Biotechnology for Agro-Industrial Residues Utilisation

Poonam Singh nee' Nigam · Ashok Pandey
Editors

Biotechnology for Agro-Industrial Residues Utilisation

Utilisation of Agro-Residues

Springer

Editors
Poonam Singh nee' Nigam
University of Ulster
Faculty of Life & Health
Sciences
Coleraine
Northern Ireland
United Kingdom BT52 1SA
P.Singh@ulster.ac.uk

Ashok Pandey
National Institute for
Interdisciplinary
Science & Technology
CSIR, Industrial Estate PO
Trivandrum-695019
India
ashokpandey56@yahoo.co.in

ISBN 978-1-4020-9941-0 e-ISBN 978-1-4020-9942-7

DOI 10.1007/978-1-4020-9942-7

Library of Congress Control Number: 2009920465

Printed on acid-free paper

9 8 7 6 5 4 3 2 1

springer.com

DEDICATED BY POONAM TO her dearest mother *ROOPA NIGAM*
And in loving memory of her late father
***MAHESH CHANDRA NIGAM*, who left too early.**

Preface

Industries from two large sectors, i.e. agriculture and food for most of its history have been environmentally benign. Industrial activity has always resulted in some kind of pollution, be it solid waste, wastewater or gaseous pollution. Even when technology began to have an impact, reliance on natural and ecological processes remained crucial. Crop residues were incorporated into the soil or fed to livestock and the manure returned to the land in amounts that could be absorbed and utilized. Since farms have become highly mechanized and reliant on synthetic fertilizers and pesticides, the crop residues, which were once recycled, are now largely wastes whose disposal presents a continuing problem for the farmer.

The agro-industrial residues consist of many and varied wastes from agriculture and food industry, which in total account for over 250 million tonnes of waste per year in the UK alone. The prospects and application of biotechnical principles facilitates these problems to be seen in a new approach, as resources, which in many cases have tremendous potential. As a result an extensive range of valuable and usable products can be recovered from what was previously considered waste. This encompasses a huge area of microbial-biotechnology with many possibilities that have been researched; and such findings have shown the massive potential when they are practically and economically applied.

Although several agricultural residues can be disposed of safely (due to biodegradable nature) in the environment, the vast quantities in which they are generated as a result of diverse agricultural and industrial practices, necessitates the requirement to look for some avenues where these could be utilized for some application. Since these are rich in organic nature, they represent one of the most energy-rich resources on the planet. Accumulation of this kind of biomass in large quantities every year results not only in the deterioration of the environment, but also in the loss of potentially valuable material which can be processed to yield a number of valuable added products, such as food, fuel, feed and a variety of chemicals. These residues include renewable lignocellulosic materials such as the stalks, stems, straws, hulls and cobs which all vary slightly in composition. Cellulose and hemicellulose, the major constituents of these materials, can be referred to as valuable resources for a number of reasons, largely due to the fact that they can be bio-converted for the production of several valuable products.

Thus, today, for better or for worse, we live in a society with a throw away attitude which often chooses in many cases to ignore the potential that is all around it. Particularly in the case of agriculture, there can be considerable damage to the environment which is already being continually put under increasing stress by waste disposal. Furthermore it is often quite expensive to dispose of these wastes this is not to mention the economic loss of not exploiting them properly in the first place. Biotechnology can offer many viable alternatives to the disposal of agricultural waste with the production of many much needed products such as fuels, feeds, and pharmaceutical products.

Therefore, this book has been presented with the up-to-date information available on a biotechnology approach for the utilisation of agro-industrial residues. The book contains twenty four chapters by the experts working in the field of Biotechnology for Agro-Industrial Residues Processing. Each of the chapters includes information on materials and suitable technology for their utilization and bioconversion methods to obtain products of economic importance. The chapters have been categorised in appropriate sections: (1) General; (2) Production of industrial products using agro-industrial residues as substrates; (3) Biotechnological potential of agro-industrial residues for bioprocesses; (4) Enzymes degrading agro-industrial residues and their production; and (5) Bioconversion of agro-industrial residues.

It is hoped that the book will provide a useful information resource for academics, researchers, and industries.

Northern Ireland, UK — Poonam Singh nee' Nigam
Trivandrum, Kerala, India — Ashok Pandey

Acknowledgements

Editors sincerely thank all contributing authors of the chapters included in this book for their cooperation in submitting and revising their manuscripts on due dates as per guidelines of the publisher Springer. Thanks for the approval of our book-proposal on this particular topic are due to Peter Butler, publishing director; Dugald MacGlashan, senior publishing editor; and Sara Huisman, publishing assistant, Springer Science & Business Media B.V. Efforts of Max Haring and Agnieszka Brodawka, are acknowledged for realising the production of this book. Finally, the support extended by our families could make this project possible.

Northern Ireland, UK — Poonam Singh nee' Nigam
Trivandrum, Kerala, India — Ashok Pandey

Contents

Contributors

Olugbenga O. AdeOluwa Department of Agronomy, Faculty of Agriculture, University of Ibadan, Ibadan, Nigeria, adeoluwaoo@yahoo.com

Ashish Anthwal Department of Earth and Environmental Sciences, Atmospheric Environment Laboratory, Sejong University, Gwangjin-Gu, Seoul 143-747, Republic of Korea, ashishaanthwal25@rediffmail.com

Sumathy Babitha Skin Bioactive Material Laboratory, Inha University, Yong-hyunong, Nam-gu, Incheon 402-751, Republic of Korea, babisp2003@yahoo.com

Soham Chattopadhyay Department of Biotechnology, Indian Institute of Technology, Kharagpur, West Bengal 721302, India, sohamcrj@gmail.com

Hongzhang Chen State Key Laboratory of Biochemical Engineering, Institute of Process Engineering, Chinese Academy of Sciences, Beijing 100190, PR China, hzchen@home.ipe.ac.cn

Paul Christakopoulos Biotechnology Laboratory, School of Chemical Engineering, National Technical University of Athens, Zografou 157 80, Athens, Greece, hristako@orfeas.chemeng.ntua.gr

Syed G. Dastager National Institute of Interdisciplinary Science and Technology (Formerly RRL), CSIR, Industrial Estate, Thiruvananthapuram-695019, Kerala, India, syedmicro@gmail.com

Dhanya Gangadharan Biotechnology Division, National Institute for Interdisciplinary Science and Technology (NIIST), CSIR, Trivandrum-695 019, Kerala, India, dhanya_28@yahoo.com

V.K. Garg Department of Environmental Science and Engineering, Guru Jambheshwar University of Science and Technology, Hisar 125001, Haryana, India, vinodkgarg@yahoo.com

Nutan Gupta School of Biomedical Sciences, Faculty of Life and Health Sciences, University of Ulster, Coleraine, BT521SA, Northern Ireland, UK, nutangupta100@rediffmail.com

Renuka Gupta Department of Environmental Science and Engineering, Guru Jambheshwar University of Science and Technology, Hisar 125001, Haryana, India, renug77@gmail.com

Linda M. Harvey Strathclyde Fermentation Centre, Strathclyde Institute for Pharmacy & Biomedical Science, University of Strathclyde, Glasgow, Scotland, UK, lm.harvey@strath.ac.uk

Nicemol Jacob Biotechnology Division, National Institute for Interdisciplinary Science and, Technology (CSIR), Trivandrum, 695019, Kerala, India, nicemariacyril@yahoo.co.in

Bo Jin School of Earth and Environmental Sciences, School of Chemical Engineering, The University of Adelaide, Adelaide, SA 5005, Australia, bo.jin@adelaide.edu.au

Rojan P. John Biotechnology Division, National Institute for Interdisciplinary Science and Technology, CSIR, Trivandrum 695 019, Kerala, India, rojanpj@yahoo.co.in

Joan M. Kelly School of Molecular and Biomedical Sciences, The University of Adelaide, Adelaide, SA 5005, Australia, joan.kelly@adelaide.edu.au

Diomi Mamma Biotechnology Laboratory, School of Chemical Engineering, National Technical University of Athens, Zografou 157 80, Athens Greece, dmamma@chemeng.ntua.gr

Brian McNeil Strathclyde Fermentation Centre, Strathclyde Institute for Pharmacy & Biomedical Science, University of Strathclyde, Glasgow, Scotland, UK, B.mcneil@strath.ac.uk

Asmita Mukerji Department of Biotechnology, Anandapur, East Kolkata Township, Kolkata 700 107, India, asmita.mukerji@gmail.com

Solange I. Mussatto Institute for Biotechnology and Bioengineering, Centre of Biological Engineering, University of Minho, Braga 4710-057, Portugal, solange@deb.uminho.pt

Poonam Singh nee' Nigam Faculty of Life and Health Sciences, School of Biomedical Sciences, University of Ulster, Coleraine BT521SA, Northern Ireland, UK, P.Singh@ulster.ac.uk

K.N Niladevi Biotechnology Division, National Institute for Interdisciplinary Science and, Technology (CSIR), Trivandrum 695019, Kerala, India, nilanandini@yahoo.co.in

Ashok Pandey National Institue for Interdisciplinary Science and Technology, CSIR, Industrial Estate PO, Trivandrum-695 019, Kerala, India, ashokpandey56@yahoo.co.in

Binod Parameswaran Bioenergy Research Centre, Korea Institute of Energy Research (KIER), Yusong, Daejon 305-343, Republic of Korea, binodkannur@yahoo.com

Antonios N. Philippoussis National Agricultural Research Foundation, I.A.A.C., Laboratory of Edible and Medicinal Fungi, 13561 Ag, Anargyri, Athens, Greece, iamc@ath.forthnet.gr

Ramkrishna Sen Department of Biotechnology, Indian Institute of Technology Kharagpur, West Bengal 721302, India, rksen@yahoo.com

Reeta Rani Singhania Biotechnology division, National Institute for Interdisciplinary Science and Technology, Trivandrum 695 019, Kerala, India, reetasinghania_123@rediffmail.com

Swetha Sivaramakrishnan Biotechnology Division, National Institute for Interdisciplinary Science and Technology (NIIST), CSIR, Trivandrum 695 019, Kerala, India, swsj79@yahoo.co.in

Mynepalli K.C. Sridhar Division of Environmental Health Sciences, Faculty of Public Health, University of Ibadan, Ibadan, Nigeria, mkcsridhar@yahoo.com

Evangelos Topakas Biotechnology Laboratory, School of Chemical Engineering, National Technical University of Athens, Zografou 157 80, Athens, Greece, vtopakas@central.ntua.gr

J. Obeta Ugwuanyi Department of Microbiology, University of Nigeria, Nsukka, Enugu, Nigeria, jerryugwuanyi@yahoo.com

Christina Vafiadi Biotechnology Laboratory, School of Chemical Engineering, National Technical University of Athens, Zografou 157 80, Athens, Greece, cvafiadi@central.ntua.gr

Erick J. Vandamme Laboratory of Industrial Microbiology and Biocatalysis, Department Biochemical and Microbial Technology, Faculty of Bioscience Engineering, Ghent University, B-9000 GENT, Belgium, Erick.Vandamme@UGent.be

Michael Ward Centre for Vision Science, Queens University of Belfast, Royal Victoria Hospital, Belfast BT12 6BA, Northern Ireland, UK, michaelward@hotmail.co.uk

Ye Yang State Key Laboratory of Biochemical Engineering, Institute of Process Engineering, Chinese Academy of Sciences, Beijing 100190, PR China, yeyang@home.ipe.ac.cn

Jianxing Zhang State Key Laboratory of Biochemical Engineering, Institute of Process Engineering, Chinese Academy of Sciences, Beijing 100190, PR China, zhangjx04@gmail.com

Part I
General

Chapter 1
Agro-Industrial Residue Utilization for Industrial Biotechnology Products

Erick J. Vandamme

Contents

Keywords Renewable-resources · Microbial-nutrition · Industrial fermentation substrates · Bio-chemicals versus petrochemicals · Submerged and solid state fermentation

1.1 Introduction

As worldwide demand for petroleum, our main fossil-resource to produce energy, chemicals and materials is steadily increasing, particularly to satisfy the fast growing economies of countries such as China and India, petroleum prices are expected to continue to rise further. The effect can be seen today, with petroleum prices over 130 $/barrel at the time of writing (May 2008). Whereas this fossil resource will certainly not become exhausted from one day to another, it is clear that its price will follow a long-term upward trend. Its scarcity and high price will not only afflict the chemical industries and energy sectors drastically all around the world, but it will impact on society as a whole (Soetaert and Vandamme 2006).

E.J. Vandamme (✉)
Laboratory of Industrial Microbiology and Biocatalysis, Department Biochemical and Microbial Technology, Faculty of Bioscience Engineering, Ghent University, Coupure Links 653, B-9000 GENT, Belgium
e-mail: Erick.Vandamme@UGent.be

P. Singh nee' Nigam, A. Pandey (eds.), *Biotechnology for Agro-Industrial Residues Utilisation*, DOI 10.1007/978-1-4020-9942-7_1,

Consequently, concerns have arisen about our future energy and chemicals supply. In the first place, this has caused an ongoing search for renewable energy sources that will in principle never run out, such as hydraulic energy, solar energy, wind energy, tidal energy, geothermal energy and also energy from renewable raw materials such as biomass. Biomass can be defined as "all organic material of vegetal or animal origin, which is produced in natural or managed ecosystems (agriculture, aquaculture, forestry), all or not industrially transformed". Bioenergy, the renewable energy released from biomass, is indeed expected to contribute significantly in the mid to long term. According to the International Energy Agency (IEA), bioenergy offers the potential to meet 50% of our world energy needs in the 21st century. The same hold true for the synthesis of fine and bulk chemicals, materials and polymers, now also mainly based on fossil resources, petroleum, gas and coal. The chemical industry will be confronted with the switch to utilize biomass sooner than anticipated.

In contrast to these fossil resources, bulk agricultural raw materials such as wheat, rice or corn have till a few years ago been continuously low (and even declining) in price because of increasing agricultural yields, a tendency that has recently drastically changed, with the competition between biomass for food use versus biomass for chemicals or biofuels use, becoming a societal issue. However, climate changes, droughts, high oil prices and the switch to non-vegetarian diets in fast developing economies such as China are actually the main underlying causes of the increasing food prices. New developments such as plant genetic engineering – specifically of industrial or energy crops (Van Beilen 2008) – and the production of bioenergy and chemicals from agricultural waste and agro-industrial residues can relieve these trends (Morris 2006, Zhang 2008). Agricultural crops such as corn, wheat, rice and other cereals, sugar cane and beet, potato, tapioca, etc. are now already processed in the starch and sugar refineries into relatively pure carbohydrate feedstocks (starch, sugars,...), primary substrates for the food industries, but also for most industrial fermentation processes and for some chemical processes (Dahod 1999, Kamm and Kamm 2004). Especially fermentation processes can convert those agro-feedstocks into a wide variety of valuable chemical products, including biofuels such as bioethanol, and organic solvents such as butanol (Demain 2000, 2007, Kunz 2008, Soetaert and Vandamme 2005, 2008, Wall et al. 2008).

Oilseeds such as soybeans, rapeseed (canola) and oilpalm seeds (but also waste vegetal oils and animal fats) are equally processed into oils that are subsequently converted into food ingredients and oleo-chemicals, but recently increasingly into biodiesel (Canakci and Sanli 2008, Vasudevan and Briggs 2008).

While these technologies are rather mature, agro-industrial residues or waste streams such as straw, bran, beet pulp, corn cobs, corn stover, oil cakes, waste wood,...all rich in lignocellulosic materials, are now either poorly valorized or left to decay on the land. (Sarath et al. 2008, Zhang 2008). These residues are now already efficiently converted into biogas and used for heat, steam or electricity generation. These waste materials attract now increasingly attention as an abundantly available and cheap renewable feedstock for chemicals, materials and biofuels production. Improved physical, chemical and biotechnological treatments must now quickly be developed to upgrade and valorize these agro-industrial side streams.

Estimates from the USA Department of Energy have shown that up to 500 million tonnes of such raw materials can be made available into the USA each year, at prices ranging between 20 – 50 $/ton.

This volume will focus on the biotechnological potential of using and upgrading renewable resource-"leftovers", especially the agro-industrial processing residues (now left unused when biomass is processed largely into food, but also into chemicals or fiber/materials). Emphasis will be put on fermentation and biocatalysis principles and processes as very suitable technologies for upgrading these agro-industrial residues in a sustainable way.

1.2 Fermentation and Biocatalysis

It is only now being fully realized by the chemical industry that microorganisms (bacteria, yeast and fungi, micro-algae) are an inexhaustible source of a wide range of useful chemical compounds: indeed, an ever increasing number of fine and bulk chemicals, solvents, food additives, enzymes, agrochemicals and biopharmaceuticals is now being produced based on microbial biotechnology via industrial fermentation or biocatalysis processes (Demain 2007, Vandamme 2007). Often, there is no alternative route for their synthesis but fermentation. Also bioconversion reactions, based on the use of (immobilised) microbial biocatalysts (cells or enzymes), yield useful interesting regio- and enantioselective molecules under mild reaction conditions, often starting from racemic precursors (Vandamme et al. 2005, 2006). Furthermore, all these microbial processes have a positive environmental impact (Table 1.1). These microbial products generally display desired chirality, are biodegradable and practically all are produced, starting from renewable (agro)-substrates, till now mainly starch and sugars. Indeed, these nutrient substrates, which are the "workhorse" ingredients in industrial fermentation processes worldwide, are mainly derived from agricultural crops, being processed in the established sugar and starch refineries. Agricultural practice as well as this industrial processing leads to agro-industrial residues, which should be considered now also as nutrient substrates, rather than as a waste!

Table 1.1 Sustainability-related properties of fermentation and bioconversion derived chemicals

- Produced from renewable agrosubstrates and agro-industrial residues
- Mild reaction conditions→ "green chemistry"
- Biodegradable
- Desired chirality

1.3 Currently Used Renewable Agrosubstrates as Industrial Fermentation Substrates

Worldwide, the feedstock for fermentation processes is provided directly or indirectly by agriculture: indeed cereal grains, plant tubers, plant oils, crop residues and agro-industrial products, side or waste streams are main sources of microbial

Table 1.2 Currently used typical carbon sources in industrial fermentation processes

Carbohydrates:	
Corn flour	Cane molasses
Starch (from various plant sources)	Beet molasses
Dextrins	Sucrose
Glucose syrups	Sulfite waste liquor
Dextrose	Wood hydrolysate
Maltose	Organic acids
Whey	Agro-industrial waste
Lactose	
Oils and Alcohols:	
Soybean oil, methyloleate	Glycerol
Corn oil, cottonseed oil, peanut oil,...	Polyols
Palm oil, ...	Hydrocarbons
Lardoil, fish oil	Methanol
	Ethanol

nutrients. With respect to the carbon and nitrogen source, most are plant derived, but certain microbial nutrients are of animal origin (i.e. peptones, lactose, whey, ...) or are derived from yeast (Dahod 1999) (Table 1.2). There is a general trend to replace these animal derived nutrients for plant derived ones, due to the threat and transfer of prion diseases.

1.3.1 Carbohydrates as Carbon Substrate

Although carbohydrates in general serve many other important functions, especially bulk carbohydrates serve as a nutrient source of carbon for the large scale cultivation of microorganisms (Table 1.3). Cheap carbohydrates such as beet and cane molasses, sucrose, starch or its hydrolysates and glucose syrups are almost universally used as renewable carbon sources in large scale fermentation processes. The worldwide total usage of carbohydrate-nature feedstock for industrial fermentation processes has been estimated at 4.10^7 tons per year. Molasses are produced both from cane or beet; the product is actually the mother liquor separated from the crystallized sucrose. The total fermentable sugar is in the range of 50–55% by weight and it is used extensively (often as a mix) in the fermentation of bulk products such as yeast, ethanol, monosodium glutamate, citric acid, industrial enzymes, and many others. It is also a source of nitrogen, minerals, vitamins and growth factors. Its varying composition is often a drawback, such that standardisation, pre-treatment and addition of further nutrients are needed, depending on the fermentation process envisaged.

Starch generally cannot be used in its native form as far as most fermentation applications are concerned, since it undergoes gelatinisation during sterilisation of the fermentation broth, resulting in high viscosity. Liquefaction with an α-amylase is needed to decrease this viscosity. Such liquefied starch can be used as carbon source, if the microorganisms involved produce the needed glucoamylases i.e. many bacilli

Table 1.3 Important functions of carbohydrates

Common:
– Diet of living organisms, including microorganisms
– Energy source
– Storage compounds
– Biological "construction" material
– Substrate for chemical derivatisation
Special:
– "Messenger" molecules: receptors, recognition sites, lectin interactions, immunostimulants, . . .
– Unusual sugars
– Chiral intermediates
– Biopolymers, bioplastics
Bulk:
Sucrose: $> 100 \times 10^6$ tons/year
Glucose: $> 10 \times 10^6$ tons/year
Carbohydrate fermentation feedstock: 4×10^7 tons/year

and fungi. Maltodextrins result from enzymatic or acid hydrolysis of starch and can be used as such i.e. for the production of antibiotics (to avoid glucose catabolite repression) i.e. penicillin, cephalosporin, streptomycin,

Glucose syrups are obtained by the action of amylases on starch liquefacts, a process called saccharification. These syrups (85–90% glucose), also known as starch hydrolysates, are most frequently used in fermentation applications: i.e. for production of citric acid, gluconic acid, itaconic acid, L-amino acids (monosodiumglutamate, L-lysine, L-threonine,. . .), xanthan, curdlan, scleroglucan, erythritol, several antibiotics, ... For the production of lactic acid and several other chemicals, pure glucose (dextrose) is often preferred to facilitate the product recovery.

Maltose syrups, obtained by β-amylase action on starch liquefacts, are suitable in fermentations, where a glucose repression effect is active, as is the case in several antibiotic fermentations.

Sulfite waste liquor, a side product of the paperpulp manufacturing process, is rich in pentose-sugars and can be utilized by *Candida* yeasts and several other microorganisms as a carbon source.

Currently, well defined – rather pure – carbon sources are preferred in industrial fermentations, due to constraints imposed by the microorganisms involved, but also with a simple downstream processing of the end product in mind.

1.3.2 Plant Oils as Carbon Substrate

Another interesting substrate for fermentation processes are the plant lipids and oils (Table 1.2), commonly used in fermentations of bulk antibiotics such as the β-lactam group (penicillins and cephalosporins), the tetracyclines, the macrolides and the antifungal polyenes.

Although carbohydrate substrates are relatively easily handled, as compared to plant oils, molasses and starch sources may need costly pre-treatment or hydrolysis. However oils contain about 2,5 times the energy content of glucose per weight basis: 8880 kcal/kg oil versus 3722 kcal/kg glucose. On a volume basis, oils display another advantage: it takes 1.24 litres of soybean oil to provide 10 kcal of energy into a fermentor, while it takes about 5 litres of sugar (50% ww) solution to reach that value (Stowell et al. 1987). Oils also display anti-foam properties and can act as a precursor in certain biosurfactant and antibiotic fermentations i.e. the polyene antifungal. On the other hand, utilization of oils necessitates handling two-phase system fermentations, demands a higher oxygen input and relies on microbial strains displaying lipase activity.

1.3.3 Nitrogen Sources, Used in Industrial Fermentation Processes

Crude pertinacious plant, animal and yeast derived products are commonly used as complex nitrogen sources in fermentation processes: in addition to nitrogen and carbon, they also supply vitamins, growth factors and minerals for microbial growth. Some examples are given in Table 1.4.

Yeast derived products are generally produced from baker's and brewer's yeast, grown themselves on molasses, malt extract or occasionally on other agro-waste substrates. Yeast extracts as well as yeast autolysates and hydrolysates are in use; all of them should be tested as to their suitability for a given microbial strain, used in a particular fermentation process.

Peptones are obtained by partial enzymatic hydrolysis of proteins of animal, dairy or plant origin (meat, gelatin, casein, whey protein, soy protein,...). The recent emergence of prion diseases among breeded animals has created a greater demand for protein hydrolysates derived from other sources such plants, fish and other marine sources.

Corn steep liquor is a fermented by-product of the cornwet-milling process; it is rich in minerals as well as in amino acids, vitamins and growth factors and is in use

Table 1.4 Typical nitrogen sources, used in fermentation processes

Plant derived	Yeast derived
Corn steep liquor (CSL)	Yeast extract
Corn gluten meal	Yeast autolysate
Cottonseed flour	Yeast hydrolysate
Peanut meal, linseed meal, rice meal, ... wheat flour,....	Distillers dried solubles
Soybean meal	
Others	Animal Derived
NH_4^+, NO_3^-, N_2	Peptones (meat, fish,...)
Urea	Lard water
	Milk proteins (casein,...)

as a nutrient source in many industrial fermentation processes for i.e. penicillin G, amino acids, enzymes, biopesticides.

Again, all these crude nitrogen sources are directly or indirectly derived from agricultural products or their industrial processing.

1.3.4 Nutrient Selection Criteria

The ultimate choice of nutrient source type for a given fermentation process is a complex decision, based on imperatives given by the microbial strain involved, or the nature of the end product (Table 1.5) and on technical and economic considerations (Table 1.6). For a single antibiotic process, as many as ten quite different carbon substrates have been used for commercial production, depending on prevailing economics and on geographical location of the production plant.

Table 1.5 Selection of starch, maltose or glucose based feedstock for fermentation processes

Fermentation product	Starch maltose glucose	Reason of preferred use (+)
Polyols: Erythritol	+	Higher yield and reduced purification steps, as compared to sucrose or molasses
Organic acids: Gluconic: Itaconic	+	Molecular structure
	+	Molecular structure
Amino acids: Lysine	+	Yield and reduced purification steps
Polysaccharides:	+	Molecular structure
Xanthan Cyclodextrins	+	Molecular structure
	+	
Enzymes:	+	Catabolite repression
Carbohydrases:	+(fedbatch)	
Proteases	+	
	+(fedbatch)	
Antibiotics:	+	Purity of product
Macrolides:	+	Molecular structure
Tetracyclines:	+	Catabolite repression
Penicillin G:	+(fedbatch)	
Vitamins: B_{12}	+	Purity of product

Table 1.6 Economic and technical considerations in the selection of fermentation nutrient sources

Availability	Consistency of nutritional quality
Cost per unit of nutrient	Flexibility in application
Transportation cost	Rheological properties
Price stability	Surface tension factors
Pre-treatment costs	Product recovery impact
Stabilization costs	Process yield
Storage costs	Product concentration and type
Safety factors	Overall productivity

Some important factors in comparing the benefits and/or disadvantages of using crude or refined carbohydrates or oils as carbon source in industrial fermentations have been compiled by Stowell et al. (1987). The key point here is that microorganisms can convert these abundantly available and renewable nutrient sources into a vast range of very complex biochemical's with often unsuspected application potential (Demain 2000). Submerged fermentation has been the mainstay industrial biotechnology production process in use, but as increasingly crude (solid) agro-industrial residues will become available, solid state fermentation processes will experience a remarkable revival in the near future (Robinson et al. 2001).

1.4 Towards Agro-Industrial Residue Utilization Technology in Industrial Biotechnology

When switching to agro-industrial residues or even agro-waste streams, the bottleneck remains to release the fermentable sugars, left in the lignocellulosic matrix, the main component of these residues (Zhang 2008, Sarath et al. 2008, Vasudevan and Briggs 2008, Canakci and Sanli 2008).

Special pre-treatments of these agro-industrial side streams is a prerequisite: mechanical (thermo) physical, chemical and enzymatic pre-treatments will be primordial in most cases, before microbial fermentation technology or enzymatic upgrading (biocatalysis) can start. An exception here is the use of solid state fermentation technology, where crude lignocellulosics are directly provided as a substrate for microbial productions (Robinson et al. 2001). The switch to agro-industrial residues will also put even more emphasis on pre-treatment (upstream) – and on downstream-processing costs in the overall economics of such "second generation" fermentation processes!

These physical, (thermo) chemical, mechanical and enzymatic pre-treatments are covered by experts in detail in the first chapters of this volume, as well as the principles of solid state versus submerged fermentation.

Subsequently, the potential of a wide range of agro-industrial residues to serve as nutrient source for industrial biotechnology processes is covered. Also the production potential of a wide range of fine and bulk chemicals, fuels and materials based on these agro-industrial residues is discussed. If these processes materialize in the near future, it will relief drastically current societal tension whether to use biomass and crops for food or for platform chemicals and biofuels (Morris 2006).

References

Canakci M and Sanli H (2008) Biodiesel production from various feed stocks and their effects on the fuel properties. J Ind Microbiol Biotechnol 35: 431–441

Dahod SK (1999) Raw materials selection and medium development for industrial fermentation processes, pp. 213–220 In "Manual of Industrial Microbiology and Biotechnology" (2nd ed); (Demain, A.L. and Davies, J.E., eds.) ASM Press, Washington DC

Demain AL (2000) Small bugs, big business: the economic power of the microbe. Biotechnol Adv 18: 499–514
Demain AL (2007) The business of biotechnology. Ind Biotechnol 3: 269–283
Kamm B and Kamm M (2004) Principles of biorefineries. Appl Microbiol Biotechnol 64: 137–145
Kunz M (2008) Bio-ethanol: Experiences from running plants, optimization and prospects. Biocat Biotransf 26: 128–132
Morris D (2006) The next economy: From dead to living carbon. J Sci Food Agric 86: 1743–1746
Robinson T, Singh D, and Nigam P (2001) Solid state fermentation: A promising microbial technology for secondary metabolite production. Appl Microbiol Biotechnol 55: 284–289
Sarath G, Mitchel RB, Satler SE, Funnell D, Pedersen JF, Graybosch RA, and Vogel KP (2008) Opportunities and roadblocks in utilizing orages and small grains for liquid fuels. J Ind Microbiol Biotechnol 35: 343–354
Soetaert W and Vandamme EJ (2005) Biofuel production from agricultural crops. In: Biofuels for fuel cells: Renewable energy from biomass fermentation, pp. 37–50 In "Series Integrated Environmental Technology" (Lens, P., Westerman, P., Haberbauer, M., and Moreno, A., eds.) IWA Publ., UK.
Soetaert W and Vandamme EJ (2006) The impact of industrial biotechnology. Biotechnol J 1(7–8): 756–769
Soetaert W and Vandamme EJ (2009) Biofuels, 242 pp. In "Renewable Resources series" (Stevens, C., Series ed.) J Wiley & Sons Ltd. ISBN 978-0-470-02674-8
Stowell JD, Beardsmore AS, Keevil CW, and Woodward JR (1987) Carbon Substrates in Biotechnology IRL-Press, Oxford-Washington DC
Van Beilen JB (2008) Transgenic plant factories for the production of biopolymers and platform chemicals. Biofuels Bioprod Bioref 2: 215–228
Vandamme EJ (2007) Microbial gems: Microorganisms without frontiers. SIM-News 57(3): 81–91
Vandamme EJ, Cerdobbel A, and Soetaert W (2005) Biocatalysis on the rise: Part 1 Principles. Chem Today 23(6): 47–51
Vandamme EJ, Cerdobbel A, and Soetaer W (2006) Biocatalysis on the rise: Part 2 Applications. Chem Today 24(1): 57–61
Vasudevan PT and Briggs M (2008) Biodiesel production: Current state of art and challenges. J Ind Biotechnol 35: 421–430
Wall JD, Harwood CS, and Demain AL (2008) "Bioenergy". ASM-Press, Washington DC
Zhang YHP (2008) Reviving the carbohydrate economy via multi-product lignocellulose biorefineries. J Ind Microbiol Biotechnol 35: 367–375

Chapter 2
Pre-treatment of Agro-Industrial Residues

Poonam Singh nee' Nigam, Nutan Gupta and Ashish Anthwal

Contents

Abstract Problem of management of agro-industrial residues complicate the farming economies. Agro-industrial residues are the most abundant and renewable resources on earth. Accumulation of this biomass in large quantities every year results not only in the deterioration of the environment, but also in the loss of potentially valuable material which can be processed to yield a number of valuable added products, such as food, fuel, feed and a variety of chemicals. The agro-industrial residues

P. Singh nee' Nigam (✉)
Faculty of Life and Health Sciences, School of Biomedical Sciences, University of Ulster, Coleraine, BT521SA, Northern Ireland, UK
e-mail: P.Singh@ulster.ac.uk

P. Singh nee' Nigam, A. Pandey (eds.), *Biotechnology for Agro-Industrial Residues Utilisation*, DOI 10.1007/978-1-4020-9942-7_2,

have alternative uses or markets. Pre-treatment is an important tool for breakdown of the structure of these residues mainly formed of cellulose, hemicellulose and lignin. Cellulose is present in large quantities in agro-industrial residues. As hemicellulose and cellulose are present in the cell wall they undergo lignification hence there is an increasing need to have an effective and economic method to separate cellulose and hemicellulose from cell wall. Various pre-treatment methods such as physical, chemical, biological (enzymatic) and combined are available. Physical and chemical treatments breakdown the materials present in the agro-industrial residues. As glucose is readily used by the microorganisms and is present in cellulose, biological pre-treatment by microrganisms is also a good method. Enzymes like phytase, laccase, LiP, MnP are produced by these microrganisms and help in delignification, bleaching, and manufacture of animal feed etc.

Keywords Agro-industrial · Pre-treatment · Physical · Enzymatic · Chemical · Microorganisms

2.1 Agro-Industrial Residues

Agro-industrial residues are directly burnt as fuel in developing world that includes crop residues, forest litter, grass and animal garbage. Crop residues are more widely burnt than animal waste and forest litter.

Agro-industrial residues are derived from the processing of a particular crop or animal product usually by an agricultural firm. Included in this category are materials like molasses, bagasse, oilseed cakes and maize milling by-products and brewer's wastes. Crop residues encompass all agricultural wastes such as straw, stem, stalk, leaves, husk, shell, peel, lint, seed/stones, pulp, stubble, etc. which come from cereals (rice, wheat, maize or corn, sorghum, barley, millet), cotton, groundnut, jute, legumes (tomato, bean, soya) coffee, cacao, olive, tea, fruits (banana, mango, coco, cashew) and palm oil.

2.1.1 Types of Agro-Industrial Residues

Agro-industrial residues are of a wide variety of types, and the most appropriate energy conversion technologies and handling protocols vary from type to type. The most significant division is between those residues that are predominantly dry (such as straw) and those that are wet (such as animal slurry).

2.1.1.1 Dry Residues

These include those parts of arable crops not to be used for the primary purpose of producing food, feed or fibre.

a. Field and Seed Crop

Field and seed crop residues are the materials remaining above the ground after harvesting, including straw or stubble from barley, beans, oats, rice, rye, and wheat, stalks, or stovers from corn, cotton, sorghum, soybeans, and alfalfa.

b. Fruit and Nut crop

Fruit and nut crop residues include orchard prunings and brushes. The types of fruit and nut crops include almonds, apples, apricots, avocados, cherries, dates, figs, grapefruit, grapes, lemons, limes, olives, oranges, peaches, pears, plums, prunes, and walnuts.

c. Vegetable Crop

Vegetable crop residues consist mostly of vines and leaves that remain on the ground after harvesting. The types of vegetable crops include such plants as artichokes, asparagus, cucumbers, lettuce, melon, potatoes, squash, and tomatoes.

d. Nursery Crop

Nursery crop residues include the prunings and trimmings taken from the plants during their growth and in the preparation for market. There are more than 30 different species of nursery crops (e.g. flowers and indoor plants, etc.) that are grown.

2.1.1.2 Wet Residues

These are residues and wastes that have high water content as collected. These include:

a. Animal Slurry
b. Farmyard manure
c. Grass silage

Silage is forage biomass harvested and fermented for use as winter fodder for cattle and sheep. Grass silage is harvested in the summer and stored anaerobically in a silage clamp under plastic sheeting.

2.2 Composition of Residues

Agro-industrial residues consist of lignocellulose that is compact, partly crystalline structure consisting of linear and crystalline polysaccharides cellulose, branched non cellulosic and non-crystalline heteropolysaccharides (hemicelluloses), and branched (non crystalline) lignin (Glasser et al. 2000).

Cellulose is made up of a linear polymer chain, which in turn consists of a series of hydroglucose units in glucan chains (Fig. 2.1). The hydroglucose units are held together by β-1-4 glycosidic linkages, producing a crystalline structure that can be

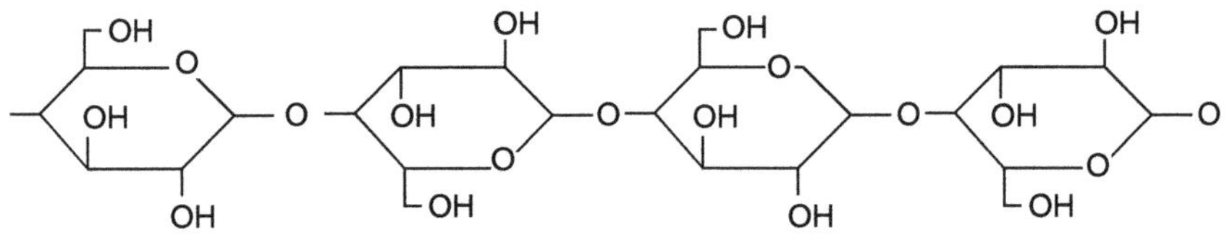

Fig. 2.1 Structure of cellulose

broken down more readily to monomeric sugars. Another major component of the lignocellulose structure is hemicelloluse, which is made up of various polysaccharides, namely, xylose, galactose, mannose and arabinose. The function of hemicellulose has been proposed as a bonding agent between lignin and cellulose. Mannose has been used as a fermentable substrate since many years, with more specific yeast being able to utilize arabinose and xylose. Hemicellulose is composed of linear and branched heteropolymers of L-arabinose, D-galactose, D-glucose, D-mannose, and D-xylose. Methyl or acetyl groups are attached to the carbon chain to various degrees. Hemicellulose and cellulose, constitute 13–39% and 36–61% of the total dry matter, respectively.

Lignin found in nature is made of three monomers which are biosynthesized in plants through shikimic acid pathway. It is made by an oxidative coupling of three major C_6–C_3 phenypropanoid units, namely sinapyl alcohol, coniferyl alcohol and *p*-coumaryl alcohol. These are arranged in a random, irregular three dimensional network that provide strength and structure and is consequently very resistant to enzymatic degradation (Fig. 2.2).

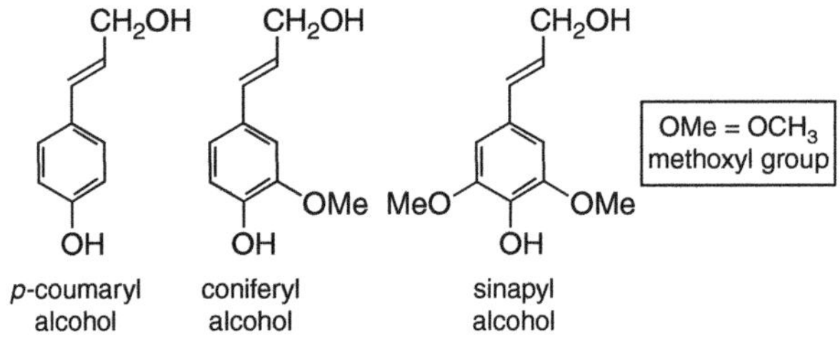

Fig. 2.2 Monomers of lignin

p-Coumaryl alcohol is a minor component of grass and forage type lignins. Coniferyl alcohol is the predominant lignin monomer found in softwoods. Both coniferyl and sinapyl alcohols are the building blocks of hardwood lignin.

2.3 Annual Yield

Agro-industrial residues are an ideal energy source if the two components can be successfully separated or treated. Over 300 million tons of lignocellulose are produced annually worldwide. In UK there are nearly 2 million ha of wheat and 1 million ha of barley. Over half a million ha of oilseed rape which is generally ploughed back, partly as it is very friable and does not lend itself conveniently to collection. Smaller areas of oats (100,000 ha), rye (9,000 ha) and triticale (13,000 ha), all of which can yield straw. In the UK, as a result of insufficient summer warmth to fully ripen the grain, most maize production (around 100,000 ha) is grown as a forage crop and used for high quality silage, with only about 2,500 ha for grain, in the far south of England.

This renewable biomass has the potential to be used for the production of fuels, chemicals, animal feed etc. Sometimes these agro-industrial residues are seen as waste and pose disposal problems for the associated industries. This can be solved through its utilization, turning a valueless waste into a valuable substrate for fermentation processes. The main components of agro-industrial residues are shown in Table 2.1.

Table 2.1 Main components of agro-industrial residues

Agro-industrial residues	Lignin (wt %)	Cellulose (wt %)	Hemicellulose (wt %)	References
Corn cobs	6.1±15.9	33.7±41.2	31.9±36.6	Ropars et al. 1992
Sugarcane baggase	10±20	40±41.3	27±37.5	Schaffeld 1994a
Wheat straw	8.9±17.3	32.9±50	24±35.5	Bjerre et al. 1996.
Rice straw	9.9±24	36.2±47	19±24.5	Patel and Bhatt 1992
Corn stalks	7±18.4	35±39.6	16.8±35	Barrier et al. 1985.
Barley straw	13.8±14.5	33.8±37.5	21.9±24.7	Fan et al. 1987
Rye straw	19.0	37.6	30.5	Fan et al. 1987
Oat straw	17.5	39.4	27.1	Fan et al. 1987
Flax	22.3	34.9	23.6	Fan et al. 1987
Soya stalks	19.8	34.5	24.8	Fan et al. 1987
Sunflower stalks	13.44	42.10	29.66	Jiménez et al. (1990)
Vine shoots	20.27	41.14	26.00	Jiménez et al. (2007)
Cotton stalks	21.45	58.48	14.38	Jiménez et al. (2007)
Sunflower seed hulls	29.40	24.10	28.60	Dekker and Wallis 1983
Thistle	22.1	31.1	12.2	JimeÂnez and Loâpez 1993

2.4 Uses of Agro-Industrial Residues

Agro-industrial residues can be used in many ways because they are cheap, abundant and their use will provide us with environmental and economic benefits:

a. Barley straw is used for animal bedding and feed
b. In UK around 40% of wheat straw is chopped and returned to the soil, 30% used on the farm (for animal bedding and feed), and 30% is sold.
c. Chopped straw can reduce phosphate and potassium needed for the following crop, and can help conserve soil moisture and structure.

d. The ash from burning or gasifying straw can be used to return minerals to the soil however cannot contribute organic matter or help soil structure.
e. Corn Stover is used as biomass which is low carbon sustainable fuel that can deliver a significant reduction in net carbon emissions when compared with fossil fuels.
f. Rice straw can be used as pulp for paper becoming an ideal solution for the California and Oregon rice burning conundrum.
g. Agro-industrial residues produce ethanol, bioethanol a product of high potential value containing minor quantities of soluble sugars, pectin, proteins, minerals and vitamins. Bioethanol produced from renewable biomass has received considerable attention in current years. Using ethanol as a gasoline fuel additive as well as transportation fuel helps to alleviate global warming and environmental pollution (Fig. 2.3).
h. They also have potential to produce biogas under anaerobic fermentation conditions.
i. For soil nutrient recycling and improvement purposes and may therefore be displacing significant quantities of synthetic fertilizers or other products.
j. In USA and Canada, the straws of wheat, barley, oats and rye, and the husks of rice have been utilised in mixture with wood fibers in the production of pulp, particleboards and fibreboards (Hesch 1978, Loken et al. 1991, Knowles 1992).

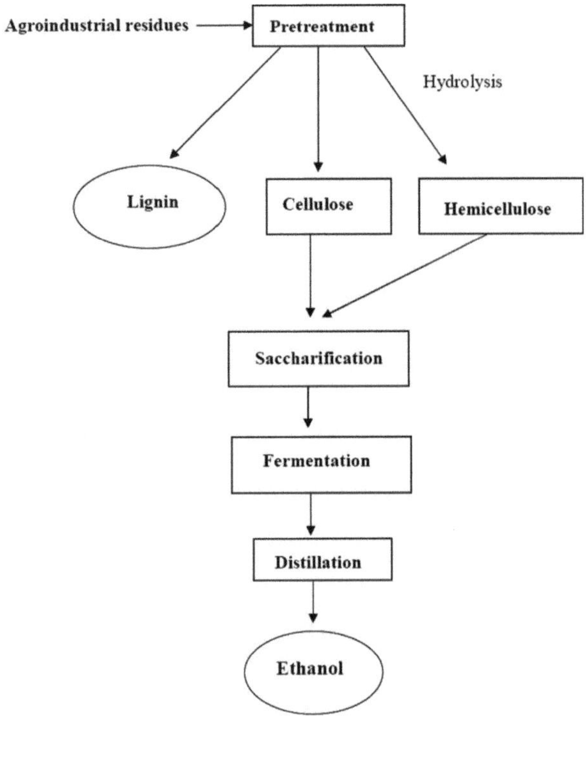

Fig. 2.3 Ethanol production from ligocellulosic material i.e agroindustrial residues. Adapted from Olsson and Hagerdal (1996)

k. In Asia, husks of rice have been used to produce cement-boards (Govindarao 1980).
l. China and Japan also have made attempts to utilize Indian cane fibers in combination with wood fibers and foamy plastics to produce various kinds of wood-boards (Wang and Joe 1983)
m. Production of charcoal and briquettes (Hulscher et al. 1992, TDRI 1983).

2.5 Pre-treatments

As glucose is readily fermented by most microorganisms, yielding a variety of products, it is very much in demand by the fermentation industries. Glucose as cellulose is present in large quantities in agro-industrial residues. Because hemicellulose and cellulose present in the cell wall undergo lignification, an effective and economic method must be used to separate cellulose and hemicellulose from cell wall. To make monomeric sugar utilization from these residues a viable option, various physical, chemical and biological pre-treatments have been explored Table 2.2.

Table 2.2 Pre-treatment of Agro-industrial residues

Pre-treatment	Examples	Effect of Pre-treatment	References
Physical	Milling	Fine, highly decrystalized structure	Li et al. 2007
	Steam Explosion, Steaming treatment	Increased pore size/hemicellulose-hydrolysis	Kokta et al. 1992
	Hydrothermal	Hemicellulose hydrolysis, alteration in properties of cellulose and lignin.	Sun and Tomkinson 2002
	Irradiation	Depolymerization	Aoyama et al. 1995
Chemical	NaOH, NH_3, $H_2 0_2$ Peroxyformic acid, Organosolvents	Lignin/ hemicellulose degradation	Singh et al. 1988
	Peroxymonosulphate	Activates deliginification	Stewart 2000
Biological	White-rot fungi (*Bjerkendra adusta, Phanerochaete chysoporium, Ceriporiopsis subvermispora*) Specific bacteria	Lignin degradation	Diana et al (2002)
Enzymatic	Lignin Peroxidases (LiP, MnP, laccase)	Selective lignin/hemicellulose degradation	Aoyama 1996

2.6 Physical Pre-treatment

Mechanical and thermal methods exist to treat agro-industrial residues, but these methods tend to require a high energy input which can increase the processing cost considerably. Product separation for fermentation purposes can also make physical pre-treatments expensive.

2.6.1 Steam Explosion

Cellulose and hemicellulose in agro-industrial wastes are natural organic resources which can be harvested for energy production. To get products like methane and alcohol, biological treatments cannot be applied directly because these agro-industrial residues have a covering of lignin in their cell walls hence one efficient physical method is steam explosion, in which agro-industrial residue is pressurised with steam for a period of time followed by a rapid decompression, producing an explosive reaction that acts on lignocellulose structure. This process is carried out at a high pressure and high temperature (180 to 240°C) and breaks up the lignocellulose structure by blowing apart the three dimensional lignin components, as well as causing the decomposition of some hemicellulose into uronic and acetic acids, which catalyzes the depolymerization of hemicelluloses and lignin.

Once the lignin and cellulose components have been separated, efficient isolation of these components is required, which is a major constraint in this technology. However it forms a necessary part of delignification or pre-treatment as lignin needs to be fully removed depending on the subsequent product. For instance, if cellulose is to be used for high quality paper or chemical production, then the amount of lignin present in the final product will affect its purity. Separating the major components (fractionation) can be carried out by solvent extraction.

Pre-treatment of agro-industrial residues such as wheat and rice straw by steam explosion has been successfully used for ethanol production (Ballerini et al. 1994), bioconversion of olive oil cake. Similarly other agro-industrial residues such as sugar cane bagasse, cassava bagasse have been pretreated by steam for producing multiple other industrial products.

Using steam explosion as a pre-treatment method for agro-industrial residues, various advantages like allowing more susceptibility to cellulose degrading enzymes. As this method is environmental friendly there are no costs for recycling as in case of chemical treatment. As compared to mechanical methods like milling there is low energy input which can further be minimized using high efficiency equipments.

But if this method is to be used on a large industrial scale then high running cost and high energy requirement must be kept in view as large amounts of energy are needed.

Energy efficient and economically viable approaches like optimizing the thickness of heat-insulating material of the explosion apparatus and also the pressure and holding times should be used.

2.6.2 Hydrothermal Processing

Hydrothermal processing of agro-industrial residues causes a variety of effects including extractive removal, hemicellulose hydrolysis and alteration of the properties of both cellulose and lignin.

Water treatments provide an interesting alternative for the chemical utilization of lignocellulose owing to the following reasons:

i. No chemicals different from water are necessary, the whole process being environment-friendly
ii. Hemicelluloses can be converted into hemicellulosic sugars at good yields with low byproduct generation (Lamptey et al. 1985), leading to solutions of sugar oligomers and/or sugars that can be utilized for a variety of practical purposes (Modler 1994, Saska and Ozer 1995, Aoyama et al. 1995, Aoyama 1996, LoÂpez-Alegret 1996)
iii. In comparison with acid pre-hydrolysis, no problems derived from equipment corrosion are expected owing to the mild pH of the reaction media
iv. Stages of sludge handling and acid recycling are avoided, resulting in a simplified process structure
v. The physico-chemical alteration provoked by treatments on lignin and cellulose facilitates the further separation and processing of these fractions
vi. Economic estimates (Schaffeld 1994b, Kubikova et al. 1996) showed advantages for water treatments over alternative technologies.

The studies on the processing of Lignocellulosic materials by water or steam have been referred to in literature as autohydrolysis (Lora and Wayman 1978, Conner 1984, Carrasco 1989, Tortosa et al. 1995), hydrothermolysis (Bonn et al. 1983, HoÈrmeyer et al. 1988, Kubikova et al. 1996), aqueous liquefaction or extraction (Heitz et al. 1986, Saska and Ozer 1995), aquasolv (Kubikova et al. 1996), water prehydrolysis (Conner 1984), hydrothermal pre-treatment or treatment (Overend and Chornet 1987, Schaffeld 1994b, Kubikova et al. 1996). All these studies are based on the same kind of reactions and are referred to as "hydrothermal treatments" in this work.

2.6.3 Irradiation

Irradiation produces delignification, depolymerization and destruction of the crystalline structure of cellulose (Lowton 1952). Pritchard et al. 1962 have reported that the solubility and digestibility of wheat straw increased by gamma irradiation.

2.7 Chemical Pre-treatment

Chemical treatment is generally used to remove lignin content of agro-industrial residues. Chemical pre-treatments by alkali or acid hydrolysis, are common in paper and pulp industries to recover cellulose for paper production. These treatments tend to be expensive hence are not used for bioconversion purposes. Caustic welling is common chemical method that has the effect of increasing the surface area of the agro-industrial residue due to the swelling and disruption of lignin.

The need for corrosion-resistant apparatus, an effective washing strategy and the capability for the safe disposal of used chemicals are the disadvantages of chemical pre-treatments for lignin removal.

2.7.1 Hydrogen Peroxide (H_2O_2)

Hydrogen peroxide is used as a pre-treatment for agro-industrial residues at operational temperatures of $\geq 100^{\circ}C$ in alkaline solution i.e. hydroperoxide anion reacts with the lignin present in pulps acting as a nucleophile as well as an oxidant and as at this high temperature decomposition of H_2O_2 takes place certain chelants are added to suppress the decomposition. In paper and pulp industries it is used for bleaching and delignification purposes (to improve the brightness of pulp as it reacts with coloured carbonyl-containing structures in the lignin structure). Decomposition of H_2O_2 in alkaline condition is rapid and as a consequence more reactive radicals such as hydroxyl radicals (HO) and superoxide anions (O_2^-) are produced which are responsible for lignin degradation. However, the H_2O_2 treatment process is expensive. It becomes particularly unstable in the presence of certain transition metals e.g. Mn, Fe and Cu, at high temperatures, necessitating the addition of chelants to reduce the rate of decomposition.

Delignification by this process on a large scale can therefore be costly. At $25^{\circ}C$ in alkaline solution of 1% H_2O_2 about half of hemicelluloses and lignin content of wheat straw and corn stover is solubilized yielding a cellulose rich insoluble residue that can be enzymatically converted to glucose (Gould 1984).

This treatment of agro-industrial residues like wheat straw increases the susceptibility of plant structural carbohydrates to fibre digesting microorganisms present in the digestive tract of ruminants. Also H_2O_2 treatment results in both partial delignification of the cell wall and at least partial decrystalization of cellulose microfibrils.

Current research has focussed on reducing energy requirements of this process and working temperatures has been reduced by over 50%, with 82% and 88% lignin dissolution occurring at temperatures of 40 and $70^{\circ}C$, respectively in 2% H_2O_2.

2.7.2 Organosolvents

The pre-treatment of lignocellulosic with organosolvents involves the use of an aqueous solvent such as ethanol, butanol, phenol, etc., in the presence of a catalyst. This hydrolyzes lignin bonds as well as lignin-carbohydrate bonds, but many of the carbohydrate bonds in the hemicellulose components are also broken. Lignin is dissolved as a result of the action of the solvent and cellulose remains in solid form.

The use of organsolvents for lignin removal is an attractive process because solvents can be recovered and recycled. It also has the advantage of being able to separate lignin from the solid cellulose with the hemicellulose hydrolysate found in liquid form.

2.7.3 *Ozone*

Ozone has restricted use as a pre-treatment for lignocellulosic substrates because it not only attacks the lignin molecule but also degrades the cellulose component. Lignin attacks as a scavenger during this pre-treatment because lignin consumes most of ozone during the degradation of the carbohydrate content of agro-industrial residues, but this pre-treatment has a positive side that it lowers the amount of ozone available for cellulose degradation hence providing a substantial hydroxyl radical reactivity. Cellulose degradation has been attributed partly to a direct reaction of ozone with the glucosidic linkage and partly to a free radical mediated oxidation of hydroxyl groups in glucose (Johansson et al. 2000). Hydroxyl radicals are formed during ozonation and although lignin is attacked more rapidly, the cellulose is also targeted when the protective layer of lignin has been removed.

Double bonds react readily with ozone hence pulp bleaching is initially fairly selective, it also has the effect of disrupting the association between carbohydrates polymers and lignin, yielding a residue that is more susceptible to attack cellulases. Due to this unspecific nature, ozonation is more widely used as a pre-treatment for pulping industry.

2.7.4 *Peroxyformic Acid*

Peroxyformic acid is generated in situ by mixing formic acid with hydrogen peroxide. The agro-industrial material is added to it and cooked at $80^{\circ}C$ for three hours. Formic acid has the ability to act as a solvent for lignin and breaks down hemicellulose chains, hence peroxyformic acid causes oxidative depolymerization of lignin increasing its solubility. The formation of peroxyformic acid causes the production of electrophilic HO^{+} ions, which react with lignin. The next stage involves an increase in the reaction temperature for a period of time and it is in this stage that most delignification occurs. At these temperatures, the cellulose component may be detrimentally affected. The third and final stage of the treatment is designed to degrade any remaining lignin. This pre-treatment is known as the Milox process, derived from "mileu pure oxidative pulping".

2.8 Biological and Enzymatic Pre-treatment

As said earlier in the chapter lignin is associated with cellulose and hemicellulose in the cell wall thus acting as a barrier preventing the availability of carbohydrates for further transformation processes. Therefore, pre-treatment becomes a necessity for utilization of agro-industrial residues to obtain a good degree of fermentable

Table 2.3 Biological treatment of agro-industrial residues by microorganisms

Lignocellulose substrate	Fungi	Product	References
Wheat straw/Rice straw	*Aspergillus niger* *A.awamori* *T. reesei* *P. chrysosporium,* *P. sajor-caju*	Production of bioethanol	Patel et al. 2007
Wood chips	*Bjerkendra adusta*	Delignification for pulp refining	Dorado et al. 2001
Rice straw	*Cyathus stercoreus*	Improving nutrional quality	Orth et al. 1993
Wheat straw	*Pleurotus ostreatus* *Trametes versicolor*	Animal feed	Eichlerova et al. 2000 Zafar et al. 1996
Kraft pulp	*Trametes versicolor*	Delignification and bleaching	Dumonceaux et al. 2001
Lignocellulosic hydrolysates	*Trametes versicolor*	Peroxidase and laccase for delignification prior to ethanol fermentation	Valmaseda et al. 1991

sugars. Pre-treatment is required for alteration in the cellular structure of cellulose containing agro-industrial residues to make more accessible to the enzymes that convert the carbohydrate polymers into fermentable sugars (Mosier et al. 2005) and to cellulase producing microorganisms. White-rot fungi and certain bacteria have been commonly used for biological pre-treatments of lignocellulosics (Table 2.3). Although lignin removal through lignin degradation is possible using biological methods, it is however, unselective as lignin is degraded only to obtain the more readily metabolized cellulose and/or hemicellulose. Solid state fermentation allows the enhancement of enzyme productivity through immobilization on agricultural residue and the reduction in cost for growth substrates of fungus because agricultural or industrial waste material can be used as cheap substrates.

Ligninolytic peroxidases gained attention by their industrial applications in pulp and paper industries such as biochemical pulping and decolorization of bleach plant effluent. The use of natural solid substrates, especially lignocellulosic agro-industrial residues, as growth substrates of fungi was done for laccase production. LiP and MnP, were used because of anticipated effects on cost reduction, waste reuse, and enhanced enzyme production. The high oxidative potential of many enzymes have a positive effect on many unusable and unwanted wastes. Enzymes obtained from agricultural sources are produced by both brown-rot and white-rot fungi. The studies about the enzyme production in brown rot fungi have been done on chemically defined liquid medium under conditions to produce that particular enzyme. But in solid state fermentation and in natural environment all wood rot fungi grow under different conditions and hence the growth of fungi in solid state fermentation (fungus immobilized) and in liquid submerged fermentation is different and therefore the enzymes produced are different. The biochemical mechanisms required for degradation of lignin have been studied on lignolytic systems of white-rot fungi while the polysaccharide degradation is efficiently done by brown-rot fungi and other ascomycetous fungi.

High amounts of laccase was produced by *C. gallica* UAMH8260 on cereal bran liquid medium (Pickard et al. 1999b). Growth of *Trametes versicolor* FPRL-28A and its laccase production on wood chips, cereal grain, wheat husk, and wheat bran was good (Ullah et al. 2000). Wheat bran was also successfully used as a solid-state medium for the laccase production by *Pleurotus pulmonarius* CCB-19 (De Souza et al. 2002) and *Fomes sclerodermeus* BAFC 2752 (Papinutti et al. 2003). Wheat straw is a better substrate of *P. ostreatus* than wheat extract for the laccase production (Morais et al. 2001b). Stimulation effect of wheat straw on the laccase production was observed in the cultures of *Lentinula edodes* 610 (Hatvani and Mecs 2002). Barley bran is a good substrate for laccase production by *T. versicolor* CBS100.29 than were grape seed and grape stalks (Lorenzo et al. 2002). Olive mill wastewater-based medium, containing large amounts of recalcitrant aromatic compounds, was used for the production of laccase and MnP in submerged and solid-state cultures of *Panus tigrinus* (Fenice et al. 2003).

White-rot and brown-rot fungi produce several enzymes when grown on *Eucalyptus grandis* using solid state fermentation technique. All fungi produced hydrolytic activities but brown-rot fungi produced higher levels of cellulose and xylanase than white rot fungi whereas phenol peroxidases were produced only by white-rot fungi (Machuca and Ferraz 2001).

Phytase is either absent or present at a very low level in the gastrointestinal tract of monogastric animals (Selle and Ravindran 2007). Dietary phytate is not digested in the intestine and consequently accumulates in faecal materials. Phytase is produced by bacteria, fungi and yeasts. Among them, strains of *Aspergillus niger* produce large amounts of extracellular phytase (Chelius and Wodzinski 1994) and show more acid tolerance than bacteria and yeasts (Kim et al. 1998). In view of its industrial importance the ultimate objective is to produce this enzyme at cost effective level and establish conditions for its industrial application. Phytase production under submerged fermentation conditions using pretreated agriculture residues is useful to remove excess inorganic phosphate which otherwise inhibit phytase production. Phytase production by SSF of agriculture residue using *A. niger* NCIM 563 which was highly active at pH 5.0 (Mandviwala and Khire 2000) and process for preparation of acidic phytase using dextrin glucose medium under submerged fermentation condition (Soni and Khire 2005, 2007).

Maximum increase of 20.3 times in phytase activity was observed in case of wheat bran as compared to de-oiled rice bran, coconut cake, peanut cake high and low oil. Phytase production under submerged fermentation conditions by *A. niger* NCIM 563, indicates that pre-treatment of agriculture residues with distilled water was useful There was substantial increase in phytase activity when this excess phosphate was removed by pre-treatment. Similarly there was increase in productivity and reduction in fermentation time when agriculture residue was used instead of dextrin in submerged fermentation hence, being more economic as the cost of any agriculture residue is much cheaper than dextrin (Bhavsar et al. 2008). More details about the enzymatic treatment is discussed in chapter 22.

2.8.1 White-Rot Fungi

Due to their ability to degrade lignin as well as polysaccharides found in cellulose and hemicellulose, white-rot fungi have the potential not only to act as a biological pre-treatment but also to degrade all the major components of lignocellulose to yield a valuable product. Although lignin can be degraded by these microorganisms, the enzymes responsible are produced only when other widely available substrates are unavailable. The purpose of lignin degradation by white-rot fungi is to allow better access to the cellulose and hemicellulose components.

Three main enzymes are thought to be involved in ligninolytic biodegradation, namely lignin peroxidise (LiP), manganese peroxidise (MnP) and laccase (Fig. 2.4). LiP has the ability to take electron from the lignin molecule to create a cation radical, which then initiates an oxidative reaction that results in the oxygenation and depolymerising of the lignin. MnP oxidizes Mn (II) to Mn (III), which has the ability to diffuse into the lignin structure and initiate the oxidation process. Laccase is phenol oxidase that differs from peroxidises in that it does not require hydrogen peroxide to directly attack lignocellulose.

White-rot fungi such as *Phanerochaete chrysosporium, Trametes versicolor, Trametes hirsuta and Bjerkandera adusta* has the ability to degrade lignin and can be used as an effective biological pre-treatment (Table 2.4). This is a cheap and effective method of delignification. As fungi grow on these agro-industrial residues, they utilize the polysaccharides after lignin degradation in order to grow and reproduce. This, in turn, has the effect of increasing the nutritional value of the agro-industrial substrates that are genrally low. After fermentation, this may be used as an animal feed or soil fertilizer. This process is mostly carried out under solid state fermentation conditions.

Recent work has concentrated on trying to make biological treatments more selective through the use of genetically manipulated fungi with cellulose promoting enzymes inactivated. This allows lignin degradation without affecting the cellulose component of the complex. Disadvantages of using mutant fungi include their high dependence on an external carbon source and an increase in hemicellulose degradation. This is true of *Sporotrichum pulverulentum* mutants.

Some processes have focussed on directly converting lignocellulosic residues to single-cell protein (SCP). In a method known as the Institut Armand-Frappier process, a *Chaetomium cellulolyticum* mutant and *Pleurotus sajor-caju* as well as strains of *Aspergillus* and *Penicillium* spp., are used. This cocultivation of fungi has the ability to utilize cellulose and hemicellulose, after lignin degradation, for SCP production. In this case, there is no need for any other pre-treatment method, as together these fungi are capable of separating lignocellulose into its individual components. The cellulose obtained may be also used for paper production or as SCP for animal or human feed.

Biological pre-treatments require a long time period in comparison to other tried and tested physical and chemical methods. A period of two to five weeks may be required for sufficient delignification. The direct application of ligninolytic enzymes has also been investigated in order to reduce the length of the treatment period, but

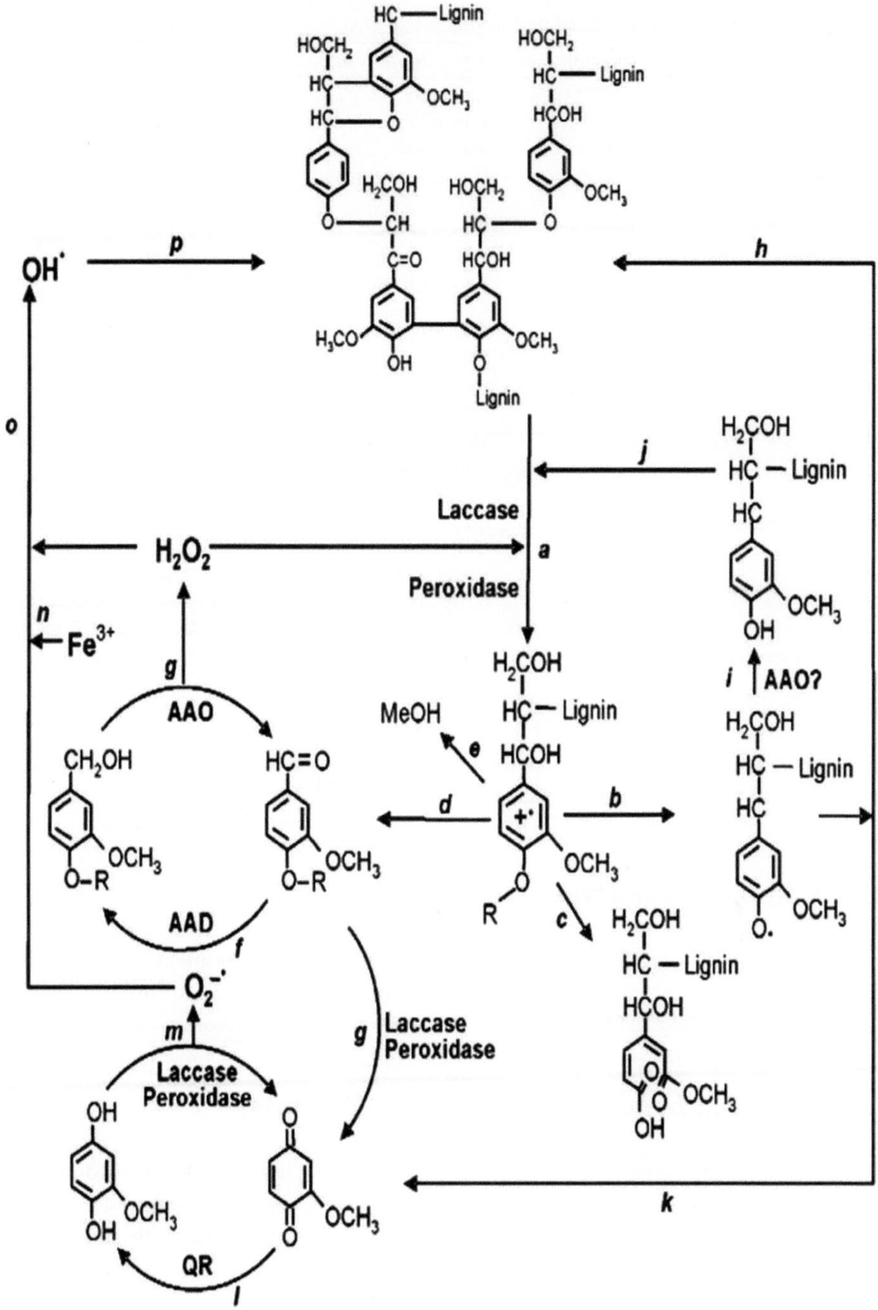

Fig. 2.4 A scheme showing lignin biodegradation showing enzymatic reactions. (Source : International Micobiology (2005) 8: 195-204 updated from Gutièrrez and Martìnez, 1996)

Table 2.4 lignin degrading Enzyme production using agricultural residues

Fungus	Enzyme	Agricultural residues	Reference
B. adusta	MnP	Rice bran	Wang et al. (2001)
C. gallica	Laccase	Wheat bran	Pickard et al. (1999b)
F. sclerodermeus	MnP, laccase	Wheat bran	Papinutti et al. (2003)
L. edodes	Laccase	Wheat straw	Hatvani and Mecs (2002)
P. ostreatus	Laccase	Wheat straw	Morais et al. (2001b)
P. ostreatus	MnP	Wood sawdust	Giardina et al. (2000)
P. ostreatus, *P. sajor-caju*	LiP (MnP), laccase	Banana waste	Reddy et al. (2003)
P. pulmonarius	Laccase	Wheat bran	De Souza et al. (2002)
P. tigrinus	MnP, laccase	Olive mill wastewater	Fenice et al. (2003)
T. hirsute	Laccase	Potato peeling with barley bran	Rosales et al. (2002)
T. versicolor, Funlia trogii	Laccase	Molasses wastewater (vinasse) with cotton stalk	Kahraman and Gurdal (2002)
T. versicolor	Laccase	Barley straw	Couto et al. (2002d)
T. versicolor	Laccase	Grape stalk, grape seed	Lorenzo et al. (2002)
T. versicolor	Laccase	Wood shaving	2002d (2002d)
T. versicolor	Laccase	Barley bran	Lorenzo et al. (2002), Couto et al. (2002d)
T. versicolor	Laccase	Wheat Straw	Couto et al. (2002d)
T. versicolor	Laccase	Wheat bran, wheat husk	Ullah et al. (2000)

the direct use of enzymes for delignification is expensive and suffers from poor enzymes on the lignocellulose material.

2.9 Combined Pre-treatments

These treatments are a combination of physical, chemical, biological and enzymatic treatments used in order to get the best of all treatments hence minimizing the disadvantages of individual treatments.

2.9.1 Gamma Irradiation and Sodium Hydroxide

The physical and chemical pre-treatment when used individually breakdown the material of the agro-industrial residues whereas a combined treatment acts on the cell wall structure. Agricultural residues when pretreated alone with sodium hydroxide increase the organic matter digestibility from 5 to 54% and gamma irradiation alone increase the organic matter digestibility from 8 to 46% but a combination of the two increases the organic matter digestibility by many folds ($p<0.05$) (Al-Masri 1999). As most of the agro-industrial residues contain high concentration of cell wall constituents therefore the increase in the apparent digestibility, as a result of alkali or irradiation pre-treatments, could be attributed to the decrease in cell wall constituents of agro-industrial residues. Individual treatments with NaOH or irradiation decrease the cellulose and hemicellulose content. Other studies have shown that alkali treatment alone decreases the lignin and hemicellulose percentage with swelling in the cellulose content.

Treatment of roughages with alkali at a higher level might upset the rumen fermentation because the high sodium intakes of unreacted alkali prevent the potential digestibility of the feed from being realized.

Irradiation can produce delignification, depolymerization and destruction of the crystalline structure of cellulose (Lowton 1952). Combined treatment of irradiation and sodium hydroxide had better effects in increasing the values of organic matter digestibility hence having a greater effect of the combined treatment on the cell-wall constituents (Al-Masri 1994). Treatment by alkali when given after irradiation of the agricultural residues, was more effective in reducing the values of cellulose and hemicellulose, which could be the cause of the degradation of cellulose and hemicellulose into soluble materials.

NaOH and irradiation on the enzymatic hydrolysis of treated rice straw increased as the irradiation dose increased (Xin and Kumakura 1993).The irradiation may have broken the structure of the lignocellulose so that the NaOH solution could enter easily into the lignocellulose complex, thus the rate of reaction was increased.

2.9.2 Sodium Hydroxide and Solid State Fermentation

In this combined treatment of solid state fermentation when the agro-industrial residue is pre-treated with sodium hydroxide, factors like pH, steam sterilization and urea are important in formulating the substrate for the growth of the microorganisms hence cellulose production which leads to commercial application in the fruit and food industries.

Abbreviations

LiP: Lignin peroxidases
MnP: Manganese peroxidases
SCP: Single cell protein
SSF: Solid State Fermentation

References

Al-Masri MR (1994) Effects of gamma irradiation and sodium hydroxide on cell-wall constituents of some agricultural residues. Second Arab Conference on the Peaceful Uses of Atomic Energy, Cairo, Egypt.

Al-Masri MR (1999) In vitro digestible energy of some agricultural residues, as influenced by gamma irradiation and sodium hydroxide. Appl Radi Iso 50: 295–301

Aoyama M (1996) Steaming treatment of bamboo grass. II. Characterization of solubilized hemicellulose and enzymatic digestibility of water-extracted residue. Cellulose Chem Technol 30: 385–393

Aoyama M, Seki K, Saito N (1995) Solubilization of bamboo grass xylan by steaming treatment. Holzforschung 49: 193–196

Ballerini D, Desmarquest JP, Pourquie ÂJ, Nativel F, Rebeller M (1994) Ethanol production from lignocellulosics: Large scale experimentation and economics. Biores Technol 50: 17–23

Barrier JW, Moore MR, Farina GE, Broder JD, Forsythe ML, Ligthsey GR (1985) Experimental production of ethanol from agricultural cellulosic materials using low-temperature acid hydrolysis. In: Third Southern Biomass Energy Research Conference, Gainesville, FA

Bhavsar K, Shah P, Soni SK, Khire JM (2008) Influence of pre-treatment of agriculture residues on phytase production by *Aspergillus niger* NCIM 563 under submerged fermentation conditions. Afr J Biotechnol 7(8): 1101–1106

Bjerre AB, Olesen AB, Fernqvist T, PloÈger A, Schmidt AS (1996) Pre-treatment of rice straw using combined wet oxidation and alkaline hydrolysis resulting in convertible cellulose. Biotechnol Bioeng 49: 568–577

Bonn G, Concin R, Bobleter O (1983) Hydrothermolysis – a new process for the utilization of biomass. Wood Sci Technol 17: 195–202

Carrasco F (1989) Fundamentos del fraccionamiento de la biomasa. Afinidad 46: 425–429

Chelius MK, Wodzinski RJ (1994). Strain improvement of *Aspergillus niger* for phytase production. Appl Microbiol Biotechnol 41: 79–83

Conner AH (1984) Kinetic modeling of hardwood prehydrolysis. Part I: Xylan removal by water prehydrolysis. Wood Fiber Sci 16: 268–277

Couto SR, Gundin M, Lorenzo M, Sanroman MN (2002d) Screening of supports and inducers for laccase production by *Trametes versicolor* in semi-solid state conditions. Process Biochem 38: 249–255

Couto SR, Moldes D, Liebanas A, Sanroman A (2003) Investigation of several bioreactor configurations for laccase production by *Trametes versicolor* operating in solid-state conditions. Biochem Eng J 15: 21

Daina S, Orlandi M, Bestetti G, Wiik C, Elegir G (2002) Degradation of 5 lignin model dimers by *Ceriporiopsis Subvermispora.* Enzyme Microb Technol 30: 499–505

Dekker RFH, Wallis AFA (1983) Enzymic saccharification of sugarcane bagasse by autohydrolysis-steam explosion. Biotechnol Bioeng 25:3027

De Souza CGM, Zilly A, Peralta RM (2002) Production of laccase as the sole phenoloxidase by a Brazilian strain of *Pleurotus pulmonarius* in solid state fermentation. J Basic Microbiol 42(2): 83–90

Dorado J, Almendros G, Field JA, Alvarez RS (2001) Infrared spectroscopy analysis of hemp (*Cannabis sativa*) after selective delignification by *Bjerkandera* sp. at different nitrogen levels. Enzyme Microb Technol 28: 550–559

Dumoceaux T, Bartholomew K, Valeanu L, Charles T, Archibald F (2001) Cellobiose dehydrogenase is essential for wood invasion and nonessential for kraft pulp delignification by *Trametes versicolor*. Enzyme Microb Technol 29: 478–489

Eichlerova I, Homolka L, Nerud F, Zadrazil F, Baldrian P, Gabriel J (2000) Screeening of *Pleurotus ostreatus* isolates for their ligninolytic properties during cultivation on natural substrates. Biodegradtion 11: 279–287

Fan TL, Gharpuray MM, Lee YH (1987) Cellulose hydrolysis. Spriner-Verlag, Berlin

Fenice M, Sermanni GG, Federici F, D'Annibale A (2003) Submerged and solid-state production of laccase and Mn peroxidase by *Panus tigrinus* on olive mill wastewater-based media. J Biotechnol 100: 77–85

Giardina P, Palmieri G, Fontanella B, RivieccioV, Sannia G (2000) Manganese peroxidase isoenzymes produced by *Pleurotus ostreatus* grown on wood sawdust. Arch Biochem Biophys 376(1): 171–179

Glasser WG, Kaar WE, Jain RK, Sealey JE (2000) Isolation options for noncellulosic heteropolysaccharides (Hetps). Cellulose 7: 299.317

Gould JM (1984) Alkaline peroxide delignification of agricultural residues to enhance enzymatic saccharification. Biotechnol Bioeng 26: 46–52

Govindarao VM (1980) Utilization of rice husk – a preliminary analysis. J Sci Ind Res 39: 495–515

Gutiérrez A, Martínez AT (1996) Biodegradación de la lignina: Una perspectiva actual. Rev Iberoam Micol 13: 18–23
Hatvani N, Mecs I (2002) Effect of the nutrient composition on dye decolorisation and extracellular enzyme production by *Lentinus edodes* on solid medium. Enzyme Microb Technol 30: 381–386
Heitz M, Carrasco F, Rubio M, Chauvette G, Chornet E, Jaulin L, Overend RP (1986) Generalized correlations for the aqueous liquefaction of lignocellulosics. Can J Chem Eng 64: 647–650
Hesch R (1978) Straw is a viable material for the particleboard industry. Plywood Panel 19(7): 26–27
HoÈrmeyer HF, Schwald W, Bonn G, Bobleter O (1988) Hydrothermolysis of birch wood as pre-treatment for enzymatic saccharification. Holzforschung 42: 95–98
Hulscher WS et al. (1992) Briquetting in South and East Asia: State of the Art Assessment. Annual Report. Technology and Development Group, The Univ. of Twente, and Densi Tech BV, The Netherlands
Jiménez L, Lopez F (1993) Characterization of paper sheets from agricultural residues. Wood Sci Technol 27: 468–474
Jiménez L, Sanchez I, Lopez F (1990) Characterization of Spanish agricultural residues with a view to obtaining cellulose pulp. Tappi J 73(8): 173–176
Jiménez L, Perez A, Torre MJ de la, Moral A, Serrano L (2007) Characterization of vine shoots, cotton stalks, Leucaena leucocephala and Chamaecytisus proliferus, and to their ethyleneglycol pulps. Bioresour Technol 98: 3487–3490
Johansson EE, Lind J, Ljunggren S (2000) Aspects of the chemistry of Cellulose degradation and the effect of ethylene glycol during ozone delignification of kraft pulps. J Pulp Paper Sci 26(7): 239–244
Kahraman SS, Gurdal IH (2002) Effect of synthetic and natural culture media on laccase production by white rot fungi. Bioresour Technol 82: 215–217
Kim YO, Kim HK, Baeks Yu JH, Oh TK (1998). Purification and properties of a thermo stable phytase from *Bacillus* spp DS11. Enzyme Microbiol Technol 22: 2–7
Knowles L (1992) A better alternative. Timberline Tribune 1: 6, USA.
Kokta BV, Ahmed A, Garceau JJ, Chen R (1992) Progress of steam explosion pulping: an overview. In: Lignocellulosics. Science, Technology, Development and Use. Ed. Ellis Horwood Limited. Chichester.
Kubikova J, Zemann A, Krkoska P, Bobleter O (1996) Hydrothermal pre-treatment of wheat straw for the production of pulp and paper. TAPPI J 79: 163–169
Lamptey J, Robinson CW, Moo-Young M (1985) Enhanced enzymatic hydrolysis of lignocellulosic biomass pretreated by lowpressure steam autohydrolysis. Biotechnol Lett 7: 531–536
Li J, Henriksson G, Gellerstedt G (2007) Lignin depolymerization/repolymerization and its critical role for delignification of aspen wood by steam explosion. Biores Technol 98: 3061–3068
LoÂpez-Alegret M (1996) FructooligosacaÂridos. Alimentaria 34: 21–23
Loken S et al. (1991) Guide to Resource Efficient Building Cements. Annual Report. CRBT. Missula, MT, USA.
Lora JH, Wayman M (1978) Delignification of hardwoods by autohydrolysis and extraction. TAPPI J 61: 47–50
Lorenzo M, Moldes D, Couto SR, Sanroman A 2002 Improving laccase production by employing different lignocellulosic wastes in submerged cultures of *Trametes versicolor*. Bioresour Technol 82: 109–113
Lowton JE (1952) Effect of high-energy cathode rays on cellulose. Ind Chem 44:2848
Machuca A and Ferraz A (2001) Hydrolytic and oxidative enzymes produced by white and brown-rot fungi during *Eucalyptus grandis* decay in solid medium. Enzmey Mircrob Technol 29: 386–391
Mandviwala TN, Khire JM (2000) Production of high activity thermostable phytase from thermo tolerant *Aspergillus niger* in solid state fermentation. J Ind Microbiol Biotechnol 24: 237–243
Martinez AT, Speranza M, Dueñas FJR, Ferreira P, Camarero S, Guillén F, Martínez MJ, Gutiérrez A, Río JCD (2005) Biodegradation of lignocellulosics: microbial, chemical, and enzymatic aspects of the fungal attack of lignin. Int Microbiol 8:195–204

Modler HW (1994) Biodogenic factors: sources, metabolism and applications. Int Dairy J 4: 383–407

Morais H, Ramos C, Forgacs E, Cserhati T, Oliviera J (2001b) Lignin-modifying enzymes of *Pleurotus ostreatus* grown on agroresidues. Acta Aliment 30(4): 363–372

Mosier N, Wyman C, Dale B, Elander R, Lee, YY, Holtzapple M, Ladisch M (2005) Features of promising technologies for pre-treatment of lignocellulosic biomass. Biores Technol 96: 673–686

Olsson, Hagerdal (1996) Fermentation of lignocellulosic hydrolysates for ethanol production. Enzyme Microbiol Technol 18: 312–331

Orth AB, Royse DJ, Tien M (1993) Ubiquity of lignin degrading peroxidases among various wood-degrading fungi. Appl Environ Microbiol 59(12): 4017–4023

Overend RP, Chornet E (1987) Fractionation of lignocellulosics by steam-aqueous pre-treatments. Phil Trans R Soc Lond A 321: 523–536

Papinutti VL, Diorio LA, Forchiassin F (2003) Production of laccase and manganese peroxidase by *Fomes sclerodermeus* grown on wheat bran. J Ind Microbiol Biotechnol 30: 157–160

Patel JS, R. Onkarappa, Shobha KS (2007) Study of ethanol production from fungal pre-treated wheat and rice straw. Internet J Microbiol 4(1)

Patel MM, Bhatt RM (1992) Optimisation of the alkaline peroxide pre-treatment for the delignification of wheat straw and its applications. J Chem Technol Biotechnol 53: 253–263

Pickard MA, Vandertol H, Roman R, Vazquez-Duhalt R (1999b). High production of ligninolytic enzymes from white rot fungi in cereal bran liquid medium. Can J Microbiol 45: 627–631

Pritchard GI, Pigden WJ, Minson DJ (1962) Effect of gamma radiation on the utilization of wheat straw by rumen microorganisms. Can J Anim Sci 42: 215

Reddy GV, Babu PR, Komaraih P, Roy KRRM, Kothari IL (2003) Utilization of banana waste for the production of lignolytic and cellulolytic enzymes by solid substrate fermentation using two *Pleurotus* species (*P. ostreatus* and *P. sajor-caju*). Process Biochem 38: 1457–1462

Ropars M, Marchal R, Pourquie ÂJ, Vandecastelee JP (1992) Largescale enzymatic hydrolysis of agricultural lignocellulosic biomass. Part 1: Pre-treatment procedures. Biores Technol 42: 197–204

Rosales E, Couto SR, Sanroman, A (2002) New uses of food waste: application to laccase production by *Trametes hirsuta*. Biotechnol Lett 24: 701–704

Saska M, Ozer E (1995) Aqueous extraction of sugarcane bagasse hemicellulose and production of xylose syrup. Biotechnol Bioeng 45: 517–523

Schaffeld G (1994a) El material lignoceluloÂsico. In: Etanol de LignoceluloÂsicos. TecnologõÂa; Perspectivas. Ed. Servicio de Publicacio Âns e Intercambio CientõÂco da Universidade de Santiago de Compostela. Santiago de Compostela. pp. 15–34

Schaffeld G (1994b) Pre-tratamiento del material lignoceluloÂsico. In: Etanol de lignoceluloÂsicos. TecnologõÂa; perspectivas. Ed. Servicio de PublicacioÂns e Intercambio CientõÂ®co da Universidade de Santiago de Compostela. Santiago de Compostela. pp. 35–60

Selle PH, Ravindran V (2007) Microbial phytase in poultry nutrition. Animal Feed Sci Technol 135: 1–41

Singh A, Abidi AB, Darmwal NS, Agrawal AK (1988) Evaluation of chemical pre-treatment for biodegradation of agricultural lignocellulosic wastes by *Aspergillus niger.* World J Microbiol Biotechnol 4(4): 473–479

Soni SK, Khire JM (2005) A Process for Preparation of Acidic Phytase. Indian patent CSIR No. NF-180/2005.

Soni SK, Khire JM (2007) Production and partial characterization of two types of phytase from *Aspergillus niger* NCIM 563 under submerged fermentation conditions. World J Microbiol Biotechnol 23: 1585–1593

Stewart D (2000) Delignification and bleaching of non-wood fibers with peroxymonosulphate.4.jute. Cellulose Chem Technol 34: 287–298

Sun RC, Tomkinson J (2002) Charaterization of hemicelluloses obtained by classical and ultrasonically assisted extractions from wheat straw. Carbohydrate Polymers 50: 263–271

Tortosa JF, Rubio M, Demetrio G (1995) AutohidroÂlisis de tallo de maõÂz en suspensioÂn acuosa. Afinidad 52: 181–188

Tropical Development and Research Institute (1983) A Profile of the Briquetting of Agricultural and Forest Residues. Report No. G181: 1–12. London, UK.

Ullah EA, Kadhim H, Rastall RA, Evans CS (2000) Evalualtion of solid substrates for enzyme production by *Coriolus versicolor*, for use in bioremediation of chlorophenols in aqueous effluents. Appl Microbiol Biotechnol 54: 832–837

Valmaseda M, Martinez JM, Martinez AT (1991) Kinetics of wheat straw solid- state fermentation with *Trametes versicolor* and *Pleurotus ostreatus* lignin and polysaccharides alteration and production of related enzymatic activities. Appl Microbiol Biotechnol 35: 817–823

Wang SY, Joe IS (1983) Studies on the properties of bamboo composite panels. For Prod Ind 2: 17–31.

WangY, Vazquez-Duhalt R, Pickard, MA (2001) Effect of growth conditions on the production of manganese peroxidase by three strains of *Bjerkandera adusta*. Can J Microbiol 47: 277–282

Xin LZ, Kumakura M (1993) Effect of radiation pre-treatment on enzymatic hydrolysis of rice straw with low concentrations of alkali solution. Biores Technol 43:13

Zafar SI, Abdullah N, Iqbal M, Sheeraz Q (1996) Influence of nutrient amendment on the biodegradation of wheat straw during solid state fermentation with *Trametes versicolor.* Int Biodeter Biodegr 83–87

Part II
Production of Industrial Products Using Agro-Industrial Residues as Substrates

Chapter 3
Production of Organic Acids from Agro-Industrial Residues

Poonam Singh nee' Nigam

Contents

P. Singh nee' Nigam (✉)
Faculty of Life and Health Sciences, School of Biomedical Sciences, University of Ulster, Coleraine BT52 1SA, Northern Ireland, UK
e-mail: P. Singh@ulster.ac.uk

P. Singh nee' Nigam, A. Pandey (eds.), *Biotechnology for Agro-Industrial Residues Utilisation*, DOI 10.1007/978-1-4020-9942-7_3,

Keywords Organic acids · Citric acid · Lactic acid · Food-acidulant · Beverage · Buffering · Chelation

3.1 Use of Agro-Wastes for Organic Acid Production

Most of the chapters in this book include information on the variety of agricultural wastes, agro-industrial by-products and residues produced globally and the authors of some of these chapters also have presented data on the annual yield of these renewable resources. Hence, not to repeat this information again here, this chapter deals with the utilization of these bio-resources for organic acid production. Couto (2008) has recently published an article on the exploitation of large scale biological wastes for the production of value-added products. The agricultural raw materials have been used for the bioconversion of their appropriate components for the production of various bio-products and in bio-refining (Koutinas et al. 2007b). The surplus agricultural materials have been used for the production of important products of industrial value such as lactic acid (Koller et al. 2005; Koutinas et al. 2007a,b,c; Reddy and Yang 2005; Venus and Richter 2006). Novel approaches have been tried for the utilisation of waste corn and corn fiber (Johnson 2006; Kalman et al. 2006; Murthy et al. 2006). Other agricultural residues and wheat straw have been studied for the production of industrial products (Chang and Ma 2007; Canilha et al. 2006; Thomsen 2005; Soccol and Vandenberghe 2003). Citric acid production has been studied using solid substrates in solid state fermentation by Prado et al. (2004) and Xie and West (2006) used corn distillers' grain for citric acid production. Oat cereal (Koutinas et al. 2007a) and wheat bran (John et al. 2007) could be utilised as substrates for Lactic acid production.

3.2 Commercial Importance of Organic Acids

Organic acids have established their important application not only in the food and beverage industry, but also in a variety of industries. These acids possess three main characteristics, which make them suitable in various formulations as one of the main ingredients. Their much sought after characteristics include their room temperature solubility, hygroscopic, buffering and chelating nature. Organic acids are used mainly as food acidulants (Taing and Taing 2007; Wang et al. 2008). The main organic acids in industrial use are citric, acetic, tartaric, malic, gluconic and lactic acids. The most utilised organic acid is citric acid (Milson et al. 1985; Moeller et al. 2007). The compound of citric acid, iron citrate is used in pharmaceutical industry as a source of iron and citric acid to preserve cosmetic preparations, ointments, stored blood, and tablets. Organic acids are obtained as the end-products or as the intermediate component of a particular biochemical cycle.

Citric acid is used as a suitable and cheaper replacement of polyphosphates in the detergent industry. The higher cost of polyphosphates formerly restricted its use, however, detergents containing polyphosphates have been prohibited in some countries and therefore their use has been completely replaced by citric acid. Citric

Table 3.1 Global production of organic, acids of industrial importance (adapted from Pandey et al. 2001)

Type of acid	Process of synthesis	Total global production*
Acetic Acid	100% by microbial Fermentaion	120,000
Citric Acid	100% by microbial Fermentation	840,000
Gluconic Acid	[i] 67% by Fermentation [ii] 33% by Chemical-synthesis	50,000
Lactic Acid	[i] 50% by Fermentation [ii] 50% by Chemical-synthesis	35,000
Malic Acid	[i] 70% by Chemical-synthesis [ii] 30% by Fermentation	30,000
Propionic Acid	100% by Chemical-synthesis	50,000
Tartaric Acid	100% by fermentation	30,000

(* measured in metric tonnes)

acid is used in chemical industry as an antifoaming agent, as a softener, and in the treatment of textiles. Pure metals are first produced as metal citrates in metal industry. Organic acids are synthesised commercially either chemically or biotechnologically. The important data has been presented in Table 3.1 on worldwide production of industrially important organic acids.

Though organic acids have wide applications in the food industry as additives and as chemical feed-stocks, but the 75% of food industry usage requires two main acids, citric and acetic acid. Citric acid has established itself as the most widely produced organic acid due to its widest range of applications. The biosynthesis of organic acids has been widely studied (Soccol et al. 2004) using agricultural residues in solid state fermentation (SSF). Citric acid has been produced through SSF technology (Shankaranand and Lonsane 1993; Hang et al. 1987; Hang and Woodams 1984, 1986) for many years, while other organic acids such as lactic, fumaric and oxalic acid have been reported to be produced in SSF only in the last few years.

Fermentation processes play an important role in the production of organic acids. All acids produced as a result of the tricarboxylic acid cycle can be produced in high yields in microbiological processes (Kapritchkoff et al. 2006). Therefore, there is increasing demand for their production. Some of the factors involved in the bulk-production of organic acids are discussed under Section 3.3.

3.3 Organic Acid Biosynthesis

3.3.1 The Production-System

The agricultural residues and by-products are the ideal substrates as the raw material. For economical production the substrates used as raw materials and fermentation medium used for the synthesis process should be cheaper and easily available globally and in large quantities. Therefore, specific medium and fermentation parameters should be simple and easy to optimise. The control of several factors including, bioreactors, oxygen, temperature, humidity and aeration have been optimised by various researchers (Moeller et al. 2007; Murthy et al. 2006).

3.3.2 *The Production-Strains*

For the production of a bio-product such as organic acid, strain selection is very important. The organism must have relatively stable characteristics and the ability to grow rapidly and vigorously (Prescott and Dunn 1959). The microorganism of choice should also be non-pathogenic and suitable for the studies related to the optimisation of parameters (Wang et al. 2008; Richter and Nottelmann 2004). The most important economic characteristic to note in the selection of an organism is its ability to produce high yields of the desired product (Zhang et al. 2007).

3.3.3 *The Product*

The fermentation end-product should be present in a heterogeneous mixture as a result of the metabolism of cultured microorganism. It should be accepted worldwide as GRAS, approved by the Joint FAO/WHO Expert committee on Food additives. At the same time the production of undesirable side-products, such as other acids, should be efficiently suppressed. An economical, easy to follow and efficient large-scale strategy for product-recovery, and purification must be developed.

There is great competition between microbiological and chemical processes for the production of various organic acids. However, the production of citric acid is the exception, which is now synthesised 100% by fermentation. Table 3.2 summarises the production of organic acids in SSF.

3.4 Citric Acid

3.4.1 *Citric Acid Biosynthesis*

Citric acid was a produced by *Mucor* and *Penicillium* sp. as fungal metabolite in media limited in phosphate. The presence of citric acid was detected as a by-product of calcium oxalate produced by a culture of *Penicillium glaucum*. A great number of problems had to overcome, before an effective fermentation process could be used commercially (Lockwood and Schweiger 1967). Other investigation showed the isolation of two varieties of fungi belonging to genus *Citromyces* (namely *Penicllium*). Initially for the production of citric acid microorganisms were cultivated in surface culture. At present over 99% of the world's output of citric acid is produced using the process of microbial fermentation. Ikramul et al. (2007) have used Sugar industry by-product molasses for citric acid production. The optimisation of citric acid production was recently studied by Moeller et al. (2007); and Maria et al. (2007).

Citric acid is a natural constituent and common metabolite of plants and animals. It is the most versatile and widely used organic acid. Citric acid is an important commercial product with a global production reaching 840,000 tons per year. Currently citric acid is produced by fermentation-technology, using the filamentous fungus *Aspergillus niger* mainly through surface (solid or liquid) and

Table 3.2 Use of agro-industrial residues and by-products for the production of organic acids (adapted from Pandey et al. (2001))

Agri-industrial waste	Production-strain	Fermentation product
Oat cereal	*Rhizopus oryzae*	Lactic acid
Molasses	*Aspergillus niger GCMC-7*	Citric acid
Sweet Potato	*Aspergillus niger*	Citric Acid
Wheat bran	*Mixed lactobacilli*	Lactic Acid
Starch	*Immobilised Rhizopus oryzae*	Lactic acid
Pineapple Waste	*Aspergillus Foetidus*	Citric acid
Pineapple Waste	*Aspergillus niger*	Citric Acid
Sugar Cane Bagasse	*Rhizopus oryzae*	Lactic Acid
Carrot-Processing Waste	*Aspergillus niger*	Citric Acid
Gallo Seeds-Cover Powder	*Aspergillus nige, A. oryzae*	Gallic Acid
Okara, Soy-Residues	*Aspergillus niger*	Citric Acid
Carob-Pods	*Aspergillus niger*	Citric Acid
Corn-Cobs	*Aspergillus niger*	Citric Acid
Sweet Potato	*Rhizopus Sp.*	Oxalic Acid
Cassava	*Aspergillus niger*	Citric Acid
Cassava Bagasse	*Aspergillus niger*	Citric Acid
Myroballan seeds	*Rhizopus oryzae*	Gallic Acid
Sugarcane Press Mud	*Aspergillus niger*	Citric Acid
Kiwifruit peel	*Aspergillus niger* NRRL 567	Citric Acid
Cassava	*Aspergillus niger*	Lactic Acid
Cassava	*Streptococcus thermophilus*	Furmaric Acid
Sweet Sorghum	*Lactobacillus helveticus*	Lactic Acid
Sugar Cane Press-Mud	*Rhizopus oryzae*	Lactic Acid
Sugar Cane Press-Mud	*Lactobacillus casei*	Lactic Acid
Coffee Husk	*Aspergillus niger*	Citric Acid
Sugar Cane Press-Mud	*Lactobacillus Paracasei*	Lactic Acid
Starch Containing root Kumara	*Aspergillus niger*	Citric Acid
Carrot-Processing Waste	*Rhizopus* Sp.	Lactic Acid
Amberlite (Inert Solid Support)	*Aspergillus niger*	Citric Acid
Polyurethane (Inert Solid Support)	*Aspergillus niger*	Citric Acid

submerged fermentation of starch or sucrose-based media, (Jianlong 2000; Vandenberghe et al. 2000). The food industry utilises about 70% of the total production of citric acid (Rohr et al. 1983), the pharmaceutical industry consumes 12% and the rest 18% has market for other applications.

3.4.2 Industrial Applications of Citric Acid

Citric acid has a long list of applications in industrial sectors such as the food, beverage, and pharmaceutical industries. It is mainly used for preservation, antioxidation, chelation and acidulation. It is used as flavour enhancer, plasticizer and synergistic as well as sequestering agent. Its rising demand is subsequently causing an increase in global production.

Citric acid is mainly used in the food industry because of its pleasant acidic taste and its high solubility in water. It is worldwide accepted as "GRAS" (generally

Table 3.3 Industrial Applications of citric acid (adapted from Pandey et al. (2001))

Properties and functions of citric acid	Commercial sector
Tartness; complementary fruits and berries flavours; effective antimicrobial preservative; pH adjuster, providing uniform acidity.	Soft drinks; canned fruit-juices; bottled beverages
Tartness; Producer of dark colour in hard sugar-candies; Acidulant; restricting sucrose inversion	confectionery
pH adjustment, antioxidant as a metallic-ion chelator, buffering agent.	Cosmetics
Emulsifier in ice creams and processed cheese; acidifying agent in many cheese products; antioxidant.	Dairy (Ice-cream & Cheese)
Synergist for other antioxidants, as sequestrant.	Fats and oils
Lowers pH to inactivate oxidative enzymes. Protects ascorbic acid by inactivating trace metals.	Frozen fruit
Sequestrant of metal ions, neutralizant, buffer agent	commercial applications
Provides tartness, pH adjustment.	Fruit-preserves; marmalade; confiture
Removes metal oxides from surface of ferrous and nonferrous metals, for preperational and operational cleaning of iron and copper oxides.	Purification of Metal oxides
As effervescent in powders and tablets in combination with bicarbonates. Provides rapid dissolution of active ingredients. Acidulant in mild astringent formulation. Anticoagulant.	Pharmaceuticals
electroplating, copper plating, metal cleaning, leather tanning, printing inks, bottle washing compounds, floor cement, textiles, photographic reagents, concrete, plaster, refractories and moulds, adhesives, paper, polymers, tobacco, waste treatment	Industrial; miscellaneous in world market

recognised as safe), approved by the joint FAO/WHO Expert Committee on Food Additives. Table 3.3 summarises the main industrial and commercial uses of citric acid.

3.4.3 Production of Citric Acid

Citric acid has always been a subject of interest and therefore of research. Solid state fermentation has been regarded as an alternative method to produce citric acid from agro-industrial residues. It is also produced from starch or sucrose-based media using liquid and surface fermentation (Milson et al. 1985; Peller and Perlman 1979). Bioproduction can be regarded as a combination of fermentation and processing techniques. The two methods should be considered as a single production unit because of the competition afforded by chemical synthesis.

The choice of the production method is dictated, amongst other things, by the effect of energy metabolism on the product formation. Production of citric acid is a good example of production of a metabolite and its downstream processing method. Citric acid is a tricarboxylic acid and is an intermediate in the Kreb's cycle. It was discovered that *Aspergillus niger* could accumulate citric acid in a media rich in carbohydrate but deficient in phosphate and trace elements like Fe^{+2}, Zn^{+2} and Mn^{+2}. The regulation of glycolysis is most important for citric acid production. The fermentation is carried out aerobically in large fermenters and a key requirement for high citric acid yields is that the medium be iron-deficient. This is because citric acid is overproduced by the fungus as a chelator to scavenge on.

3.4.4 Process-Technology

The process depends on the type of substrate used for organic acid production, therefore different workers have used different methods (Pandey et al. 2001; Soccol et al. 2004). Wheat bran agricultural residue has been used as a fermentation raw material in the Koji-process (Yamada 1977). Koji process was first developed in Japan and is used for commercial production that accounts for 20% of all the citric acid produced in the country annually. The pH of the bran is adjusted to 4.5 before sterilization, and is then steam-sterilised. During steaming, wheat bran gathers moisture so the final water-saturation becomes 70–80%. The sterilised bran at pH 4.5 is converted into a mash. The bran-mash is then cooled to 30–36°C prior to inoculation. The mash is mixed well and inoculated with "Koji", which is a preparation of fermented grains covered with fungal growth.

A variety of different cereal grains such as soybean, wheat and rice can be used. Koji is used to inoculate the wheat bran mash. A special strain of *Aspergillus niger* is used for this fermentation process because wheat bran contains a high concentration of undesirable trace elements such as like Fe^{+2}, Zn^{2+} and Mn^{2+}. This special strain is capable of producing citric acid unaffected in the presence of these ions. Koji provides as a source of important enzymes such as amylase, protease and lipases, which are used in the hydrolysis process required for cereal grains containing considerable percentage of carbohydrate-starch. Amylase helps to hydrolyse starch in the wheat bran to glucose, which is further fermented to citrate by the fungus. The inoculated wheat bran is then spread into trays at a depth of 3 to 5 cm or is arranged in long rows on the surface of fermenting units. Solid state cultivation is carried out for a week at a temperature of 30–36°C. The pH in the mash drops to 1.8–2.0. After the fermentation is over, the fermented wheat bran is harvested and placed in percolators. The fermented mass is extracted with water for citric acid. The aqueous extract is subjected to modern downstream processing, for purification and concentration of the organic acid.

3.4.5 Commercial Strains for the Synthesis of Citric Acid

Some fungal strains are capable of producing citric acid as a metabolite in their primary metabolism. Citric acid is produced by several moulds such as *Penicillium luteum, P. citrinum, Aspergillus niger, A. wentii, A. clavatus, Mucor piriformis, Citromyces pfefferianus, Paecilomyces divaricatum, Trichoderma viride* etc. Citric acid can also be produced by yeasts, such as *Yarrowia lipolytica* and *Candida guilliermondii*. However, for commercial production only mutants of *Aspergillus niger* should be used. The reason being that compared to *Penicillium* strains, the aspergilli are capable of producing higher yield of citric acid per unit time. Moreover, the production of undesirable side products such as oxalic acid, isocitric acid and gluconic acid, could be easily and efficiently suppressed in the mutants of *Aspergillus niger*.

Production strains are optimised based on the carbon sources used. A strain, which produces citric acid in good yields using one substrate, may not be able to produce the same product yield with another carbon source. If the raw material used is starch, the employed fungus should be able to form amylases for the hydrolysis of starch. Using non-amylase producing strains with starch as the starting substrate, amylase enzymes need to be supplemented in the fermentation medium (Wang et al. 2008). Since starch forms the largest part of food waste, external enzymes, especially glucoamylase (EC 3.2.1.3) which could hydrolyse starch to produce glucose would be necessary for the bioconversion of food waste into organic acids (Pandey et al. 2000). Industrial strains that produce commercial citric acid are not freely available. A few can be obtained from international culture collections. These are *Aspergillus niger* NRRL 2270, NRRL 599, ATCC 11414 and ATCC 9142; and *Y. lipolytica* ATCC 20346, ATCC 20390, NRRL Y-7576 AND NRRL Y-1095. Improvement of strains for citric acid production is focused on obtaining strains with reduced levels of activity of isocitrate dehydrogenase and aconitase.

3.5 Preparation of Production Medium

3.5.1 Pre-culture for Fermentation

The fungal-preculture used as inoculum for citric acid production is in the form of a spore-suspension. Spores are produced in glass bottles on solid substrates at optimum temperatures. The incubation period for the formation of spores is 8–14 days. Besides the total number of spores, the viability of spore crops is critical. *Aspergillus niger* spores are produced in Czapeck Dox Broth solidified with agar in bottles or flasks. Medium is incubated at 28°C for minimum of 8 days. Spores are removed in the form of a suspension using distilled water with 0.1% Tween-80. This spore-suspension can then be stored at 4°C for a maximum of two weeks. A spore-suspension used to inoculate the solid culture medium should contain at least 10^8 spores per ml suspension.

3.5.2 Medium and Culture Conditions for SSF

SSF can be carried out using several raw materials (Shankaranand and Lonsane 1993, 1994) and agricultural wastes (Tongwen and Weihua 2002). A variety of substrates can be used for citric acid production especially in SSF. Various agro-industrial by-products and residues such as sugarcane press mud, coffee-husk, wheat-bran, cassava fibrous residue, rice bran and de-oiled rice bran have been evaluated as the potential substrates for citric acid production in SSF. The list of other potential substrates includes molasses distributed on spaghnum moss with calcium carbonate, saw dust impregnated with molasses, beet pulp impregnated with pine apple juice, sweet potato residue and wheat bran, sugar-free sugarcane bagasse with molasses, sugarcane bagasse impregnated with concentrated liquor of pine apple juice, rice-bran, apple-pomace, grape-pomace, kiwi-fruit peel, sugarcane bagasse impregnated with cellulose hydrolysate, cassava waste, citrus waste, mandarin wastes, polyurethane foam, starch cake with rice-bran as well as cassava and coffee wastes.

Generally, the substrates are moistened to about 70% depending on the substrate absorption capacity. The initial pH is normally adjusted to a lower value before sterilisation. After sterilisation the material is inoculated with spore-suspension and spread on trays in layers of 5 cm maximum thickness. The temperature of incubation varies according to the microorganisms used. Growth can be accelerated by adding (x-amylases), although *Aspergillus niger* can hydrolyse starch with its own amylase. The solid culture process takes 90 hours, at the end of fermentation period the fermented mass is extracted with hot water to isolate citric acid from the solid medium.

One of the important advantages of the SSF process is that the presence of trace elements may not affect the citric acid production as it does in submerged liquid-fermentation. Consequently, substrate pre-treatment is not required. Citric acid production is directly influenced by the nitrogen source. Physiologically, ammonium salts are preferred such as urea, ammonium chloride and ammonium sulphate. Nitrogen consumption leads to a pH decrease, which is very favourable for citric acid fermentation. Urea has a tampon effect that assures pH control. A high nitrogen concentration increases fungal growth and the consumption of substrate, but this result in a decrease yield.

The presence of phosphate in the fermentation medium has a great influence on the yield of citric acid. Potassium dihydrogen phosphate has been found to be the most suitable and effective phosphorus source. Low levels of phosphorus favour citric acid production. The higher concentrations of phosphorus in the fermentation medium results in the formation of certain sugar-acids, a decrease in the fixation of carbon dioxide and the stimulation of culture-growth. Phosphates act at the level of enzyme activity. Different strains require distinct nitrogen and phosphorus concentrations in the medium. In fact, nitrogen and phosphorus limitation is the crucial factor in citric acid production as there is interaction between them.

The specificity of solid culture is largely due to a lower diffusion rate of nutrients and metabolites occurring at lower water activity in SSF. Consequently, the strains with large requirements of nitrogen and phosphorus are not the ideal organisms for SSF. The production process employing solid culture medium is in use solely in small production-plants and a maximum of 500 tons per year are produced in these plants. The other production processes used are surface-fermentation, which are performed using solid and liquid media. Several factors affect the choice of fermentation processes or production-type. Important ones are: availability of investment capital, energy availability, cost of labour and training, and the availability of techniques for the measurement and regulation of the process.

3.5.3 Bioreactors for Citric Acid Production

Various types of fermenters can be used for the production of citric acid in SSF (Prado et al. 2004; Hang, 1988). The most common fermenters for this purpose are glass-incubators, glass-columns, trays, Erlenmeyer conical flasks, rotating drum bioreactors, single-layer and multi-layer packed-bed bioreactors and packed-bed column bioreactors (Hang and Woodams 1985; Raimbault and Alazard 1980). Production is usually better in flasks while the yields could be lower in tray and rotating-drum bioreactors. A multi-layered packed-bed bioreactor is capable of considerably improved mass-transfer compared to a single-layer packed-bed bioreactor operated under similar conditions. Packed-bed bioreactors show superior production of citric acid than the production in flask culture under similar conditions. Higher yields of citric acid could be obtained using cassava bagasse as a SSF-substrate in a packed-bed column bioreactor than in flasks. Aeration, heat and mass-transfer effects can be controlled and improved in bioreactors. Different mechanisms of heat removal such as conductive, convective and evaporative from packed-bed bioreactors in SSF may also affect citric acid production. The conductive heat transfer could be the least efficient mechanism compared to convective and evaporative mechanisms. Evaporative heat-transfer is the best mechanism of heat removal from packed-bed bioreactors is SSF.

3.6 Optimisation of Factors

One of the most important factors is the employment of a suitable microorganism in fermentation. The strain of microorganism should have the ability to accumulate in the presence of high concentrations of various trace elements (Sikander and Ikramul 2005). The ability of microorganisms to produce amylolytic enzymes is of great importance if the SSF-substrates used are starchy in nature. The selection of strain is done for the utilisation and fermentation of specific substrates in SSF. The rate of citric acid production may become slower if the sufficient enzymes are not produced by the culture for the fermentation of substrates (Maria et al. 2007). In such cases the saccharification of the substrate can be performed separately, or by adding commercial preparations of enzymes to the fermentation medium at the time of culture-inoculation the rate of saccharification can be enhanced.

3.6.1 Influence of Environmental Factors

The pH of a culture may change in response to microbial-metabolic activities. The most obvious reason is the secretion of organic acids such as citric acid, acetic or lactic acids. Production and accumulation of organic acids in the fermentation medium will cause the pH of the medium to decrease. Changes in pH kinetics depend mainly also on the microorganisms employed in SSF. The pH can drop very quickly to less than 3.0 with the organisms such as *Aspergillus* sp., *Penicillium* sp., and *Rhizopus* sp. The pH is more stable between 4.0 and 5.0 for other groups of fungi such as *Trichoderma, Sporotrichum*, and *Pleurotus* sp. The nature of the substrate may also influence the pH kinetics in the fermentation medium. Generally, a pH below 2.0 is required for optimum production of citric acid. A low initial pH has the advantage of preventing bacterial contamination and also helps in inhibiting the formation of oxalic acid. A lower pH is the optimum for the growth of the fungal cultures and at the same time for accumulation of citric acid (Kolicheski 1995).

The synthesis of citric acid by *Aspergillus niger* using different agro-industrial residues of variable composition requires a number of problems to be encountered due to high sensitivity of the culture to a number of medium-ingredients. Citric acid synthesis is directly influenced by the nitrogen source and its consumption leads to a pH decrease, which is very important point in citric acid fermentation. The presence of phosphate and trace elements also has a significant influence. Another important factor is the presence of lower alcohols, which has been found to enhance the citric acid synthesis. Appropriate alcohols being methanol and ethanol and their optimal amount depend on the strain being used and the composition of the fermentation medium.

Relationship between citric acid production and respiration rate of the cultivated microorganism has been studied in detail by Prado et al. (2004). Glass column fermenters have been used in order to study the influence of aeration on citric acid production. On-line respirometry monitoring is used to evaluate the performance of SSF-process. A nutritive solution composed of urea, potassium dihydrogen phosphate, methanol and $FeSO_4$. $7H_2O$ has been found to increase the production of citric acid to 347.7 g per kg dry substrate. The yield of citric acid was achieved 81% based on starch-consumption.

Other factors can also influence the citric acid production in SSF. The presence and concentration of lower alcohols affect the yield. A list of parameters affecting the yield of citric acid includes: weight of the substrate in the bioreactor, the availability of surface-area of the SSF-substrate for microbial-colonisation, pre-culture condition such as the type, age and size of inoculum, initial moisture content of the prepared SSF-medium, various fermentation-parameters such as incubation-temperature, rate of aeration and mixing, concentration and nature of the carbohydrates in SSF-substrate used and the pH of the SSF-medium. It has been generally found that the addition of methanol may increase production. Glycolytic rate can influence the production of citric acid and oxalic acid in SSF. The overproduction of citric acid is found to be related to an increased glucose-flux through glycolysis. At a lower glucose-flux oxalic acid is found to accumulate in the medium. The specificity of solid state culture is largely due to a lower diffusion rate of nutrients and metabolites, which occurs in lower water activity conditions. Consequently, strains with large requirements of nitrogen and phosphorus are not favoured, due to the restriction of accessibility to the nutrients in SSF-medium.

3.6.2 Influence of Micronutrients

Trace-element nutrition is the main factor influencing the yield of citric acid. A number of divalent metals such as zinc, manganese, iron, copper and magnesium have been found to affect the citric acid yield in fermentation by *Aspergillus niger* (Shankaranand and Lonsane 1994). Different additives and metallic micro minerals have been reported for the enhanced production of citric acid by *Aspergillus niger* cultivated on different carbohydrate materials (Sikander and Ikramul 2005). It is crucial to take into account the interdependence of medium-constituents probably in SSF. Zinc favours the production of citric acid when added with potassium hydrogen phosphate. The presence of manganese ions and iron and zinc in high concentrations could cause the reduction of citric acid yields only in phosphate-free medium. Some microorganism such as *Aspergillus niger* show a difference in response to metal ions and minerals in different fermentation systems, therefore the response of the strain may be different in SSF compared to submerged fermentation. SSF-systems are able to overcome the adverse effects of high concentrations of metals and minerals in the medium. As a consequence of this, the addition of chelating agents such as potassium ferrocyanide to the solid medium proves to be of no use.

Copper is found to complement the ability of iron at optimum level, to enhance the biosynthesis of citric acid. Manganese deficiency results in the repression of the anaerobic and TCA cycle enzymes with the exception of citrate synthetase. This leads to overflow of citric acid as an end-product of glycolysis. A low level of manganese (ppm) is capable of reducing the yield of citric acid by 10%. Citric acid accumulation decreases with the addition of iron, which also has some effect on mycelial growth and the colonisation of the fungus. Presence of different copper concentrations in the pellet formation medium could be very important in order to enhance a suitable structure, related to the cellular physiology and for citric acid production. Magnesium is required for growth as well as for citric acid production.

3.6.3 Influence of Various Supplements

Some compounds that can inhibit metabolism, such as calcium fluoride, sodium fluoride and potassium fluoride have been found to accelerate the citric acid production, but potassium ferrocyanide decreases the yield. There are many compounds, which could act in many ways to favour the production of citric acid. These favouring compounds are capable of impairing the action of metal-ions and other toxic compounds, which influence the growth during the initial phase. Some of the favouring compounds are: 4-methyl-umbelliferone, 3-hydroxi-2-naphtoic, benzoic acid, 2-naphtoic acid, iron cyanide, quaternary ammonium compounds, amine oximes, starch, EDTA and vermiculite etc.

Addition of lower alcohols enhances citric acid production from crude carbohydrates. Appropriate alcohols could be methanol, ethanol, isopropanol or methyl acetate. The optimal amount of methanol/ethanol depends upon the strain and the composition of the fermentation medium. The addition of ethanol may result in two-fold increase in citrate synthetase activity and a 75% decrease in aconitase activity. Whereas the activities of other TCA cycle enzymes increase slightly. Alcohols have been shown to principally act on membrane permeability in microorganisms by affecting phospholipid composition on the cytoplasmatic membrane. The role of membrane permeability may be argued in the citric acid accumulation. The alcohols stimulate citric acid production by affecting the growth and sporulation through by not only acting on the cell-permeability but also on the spatial organisation of the membrane, or changes in lipid composition of the cell wall.

3.7 Mechanism of Citric Acid Synthesis

Recently advances in citric acid fermentation including biochemical aspects, membrane transport system and the modelling of the process are well documented by Maria et al. (2007). Citric acid (2-hydroxy propane-1,2,3, tricarboxylic acid) is a primary metabolite, which is formed in the tricarboxylic acid cycle. Glucose is the starting carbon-source in metabolism, which is released from the enzymatic-hydrolysis of various substrates used as raw material in the citric acid fermentation. A part of glucose coming from the raw materials used in the fermentation is consumed in the trophophase for the production of fungal-mycelium and is converted through respiration into CO_2. The rest of the glucose, in the idiophase, is converted into organic acids. There is a minimal loss through respiration during this phase. The theoretical yield is 123 g citric acid-1 -hydrate or 112 g anhydrous citric acid per 100 g sugar. These yields are theoretical, and are not obtained in practice because of loss during the trophophase.

3.7.1 Pathways of Citric Acid-Biosynthesis

Biochemical aspects, membrane transport and modelling of citric acid fermentation have been recently published by Maria et al. (2007). Glucose used for the citric acid synthesis is metabolised in two pathways; 80% is broken down by reactions of the Embden-Meyerhof-Parnas (EMP) pathway and the rest 20% goes through the reactions of the Pentose-phosphate cycle. The relationship between these two pathways is 2:1 during the growth phase. The enzymes of the EMP-pathway are present throughout the fermentation process in the strains of citric acid production. The activity of the pathway is regulated in a positive manner with phosphofructokinase but pyruvate kinase regulates the activity in a negative manner. The acetate residue is channelled into the tricarboxylic acid cycle after pyruvate is decarboxylated with the formation of acetyl-CoA. All enzymes of the Krebs cycle are expressed during the idiophase except α-ketoglutarate dehydrogenase. During the production of citric acid, the citrate synthase activity, a condensing enzyme, is increased by a factor of 10. Those enzyme activities, which catabolise citric acid, are aconitase and isocitrate dehydrogenase, both activities are sharply reduced in the trophophase. A mitochondrial enzyme, one of the three isocitrate dehydrogenase isozymes, is specific for

NADP, which is inhibited by glycerol that accumulates during the spore germination process. In addition, citric acid production is inhibited by high intracellular concentrations of ammonium ion.

Citrate synthase can not be solely responsible for maintaining the activity of the tricarboxylic acid cycle, because the cycle would cease if citric acid were removed. The tricarboxylic acid cycle intermediates, replenished by distinct sequences (anaplerotic sequences), must exist in the production phase. The first anaplerotic enzymes present in *Aspergillus* is a pyruvatecarboxylase. Pyruvate and carbon dioxide are converted into oxalacetate, inorganic phosphate, and ADP with the consumption of ATP. The reaction is dependent on Mg^{2+} and K^+ ions. Acetyl CoA is not required for this reaction, although this is required in the metabolic reactions of other microorganisms. Therefore, pyruvate carboxylase is the essential enzyme for citric acid biosynthesis. The second anaplerotic sequence involves a phosphoenol pyruvate (PEP) carboxykinase enzyme. This enzyme converts PEP and CO2 into oxalacetate and ATP in the presence of ADP. This system needs Mg^{2+} or Mn^{2+} and K^+ or NH_4^+.

3.7.2 Biochemistry of Citric Acid-Overproduction

The metabolic pathways involved in the overproduction of citric acid by *Aspergillus niger* have been studied (Moresi and Parente 2000). Essential for overproduction are a high flux of metabolites through the glycolysis (5–10 mmol/min per mg of protein), a block of reactions of the tricarboxylic acid (TCA) cycle that degrade citrate, and an anaplerotic sequence, to replenish the oxaloacetate (OAA) used for the synthesis of citric acid.

Although the regulation of enzyme activity is critical in controlling the metabolic flux toward citric acid overproduction, the most important steps in controlling the rate of the pathway are glucose transport and hexokinase activity (HK). Ikramul et al. (2002) have studied that the improvement of strains of micro-organisms for the production of citric acid has been and can be traditionally achieved by mutagenesis and screening. But it has been postulated that the overexpression of HK and the glucose-transport systems would result in the maximum possible increase in citric acid synthesis. This strategy appears to be viable. The overexpression of about 20–30-fold of pyruvate kinase (PK) and phosphofructokinase (PFKI) has been achieved in *Aspergillus niger*. The metabolic changes necessary for the overproduction of citric acid by *Aspergillus niger* and *Yarrowia lipolytica* are induced by high sugar concentrations (Moeller et al. 2007) and manganese deficiency, although other factors are also important mainly the concentrations of phosphate, nitrogen and trace metals, a low pH, and high dissolved oxygen concentration.

3.8 Extraction of Citric Acid from Fermentation

Citric acid from the fermented solid material is leached out using water. The extract obtained contains citric acid, which is recovered following downstream processing by conventional methods. Recovery from the extracts is generally accomplished in three processes, precipitation, extraction and adsorption and absorption using ion-exchange resins. The Food and Drug Administration of the United States have described citric acid extraction. Citric acid extracted by this method has been recommended suitable for use in food and drugs. Precipitation is the classical method and it is performed by the addition of calcium oxidehydrate (milk of lime) to form the slightly soluble tri-calcium citrate tetrahydrate. The precipitated tri-calcium citrate is removed by filtration and washed several times with water. The precipitate is then treated with sulphuric acid to form calcium sulphate, which is removed by filtration. Mother liquor contains citric acid is treated with active carbon and passed through cation and anion exchangers. Several anion-exchange resins are commercially available. Finally the liquor is concentrated in vacuum crystallisers at 20–25°C, forming citric acid monohydrate. Anhydrous citric acid is prepared following the process of crystallisation at higher temperature than 25°C.

3.9 Productivity of Citric Acid

Few examples of citric acid production using various agricultural residues and industrial by-products have been presented in Table 3.4

Different agro-industrial residues have been investigated (as presented above) for citric acid production (Kolicheski et al. 1997). In order to achieve economic development, focus has shifted to the industrial application of cassava for value addition. Industrial processing of cassava tubers is mainly done to isolate flour and starch. Processing for flour generates solid-residues including brown peel, inner peel, unusable roots, crude bran and bagasse. Cassava bagasse is a fibrous residue from the extraction process, which is generated during the separation stage. This bagasse contains about 40–70% starch that physically could not be extracted. It has a large absorption capacity and may hold up to 70% moisture. This waste can be used for citric acid production along with two other potential substrates, sugarcane bagasse and coffee-husk.

Citric acid production from cassava bagasse can be carried out using a culture of *Aspergillus niger* NRRL 2001 in solid state fermentation. SSF is performed at 30°C for 120 h. These three substrates are ground to a particle size of 0.8–2.0 mm and dried at 55-60°C for 12 h. The fermentation-nutrient medium containing $ZnSO_47H{\sim}O$ (0.2 g/l) and $FeCl_3$. H_2O (0.014 g/l) are sterilised at 121°C for 15 min. After cooling methanol (4%) is added aseptically to the medium. Sugar cane bagasse and coffee-husk are supplemented with 40% of glucose (corresponding to the starch content of cassava bagasse). Solid substrates are inoculated with the spore-suspension, mixed with the nutrient medium. The inoculum size must contain 10^7 spores per g of dry substrates. The initial moisture contents of the three substrates vary due to their structure and composition. Solid culture medium should contain 90, 65 and 62% of moisture contents for sugar cane bagasse, coffee-husk and cassava bagasse, respectively. Fermentation of these solid substrates can be carried out in vertical column fermenters, further details and the kinetics of this process are discussed by Pandey et al. (2001).

Table 3.4 Biosynthesis of citric acid from agro-industrial substrates (adapted from Pandey et al. 2001)

Agro-industrial material	Culture used for Biosynthesis	Recovery
Food wastes	*A. niger* UV60	45.5 g/l
Wheat bran	*A. niger* CFTRI30	85 g/kg
Apple-pomace	*A. niger* NRRL2001	766 g/kg
Cassava residue	*A. niger* CFTRI30	234 g/kg
Apple-pomace	*A. niger*NRRL2270	816 g/kg
Rice-bran	*A. niger* CFTRI30	127 g/kg
Apple pomace	*A. niger* NRRL2270	771 g/kg
De-oiled rice bran	*A. niger* CFTRI30	92 g/kg
Sugar cane press-mud+ Wheat bran (4: 1)	*A. niger* CFTRI30	116 g/kg
Apple pomace	*A. niger* NRRL 328	789 g/kg[a]
Apple pomace	*A. niger* NRRL 567	883 g/kg[a]
Grape pomace	*A. niger* NRRL 2001	413 g/kg[a]
Grape pomace	*A. niger* NRRL 2270	511 g/kg [a]
Grape pomace	*A. niger* NRRL 599	498 g/kg [a]
Grape pomace	*A. niger* NRRL 328	523 g/kg [a]
Grape pomace	*A.. niger* NRRL 567	600 g/kg [a]
Kiwifruit peel	*A. niger* NRRL 567	100 g/kg [a]
Sugar cane	*A. niger*	29 g/kg
Orange waste	*A. niger*	46 g/kg
Beet molasses	*A. niger* ATCC 942	35 g/l
Sugar cane Bagasse	*A. niger* CFTRI 30	174 g/kg[b]
Coffee husk	*A.. niger* CFTRI 30	150 g/kg[b]
Carrot waste	*A. niger* NRRL 2270	29 g/kg[a]
Okara (soy residue)	*A. niger*	96 g/kg[a]
Pineapple waste	*A. niger* ATCC 1015	132 g/kg[b]
Pineapple waste	*A. niger* ACM 4942	194 g/kg[b]
Glucose (Sugar can bagasse)	*A. niger* CBS 733.88	21.24 g/l
Kumara (starch containing)	*A. niger*	103 g/kg[b]
Cassava bagasse	*A. niger LPB*– 21	347 g/kg[b]

a – based on sugar consumed; b - based on dry matter

3.10 Prospects of R & D in Citric Acid Biosynthesis

The production of citric acid has not received much attention in the form of modern methods of molecular biology, presumably because it is considered a well-established area. Mutagenesis and selection have carried out the improvement of strains of *Aspergillus niger*. The two principal methods of selecting populations, namely, "the single-spore technique" and the "passage method" have been used for selecting citric acid producing strains. The single-spore technique has the disadvantage in that mineral acid and organic acids (gluconic and oxalic acids) simulate the presence of citric acid; this method has been improved by incorporating a specific stain para-di-methylamino benzaldehyde instead of using the indicator for citric acid. The most employed technique to improve citric acid producing strains has been by inducing mutations in parental strains using mutagenesis. Among physical mutagens, γ-radiation and UV-radiation have often used. Hyperproducer strains are obtained by combining UV treatment with some chemical mutagens such as aziridine, N-nitrosoN-methylurea, or ethyl methane-sulphonate.

Another approach for strain improvement has been the para-sexual cycle. Diploid strains display higher citric acid yields compared to their parent haploids, but they tend to be less stable. Protoplast fusion appears to be a promising tool to extend the range of genetic manipulation of *Aspergillus niger* with respect to citric acid production. Protoplast fusion techniques have produced fusant-strains that have acid production capacities exceeding those of the parent strains in solid state fermentation. The other important aspects of strain improvement could be the resistance to detrimental constituents of fermentation raw-materials, capability of utilizing a variety of raw materials e.g. starch, cellulose, pectin-containing and other waste materials. However, there is no single effective technique to achieve the hyper-producing mutants and much remains to be done in this area.

Metabolic engineering is now possible, in the light of increased knowledge about the regulation of acidogenesis. The biochemistry of the citric acid biosynthesis has been studied in detail and knowledge of pathways is necessary to regulate the biosynthesis.

3.11 Lactic Acid

Food industry wastes have been used for lactic acid synthesis using integrated glucoamylase production (Wang et al. 2008). Development of an oat-based biorefinery for the production of lactic acid by *Rhizopus oryzae* has been recently reported by Koutinas et al. (2007a). Lactic acid ($C_3H_6O_3$: 2-hydroxypropionic acid) is produced by fermentation (50%) and by chemical-synthesis (50%). Lactic acid is common in nature; it is present in plants and microorganisms. Lactic acid has two enantiomers, L-(+) and D-(−)-lactic acid. The L-(+) isomer is used by human metabolism and is preferred for food use because the D-(−) isomer is slightly toxic. Nevertheless, synthetic racemic (DL) lactic acid is the primary commercial form. Lactic acid is widely used in the food industry as taste-enhancing additive. Ferrous salts and the various L(+)-lactic acid salts are used in the pharmaceutical industry for their therapeutic qualities.

3.11.1 Industrial Applications of Lactic Acid

The free acid is used as an acidulant and preservative in several food products such as cheese, meat, beer and jellies. Ammonium lactate is used as a source of non-protein nitrogen in feeds; sodium and calcium stearoyl lactylates are used as emulsifiers and dough conditioners. Only the L(+) form of the lactic acid is metabolised in animal and human cells, because these cells can synthesize only L(+)-lactate dehydrogenase enzyme for the metabolism. Therefore, consumption of large quantity of D(−)-lactic acid will result in its accumulation in the blood. As a result of D(−)lactic acid accumulation hyper-acidity of urine and decalcification may occur.

In the pharmaceutical industry lactic acid is used for the adjustment of pH of pharmaceutical preparations and topical wart preparations. Other applications are as a blood coagulant and dietary calcium source. Ethyl lactate is used in the preparation of anti-inflammatory drugs.

Lactic acid also is a good solvent and provides acidity in foods and beverages. Ethyl and butyl lactates are used as flavour ingredients. Lactic acid has some important applications in food industry such as for the production of fermented foods, in the pickling-process of gherkins, dill, olives, sauerkrauts, carrot and some leafy vegetables, and in the processing of oriental-foods. Lactic acid finds its application for the production of dairy products such as yoghurt, buttermilk, acidophilus milk, cottage cheese, creamy-cheese and fermented cheese etc.

Lactic acid production has been studied recently with increased interest because of its application in the synthesis of biodegradable, biocompatible plastics and coatings (Koutinas et al. 2007a). L(+)-lactic acid can be polymerised to form polylactic acid (PLA). This polymer can be used in the manufacture of new biodegradable plastics. The plastics prepared in such way are increasingly used in surgery for self-dissolving suture thread. Biodegradable plastics could play an increasing role in the industry by replacing or minimizing the use of non-degradable ordinary plastics to solve environmental pollution problems. These biodegradable plastics synthesized from lactic acid could be considered as a substitute for plastics manufactured from the petroleum products. Lactic acid has been used in the manufacture of cellophane, resins and some herbicides and pesticides. Another important application of lactic acid is in textile and tanning industries.

3.11.2 Lactic Acid Production

Lactic acid using solid state fermentation can be carried out using fungal as well as bacterial cultures. Strains of *Rhizopus* sp. (Koutinas et al. 2007a; Tay and Yang 2002) have been common among the fungal cultures and that of *Lactobacillus* sp. among the bacterial cultures (John et al. 2007). The continuous fermentation systems using total microbial cell retention have been used as an empiric steady state model for lactic acid production (Richter and Nottelmann 2004).

Different crops such as cassava and sweet sorghum and the various crop-residues such as sugar cane bagasse, sugar cane press-mud, carrot-processing waste and starch (Tay and Yang 2002) can be used as substrates. A strain of *Rhizopus oryzae* has the potential of producing lactic acid in solid state fermentation. This strain has been used to evaluate the lactic acid production in different fermentation systems. An inert solid support, sugar cane bagasse impregnated with a nutrient solution has been used for solid state fermentation. Both production level and productivity of lactic acid have been found higher in SSF with a yield of 137.0 g/l and the productivity of 1.43 g/l per hour.

Food wastes produced in large amounts in food and catering industry have been studied for lactic acid production by Kim et al. (2003) and Wang et al. (2005). One bacterial strain *Lactobacillus paracasei* has been found to produce lactic acid in SSF. Lactate concentration and yield of 90 g/kg and 91–95% has been achieved in SSF using bacterial culture. The time required to complete SSF for lactic acid production is usually 120–200 h. Sugar cane press-mud has been used as substrate in SSF employing three bacterial cultures. A strain of *Lactobacillus casei* has been found to produce a higher concentration of lactic acid in comparison to *Lactobacillus helveticus* and *Streptococcus thermophilus*.

Production of L(+)-lactic acid by *Rhizopus oryzae* can be carried out in SSF using a solid medium. This solid medium is prepared by impregnating sugar cane bagasse with a nutrient solution containing glucose and calcium carbonate. The optimal concentration of glucose is 18%, producing 13.7% lactic acid. In such process the productivity can be achieved up to 1.43 g/l per h and the fermentation yield is in the range of 77%. These data are significant for L(+)- lactic acid biosynthesis.

A process can be designed for example to carry out SSF for the maximum production of lactic acid. Sugar cane bagasse, a lignocellulosic waste can be used as the support for solid culture. Crushed, moulded, carefully washed and 0.8–2.0 mm size particles are used after sterilisation. Standard fermentation medium is prepared using 200 g/l glucose. The fermentation medium is sterilised at 110°C for 30 min and the pH is adjusted to 7.5 using a 4% solution of NaOH. A solid-fermentation medium of 1200 g-wet weight, with moisture content of 70%, consists of 100-g sugar cane bagasse, 100-g calcium carbonate, and 800-ml fermentation medium. The fermentation medium contains glucose, ammonium sulphate, potassium dihydrogen phosphate, zinc and magnesium sulphates. Inoculum size of 200 ml is used and is included on 1200g wet weight. The inoculum of fungal culture is prepared by growing a sporangiospore suspension using 10^7 spores per g glucose in a liquid fermentation medium for about 14–15 h under shaking condition. Inoculum mixed solid fermentation medium can be incubated in various fermenters such as Erlenmeyer flasks or glass column reactors taking about $0.45\ g/cm^3$. If the fungal culture used is *Aspergillus oryzae*, the SSF should be run for at least 96-98 h at 35°C and the aeration rate is fixed at 1.2 l/h in each glass column reactors.

3.11.3 Parameters Influencing Lactic Acid-Synthesis

Effects of cultivation parameters on morphology of *Rhizopus arrhizus* and the lactic acid production in a bubble column reactor has been studied by Zhang et al. 2007). In solid state fermentation, aeration of the moistened medium is important. It controls the humidity of the solid support and simultaneously the heat-release of the fermentation metabolism. Aeration also provides the oxygen required for the fungal colonisation in solid medium. Therefore, the optimisation of aeration rate is very important in SSF and every SSF-process is different due to the use of different substrate and microorganism. In one example of lactic acid production as described above, aeration rates of 0 to 100 ml/min per column reactor were tried; the maximal L(+)- lactic acid yield was obtained with the aeration rate of 20 ml/min. A decrease of 35% was caused in lactic acid yield at the aeration rate of 100 ml/min. A higher aeration may cause a significant reduction in the lactic acid yield, which is directly related to an increase in the aerobic respiration rate of the fungus. Similarly, at very low or with a poor aeration a significant amount of glucose may not be consumed in the fermentation that may result in decreased biosynthesis of lactic acid.

A second important factor influencing the SSF is the inoculation rate of fungal spores. The preculture for a SSF is mostly prepared by growing the fungal spores in liquid fermentation medium, which is used to inoculate the solid fermentation medium. Various sizes of spore-inoculum using 10^5 to 10^8 spores in a suspension form have been used to produce a preculture of *Aspergillus oryzae*. The optimal seed-inoculum has been found to be 10^6 spores per g glucose used. This inoculum size produced 120 g/l lactic acid with complete glucose uptake and 75% yield of lactic acid. Inoculation rate and the lactic acid production are parallel. A similar variation in lactic acid yield has been observed as any variation occurred in inoculation rate.

Lactic acid synthesis has been found to be affected by the glucose concentration in the fermentation medium. SSF using 12 to 24% glucose has been performed to optimize sugar concentration for an enhanced yield of lactic acid. The optimum glucose concentration, as the initial intake in the medium, was found to be the 18% producing 137 g/l lactic acid with a 76% yield. A higher concentration of glucose as 24% led to a decrease in lactic acid biosynthesis. Because of a partial consumption of glucose in the fermentation medium, a glucose concentration of about 16% produced 117 g/l of lactic acid with a fermentation yield of 74%. If the glucose concentration used was lower as 12%, lactic acid production was reduced to 75 g/l, with a lower fermentation yield of 63%.

During the course of fermentation various parameters affect the yield and the fermentation efficiency. Simultaneously with lactic acid biosynthesis, variation in pH and the moisture content have been recorded. The pH normally falls by more than two units from an initial value of 6.96 to 4.80. A slight increase in relative humidity is observed throughout the fermentation; it changes 70–76% after 96 h of fermentation. In SSF of lactic acid, fumaric acid has also been found to be produced at the concentration of 20 g/l after 96 h.

The type of fermenter used is another factor affecting the product-yield. Rotating fibrous bed bio-reactor have been found useful for lactic acid production by immobilised *Rhizopus oryzae* (Tay and Yang 2002). Bubble column reactor has been used for lactic acid synthesis through the cultivation of *Rhizopus arrhizus* (Zhang et al. 2007). A comparative study of SSF in column-bioreactor and culture in Erlenmeyer flasks showed 77 and 74% fermentation yields, respectively. Fumaric acid production (20 g/l) has been found important in SSF, while this acid has been noticed in very small quantity of 2.1 g/l in liquid fermentation of lactic acid. The culture conditions cause the synthesis of fumaric acid. A closer investigation of the biochemical processes generating fumaric acid is necessary for the understanding of the metabolic pathways involved. Lactate-dehydrogenase activity is related to the presence of its substrate, pyruvic acid as well as to the co-factors that are indispensable in the reaction it catalyses. The absence of reduced co-factors such as $NADH^2$ or the degree of cytochrome oxidation may limit the reaction.

The decrease in pH during the SSF of lactic acid was noticed from 6.96 to 4.8. Membrane processes are involved in proton-translocation between the cell medium and the outside medium. The nature of the proton-pumping activities concerned should be characterized in relation to L(+)-lactic acid production. This type of approach has been used in numerous studies of lactic acid bacteria.

3.11.4 Downstream Processing of Lactic Acid

The solid-fermented mass obtained from the fermenters is used for the extraction of lactic acid. A solution of 1M H_2SO_4 is mixed with the fermented material. The whole content is then placed in a press cell with 0.5–1.5 mm orifices over its whole surface. A hydraulic pressure of 500 to 2,100 kg/cm^3 is used to extract concentrated liquid from the fermented material. The extract obtained consists of L(+)-lactic acid unfermented sugar, salts and other metabolites formed such as fumaric acid. The insoluble fraction contains the support used in SSF, calcium phosphate, and mycelial biomass. Recovery of lactate is complicated by the high solubility of its salts.

The traditional process involves precipitation of calcium lactate and regeneration of lactic acid by addition of sulphuric acid followed by further purification steps, which are ion exchange and decolourisation. The extract is treated with hydrated lime for four purposes; it kills lactic acid bacteria, coagulates proteins, degrades residual sugars and solubilises calcium lactate precipitated due to the reaction of lactic acid with calcium hydroxide and calcium carbonate present in fermenter. The solution is then filtered and the filtrate containing soluble calcium lactate is treated with sulphuric acid to precipitate calcium sulphate liberating lactic acid into the solution. The lactic acid solution is filtered to remove calcium sulphate precipitates and then bleached. The bleaching is performed by treatment with activated carbon. The bleached lactic acid solution is then concentrated to 35–40% acidity in evaporator. The process of bleaching and evaporation is repeated 2–3 times to obtain lactic acid of different grades, such as technical-grade and food-grade lactic acid. Alternative processes are the extraction by liquid membranes, electro dialysis and ion exchange.

3.12 Prospects of R & D in Lactic Acid Biosynthesis

The concept of biorefinery using agricultural substrate such as oat is certainly has good prospects (Koutinas et al. 2007a). A physiological and biochemical approach is required for a better understanding of the contribution of the enzymes of the metabolic pathway or pathways leading to the synthesis of L(+)-lactic acid. An NAD^+-dependent lactate dehydrogenase (L-lactate: NAD^+ oxidoreductase, EC 1.1.1.27) catalysing the reduction of pyruvate into lactate is certainly one of the fundamental enzymes involved. A shuttle between pyruvic acid and oxaloacetic acid involving a carboxylation reaction often forms the furmaric acid. Better understanding of formation of both ethanol and fumaric acid is required.

References

Canilha L, Carvalho W, Almeida E and Silva JB (2006) Xylitol bioproduction from wheat straw: Hemicellulose hydrolysis and hydrolyzate fermentation. J Sci Food Agric 86: 1371–1376

Chang C, Cen P and Ma X, (2007) Levulinic acid production from wheat straw. Bioresour Technol 98: 1448–1453

Couto SR (2008) Exploitation of biological wastes for the production of value-added products under solid-state fermentation conditions. Biotechnol J 3(7): 859–870

Hang YD and Woodams EE (1984) Apple pomace: A potential substrate for citric acid production by *Aspergillus niger*. Biotechnol Lett 6: 253–254

Hang YD, Luh B S and Woodams EE (1987) Microbial production of citric acid by solid state fermentation of kiwi fruit peel. J Food Sci 52: 226–227

Hang YD (1988) Microbial production of citric acid in fixed-bed column bioreactors. Biotechnol Lett 10: 421–426

Hang YD and Woodams EE (1985) Grape pomace: A novel substrate for microbial production of citric acid. Biotechnol Lett 7: 253–254

Hang YD and Woodams EE (1986) Utilisation of grape pomace for citric acid production by solid state fermentation. Am J Enol Vitic 34: 426–428

Ikramul H, Sikander A, Qadeer MA and Javed I (2002) Citric acid fermentation by mutant strain of *Aspergillus niger* GCMC-7 using molasses based medium. Process Biotechnol 5(2): 5–12

Jianlong W (2000) Production of citric acid by immobilized *Aspergillus niger* using a rotating biological contactor (RBC). Bioresour Technol 75: 245–247

John RP, Nampoothiri MK and Pandey A (2007) Simultaneous saccharification and L-(+)-lactic acid fermentation of protease-treated wheat bran using mixed culture of lactobacilli. Biotechnol Lett 28(22): 1823–1826

Johnson DL (2006) The corn wet milling and corn dry milling industry – a base for biorefinery technology developments. In Biorefineries – Industrial Processes and Products, ed. by Kamm B, Gruber PR and Kamm M. Wiley-VCH, New York, pp. 345–353

Kalman G, Recseg K, Gaspar M and Reczey K (2006) Novel approach of corn fiber utilization. Appl Biochem Biotechnol 129–132: 738–750

Kapritchkoff FM, Viotti AP, Alli RCP, Zuccolo M, Pradella JGC, Maiorano AE, et al. (2006) Enzymatic recovery and purification of polyhydroxybutyrate produced by *Ralstonia eutropha*. J Biotechnol 122: 453–462

Kim KI, Kim WK, Seo DK, Yoo IS, Kim EK and Yoon HH (2003) Production of lactic acid from food waste. Apple Biochem Biotechnol 105–108: 637–647

Kolicheski MB (1995) Prodacao de acido citrico per fermentacao no estado solido utilizando como substrato bagaco de mondioca. Tese de Mestrado, Universidade Federal do Parana, Curitiba, Brazil, p. 115

Kolicheski MB, Soccol CR, Martin B, Medeiros and Rainbault, M (1997) Citric acid production on three cellulosic supports in solid state fermentation. In Advances in Solid State Fermentation, eds. Roussos S, Lonsane BK, Raimbault M and Viniegra-Gonzalez G. Kluwer Academic Publishers, The Netherlands, pp. 449–462

Koller M, Bona R, Braunegg G, Hermann C, Horvat P, Kroutil M, et al. (2005) Production of polyhydroxyalkanoates from agricultural waste and surplus materials. Biomacromol 6: 561–565

Koutinas AA, Xu Y, Wang R-H and Webb C (2007a) Polyhydroxybutyrate production from a novel feedstock derived from a wheat-based biorefinery. Enzyme Microbiol Technol 40: 1035–1044

Koutinas, AA, Malbranque F, Wang R-H, Campbell GM and Webb C, (2007b). Development of an oat-based biorefinery for the production of lactic acid by *Rhizopus oryzae* and various value-added co-products. J Agric Food Chem 55: 1755–1761

Koutinas, AA, Wang R and Webb C (2007c) The biochemurgist -Bioconversion of agricultural raw materials for chemical production. Biofuels Bioprod Biorefin 1(1): 24–38

Lockwood LB and Schweiger LB (1967) Citric acid and itaconic acid fermentation. In Microbial Technology, ed. Peppler HJ. Reinhold Publishing Corp., New York,

Milson PJ and Meers JL (1985) Citric acid. In Comprehensive Biotechnology, Volume 3 The Practice of Biotechnology, eds. Blanch HW, drew S and Wang DIC. Pergamon Press, Oxford, pp. 665–680

Maria P (2007) Advances in citric acid fermentation by *Aspergillus niger*: Biochemical aspects, membrane transport and modeling. Biotechnol Adv 25(3): 244–263

Milson PE, Meers JL (1985) Gluconic and Itaconic acid. Comprehensive Biotechnology 3: 672–700

Moeller L, Strehlitz B, Aurich A, Zehnsdorf A, Bley T (2007) Optimization of citric acid production from glucose by *Yarrowia lipolytica*. Eng Life Sci 7(5): 504–511

Moresi M and Parente E (2000) Production of organic acids. In Encyclopedia of Food Microbiology, eds. Robinson RK, Batt CA and Patel PD. Academic Publishers, London, pp. 705–717

Murthy GS, Singh V, Johnston DB, Rausch KD and Tumbleson ME (2006) Evaluation and strategies to improve fermentation characteristics of modified dry-grind corn processes. Cereal Chem 83: 455–459

Pandey A, Nigam P, Soccol CR, Soccol VT, Singh D and Mohan R, (2000) Advances in microbial amylases. Biotechnol Appl Biochem 31:135–152

Pandey A, Soccol CR, Rodriguez-Leon JA and Nigam P (2001) Production of organic acids by solid state fermentation. In Solid state fermentation in Biotechnology-Fundamentals and applications. Asitech Publishers N. Delhi, pp. 113–126

Peller H J and Perlman D (1979) Microbial Technology, 2nd edition. Academic Press, New York

Prado FC, Vandenberghe LPS, Lisboa C, Paca J, Pandey A and Soccol CR (2004) Relation between citric acid production and respiration rate of *aspergillus niger* in solid-state fermentation. Eng Life Sci 4(2): 179–186

Prescott SC and Dunn CG (1959) Industrial Microbiology, 3rd edition. McGraw-Hill Book Co., New York

Raimbault M and Alazard D (1980) Culture method to study fungal growth in solid fermentation. Eur J Appl Microbiol Biotechnol 9: 199–209

Reddy N and Yang Y, (2005) Biofibers from agricultural byproducts for industrial applications. Trends Biotechnol 23: 22–27

Richter K and Nottelmann S (2004) An empiric steady state model of lactate production in continuous fermentation with total cell retention. Eng Life Sci 4(5): 426-432

Rohr M, Kubicek CP, Kowinek J (1983) Citric acid. Biotechnology 3: 419–454

Shankaranand VS and Lonsane BK (1993) Sugarcane press mud as a novel substrate for the production of citric acid by sopild state fermentation. World J Microbiol Biotechnol 9: 377–380

Shankaranand VS and Lonsane BK (1994) Coffee husk: An inexpensive substrate for production of citric acid by *Aspergillus niger* in solid-state fermentation system. World J Microbiol Biotechnol 10: 165–168

Sikander A and Ikramul H (2005) Role of different additives and metallic micro minerals on the enhanced citric acid production by *Aspergillus niger* MNNG-115 using different carbohydrate materials. J Basic Microbiol 45(1): 3–11

Soccol C R, Prado FC, Vandenberghe LPS, and Pandey A (2004) Organic acids: Production and application. Citric acid. In Concise Encyclopedia of Bioresource Technology, ed. Pandey A. The Howarth Reference Press, pp. 617–628

Soccol C R and Vandenberghe LPS (2003) Overview of applied solid-state fermentation in Brazil. Biochem Eng J 13: 205–218

Taing O and Taing K (2007) Production of malic and succinic acids by sugar-tolerant yeast *Zygosaccharomyces rouxii*. Eur Food Res Technol 224: 343–347

Tay A and Yang S (2002) Production of L(+)-lactic acid from glucose and starch by immobilized cells of *Rhizopus oryzae* in a rotating fibrous bed bioreactor. Biotechnol Bioeng 80(1): 1–12

Thomsen M H (2005) Complex media from processing of agricultural crops for microbial fermentation. Appl Microbiol Biotechnol 68: 598–606

Tongwen X U and Weihua Y (2002) Citric acid production by electrodialysis with bipolar membranes. Chem Eng Process 41: 519–524

Vandenberghe LPS, Soccoi CR, Pandey A and Lebault JM (2000) Solid state fermentation for the synthesis of citric acid by *Aspergillus niger*. Bioresource Technol 74(2): 175–178

Venus J and Richter K (2006) Production of lactic acid from barley: Strain selection, phenotypic and medium optimization. Eng Life Sci 6(5): 492–500

Wang XQ, Wang QH, Ma HZ and Yin W (2008) Lactic acid fermentation of food waste using integrated glucoamylase production. J Chem Technol Biotechnol online publication (www.interscience.com) DOI 10.1002/jctb.2007

Wang XM, Wang QH, Ren NQ and Wang XQ (2005) Lactic acid production from kitchen waste with a newly characterized strain of lactobacillus plantarum. Chem Biochem Eng Q 19: 383–389

Xie G and West TP (2006) Citric acid production by *Aspergillus niger* on wet corn distillers grains. Lett Appl Microbiol 43(3): 269–273

Yamada K (1977) Bioengineering report. Recent advances in industrial fermentation in Japan. Biotechnol Bioeng 19: 1563–1621

Zhang Y, Jin B and Kelly JM (2007) Effects of cultivation parameters on the morphology of *Rhizopus arrhizus* and the lactic acid production in a bubble column reactor. Eng Life Sci 7(5): 490–496

Chapter 4
Biofuels

Soham Chattopadhyay, Asmita Mukerji and Ramkrishna Sen

Contents

Abstract It is the cost and abundant availability of raw materials that determine the economic feasibility of biofuel production. Considering these constrains, agro-industrial residues may offer cheaper options as raw materials for biofuel production. This chapter thus aims at presenting the current status and future directions of biofuel production using both conventional substrates and agro-industrial residues

R. Sen (✉)
Department of Biotechnology, Indian Institute of Technology, Kharagpur,
West Bengal 721302, India
e-mail: rksen@yahoo.com

P. Singh nee' Nigam, A. Pandey (eds.), *Biotechnology for Agro-Industrial Residues Utilisation*, DOI 10.1007/978-1-4020-9942-7_4,

as raw materials and critically analyzing the prospect of agro-industrial residue based production of biofuels. Utmost care has been taken to address all the critical economic and environmental issues related to the production of each of the gaseous and liquid biofuels namely biomethane, biohydrogen, bioethanol and biodiesel in the light of available published literature information. While the important process parameters involved in anaerobic digestion and co-digestion of agro-industrial residues in various judicious combinations have been discussed for biomethane production, roles of different reactor configurations, designs and various types of cultivation processes including photo and dark fermentation employing agro-industrial waste as substrates for biohydrogen production have been analyzed. Similarly, a comparative study of bioethanol production using lignocellulosic and non-lignocellulosic wastes has been presented and discussed. Though the oil cake as an agro-industrial waste holds some promise, biodiesel production using agro-industrial residues has not been reported in literature.

Keywords Biofuels · Agro-industrial residues · Biomethane · Biohydrogen · Bioethanol

4.1 Introduction

Agro-industrial wastes are those end-products of primary production that have not been reused, recycled or salvaged. Recycling, reprocessing and eventual utilization of agro-industrial residues can be accomplished for the benefit of man and his amenities rather than their discharge to the environment which may cause detrimental effects (Hamza 1989). Extensive research has been done on the utilization of agro-industrial residues and a good number of publications are available on this (Fernández et al. 2005; Khardenavis et al. 2007; Sellami et al. 2007; Gañán et al. 2008). Table 4.1 shows chronological use of agro-industrial residues as raw materials for various industrial applications.

4.2 Bio-Fuels

Bio-fuels can be broadly classified into two major types, gaseous and liquid biofuels. Purification of the conventional biogas into methane-enriched biofuel led to the development biomethane. Biohydrogen is a relatively new type of gaseous biofuel, which is produced by anaerobic fermentation of agro-industrial wastes by the synergistic action of a consortium of methanogenic, acidogenic and hydrogenic bacteria. On the other hand, liquid biofuels have recently been classified into bioethanol and biodiesel. While bioethanol has recently gained rejuvenated importance on the wake of present energy crisis worldwide, biodiesel occupied the centre stage as a potential

Table 4.1 Various uses of agro-industrial residues

Agro-industrial residues	Uses	Reference
Pressmud (waste of sugar industry), Cow dung, Neem and Mahua cakes (soil-expelled seed material), and Neem leaf	Improving the nodulation and dry matter production of soybean	Jauhri 1989
Sugarbeet pulp, Potato pulp, Brewery grain	Methane production	Kang and Wetland 1993
Grape skin pulp extract, Starch waste, Olive oil waste effluents and Molasses	Extracellular polysaccharide (pullulan) production	Israilides et al. 1994
Wheat bran	Production of ferulic acid.	Faulds et al. 1997
Sugarcane Bagasse	Production of protein-enriched cattle feed and enzymes	Pandey et al. 2000a
Cassava Bagasse	Production of organic acids, flavour and aroma compounds, and mushrooms	Pandey et al. 2000b
Coffee husk	Production of Aroma compounds	Soares et al. 2000
Cassava bagasse, Apple pomace, Soyabean, Amaranth grain and Soyabean oil	Cultivation of edible Rhizopus strains.	Christen et al. 2000
Technical oleic acid (TOA), waste frying oil (WFO) and waste-free fatty acids from soybean oil (WFFA)	Production of poly(3-hydroxyalkaonates) (PHA)	Fernández et al. 2005
Wheat bran (WB), Rice bran (RB), Rice straw (RS), Sawdust (SD), Coconut pith (CP)	Production of endoxylanase.	Poorna and Prema 2006
Biomass (corn cobs and stalks, sugarcane waste, wheat or rice straw)	Ethanol production	Lin and Tanaka 2006
Olive-cake with Poultry manure and Sesame shells	Composting and soil amendment	Sellami et al. 2007
Cotton stalk, Rice straw, Bagasse, and Banana plant waste	Extraction of Lignocellulosic fibers	Habibi et al. 2008
Sugar cane, starchy materials and lignocellulosic biomass	Bioethanol production	Sánchez and Cardona 2008

substitute for petroleum diesel in the last two decades. Various types of the gaseous and liquid biofuels are discussed below.

4.3 Biogas

The conventional biogas, which is produced in biogas plants employing anaerobic digestion of organic wastes including manures by mixed microbial cultures, is composed primarily of methane (typically 55%–70% by volume) and carbon dioxide (typically 30%–45%) and may also include smaller amounts of hydrogen sulfide (typically 50–2000 ppm), water vapor (saturated), oxygen, and various trace hydrocarbons (Amigun et al. 2008; Kashyap et al. 2003; Siso 1996). Due to its lower methane content (and therefore lower heating value) compared to natural gas, biogas use is generally limited to engine-generator sets and boilers (Krich et al. 2005).

4.3.1 Biomethane

Biomethane is upgraded or sweetened biogas after the removal of the bulk of the carbon dioxide, water, hydrogen sulfide and other impurities from raw biogas. From a functional point of view, biomethane is extremely similar to natural gas (which contains 90% methane) except that it comes from renewable sources. (Krich et al. 2005). Biogas can also be purified and upgraded and used as vehicle fuel. Over a million vehicles are now using biogas and fleet operators have reported savings of 40–50% in vehicle maintenance costs (Parawira 2004). Table 4.2 compares wastes generated by various agro-industries on the basis of their annual production rate, yield, power generation capacity and the amount of fossil fuel they can replace.

Anaerobic digestion has proved to be the most feasible strategy for biogas production from agro-industrial wastes. The potential substitution of fossil fuel with

Table 4.2 Various agro-industries and their annual production rate, yield, power generation capacity and the amount of fossil fuel they can replace

Industry	Annual production rate of primary product	Annual production rate of secondary product	Yield of Methane m^3CH_4/unit	Power generation capacity (million kWh)	Amount of Fossil fuel replaced (tons)	References
Olive oil	377.75 Kt of olive oil	2.266 Mt of olive oil wastewater	–	–	–	Adapted from Fountoulakis et al. 2008
Wine	348.43 Kt of must	135.50 Kt of wine-grape residues	–	–	–	
Meat	315 Kt of meat annually	4.185 Mt of slaughter house waste water	–	–	–	
Cereals	3,733,700	336,060 Bran Residue	450	454	127333	Adapted from Kivaisi and Rubindamayugi 1996
Sugar	123,620	20,000	230	14	3873	
Coffee	56,053	199,200	650	190	53198	

biogas represents an annual reduction in the atmosphere of 1.05 million m^3 of CO_2. (Kivaisi and Rubindamayugi 1996).

4.3.2 The Phases of Anaerobic Methane Production

The anaerobic digestion process encompasses mainly four interdependent, complex sequential and parallel biological reactions. The phases are hydrolysis/liquefaction, acidogenesis, acetogenesis and methanogenesis. Figure 4.1 represents how complex polymers like carbohydrates, proteins and fats are hydrolyzed into simple sugars, amino acids, fatty acids and alcohols, respectively which are in turn converted into CH_4 and CO_2 on acetogenesis and methanogenesis.

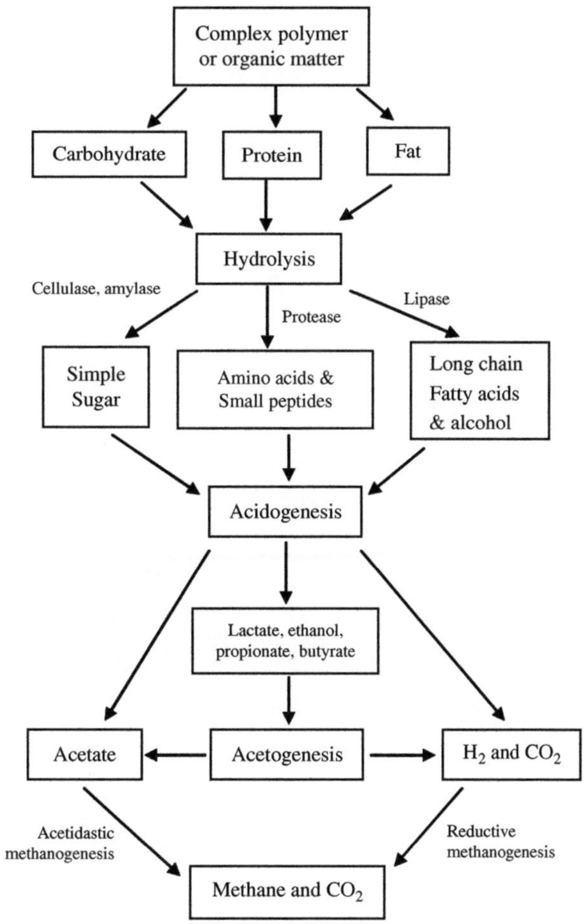

Fig. 4.1 Methane and carbon di-oxide production from complex polymers or organic matter

4.3.3 The Strategy of Co-digestion

Co-digestion of different types of organic by-products has been increasingly applied in order to improve plant profitability through easier handling of mixed wastes; common access facilities and the known effect of economy scale are some of the advantages of co-digestion. LHL (Laying Hen Litter), CW (Cheese Whey), SW (Slaughter house Wastewater), cattle manure, swine manure or piggery effluent have been utilized as biogas yield ameliorating agents (Azbar et al. 2008). Co-digestion of OMW (Olive Mill Wastewater) and WGR (Wine Grape Residues) with SW yields much improved results, with a 30–57% increase in methane yields as compared to individual digestion of the substrates (Table 4.3).

Biogas releases by OME increases 90% on co-digestion with LHL, but only 22% increases with CW (Azbar et al. 2008). Methane yields during thermophilic digestion are 14–35% higher than mesophilic digestion (Table 4.4). Thermophilic digestion, intrinsically, has higher degrading capability and methanogenic activity in biogas production. Results of comprehensive studies suggest that thermophilic anaerobic digestion may be attractive for treating high-temperature industrial effluents and specific types of slurries (Parawira 2004).

Table 4.3 Increase of methane yield and energy yield with co-digestion (adapted from Fountoulakis et al. 2008)

Agro-industrial wastes	Methane Yield (L CH_4/kg COD)	% Increase in comparison to individual digestion.	Energy yield (MJ/kg COD)
OMW	108	–	3.89
WGR	147	–	5.3
SW + OMW	170	57.4	6.13
SW + WGR	191	29.93	6.87
OMW + WGR	163	50.9 (w.r.t. OMW) 10.88 (w.r.t. WGR)	5.87

Table 4.4 Percentage increase in Methane yield in thermophilic condition (adapted from Fountoulakis et al. 2008)

Materials	Temperature	Methane yield ($l\ g^{-1}$ COD added)	Percentage increase
SW+OMW	35	0.184	34.5
	55	0.282	
SW+WGR	35	0.188	14.1
	55	0.219	
OMW+WGR	35	0.214	28.9
	55	0.301	

4.3.4 *Influence of Process Parameters and Steps*

4.3.4.1 Effect of pH

A pH range of 6.4–7.6 ensures normal functioning of a digester, beyond which inhibitory effects may be due to the toxic effects of the hydrogen ions, which are closely related to the accumulation of VFAs (Volatile Fatty Acids) (Anderson and Yang 1992).

4.3.4.2 Effect of VFA: Alkalinity Ratio

It is essential that the reactor contents provide enough buffering capacity to neutralize any possible VFA accumulation in the reactor and maintain pH 6.7 to 7.4, for stable operation. The VFA: alkalinity ratio $< 0:4$ indicates digester is stable; a range from 0:4 to 0:8 shows some instability, whereas $\geq 0:8$ shows marked instability (Switzenbaum and Jewell 1980). The buffering capacity is an added advantage of using proteins during methane generation and eliminating the requirement of an external buffer (Mshandete et al. 2004).

4.3.4.3 Effect of C: N Ratio

The C: N ratio in the range of 25–30 has been suggested as optimum for anaerobic digestion, there has been a contradiction with the ranges of 16–19 (Nyns 1986) and 16.8–18 (Kivaisi and Mtila 1998) having been proposed as ideal when lignin are taken into account.

4.3.4.4 Effect of VFA's

The major VFA's usually produced are acetic acid (A), propionic acid (P) and butyric acid. It has been seen that a P/A ratio exceeding 1.4 and acetate and butyrate build-up above 200 Mm and that of propionate above 100 Mm leads to inhibition which culminates into digester failure (Hill et al. 1987; Ahring et al. 1995).

4.3.4.5 Pre-treatment of Lignocellulosics

The pre-treatment of wastes on alkaline hydrolysis before co-digestion with activated sludge demonstrates a steep rise in methane yield from 25L to 222L CH_4 (STP)/kg $VS_{initial}$. Consequently, the total solids and volatile solids were reduced by 67% and 84%, respectively. Acidogenic bacteria can ferment the pre-treated lignocellulose even though no delignification or cellulose hydrolysis occurs during the pre-treatment (Neves et al. 2006).

4.3.4.6 Effect of Temperature

Anaerobic digestion reactors are normally operated within the mesophilic and thermophilic ranges. Methanogenesis is also possible under psychrophilic ($< 20°C$) conditions but occurs at lower rates. Bacterial activity and growth decrease by one half for every 10°C decrease in temperature below 35°C (Kashyap et al. 2003).

4.3.4.7 Accumulation of Inhibitory Compounds

It has been observed that biogas production decreases significantly with time during OMW digestion. The primary reason, appears to be OMW's high content of polyphenols inhibit the anaerobic process (Beccari et al. 1998). It was found that the individual chemical structure of compounds also greatly influences and determines the rate and mechanisms of methanogenic degradation (Fountoulakis et al. 2008).

4.4 Biohydrogen

Hydrogen is a very high energy (122 kJ/g) yielding fuel in comparison to methane or ethanol, produces water instead of greenhouse gases when combusted. Photo-autotrophically growing bacteria and (micro)-algae, utilize light as primary energy source to split water into hydrogen and oxygen by the enzyme hydrogenase. The basic reactions are (de Vrije and Claassen 2003):

$$C_6H_{12}O_6 + 2H_2O \rightarrow 4H_2 + 2CO_2 + 2C_2H_4O_2 \ \Delta Go' = -206\,\text{kJ}$$

$$C_2H_4O_2 + 2H_2O \rightarrow 2CO_2 + 4H_2 \ \Delta Go' = 104\,\text{kJ}$$

Figure 4.3 shows how biomass containing carbohydrates is converted into organic acids and hydrogen by the process of thermophilic heterotrophic fermentation. The organic acids are subsequently converted into hydrogen by photoheterotrophic fermentation process. During growth of thermophilic bacteria, hydrogen production is directly linked to central metabolic pathways unlike the case during photo-heterotrophic growth (de Vrije and Claassen 2003).

4.4.1 Hydrogen from Agro-Industrial Residue

Several forms of organic waste streams, ranging from solid wastes like rice straw, black strap molasses (Nath et al. 2005) to waste water from a sugar factory and a rice winery have been successfully used for hydrogen production. Most experiments have shown considerable hydrogen production with the limited number of thermophilic strains used. The utilization of potato steam peels by the two phase approach has been examined. The organic acids, already present in the initial substrate and additionally produced in the first fermentation step, are the substrates of choice for the photo-heterotrophic fermentation (de Vrije and Claassen 2003). Assessment of the total production of hydrogen and acetate from glucose and from equivalent

Table 4.5 Hydrogen and Acid Production from different carbon sources

Carbon source	Hydrogen production (mM)	Maximum Hydrogen productivity (mmol/L.h)
Glucose	130	10.7
Potato steam peel hydrolysate	218	11.7

amount of sugars in potato steam peel hydrolysate (prepared by the action of amylase and glucoamylase) revealed that higher hydrogen production occurred from the peels as reflected by the data in Table 4.5. (adapted from Claassen et al. 2004).

4.4.2 Production of Biohydrogen in Fermenter

Among three different reactors, namely, Continuously Stirred Tank Reactor (CSTR), an Up flow Fixed Bed Reactor (UFBR) and an Upflow Anaerobic Sludge Blanket (UASB) reactor, the CSTR gives the best performance with a yield of 0.30 L H_2/g carbohydrate and production rate of 4.50 mmol H_2/L reactor, h. Its superior performance has been attributed to the mechanical stirring that promoted both H_2 and CO_2 removal from the fermentation broth, therefore reducing feed-back inhibition phenomena (Camilli and Pedroni 2005).

4.5 Coupling of Biohydrogen and Biomethane Production

The effluent from a hydrogen producing reactor containing high concentrations of fatty acids may be subjected to a subsequent anaerobic digestion step with the conversion of the remaining organic content to biogas, mainly methane and carbon dioxide (Antonopoulou et al. 2008). Methane yields from the mesophilic reactor receiving effluent from the hydrogenogenic reactor emulate the yields of the original feed.

4.6 Bio Ethanol

4.6.1 Fermentative Production of Bioethanol

Bioethanol is a biofuel used as a petrol substitute, produced by simple fermentation processes involving cheaper and renewable agricultural carbohydrate feedstock and yeasts as biocatalysts. A variety of common sugar feedstocks including sugarcane stalks, sugar beet tubers and sweet sorghum are used. The fermentation process is mediated by two enzymes invertase and zymase, produced by the yeast cells. The overall process steps are as follows (Zanichelli et al. 2007):

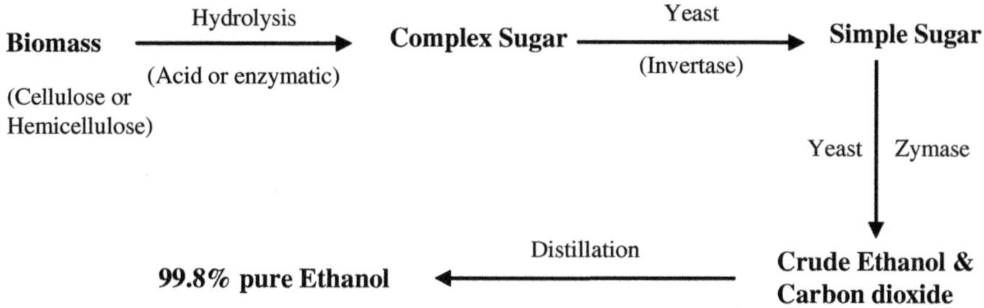

Fig. 4.2 The flow chart of basic ethanol production process

$$(C_6H_{10}O_5)n + nH_2O \xrightarrow{\text{invertase}} nC_6H_{12}O_6 \xrightarrow{\text{zymase}} 2nC_6H_{12}OH + 2nCO_2$$

Figure 4.2 describes an overview of bioethanol production from cellulosic or hemi cellulosic biomass. The first step is a pre-treatment step for the conversion of cellulosic and hemicellulosic biomass into complex sugars by acid catalysts or enzyme. The bioconversion of the complex sugars into bioethanol is mediated by invertase and zymase as discussed above. But the ethanol produced from this fermentation process contains significant amount of water in it. To remove water, fractional distillation process is used, wherein the ethanol-water mixture is vaporized. Bioethanol gets separated from water due to lower boiling point (78.3°C). After distillation step the final product is enriched with 95% to 99.8% ethanol.

The discovery of continuous mode during 1960s permits recycling of yeast; increase the speed of the process and reducing the cost. The Yield % of theoretical max in continuous process is around 95% as compared to less than 90% for batch or fed-batch (Sánchez and Cardona 2008).

4.6.2 Bioethanol from Agro-Industrial Residues

Ethanol produced from renewable and cheap agricultural products reduces the green house gas emissions like CO_X, NO_X and SO_X and eliminate smog from the environment. Agricultural residues and wastes have several advantages as they do not require any additional lands because they are collected into piles at large agricultural and forestry facilities. Some of the agro-industrial residues and waste materials abundantly available are mentioned in Table 4.6, which are used as potential substance for ethanol production in various countries.

Table 4.6 Agro-industrial residues and plant waste materials used for bioethanol production in various countries

Agro-industrial residues and plant wastes	Country	Bioethanol yield (% of theoretical maximum)	Reference
Thippi	India (Tamilnadu)	93.18	Patle and Lal 2008
Switchgrass	USA	72	Asli et al. 2008
Corn steep liquor (CSL)	USA and Brazil	–	Ruanglek et al. 2006
Ami-ami solution, Brewer's yeast autolysate and Fish soluble waste	Thailand	88	Ruanglek et al. 2006
Waste Potato	Finland	87	Liimatainen et al. 2004
Rice straw, oat straw, wheat straw	Not specified	–	Kim and Dale 2004

4.6.3 Major Raw Materials Used in Bioethanol Production

4.6.3.1 Lignocellulosic Materials

Removal of lignin from lignocellulosic raw materials is the most critical step. Among various methods physicochemical and biological methods are mainly used for pre-treatment. Saturated steam at 160°C to 290°C and at high pressure (0.69 to 4.65 MPa) is used to convert hemicelluloses into soluble oligomers (Hamelinck et al. 2005; Ballesteros et al. 2004). Lignin is not solubilized but redistributed. Ammonia soaking of corn stover at room temperature can remove as much as 74% of the lignin (Asli et al. 2008). The fungus *Phanerochaete chrysosporium* can also be used for degrading lignin (Sánchez and Cardona 2008).

4.6.3.2 Non-lignocellulosic Materials

Thippi

Thippi that is an agro-industrial waste composed of starch, pectin, fiber and protein. After pre-treatment at 121°C for 20 minutes, acid or enzymatic treatment is done. Acid treatment is done with 0.75% H_2SO_4 at 55°C for 3 h. Enzymatic treatment is performed with various enzymes like amylase, pectinase and cellulase at 55°C for 3 h at pH 5. Fermentation is carried out to use the reducing sugars obtained from above process to produce ethanol (Patle and Lal 2008). Because of higher yield % of maximum theoretical value (>90%) as compared to other available non-lignocellulosic materials, thippi can be used as preferred substrate with great potential for bioethanol production.

Switchgrass

Switchgrass (*Panicum virgatum*) is a perennial grass grown in warm season and resistant to harsh conditions, pests, and diseases. It is capable of producing high biomass yields at low fertilizer application rates. Untreated switch grass contained 42% cellulose, 31% hemi-cellulose, 6% acid detergent lignin (ADL), 22% klason lignin and 0.7% ash. The yield % of theoretical max is 72% with simultaneous saccharification and fermentation (Asli et al. 2008).

4.7 Biodiesel

Finite resources and gradually increasing demand for diesel all over the world lead researchers to find some alternative sources. Emission of toxic green house gases from the combustion of petroleum diesel is also a major contributing factor for this. Right from the experiment of Rudolf Diesel using pea nut oil in his self-designed engine (World Exhibition in Paris in 1900), numerous studies have followed to establish the potential of triglycerides as alternative sources of diesel. But using triglycerides directly into a diesel engine leads to some operational difficulties due to its high viscosity and poor low temperature properties like pour point and cloud point (Fukuda et al. 2001). This problem could be fixed by developing vegetable oil derivatives that resemble properties of petrodiesel. Transesterification is the most widely used process in which triglycerides react with an alcohol (mainly methanol) in presence of chemical (acid or alkali) or biological (enzyme) catalysts to produce mono alkyl esters, popularly known as biodiesel. Some alkali catalyzed batch processes have been commercialized.

4.7.1 Chemical Catalyzed Method

In this method, acid (H_2SO_4) or alkali (NaOH or KOH) is used for transesterification of triglycerides. Acid catalyzed reaction has some disadvantages which include very high temperature and pressure and incomplete conversion. On the other hand, alkali catalyzed process, though requires high temperature (about 80°C) is very quick with higher conversion rates (Chongkhon et al. 2007; Vicente et al. 2004). Alkali catalyzed methods have been commercialized. This process has certain disadvantages and transesterification with biocatalyst have been tried. Figure 4.3 shows biodiesel production using alkali as catalyst.

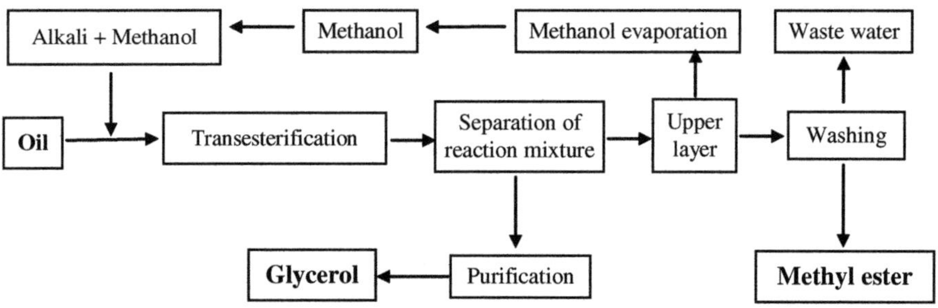

Fig. 4.3 The flowchart of Alkali catalyzed Transesterification

4.7.2 Enzymatic Transesterification

Mostly lipases from different sources are used for this process. To make the process cost effective immobilization of lipase has been done. This process can be performed under normal temperature and pressure, but takes more time than alkali catalyzed reaction. Percentage conversion is quite high and glycerol obtained in this method is of good quality and can be directly used to produce some valuable products (Fukuda et al. 2001). The enzyme catalyzed transesterification process is summarized in Fig. 4.4. Methanol is preferred to other alcohols for its abundant supply and diesel like quality of the transesterified product. By process optimization over 90% conversions can be obtained. (Shimada et al. 2002).

4.7.3 Biodiesel from Agro-Industrial Residues

Biodiesel is generally produced from vegetable oils or animal fats. Various oils like palm oil, soybean oil, sunflower oil, rice bran oil, rapeseed oil etc. are used. The choice of vegetable oil used depends on its abundant availability in the country where biodiesel is produced. To our knowledge, there are no reports available on the use of agro-industrial residues for biodiesel production. However, bioethanol produced from agro-industrial residues can in turn be used for the transesterification of vegetable oils to produce mono ethyl esters of fatty acids as biodiesel. Some residues can be successfully utilized as carbon sources for single cell oil production. It reduces the fermentation costs (Peng and Chen 2007). Whey concentrate and tomato waste hydrolysate, which contains more than 1 g/l total organic nitrogen, in turn produce 14.3 % and 39.6 % lipid respectively can be used for gamma-linolenic acid production. The amount of gamma-linolenic acid produced from these wastes is 14.1% and 11.5%, which makes whey concentrate and tomato waste hydrolysate, two good raw materials for biodiesel production (Fakas et al. 2008).

4.8 Future Prospectives

Rapidly spiraling crude oil prices and cost ineffectiveness of most of the biofuel technologies, mainly due to expensive raw materials and manufacturing processes, have fueled extensive worldwide search and utilization of agro-industrial residues for the cost competitive production of alternative biofuels. The major gaseous biofuels, namely, biomethane and biohydrogen and the major liquid biofuels, namely, bioethanol and biodiesel have evolved as potential alternative to the dwindling fossil fuel resources. Bioethanol and biodiesel are gaining importance as alternative fuels to petrol and diesel respectively. Bioethanol, which is conventionally produced from cane molasses by yeast fermentation, can also be produced from various agro-industrial residues and plant wastes. Efficient process optimization and integration by combining production and recovery processes may lead to economic production of bioethanol. Switchgrass that grows mainly in the USA in drastic climatic conditions and contains high percentage of cellulose and hemicelluloses generated some excitement in the field of bioethanol production. Biodiesel on the other hand

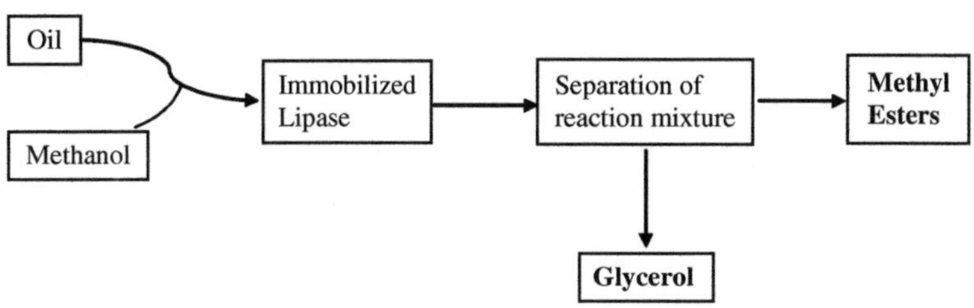

Fig. 4.4 The flowchart of Enzymatic Transesterification

is generally produced from vegetable oils. Agro-industrial residues are still not used as the substrate for biodiesel, though the residual oil present in oil cake, a waste product of oil extraction units, hold some hope. But in future, suitable residues with high lipid content may be used as potential raw materials for biodiesel production. Biotechnological techniques to produce biofuels from agro-industrial wastes and residues are potentially effective in reducing the emission of toxic pollutants and greenhouse gases, saving our environment and partly solving the worldwide fuel crisis. By focusing the transformative power of biotech on challenges in biofuel production, while considering sustainability in all its dimensions, one can reasonably hope to enable the 'second industrial revolution' that our society now requires.

References

Ahring BK, Sandberg M, and Angelidaki I (1995) Volatile acids as indicators of process imbalance in anaerobic digesters. Appl Microbiol Biotechnol 43: 559–565.

Amigun B, Sigamoney R, and von Blottnitz H (2008) Commercialisation of biofuel industry in Africa: A review. Renewable Sustainable Energy Rev 12: 690–711.

Anderson GK and Yang G (1992) Determination of bicarbonate and total volatile acid concentration in anaerobic digesters using a simple titration. Water Environ Res 64: 53–59.

Antonopoulou G, Gavala HN, Skiadas IV et al. (2008) Biofuels generation from sweet sorghum: Fermentative hydrogen production and anaerobic digestion of the remaining biomass. Bioresour Technol 99: 110–119.

Asli I, Jennifer N H, Anthony L et al. (2008) Aqueous ammonia soaking of switchgrass followed by simultaneous saccharification and fermentation. Appl Biochem Biotechnol 144: 69–77

Azbar N, Keskin T, and Yuruyen A (2008) Enhancement of biogas production from olive mill effluent (OME) by co-digestion. Biomass Bioenergy doi:10.1016/j.biombioe.2008.03.002.

Beccari M, Majone M, and Riccardi C (1998) Two reactor system with partial phase separation for anaerobic treatment of olive mill effluents. Water Sci Technol 38: 53–60.

Ballesteros M, Oliva JM, Negro MJ et al. (2004) Ethanol from lignocellulosic materials by a simultaneous saccharification and fermentation process (SFS) with K. marxianus CECT 10875. Proc Biochem 39:1843–1848.

Camilli M and Pedroni PM (2005) Comparison of the performance of three different reactors for Biohydrogen production via dark anaerobic fermentations. Proceedings International Hydrogen Energy Congress and Exhibition IHEC 2005 Istanbul, Turke.

Christen P, Bramorski A, Revah S et al. (2000) Characterization of volatile compounds produced by Rhizopus strains grown on agro-industrial solid wastes. Bioresour Technol 71: 211–215

Chongkhon S, Tongurai C, Chetpattananondh P et al. (2007) Biodiesel production by esterification of palm fatty acid distillate. Biomass Bioenergy 31: 563–568.

Claassen PAM, Budde MAW, van Noorden GE et al. (2004) Biological Hydrogen Production from Agro-Food By-Products. http://www.totalfood2004.com/TotalFood.pdf.

de Vrije T and Claassen PAM (2003) Dark hydrogen fermentations. In: J.H. Reith, R.H. Wijffels, H. Barten (eds) Bio-Methane and Bio-Hydrogen. Dutch Biological Hydrogen Foundation. Smiet Offset, The Hague, ISBN 90-9017165-7: 103–123.

Fakas S, Certik M, Papanikolaou S et al. (2008) γ-Linolenic acid production by Cunninghamella echinulata growing on complex organic nitrogen sources. Bioresour Technol 99: 5986–5990

Faulds CB, Bartolome B, Williamson G et al. (1997) Novel biotransformation of agro-industrial cereal waste by ferulic acid esterase. Ind Crops Prod 6: 367–374

Fernández D, Rodríguez E, Bassas M et al. (2005) Agro-industrial oily wastes as substrates for PHA production by the new strain P. aeruginosa NCIB 40045: effect of culture conditions. Biochem Eng J 26: 159–67

Fountoulakis MS, Drakopoulou S, Terzakis S et al. (2008) Potential for methane production from typical Mediterranean agro-industrial by-products. Biomass Bioenergy 32: 155–161
Fukuda H, Kondo A, and Noda H (2001) Biodiesel fuel production by transesterification of oils. J Biosci Bio Eng 92: 405–416
Ganan P, Zuluaga R, Restrepo A et al. (2008) Plantain fibre bundles isolated from Colombian agro-industrial residues. Bioresour Technol 99: 486–491
Habibi Y, Zawawy WKE, Ibrahim MM et al. (2008) Processing and characterization of reinforced polyethylene composites made with lignocellulosic fibers from Egyptian agro-industrial residues. Compos Sci Technol 68: 1877–1885
Hamelinck CN, van HG, and Faaij APC (2005) Ethanol from lignocellulosic biomass: techno-economic performance in short-, middle-and long-term. Biomass Bioenergy 28: 384–410.
Hamza A (1989) Utilization of Agro-industriai Residues in Alexandria: Experience and Prospects. Biol Wastes 29: 107–121
Hill DT, Cobb SA, and Bolte JP (1987) Using volatile fatty acid relationships to predict anaerobic digester failure. Trans ASAE 30: 496–501.
Israilides C, Scanlon B, Smith A et al. (1994) Characterization of pullulans produced from agro-industrial wastes. Carbohydr Polym 25: 203–209
Jauhri KS (1989) Evaluation of certain agro-industrial wastes as seed pelletants under adverse soil pH conditions. Biol Wastes 28: 233–236
Kang H and Wetland P (1993) Ultimate anaerobic biodegradability of some Agro-industrial residues. Bioresour Technol 43: 107–111
Kashyap DR, Dadhich KS, and Sharma SK (2003) Biomethanation under psychrophilic conditions: a review. Bioresour Technol 87: 147–153
Khardenavis AA, Kumar MS, Mudliar SN et al. (2007) Biotechnological conversion of agro-industrial wastewaters into biodegradable plastic, poly b-hydroxybutyrate. Bioresour Technol 98: 3579–3584
Kim S and Dale BE (2004) Global potential bioethanol production from wasted crops and crop residues. Biomass Bioenergy 26: 361–375
Kivaisi AK and Rubindamayugi MST (1996) The potential of agro-industrial residues for production Of biogas and electr. ictty in Tanzania. WREC.
Kivaisi AK and Mtila M (1998) Production of biogas from water hyacinth (Eichhornia crassipes) (Mart) (Solms) in a two stage bioreactor. World J Microbiol Technol 14: 125–131.
Krich K, Augenstein D, Batmale J P et al. (2005) Biomethane from Dairy Waste: A Sourcebook for the Production and Use of Renewable Natural Gas in California.
Liimatainen H, Kuokkanen T, Kääriäinen J et al. (2004) Development of bio-ethanol production from waste potatoes. Proceedings of the Waste Minimization and Resources Use Optimization Conference, June 10, 2004, University of Oulu, Finland. Oulu University Press, Oulu, pp. 123–129
Lin Y and Tanaka S (2006) Ethanol fermentation from biomass resources: current state and prospects. Appl Microbiol Biotechnol 69: 627–642.
Mshandete A, Kivaisi A, Rubindamayugi M et al. (2004) Anaerobic batch co-digestion of sisal pulp and fish wastes. Bioresour Technol 95: 19–24
Nath K, Chittibabu G, Das D (2005) Hydrogen production by Rhodobacter sphaeroides strain O.U.001 using spent media of enterobacter cloacae strain DM11. Appl Microbiol Biotechnol 68: 533–541.
Neves L, Ribeiro R, Oliveira R et al. (2006) Enhancement of methane production from barley waste. Biomass Bioenergy 30: 599–603
Nyns EJ (1986) Biomethanation processes. In: Schonborn, W. (ed), Microbial Degradations. Wiley-VCH Weinheim, Berlin 207–267.
Pandey A, Soccol C R, Nigam P et al. (2000a) Biotechnological potential of agro-industrial residues. I: sugarcane bagasse. Bioresour Technol 74: 69–80
Pandey A, Soccol C R, Nigam P et al. (2000b) Biotechnological potential of agro-industrial residues. II: cassava bagasse. Bioresour Technol 74: 81–87
Parawira W (2004) Doctoral Dissertation, Biotechnology Department, Lund University, Sweden

Patle S and Lal B (2008) Investigation of the potential of agro-industrial material as low cost substrate for ethanol production by using Candida tropicalis and zymomonas mobilis. Biomass Bioenergy 32: 596–602

Peng XW and Chen HZ (2007) Single cell oil production in solid-state fermentation by Microsphaeropsis sp. from steam-exploded wheat straw mixed with wheat bran. Bioresour Technol doi:10.1016/ j.biortech.2007.08.015.

Poorna CA and Prema P (2006) Production and partial characterization of endoxylanase by Bacillus pumilus using agro-industrial residues. Biochem Eng J 32: 106–112

Ruanglek V, Maneewatthana D, Tripetchkul S (2006) Evaluation of Thai agro-industrial wastes for bio-ethanol production by zymomonas mobilis. Process Biochem 41: 1432–1437

Sánchez OJ and Cardona CA (2008) Trends in biotechnological production of fuel ethanol from different feedstocks. Bioresour Technol 99: 5270–5295

Sellami F, Jarboui R, Hachicha S et al. (2007) Co-composting of oil exhausted olive-cake, poultry manure and industrial residues of agro-food activity for soil amendment. Bioresour Technol 99: 1177–1188

Shimada Y, Watanabe Y, Sugihara A et al. (2002) Enzymatic alcoholysis for biodiesel fuel production and application of the reaction to oil processing. J Mol Catal B: Enzym 17: 133–142

Siso MIG (1996) Biotechnological utilization of cheese whey: review. Bioresour Technol 57:1–11.

Soares M, Christen P, Pandey A et al. (2000) A novel approach for the production of natural aroma compounds using agro-industrial residue. Bioproc Eng 23: 695–699.

Switzenbaum MS and Jewell WJ (1980) Anaerobic attached-film expanded-bed reactor treatment. J Water Pollut Control Fed 52: 1953–1965.

Vicente G, Martinez M, Aracil J et al. (2004) Integrated biodiesel production: a comparison of different homogeneous catalysts systems. Bioresour Technol 92: 297–305.

Zanichelli D, Carloni F, Hasanaj E et al. (2007) Production of Ethanol by an Integrated valorization of olive oil byproducts. Envron Sci Pollut Res 14: 5–6

Chapter 5
Production of Protein-Enriched Feed Using Agro-Industrial Residues as Substrates

J. Obeta Ugwuanyi, Brian McNeil and Linda M. Harvey

Contents

Abstract Agricultural and food industry residues, refuse and wastes constitute a significant proportion (estimated to amount to over 30%) of world wide agricultural productivity. These wastes, include lignocellulosic materials, fruit and vegetable wastes, sugar industry wastes as well as animal and fisheries operations refuse and wastes. They represent valuable biomass and potential solutions to problems of animal nutrition and world wide supply of protein and calories if appropriate technologies can be deployed for their valorization by protein enrichment. Technologies available for protein enrichment of these wastes include solid substrate fermentation, ensiling and high solid or slurry processes. Technologies to be deployed for the reprocessing of these wastes will need to take account of the peculiarities of individual wastes and the environment in which they are generated, reprocessed and

J.O. Ugwuanyi (✉)
Department of Microbiology, University of Nigeria, Nsukka, Enugu, Nigeria
e-mail: jerryugwuanyi@yahoo.com

P. Singh nee' Nigam, A. Pandey (eds.), *Biotechnology for Agro-Industrial Residues Utilisation*, DOI 10.1007/978-1-4020-9942-7_5,

used. In particular such technologies need to deliver products that are safe not just for animal feed use but also from the perspective of human feeding. The use of organisms that are generally recognized as safe (GRAS) for the protein enrichment and reprocessing of waste will enhance user confidence

Keywords Protein-enrichment · Solid state fermentation · Silage-making · Liquid process · Slurry-process

5.1 Introduction

Agricultural and food industry residues, refuse and wastes constitute a significant proportion of world wide agricultural productivity. It has been variously estimated that up to 30% of global agricultural produce are left behind in the farms as residues and refuse. This is in addition to the proportion of photosynthetically produced energy that ends up as unused but potentially usable biomass (Robinson and Nigam 2003; Graminha et al. 2007). Large volumes of wastes, both solids and liquids, are also generated from the food processing industries. Together, these represent potential solutions to problems of animal nutrition if appropriate technologies can be deployed for their valorization. The problems posed by the continuing accumulation of these wastes have been compounded by the increasing concentration of food processing activities in large industries, particularly in the developed countries. Even in the less developed nations, the need to transport large quantities of food to the cities has led to increasing movement of processed food. In addition, emphasis on intensive agriculture to meet growing demand, as well as the concentration of food processing facilities in small land areas have resulted in the production of agro-food wastes in much higher concentrations than the available land space can take for disposal. The non-utilization of these vast resources constitutes significant loss of value.

Growing global demand for environmentally sustainable methods of production, pollution prevention and also economic motives have changed the way wastes and refuse are looked at. In other words, wastes have come to be seen more as resources in the wrong location and form than as a problem to be safely disposed of. Aside from their capacity to cause pollution, most food processing wastes are potentially of good enough quality to be recycled as raw materials for other applications or/and may be reprocessed to higher value products with relative ease. Even without any biotechnological improvement or upgrading, a wide variety of these wastes are already in use, albeit in small quantities in animal nutrition and other biotechnological processes (Martin 1998).

Developing and deploying appropriate technologies for the reprocessing and reuse of these abundant energy rich resources in human or animal feeding will go a long way in reducing the pressure on agricultural productivity and increase resource utilization efficiency. In particular, it will help to fight protein-energy malnutrition in those areas of the world where humans and animals compete for the same sources of protein and calories. Besides reducing pressure on productivity and improving global food security, efficient utilization of agricultural wastes will help to improve environmental health particularly with respect to those wastes whose accumulation constitute significant health hazards, and whose treatment and disposal incur considerable cost penalties.

Ultimately, some level of waste will arise from human productive activity, even with the growing application of clean technology. However, the idea behind waste minimization, which is the emerging face of waste management, requires that much of what is considered as waste and refuse, particularly those of considerable calorific value be seen first as valuable raw material for the production of value added product than as waste for disposal. This approach is likely to lead to ever decreasing volumes of waste meant for ultimate disposal. As transport and logistic costs will always put constraints on the movement of the vast quantities of agro-food wastes away from their point of production, reprocessing options need necessarily be designed to handle the wastes at the point of production. Technologies based on small and medium scale farm and factory based digesters and reactors should go a long way in making the processes attractive and economically viable. This work will attempt to review the current trend in biotechnological reprocessing by protein enrichment of agricultural wastes for feed use.

5.2 Technologies Applied to Waste Reprocessing for Protein Enrichment

5.2.1 Solid Substrate Fermentation

The term solid substrate fermentation (SSF) has been variously defined as the cultivation of microorganisms on solid, moist substrates in the absence of free aqueous phase (Pandey 2003), or cultivation of microorganisms in the presence of a liquid phase at maximal solid substrate concentrations (Mitchell et al. 2000), or on inert carriers supporting moist substrate (Ooijkaas et al. 2000). SSF has been historically used for the cultivation of microorganisms as it approximates the natural growth condition of most microorganisms (particularly of filamentous fungi). In Asia and Latin America, SSF has a traditional presence in the production of food and condiments such as tempeh, shoyu, and miso (and the fermented oil seeds common in Africa). On the contrary SSF seems to be employed much less frequently in Europe, except perhaps for its role in cheese and bread making.

SSF has continued to thrive on account of its labour convenience and generally low technological level. The majority of studies on the application of SSF have concentrated on its use for production of different enzymes for mostly environmental and industrial application (Couto and Sanroman 2006). But of particular interest, and one which appears to have immediate commercial potential is the application of SSF for the re-processing of bulk agricultural wastes for animal feed use (Laufenberg et al. 2003; Pandey et al. 2000a,b). SSF simulates the living conditions of filamentous fungi in their natural habitat. It is therefore convenient to grow them in such environments. In addition, in reaction environments, such as all SSF processes, where the ability to penetrate into the tissue or mass of solid substrates is important these organisms are more suited for cultivation (Santos et al. 2004) than unicellular ones, even though these have also been successfully employed in SSF (Virupakshi et al. 2005; Prakasham et al. 2006).

In theory SSF is versatile enough to be applied in a wide variety of biotechnological processes. In practice however, it is currently most suited for processes that generate low value bulk products such as applications for environmental control, including the production of compost and animal feed from solid wastes, bioremediation and biodegradation of hazardous compounds, detoxification of industrial wastes; nutritional enrichment of crops or crop-residues by biotransformation, biopulping, production of fermented foods, industrial enzymes, pigments, biopesticides, organic acids and flavor compounds. Cultivation of microbial consortia can also be more easily achieved in SSF than in submerged processes. Over the last few years, several topical reviews have been written on the available SSF technologies for the reprocessing and reuse of different agricultural and food industry wastes particularly lignocellulosic and vegetable wastes (Nigam and Singh 1994; 1996; Singh et al. 1996; Pandey and Soccol 2000; Pandey et al. 2000a,b; Pandey 2003; Raghavarao et al. 2003; Tengerdy and Szakacs 2003; Das and Singh 2004; Singh 2004; Couto and Sanroman 2006; Graminha et al. 2007).

5.2.2 Silage Making (Ensiling)

Silage making is the (lactic) fermentation/storage of (forage) for use in animal (ruminant) feeding. Ensiling is a multistage process which ultimately results in low pH (< 4.0) products that have extended resistance to spoilage. During ensiling, some bacteria are able to break down some cellulose and hemicellulose to their components sugars which are subsequently metabolized to low molecular weight acids, mostly lactic acid. This can also be encouraged by the use of appropriate mix of enzymes and microbial (lactic acid bacteria) silage inoculants (Okine et al. 2005; Colombatto et al. 2004. The lactic acid bacteria are also believed to produce bacteriocins that discourage the growth of and spoilage by unwanted populations. Efficient fermentation ensures a palatable and digestible feed. Production of good quality silage requires that anaerobiosis be achieved quickly to enable the lactic acid bacteria to develop and predominate, and in the process further bring down the pH of the mass. This discourages spoilage of the silage by putrefactive aerobic populations and ensures the retention of the most nutrients in the final product (Arvidsson et al. 2008).

Silage making starts with the impounding of the biomass and is initiated by aerobic populations. During this stage the aerobic organisms scavenge oxygen and bring about anaerobiosis. This phase is undesirable, because the aerobic bacteria consume soluble carbohydrates that should otherwise be available for the beneficial lactic acid bacteria. It also leads to the production of moisture, and heat generation, which if not properly managed are capable of destroying the process. Proteinaceaous materials may also be rapidly broken down during this phase and this can lead to loss of nutrients and the accumulation of ammonia (Slottner and Bertilsson 2006). To encourage rapid acidification during ensiling, fermentable sugars and lactic acid bacteria inoculants are often added to the silage (Okine et al. 2005; Yang et al. 2006). This is common during ensiling of protein rich feeds such as manure, slaughter house and fish wastes as well as many agricultural residues such as wheat straw, tomato or apple pomace and citrus waste (Chaudhry et al. 1998; Shaw et al. 1998; Scerra et al. 2001; Vidotti et al. 2003; Bampidis and Robinson 2006; Yang et al. 2006; Volanis et al. 2006; Santana-Delgado et al. 2008; Vázquez et al. 2008).

In the anaerobic stage of ensiling, a mixed population of lactic acid bacteria predominates and metabolizes fermentable sugars, producing lactic acid and reducing the pH of the mass to acidic levels. As the pH drops, minor acetic fermentation (if present) ends. This process continues until most of the available sugars have been consumed, and the pH has dropped to a level low enough to discourage bacterial activity. The duration of this stage varies with the nature of the biomass being ensiled, particularly, the initial concentration of fermentable sugars and the population of lactic acid bacteria. Ensiling by itself hardly leads to protein enrichment of the biomass except if mineral nitrogen is included. However, its capacity to achieve conservation of waste protein for use in animal feeding makes it important in schemes for the reuse of agricultural refuse.

5.2.3 Waste Enrichment in Liquid and Slurry Processes

A number of agricultural wastes are produced with considerable moisture content giving them a slurry consistency. In general the evaporation of such wastes to a solid consistency prior to biotechnological improvement (by SSF) would increase the cost of reprocessing and make them economically unattractive. Reprocessing of such residues as citrus pulp, farm house slurries, potato process slurries, cane process wastes has to be implemented as they are generated to be economically viable (Stamford and Decamargo 1992; De Gregorio et al. 2002). Production of protein enriched food (Teniola and Odunfa 2001; Nguyen et al. 2007a,b) and feed (Ugwuanyi 2008a; Ugwuanyi et al. 2006; 2008) in slurry state fermentation have been reported. The process has advantages in its capacity to operate at self heating elevated temperatures, achieve simultaneous protein enrichment and pasteurization of reprocessed waste, convert mineral nitrogen sources to protein rich microbial biomass and produce high protein content waste for use in animal feeding, and in employing high protein accumulating thermophiles to drive the process. Slurry reactions may be very rapid (Ugwuanyi et al. 2008) and can be adapted for use in the treatment of vast range of waste biomass, particularly those generated at elevated temperatures. As slurry reactions are a midpoint between SSF and SmF they may be expected to enjoy borderline advantages.

5.3 Agro-Food Industry Wastes and Residues and Their Reuse Potentials

The nature and quantity of waste generated from agricultural and food manufacturing practices vary with the predominant crop types and processing technology. Similarly, the need to reuse available waste varies with the pressure on calorie and the environment. In economies that produce mostly grains, large quantities of straw are produced, while economies that produce mainly root and tuber crops contend with high starch containing wastes. Although it is known that reuse of these wastes can significantly improve the economics of crop production, their effective reuse has often been constrained by the available technology. In general, the major wastes from agricultural and food industry operations include a variety of lignocellulosic materials, such as rice straw, wheat straw, maize stalk and cobs, barley straw, cane bagasse, cassava bagasse, vegetable process wastes including starch wastes, sugar industry wastes as well as farm animal refuse and waste from slaughter house and fisheries operations. Food processing industries also generate large quantities of rejects, trimmings and other substandard food materials that do not make it into the production chain. Fruit processing industries produce vast quantities of waste such as pomace and pulp that present disposal problems related to bulk and nutritional insufficiency making them unsuited for large scale use animal feeding and expensive to dispose. From the fermentation industries, vast amounts of spent media and microbial biomass are also generated.

5.3.1 Lignocellulosic Wastes

Agricultural refuse in this class are composed principally of cellulose, hemicellulose and lignin, and include cereal and vegetable wastes such as straw, bagasse, stover, cobs, cotton husk, groundnut husk, fibrous remnants of forage grass among others. They are arguably the most abundant agricultural wastes (Tengerdy and Szakacs 2003). In general, wastes in this class are composed of nearly 50% cellulose on a dry weight basis while hemicellulose and lignin account for the balance on a nearly equal basis (Bisaria 1998; Pandey et al. 2000a,b). On a global scale the quantity of cellulosic wastes available varies with the predominant agricultural and industrial crop produced in a given society. Although lignocellulosic wastes have found significant application as sources of (heat and electric) energy, it is believed that considerable value addition may be achieved by using these wastes for animal nutrition (Pandey et al. 2000a,b). Unfortunately their use in animal feeding is constrained by very low content of protein, vitamin, oil and other nutrient and limited digestibility and palatability to ruminants. However, they may be applied for animal nutrition following protein enrichment by using a variety of micro and macro fungi and bacteria. The predominant efforts in this direction have emphasized the use of SSF. In addition to the protein enrichment of wastes for use in ruminant nutrition, they have also been employed as principal raw materials, as a carbon source, for the fermentative production of feed related products such as enzymes and organic acids, single cell oil and flavour compounds, etc. The use of these wastes as principal carbon and energy sources for the production of microbial biomass protein such as single cell protein and mushrooms has also received considerable attention and a number of topical reviews exist in the literature (See Table 5.1).

The choice of microbial type to grow on lignocellulosic wastes depends to a large extent on the desired end product, and on whether or not a pre-treatment step is included or needed in the cultivation process. Pre-treatment processes that have been applied include steam explosion, acid, alkali, peroxide treatment, gamma irradiation, and combination of two or more of these processes as well as the usually more expensive but environmentally friendlier enzymatic treatment using a variety of cellulases hemicellulases and ligninases, (See Chapters 20 and 22 of this volume). Table 5.1 shows some processes and organisms that have been employed at both laboratory and pilot scale to achieve protein enrichment of some agricultural produce of mostly lignocellulosic nature and other wastes for food use and animal (ruminant) nutrition. In many instances, protein enrichment of agro-food waste may be accompanied by the economic extraction of valuable biochemicals such as food/feed grade enzymes and organic acid. In most cases the production of protein enriched lignocellulosic waste has been associated with the reduction in the content of lignocellulose (associated with loss of biomass via microbial respiration as carbon dioxide). Of particular importance in the protein enrichment of lignocellulosic waste for feed use is the improvement in palatability/acceptability/digestibility of such treated wastes for ruminants (Misra et al. 2007). This is due to the enzymatic disintegration of the lignocellulosic structure of plant cell wall. In addition, the low technology and reduced reactor volume employed in the SSF process (Nigam and Singh 1996) means that the process may be easily adapted for use in less affluent farm communities.

Although agricultural and food industry waste can be cheap raw materials for the production of enriched feed, the deployment of this technology may be constrained by logistic costs, particularly in the less developed economies. As the feed so produced is bulky and of low economic value, a key "selling point" of the process is the reduction of the pollution potential of the reprocessed waste. The implication of this for cost of the final product will be central to the economics of the reprocessing technology. Whatever options are adopted, a direct local need, such as the utilization of the produced feed for animal nutrition in the vicinity of the producing facility will go a long way in increasing the competitiveness of the process and product. The challenge is to develop on-site applicable and scalable technologies for the enrichment of agro-food waste in the vicinity of waste generating facilities.

Table 5.1 Microorganism grown on lignocellulose and other wastes in solid substrate fermentation and the products

Product/process/	Principal substrate	Microorganism used	Reference
Animal feed and food; protein enriched biomass, SCP; edible mushroom; cyanogenic glycoside detoxification; Protein enriched flour Glutamic acid; citric acid; volatile compounds	Cassava wastes (peels; slurry; bagasse; waste water); cassava tubers Cassava starch; wastewater	*Saccharomyces cerevisae; Lactobacillus spp; Rhizopus oryzae; Rhizopus spp; Aspergillus niger; Aspergillus spp.; Cephalosporium eichhorniae; Pleurotus spp; Lentinus spp, Brevibacterium divaricatum; Geotricum fragrans*	Ubalua 2007; Oboh and Elusiyan 2007; Oboh 2006; Obadina et al. 2006; Fagbemi and Ijah 2006; Srirotha et al. 2000; Soccol 1996; Balagopalan 1996; Daubresse et al. 1987; Brook et al. 1969; Zvauya and Muzondo 1994; Nguyen et al. 1992; Noomhorm et al. 1992 Jyothi et al. 2005; Damasceno et al. 2003
Protein enrichment; anti-nutrient removal; protein rich biomass	Coffee pulp; coffee husk; other coffee wastes	*Streptomyces; Pleurotus spp*	Orozco et al. 2008; Salmones et al. 2005; Brand et al. 2001; Fan et al. 2000
Single cell oil; protein enriched straw/feed; single cell protein; mushroom; gamma linoleic acid; citric acid; vitamins; essential amino acids Medicinal fungus; feed	Wheat bran/straw/corn stover/buckwheat/ millet/sugar beet pulp/citrus waste/water hyacinth; Mustard straw; bean straw; agave bagasse Agro-residues; Perennial grass	*Microsphaeropsis sp; Streptomyces cyaneus; various Basidiomycete fungi; Coprinus-fimetarius; Micromycetes; Phanerochaete chrysosporium; Pleurotus ostreatus; Thamnidium elegans; cellulolytic bacteria; Neurospora sitophila; Rhodotorula gracilis; Trametes spp Ganoderma* spp, *Coriolus versicolorTrichoderma* spp *Lentinus edodes; Cellulomonas biazoteain*	Peng and Chen 2008; Certik et al. 2006; Chaudhary and Sharma 2005; Varnayte and Raudonene 2004; Basu et al. 2002; Berrocal et al. 2000; Woomer et al. 2000; Dorado et al. 1999; Zadrazil 1997; Singh et al. 1995; Tripathi and Yadav 1992; Jacob 1991; Vladimirova et al. 1995; Iconomou et al. 1998; Mukherjee and Nandi 2004 Tripathi et al. 2008; Misra et al. 2007; Ortiz-Tovar et al. 2007, Philippoussis et al. 2007; Gaitan-Hernandez et al. 2006; Rajoka 2005
Protein rich fungi and feed; single cell protein	Apple pomace; apple waste; apple pulp; grape waste; carob pod; pineapple waste	*Rhizopus oligosporus; Candida utilis and Pleurotus ostreatus; Kloeckera-apiculata; Penicillium funiculosum Myrothecium verrucaria Aspergillus niger; Saccharomyces spp*	Rodriguez-Ramirez et al. 2007; Albuquerque et al. 2006; Villas-Boas et al. 2003; Nigam and Singh 1996; Nigam 1998; Rahmat et al. 1995; Bhalla and Joshi 1994; Kuzmanova et al. 1991; Correia et al. 2007

Table 5.1 (Continued)

Product/process/	Principal substrate	Microorganism used	Reference
Protein enriched feed	Cactus pear; cactus waste fibre	*Saccharomyces cerevisae; Aspergillus niger*	Araujo et al. 2005; Oliveira et al. 2001
Protein rich biomass/feed; Protein rich mushrooms	Rice polishing/rice bran/straw/chaff; sago fibre; saw dust; paddy straw; Lignocellulosic waste	*Candida utilis; Aspergillus niger; Trichoderma viride; Pleurotus sajor-caju; Pleurotus ostreatus Trichoderma reesei, Saccharomyces cerevisiaes; A. oryzae*	Rajoka et al. 2004; Bonatti et al. 2004; Ravinder et al. 2003; Vadiveloo 2003; Yang et al. 2003; Anupama 2001; Banik and Nandi 2004; Youssef and Aziz 1999; Patrabansh and Madan 1997
Protein rich food/feed	Viticulture waste	*Pleurotus spp*	Sanchez et al. 2002; Zhang et al. 2008
Protein enriched silage	Corn straw	*Silage population; silage* and *ssf*	Yang et al. 2001;
Fruit aroma	Various agro waste	*Ceratocystis fimbriata*	Bramorski et al. 1998
Protein rich feed	Cane bagasse and residues; other cane wastes in solid and slurry	*Trichoderma reesei* and *Trichoderma viride; Aspergillus niger; white rot fungi; Pleurotus spp*	Valino et al. 2002; 2003; Gutierrez-Correa et al. 1999; Zadrazil and Puniya 1995; Ortega et al. 1992; Echevarria et al. 1991
Protein enriched waste feed	Saw dust	*Pleurotus spp*	Lal and Panda 1995
Protein rich waste/feed/single cell protein	Mango waste; date industry waste	*Pleurotus spp*	Jwanny et al. 1995
Protein enrichment	Sugar beet pulp	*Trichoderma reesei; Trichoderma aureoviride*	Israilides et al. 1994; Illanes et al. 1992
Protein enrichment; scp production	Cashew waste	*Various yeasts*	De Holanda et al. 1998
Protein enrichment; Cellulose degration	Palm kernel cake	*Sclerotium rolfsii, Trichoderma harzianum, Trichoderma longiobrachiatum, Trichoderma koninggi and Aspergillus niger*	Iluyemi et al. 2006
Protein enriched waste; hydrolytic enzymes; single cell protein	Cabbage waste; Chinese cabbage	*Candida utilis, Pichia stipitis, Kluyveromyces marxianus, Saccharomyces cerevisiae; Indigenous microbes*	Choi and Park 2003; Krishna and Chandrasekaran 1995

5.3.2 Slaughter House Wastes and Manure

Large scale production of animal products including meat and poultry, and the processing and packaging of these in large scale facilities has resulted in the generation of large volumes of wastes including blood, feather, hoofs, horns, poultry intestines among others. The world wide production of chicken intestine runs into several million tons (Shaw et al. 1998). Often these wastes are treated on site for disposal, resulting in considerable loss of otherwise useful biomass. The high protein content of these wastes makes them attractive for use in animal nutrition. However, reprocessing them for use in animal nutrition can be a considerable challenge due to their ease of spoilage. Consequently, techniques applied to these wastes are essentially preservative. Fermentative ensiling has been studied extensively as an economical process to reprocess these wastes for use as ingredients in animal feeds as alternative to the conventional, but more expensive fish and soy meal. As animal offals are poor in carbohydrates, preservation is usually effected following addition of fermentable sugars (Shaw et al. 1998). Ensiling of animal wastes also has the advantage of causing reduction in the level of pathogenic organisms present in the wastes.

Reuse of poultry litter and manure for animal feeding has also attracted considerable attention due to the inefficiency of feed utilization by poultry. Pure and mixed culture lactic acidification and silage type reactions have been variously applied to improve the protein content and preservation quality and improve the smell of poultry manure for use in the production of poultry layers, beef cattle and pigs without impairing performance (Lallo et al. 1997; El Jalil et al. 2001). The process is energy efficient and has enabled the use of much higher concentration of animal waste in feed formulation than could be achieved with chemically or physically modified litter and offal (Kherrati et al., 1998).

Lactic acidification of animal wastes is particularly interesting in the tropics where the process can be very rapid and the storage life of this form of waste can be very short, with putrefaction starting only a few hours following the collection of offal and litter. Lactic acidification also leads to rapid elimination of spoilage and pathogenic organisms Shaw et al. 1998. A slight decline in protein content of the waste (if it happens) can be compensated for by the fact of reuse of the waste, rather than incurring cost penalties in its disposal.

Long term preservation by lactic preservation of slaughter house sludge for use in animal feeding has been demonstrated in processes that also achieve rapid inactivation of potential pathogens and spoilage organisms (Skrede and Nes 1988; Urbaniak and Sakson 1999; De Villiers and Pretorius 2001). As regulatory control of sludge disposal tightens, the process could play pivotal roles in waste minimization during animal production. A slightly different approach to the protein enrichment of (poultry) manure was reviewed by El Boushy (1991), which involves the use of house fly pupae as the protein enriching principle. Although this technology appears promising for the protein enrichment/extraction from poultry manure and other protein rich wastes, the necessity for sterilization at elevated temperature and its attendant energy cost may be a disincentive for the use of this technology, besides its aesthetic drawbacks.

5.3.3 *Fish and Fisheries Industries Wastes*

In the fish and fisheries industries including shrimp and crustacean processing, large amounts of wastes including rejects, discards and by-products, are produced worldwide. It is estimated that up to 30% of the total landings in the fisheries industries are considered as underutilized, by-catch and unconventional or unexploited (Venugopal and Shahidi 1995; Evers and Carroll 1996, 1998). It has also been estimated that over 32 million tones of fish wastes accumulate annually from the processing of fish (Kristinsson and Rasco 2000). Although a proportion of these get reprocessed into fish meal, oil and cake, several tons end up as waste requiring disposal. (Arvanitoyannis and Kassaveti 2008; Arvanitoyannis and Ladas 2008) provide a current review on the environmental position of fishery and meat wastes and management processes that are being considered for handling these, particularly within the context of the European Union where several directives and legislations aim to control their disposal. A prominent approach to the valorization of these wastes involves their biotechnological reprocessing (particularly by ensiling) for use in animal nutrition.

The ensiling of fish and shrimp wastes for use in fishery, as well as other animal nutrition, is receiving considerable attention and several studies have reported on the optimization of processes for the reuse of these wastes. In addition to using the ensiled feed for fish culture the processes lead to effective management of the waste. De Arruda et al. (2007) have shown that the use of ensiled fish waste in fish feeding can significantly improve the economics of fish production, considering that feeding accounts for up to 60% of cost. Dong et al. (2005) and Ngoan et al. (2000) used ensiled shrimp waste to replace soy meal in the feed of duck and pigs with comparable efficiency. Coello et al. (2002) optimized fish waste ensiling for the production of L-lysine using *Corynebacterium glutamicum*. The lactic preservation of these wastes by ensiling alone or with a variety of straw, forage and molasses has been shown to increase storage life of the product and increase acceptability, intake and digestibility by poultry, cattle, fish and other animals with the possibility of the silage serving a probiotic function (Faid et al. 1997; Hammoumi et al. 1998; Gerona et al. 2007; Goncalves and Viegas 2007). Importantly, rapid lactic preservation has enabled the reuse of an abundant waste, the disposal of which would otherwise attract considerable cost penalties.

5.3.4 *Microbial Biomass (Single Cell Protein) from Animal Wastes*

A slightly different approach to the valorization of animal and fishery waste is the hydrolysis and conversion wastes to single cell protein. Horn et al. (2005) used hydrolysate of cord viscera which constitutes about 17% of the fish biomass to grow *Lactobacillus* spp. and demonstrated that the medium so formulated was as effective as commercial peptone based media in the cultivation of the organism. This underscores the potential for the use of fishery waste of this kind for the cultivation of even fastidious organisms for the production of microbial biomass. Kuhn et al., (2008) fed microbial biomass produced from fish effluent to shrimps and demonstrated that the process improved the economics of shrimp production. In addition, the process led to the effective treatment of the resulting effluent. Single cell protein production for feed use has been achieved by cultivation of organisms on ram horn hydrolysate (Kurbanoglu and Algur 2002; Kurbanoglu 2003), poultry process waste Najafpour et al. 1994 and acid hydrolysed shrimp waste (Ferrer et al. 1996). Composted fish waste has been used for the production of *Scytalidium aciabphilum* biomass in submerged fermentation with good protein yield for animal feeding. Amar et al. (2006) also employed bacterial digestion of fish waste to produce feed for the production of Indian white prawn and in the process achieved both treatment and reuse the fish waste. Schneider et al. (2006) produced protein enriched bacterial biomass for animal feed use from a suspended growth process using aquaculture waste and in the process achieved treatment of a particularly recalcitrant waste stream. Viera et al. (2005) used microalgae to treat fish pond waste water effluent, and demonstrated that the protein rich algal biomass could be used as feed for the production of abalone.

5.3.5 Wastes from Cassava and Other Roots and Tuber Crops

Root and tuber crops including cassava, potato, cocoyams and yams are the principal sources of calories in many countries. The processing of these crops for human nutrition often results in the generation and disposal of several tones of carbohydrate rich wastes. Cassava, which is acknowledged to be a most important source of calories for large populations in the tropics, ranks as the world's sixth most important food crop (Soccol 1996). Besides its significant place in tropical and global food security, cassava has recently become recognized as an industrial crop in many countries where it is playing significant roles in animal nutrition and supply of industrial starch (Obadina et al. 2006).

5.3.5.1 Protein Enrichment of Cassava Wastes

Estimates vary considerably, but in the processing of cassava into food and starch, waste biomass may account for up to 30% of total produce (Antai and Mbongo 1994). Cassava wastes are very toxic due to the disproportionate partitioning of cyanogenic glycoside into the waste. As a result, without treatment, the waste can only be used in limited quantities in animal nutrition, while the capacity of cassava waste to cause pollution limits the disposal of the waste to land. Apart from the problem of toxicity, use of the waste in animal nutrition is also constrained by its limited protein content. Yet cassava waste remains a valuable resource which if widely used in animal nutrition can reduce pressure on food crops. In order to achieve the reuse of cassava wastes, a number of processing methods have been reported which result in detoxification and improvement in the protein content of the waste thereby converting this strong environmental pollutant to a value added product.

A number of processes have been implemented using cassava waste alone or in combination with other waste types, including poultry droppings to achieve reprocessing and protein enrichment of wastes. Organisms that have been employed in the protein enrichment and detoxification of cassava process wastes for use in ruminant nutrition include *Aspergillus* spp, *Trichoderma* spp. as well as a variety of bacteria, yeasts and ruminal microflora (Noomhorm et al. 1992; Oboh 2006; Adeyemi et al. 2007. See also Table 5.1). The carbohydrate content of cassava waste has also been variously exploited for the production of various food additives and ingredients, including citric acid and lactic acid (John et al. 2006; Pandey and Soccol, 2000; Ghofar et al. 2005). Although room exists for improvement of the protein content of the product, it is interesting that by the application of relatively inexpensive solid substrate processes cassava waste could be converted into useful products rather than being disposed of by expensive waste treatment strategies.

5.3.5.2 Protein Enrichment of Cassava

Various approaches have been studied in small laboratory and pilot scales for the protein enrichment of cassava for food use ((Brook et al. 1969; Daubresse et al. 1987; Oboh and Akindahunsi 2003). The processes led to slight drop in the carbohydrate content of the food. However, it was considered that the drop in carbohydrate was compensated for by the increase in the protein content of the resulting foods. Cassava starch has found considerable application in the production of food and feed grade single cell protein using various microorganisms (Ejiofor et al. 1996; Srirotha et al. 2000).

5.3.5.3 Protein Enrichment of Cocoyam Waste

Cocoyams (*Colocasia* and *Xanthosoma* spp) are widely cultivated for food in West Africa, Asia and the Oceania, with Nigeria, China and Ghana leading in world production (Onwueme and Charles 1994). In 1999, world wide production of cocoyams topped 6.5 million tones with Africa producing over half of the total (Onwueme 1999). In producing countries, cocoyams account for a significant proportion of the total energy intake, and this varies from about 7% in Ghana to about 18% in parts of the Oceania (Horton 1988). The processing of cocoyams to food and starch is associated with the generation of vast quantities of waste and residue that account for a significant proportion of the entire cocoyam produce (Ugwuanyi 2008b). The preservation and reuse of these vast wastes in animal nutrition will enhance food security in areas where cocoyams are abundant. Duru and Uma (2003a,b) have demonstrated the potential of using SSF to achieve over 50% increase in the protein content of cocoyam process waste using *Aspergillus oryzae*. The protein enriched waste could be used for the feeding of both ruminants and monogastric animals.

5.3.5.4 Protein Enrichment of Potato and Sweet Potato Wastes

Gelinas and Barrette (2007) employed *Candida utilis* to improve the protein content of waste potato starch from a chip manufacturing facility. Up to 11% protein was accumulated in a process that yielded 8% yeast protein in a submerged fermentation.

In a process that mixed sweet potato and sugar cane, Rodriguez et al. (2005) achieved improvement in the protein content of waste digest. Other processes in which potato process pulp and waste water, and sweet potato wastes have been reprocessed for protein enrichment including production of food and feed grade SCP have been reported (Yang 1993; Yang et al. 1993; Abu et al. 2000; Okine et al. 2005). The processes have been operated as SSF reactions, silage and as submerged fermentations.

5.3.6 Protein Enrichment of Fruit and Vegetable Wastes

5.3.6.1 Fruit Industry Wastes

Growing international production and marketing of fruits has led to increasing accumulation of fruit wastes such as citrus pulp, seeds and peels, grape pommace among others (Volanis et al. 2006; Bampidis and Robinson 2006). Disposal of these wastes can be a major cost component of fruit production since they may account for up to 50% by weight of fruits (Scerra et al. 2000; Graminha et al. 2007). These materials are very high in cellulosic materials (cellulose and hemicellulose), but low in lignin, making them potentially good feed sources for ruminants and promising substrates for the production of microbial protein. In countries with inadequate supplies of conventional ruminant feeds the use of fruit industry waste can impact quite positively on the supply of feed for animal nutrition while reducing environmental pollution. Unfortunately, fruit wastes have only minimal protein content which limits their value in animal nutrition. So, exploitation of these wastes in animal nutrition will depend on the deployment of processes for their protein enrichment by biotechnological means (Hang and Woodams 1986; Hang et al. 1987; Shojaosadati and Babaripour 2002; Volanis et al. 2006).

Fermentative processes in both SSF and slurries employing both filamentous and unicellular microorganisms have been employed for the protein enrichment of a variety of fruit industry wastes for animal feed use. (Shojaosadati et al. 1999; Scerra et al. 2000; De Gregorio et al. 2002; Correia et al. 2007; Plessas et al. 2008; Vendruscolo et al. 2008). Up to 500% protein enrichment of apple pomace using a combination of *Candida utilis* and *Pleurotus ostreatus* has been reported (Villas-Boas et al. 2003) and these wastes have also been employed to produce food and feed grade SCP. Protein enrichment and detoxification of coffee pulp for animal feed use has been reported (Orozco et al. 2008). Sunita and Rao (2003) used mango processing waste to produce blue green algal biomass for the production of Tilapia.

5.3.6.2 Vegetable Waste

Vegetable waste including trimmings, pressing fluids and rejects account for significant proportions of vegetable produce world wide. These wastes have high content of fermentable sugars and are very perishable. As a result, they have been treated for protein enrichment by a number of processes including ensiling and solid substrate fermentation. Vegetable wastes that have been reprocessed using food grade yeasts include Chinese cabbage juice, waste brine generated from kimchi production, deproteinized leaf juices, corn silage juice, date waste, tea process waste (Chanda and Chakrabatri 1996; Nancib et al. 1997; Choi and Park 1999; Choi et al. 2002; Hang et al. 2003; Murugesan et al. 2005). Stabnikova et al. (2005) produced specialty selenium enriched *Saccharomyces cerevisae* biomass by growing the organism in extracts of cabbage, watermelon, a mixture of residual biomass of green salads and tropical fruits.

5.3.7 Other Wastes

5.3.7.1 Olive Mill and Other Lipid Wastes

Production of olive oil, an important produce in some Mediterranean countries, particularly Spain, Italy, Greece and Tunisia, results in the production and disposal of large volumes of strongly polluting and toxic olive mill wastewater (OMW) (Israilides et al. 1997; Christodoulou et al. 2008). The considerable biomass content of OMW has necessitated efforts being made to develop biotechnological processes for the valorization of the waste. Fermentation and composting of OMW mixed with a variety of agricultural wastes as bulking agents have been practiced as means of detoxifying the waste (Haddadin et al. 1999; Garrido Hoyos et al. 2002; Mantzavinos and Kalogerakis 2005; Laconi et al. 2007; Hachicha et al. 2008; Cayuela et al. 2008). The use of OMW for production of microbial biomass as part of the detoxification process, and use of the detoxified waste/produced biomass as source of vitamins and mineral in animal nutrition has also been studied (Gharsallah 1993; DeFelice et al. 1997; Haddadin et al. 1999; Sampedro et al. 2005, 2007; Aloui et al. 2007; Christodoulou et al. 2008). Recently, several workers have studied the possibility of using OMW in media for the production of edible mushrooms with promising results (Zervakis et al. 1996; Tsioulpas et al. 2002; Roig et al. 2006; Soler-Rivas et al. 2006; Kalmis et al. 2008. See also Table 5.1).

5.3.8 Feed Grade Enzymes from Agro-Food Wastes

A variety of agricultural products contain anti-nutrients that limit their value as feedstuffs. Reduced nutrient utilization due to these anti-nutrients can contribute to environmental pollution due to excessive excretion of unabsorbed nutrients, especially nitrogen (N) and phosphorus (P), and increase production costs due to inefficient feed utilization (Woyengo et al. 2008). The use of enzymes to increase digestibility and nutritional value of feed has increased recently, and various depolymerizing enzymes have found their way into animal feed as supplement and additives (Selle et al. 2003; Roopesh et al. 2006; Nortey et al. 2007; Sands and Kay 2007). The use of solid substrate processes for the production of various feed grade enzymes including xylanases, cellulases, pectinases, chitinases, phytases, and ligninases is attractive, because in addition to its process engineering advantages, the enzyme may be produced and used (in situ) in the feed, in situations where the substrate is also the feed undergoing (protein and enzyme) enrichment leading to enhanced digestibility (Karunanandaa and Varga 1996; Kang et al. 2004; Carmona et al. 2005; Couto and Sanroman 2005; 2006; Mazutti et al. 2006; Roopesh et al. 2006; Mamma et al. 2008). The application of organisms with GRAS status for this process should be valuable and promising in the reuse of lignocellulosics for animal feeding. The use of exogeneous enzymes to improve the nutrient availability of feed has found most application in silage making, where a number of depolymerising enzymes are commercially available as silage additives (Schimidt et al. 2001; Colombatto et al. 2004).

A number of studies and reviews on SSF production, and use of depolymerising enzymes in animal nutrition, have been published recently (Juanpere et al. 2004; Titi and Tabbaa 2004; Eun et al. 2006; Cao et al. 2007; Graminha et al. 2007).

5.4 Conclusion and Safety Considerations

Global perception of wastes in general and agricultural wastes in particular is changing rapidly in response to need for environmental conservation sustainable agricultural productivity and global food security. Consequently, wastes are more currently seen as resources in the wrong form and location that needs to be reprocessed and reused than as wastes to be disposed of. This has increased the need for appropriate technologies for the reprocessing of such wastes. Protein enrichment of wastes for use in animal nutrition offers opportunities for the reuse of abundant agricultural wastes and refuse. Biotechnological processes such as solid substrate fermentation and ensiling offer great opportunities for the reuse of abundant agricultural wastes in animal nutrition. The use organisms with GRAS status to effect the protein enrichment, enzyme production and biomass production and detoxification reactions will help improve confidence in the final products derived from these processes, and help drive development of the application of biotechnology for the valorization of agricultural waste.

References

Abu OA, Tewe OO, Losel DM, Onifade AA (2000) Changes in lipid, fatty acids and protein composition of sweet potato (*Ipomoea batatas*) after solid-state fungal fermentation. Bioresource Technology 72: 189–192

Adeyemi AO, Eruvbetine D, Oguntona T, Dipeolu MA, Ogunbiade JA (2007) Enhancing the nutritional value of whole cassava root meal by rumen filtrate fermentation. Archivos de Zootecnia 56: 261–264

Albuquerque PM, Koch F, Trossini TG, Esposito E, Ninow JL (2006) Production of *Rhizopus oligosporus* protein by solid state fermentation of apple pomace. Brazilian Archives of Biology and Technology 49: 91–100

Aloui F, Abid N, Roussos S, Sayadi S (2007) Decolorization of semisolid olive residues of "alperujo" during the solid state fermentation by *Phanerochaete chrysosporium*, *Trametes versicolor*, *Pycnoporus cinnabarinus* and *Aspergillus niger*. Biochemical Engineering Journal 35: 120–125

Amar B, Philip R, Bright SIS (2006) Efficacy of fermented prawn shell waste as a feed ingredient for Indian white prawn, *Fenneropenaeus indicus*. Aquaculture Nutrition 12: 433–442

Antai SP, Mbongo PM (1994) Utilization of cassava peels as substrate for crude protein formation. Plant Foods for Human Nutrition 46: 345–51

Anupama RP (2001) Studies on production of single cell protein by *Aspergillus niger* in solid state fermentation of rice bran. Brazilian Archives of Biology and Technology 44: 79–88

Araujo LD, Medeiros AN, Neto AP, Oliveira LD, da Silva FLH (2005) Protein enrichment of cactus pear (*Opuntia ficus-indica* Mill) using *Saccharomyces cerevisiae* in solid-state fermentation. Brazilian Archives of Biology and Technology 48: 161–168

Arvanitoyannis IS, Kassaveti A (2008) Fish industry waste: treatments, environmental impacts, current and potential uses. International Journal of Food Science and Technology 43: 726–745

Arvanitoyannis IS, Ladas D (2008) Meat waste treatment methods and potential uses International Journal of Food Science and Technology 43: 543–559

Arvidsson K, Gustavsson A-M, Martinsson K (2008) Effect of conservation method on fatty acid composition of silage. Animal Feed Science and Technology doi:10.1016/j.anifeedsci.2008.04.003

Balagopalan C (1996) Improving the nutritional value of cassava by solid state fermentation: CTCRI experiences. Journal of Scientific and Industrial research 55: 479–482

Bampidis VA, Robinson PH (2006) Citrus by-products as ruminant feeds: A review. Animal Feed Science and Technology 128: 175–217

Banik S, Nandi R (2004) Effect of supplementation of rice straw with biogas residual slurry manure on the yield, protein and mineral contents of oyster mushroom. Industrial Crops and Products 20: 311–319

Basu S, Gaur R, Gomes J, Sreekrishnan TR, Bisaria VS (2002) Effect of seed culture on solid-state bioconversion of wheat straw by *Phanerochaete chrysosporium* for animal feed production. Journal of Bioscience and Bioengineering 93: 25–30

Berrocal M, Ball AS, Huerta S, Barrasa JM, Hernandez M, Perez-Leblic MI, Arias ME (2000) Biological upgrading of wheat straw through solid-state fermentation with *Streptomyces cyaneus*. Applied Microbiology and Biotechnology 54: 764–771

Bhalla TC, Joshi M (1994) Protein enrichment of apple pomace by coculture of cellulolytic molds and yeasts. World Journal of Microbiology and Biotechnology 10: 116–117

Bisaria VS (1998) Bioprocessing of agro-residues to value added products. In A.M. Alexander (ed.) Bioconversion of waste materials to industrial products Blackie Academic and Professional, London.

Bonatti M, Karnopp P, Soares HM, Furlan SA (2004) Evaluation of *Pleurotus ostreatus* and *Pleurotus sajor-caju* nutritional characteristics when cultivated in different lignocellulosic wastes. Food Chemistry 88: 425–428

Bramorski A, Soccol CR, Christen P, Revah S (1998) Fruity aroma production by *Ceratocystis fimbriata* in solid cultures from agro-industrial wastes. Revista de Microbiologia 29: 208–212

Brand D, Pandey A, Rodriguez-Leon JA, Roussos S, Brand I, Soccol CR (2001) Packed bed column fermenter and kinetic modeling for upgrading the nutritional quality of coffee husk in solid-state fermentation. Biotechnology Progress 17: 1065–1070

Brook EJ, Stanton WA, Wallbridge A (1969) Fermentation methods for protein enrichment of cassava by solid substrate fementation in rural conditions fermentation. Acta Horticultural 375: 217–224.

Cao L, Wang W, Yang C, Yang Y, Diana J, Yakupitiyage A, Luo Z, Li D (2007) Application of microbial phytase in fish feed Enzyme and Microbial Technology 40: 497–507

Carmona, E.C., Fialho, M.B., Buchgnani, E.B., Coelho, G.D., Brocheto-Braga, M.R., Jorge, J.A., 2005. Production, purification and characterization of a minor form of xylanase from *Aspergillus versicolor*. Process Biochemistry 40: 359–364

Cayuela ML, Mondini C, Sanchez-Monedero MA, Roig A (2008) Chemical properties and hydrolytic enzyme activities for the characterisation of two-phase olive mill wastes composting. Bioresource Technology 99: 4255–4262

Certik M, Slavikova L, Masrnova S, Sajbidor J (2006) Enhancement of nutritional value of cereals with gamma-linolenic acid by fungal solid-state fermentations. Food Technology and Biotechnology 44: 75–82
Chanda S, Chakrabatri S, (1996) Plant origin liquid waste: a resource for single cell protein production by yeast. Bioresource Technology 57: 51–54
Chaudhary N, Sharma CB (2005) Production of citric acid and single cell protein from agrowaste. National Academy Science Letters-India 28: 187–191
Chaudhrya SM, Fontenot JP, Naseer Z (1998) Effect of deep stacking and ensiling broiler litter on chemical composition and pathogenic organisms. Animal Feed Science and Technology 74: 155–167
Choi MH, Ji GE, Koh KH, Ryu YW, Jo DH, Park YH (2002) Use of waste Chinese cabbage as a substrate for yeast biomass production. Bioresource Technology 83: 251–253
Choi MH, Park YH (1999) Growth of *Pichia guilliermondii* A9, an osmotolerant yeast, in waste brine generated from kimchi production. Bioresource Technology 70: 231–236.
Choi, MH, Park YH (2003) Production of yeast biomass using waste Chinese cabbage. Biomass and Bioenergy 25: 221–226
Christodoulou V, Bampidis VA, Israilides CJ, Robinson PH, Giouzelyiannis A, Vlyssides A (2008) Nutritional value of fermented olive wastes in growing lamb rations. Animal Feed Science and Technology 141: 375–383
Coello N, Montiel E, Concepcion M, Christen P (2002) Optimization of a culture medium containing fish silage for L-lysine production by *Corynebacterium glutamicum*. Bioresource Technology 85: 207–211
Colombatto D, Mould FL, Bhat MK, Phipps RH, Owen E (2004) In vitro evaluation of fibrolytic enzymes as additives for maize (*Zea mays* L) silage: III. Comparison of enzymes derived from psychrophilic, mesophilic or thermophilic sources. Animal Feed Science and Technology 111: 145–159
Correia R, Magalhaes M, Macedo G (2007) Protein enrichment of pineapple waste with *Saccharomyces cerevisiae* by solid state bioprocessing Journal of Scientific and Industrial Research 66: 259–262
Couto SR, Sanroman MA (2005) Application of solid-state fermentation to ligninolytic enzyme production- review. Biochemical Engineering Journal 22: 211–219
Couto SR, Sanroman MA (2006) Application of solid-state fermentation to food industry-a review. Journal of Food Engineering 76: 291–302
Damasceno S, Cereda MP, Pastore GM, Oliveira JG (2003) Production of volatile compounds by *Geotrichum fragrans* using cassava wastewater as substrate. Process Biochemistry 39: 411–414
Das H, Singh SK (2004) Useful byproducts from cellulosic wastes of agriculture and food industry – A critical appraisal. Critical Reviews in Food Science and Nutrition 44: 77–89
Daubresse PS, Gheyen NS, Meyer JA (1987) A process for protein enrichment of cassava by solid substrate fermentation in rural conditions. Biotecthnology and Bioenergy 29: 962–968
De Arruda LF, Borghesi R, Oetterer M (2007) Use of fish waste as silage – A review. Brazilian Archives of Biology and Technology 50: 879–886
De Gregorio A, Mandalari G, Arena N, Nucita F, Tripodo MM, Lo Curto RB (2002) SCP and crude pectinase production by slurry-state fermentation of lemon pulps. Bioresource Technology 83: 89–94
De Holanda JS, De Oliveira AJ, Ferreira AC (1998). Protein enrichment of cashew waste with yeast for animal alimentation. Pesquisa Agropecuaria Brasileira 33: 787–792
De Villiers GH, Pretorius WA (2001) Abattoir effluent treatment and protein production. Water Science and Technology 43: 243–250
DeFelice B, Pontecorvo G, Carfagna M (1997) Degradation of waste waters from olive oil mills by *Yarrowia lipolytica* ATCC 20255 and *Pseudomonas putida*. Acta Biotechnologica 17: 231–239
Dong NTK, Elwinger K, Lindberg JE, Ogle RB (2005) Effect of replacing soybean meal with soya waste and fish meal with ensiled shrimp waste on the performance of growing crossbred ducks. Asian-Australasian Journal of Animal Sciences 18: 825–834

Dorado J, Almendros G, Camarero S, Martinez AT, Vares T, Hatakka A (1999) Transformation of wheat straw in the course of solid-state fermentation by four ligninolytic basidiomycetes. Enzyme and Microbial Technology 25: 605–612

Duru CC, Uma NU (2003a) Protein enrichment of solid waste from cocoyam (*Xanthosoma sagittifolium* (L.) Schott) cormel processing using *Aspergillus oryzae* obtained from cormel flour. African Journal of Biotechnology 2: 228–232

Duru CC, Uma NU (2003b) Production of fungal biomass from cormel process waste-water of cocoyam (*Xanthosoma sagittifolium* (L) Schott) using *Aspergillus oryzae* obtained from cormel flour. Journal of the Science of Food and Agriculture 83: 850–857

Echevarria J Leon JAR, Espinosa ME, Delgado G (1991) Optimization of solid-state fermentation of sugar-cane by *Aspergillus niger* considering particles size effect. Acta Biotechnologica 11: 15–22

Ejiofor AO, Chisti Y, Moo-Young M (1996) Culture of Saccharomyces cerevisiae on hydrolyzed waste cassava starch for production of baking-quality yeast. Enzyme Microbial Technology 18: 519–525

El Boushy AR (1991) House-fly pupae and poultry manure converters for animal feed: A review. Bioresource Technology 38: 45–49

El Jalil MH, Faid M, Elyachioui M (2001) A biotechnological process for treatment and recycling poultry wastes manure as a feed ingredient. Biomass and Bioenergy 21: 301–309

Eun JS, Beauchemin KA, Hong SH, Bauer MW (2006) Exogenous enzymes added to untreated or ammoniated rice straw: effects on in vitro fermentation characteristics and degradability. Animal Feed Science and Technology 131: 87–102

Evers DJ, Carroll DJ (1996) Preservation of crab or shrimp waste as silage for cattle. Animal Feed Science and Technology 59: 233–244

Evers DJ, Carroll DJ (1998) Ensiling salt-preserved shrimp waste with grass straw and molasses. Animal Feed Science and Technology 71: 241–249

Fagbemi AO, Ijah UJJ (2006) Microbial population and biochemical changes during production of protein-enriched fufu. World Journal of Microbiology and Biotechnology 22: 635–640

Faid M, Zouiten A, Elmarrakchi A, AchkariBegdouri A (1997) Biotransformation of fish waste into a stable feed ingredient. Food Chemistry 60: 13–18

Fan L, Pandey A, Mohan R, Soccol CR (2000) Use of various coffee industry residues for the cultivation of *Pleurotus ostreatus* in solid state fermentation. Acta Biotechnologica 20: 41–52

Ferrer J, Paez G, Marmol Z, Ramones E, Garcia H, Forster CF (1996) Acid hydrolysis of shrimp-shell wastes and the production of single cell protein from the hydrolysate. Bioresource Technology 57: 55–60

Gaitan-Hernandez R, Esqueda M, Gutierrez A, Sanchez A, Beltran-Garcia M, Mata G (2006) Bioconversion of agrowastes by *Lentinula edodes*: the high potential of viticulture residues. Applied Microbiology and Biotechnology 71: 432–439

Garrido Hoyos SE, Martinez Nieto L, Rubio FC, Cormenzana AR (2002) Kinetics of aerobic treatment of olive-mill wastewater (OMW) with *Aspergillus terreus*. Process Biochemistry 37: 1169–1176

Gelinas P, Barrette J (2007) Protein enrichment of potato processing waste through yeast fermentation. Bioresource Technology 98: 1138–1143

Gerona LJ, Zeoula LM,Vidotti RM, Matsushita M, Kazama R, Caldas Neto SF, Fereli F (2007) Chemical characterization, dry matter and crude protein ruminal degradability and in vitro intestinal digestion of acid and fermented silage from tilapia filleting residue. Animal Feed Science and Technology 136: 226–239

Gharsallah N (1993) Production of single cell protein from olive mill waste-water by yeasts. Environmental Technology 14: 391–395

Ghofar A, Ogawa S, Kokugan T (2005) Production of L-lactic acid from fresh cassava roots slurried with tofu liquid waste by streptococcus bovis. Journal of Bioscience and Bioengineering 100: 606–612

Goncalves LU, Viegas EMM (2007) Production, characterization and biological evaluation of shrimp waste silage for Nile tilapia. Arquivo Brasileiro de Medicina Veterinaria e Zootecnia 59: 1021–1028

Graminha EBN, Goncalves AZL, Pirota RDPB, Balsalobre MAA, Da Silva R, Gomes E (2007) Enzyme production by solid-state fermentation: Application to animal nutrition. Animal Feed Science and Technology doi:10.1016/j.anifeedsci.2007.09.029

Gutierrez-Correa M, Portal L, Moreno P, Tengerdy RP (1999) Mixed culture solid substrate fermentation of *Trichoderma reesei* with *Aspergillus niger* on sugar cane bagasse. Bioresource Technology 68: 173–178

Hachicha S, Cegarrab J, Sellamia F, Hachichaa R, Drirac N, Medhiouba K, Ammara E (2008) Elimination of polyphenols toxicity from olive mill wastewater sludge by its co-composting with sesame bark. Journal of Hazardous Material doi:10.1016/j.jhazmat.2008.04.066

Haddadin MS, Abdulrahim SM, Al-Khawaldeh GY, Robinson RK (1999) Solid state fermentation of waste pomace from olive processing. Journal of Chemical Technology and Biotechnology 74: 613–618

Hammoumi A, Faid M, El Yachioui M, Amarouch H (1998) Characterization of fermented fish waste used in feeding trials with broilers. Process Biochemistry 33: 423–427

Hang YD, Woodams EE (1986) A solid state fermentation of apple pomace for citric acid production using *Aspergillus niger*. Journal of Applied Microbiology and Biotechnology 2: 283–287

Hang YD, Woodams EE, Hang LE (2003) Utilization of corn silage juice by *Klyuveromyces marxianus*. Bioresource Technology 86: 305–307

Hang YD, Woodams EE, Luh BS (1987) Microbial production of citric acid by solid state fermentation of kiwi fruit peel. Journal of Food Science 52: 226–227

Horn SJ, Aspmo SI, Eijsink VGH (2005) Growth of *Lactobacillus plantarum* in media containing hydrolysates of fish viscera. Journal of Applied Microbiology 99: 1082–1089

Horton D (1988) Underground crops: long-term trends in production of roots and tubers. Morrilton AR: Winrock International, USA. 132pp

Iconomou D, Kandylis K, Israilidis C, Nikokrys P (1998) Protein enhancement of sugar beet pulp by fermentation and estimation of protein degradability in the rumen of sheep. Small Ruminant Research 27: 55–61

Illanes A, Aroca G, Cabello L, Acevedo F (1992) Solid substrate fermentation of leached beet pulp with *Trichoderma-aureoviride*. World Journal of Microbiology and Biotechnology 8: 488–493

Iluyemi FB, Hanafi MM, Radziah O, Kamarudin MS (2006) Fungal solid state culture of palm kernel cake. Bioresource Technology 97: 477–482

Israilides CJ, Iconomou D, Kandylis K, Nikokyris P (1994) Fermentability of sugar-beet pulp and its acceptability in mice. Bioresource Technology 47: 97–101

Israilides CJ, Vlyssides AG, Mourafeti VN, Karrouni G (1997) Olive oil wastewater treatment with the use of an electrolysis system. Bioresource Technology 61: 163–170

Jacob Z (1991) Enrichment of wheat bran by Rhodotorula gracilis through solid state fermentation. Folia Microbiologica 36: 86–91

John RP, Nampoothiri KM, Pandey A (2006) Solid-state fermentation for L-lactic acid production from agro wastes using *Lactobacillus delbrueckii*. Process Biochemistry 41: 756–763

Juanpere J, Pérez-Vendrell AM, Brufau J (2004) Effect of microbial phytase on broilers fed barley-based diets in the presence or not of endogenous phytase. Animal Feed Science and Technology 115: 265–279

Jwanny EW, Rashad MM, Abdu HM (1995) Solid-state fermentation of agricultural wastes into food through *Pleurotus* cultivation. Applied Biochemistry and Biotechnology 50: 71–78

Jyothi AN, Sasikiran K, Nambisan B, Balagopalan C (2005) Optimisation of glutamic acid production from cassava starch factory residues using *Brevibacterium divaricatum*. Process Biochemistry 40: 3576–3579

Kalmis E, Nuri Azbar N, Yıldız H, Kalyoncu F (2008) Feasibility of using olive mill effluent (OME) as a wetting agent during the cultivation of oyster mushroom, Pleurotus ostreatus, on wheat straw. Bioresource Technology 99: 164–169

Kang SW, Park YS, Lee JS, Hong SI, Kim SW (2004) Production of cellulose and hemicellulases by *Aspergillus niger* KK2 from lignocellulosic biomass. Bioresource Technology 91: 153–156
Karunanandaa K, Varga GA (1996) Colonization of rice straw by white-rot fungi (*Cyathus stercoreus*): effect on ruminal fermentation pattern, nitrogen metabolism, and fiber utilization during continuous culture. Animal Feed Science Technology 61: 1–16
Kherrati B, Faid M, Elyachioui M, Wahmane A (1998) Process for recycling slaughter-houses wastes and by-products by fermentation. Bioresource Technology 63: 75–79
Krishna C, Chandrasekaran M (1995) Economic utilization of cabbage wastes through solid-state fermentation by native microflora. Journal of Food Science and Technology-Mysore 32: 199–201
Kristinsson HG, Rasco BA (2000) Fish protein hydrolysates: production, biochemical, and functional properties. Critical Reviews in Food Science and Nutrition 40: 43–81
Kuhn, DD, Boardman GD, Craig SR, Flick GJ Jr., Mclean E (2008) Use of microbial flocs generated from tilapia effluent as a nutritional supplement for shrimp, *Litopenaeus vannamei*, in recirculating aquaculture systems. Journal of the World Aquaculture Society 39: 72–82
Kurbanoglu EB (2003) Investigation of the use of ram horn hydrolysate for fungal protein production. Journal of the Science of Food and Agriculture 83: 1134–1138
Kurbanoglu EB, Algur OF (2002) Single-cell protein production from ram horn hydrolysate by bacteria. Bioresource Technology 85: 125–129
Kuzmanova S, Vandeska E, Dimitrovski A (1991) Production of mycelial protein and cellulolytic enzymes from food wastes. Journal of Industrial Microbiology 7: 257–261
Lallo CHO, Singh R, Donawa AA, Madoo G (1997) The ensiling of poultry offal with sugarcane molasses and *Lactobacillus* culture for feeding to growing/finishing pigs under tropical conditions. Animal Feed Science Technology 67: 213–222
Laconi S, Molle G, Cabiddu A, Pompei R (2007) Bioremediation of olive oil mill wastewater and production of microbial biomass. Biodegradation 18: 559–566
Lal N, Panda T (1995) Studies on protein enrichment in sawdust by *Pleurotus-sajor-caju*. Bioprocess Engineering 12: 163–165
Laufenberg G, Kunz B, Nystroem M (2003) Transformation of vegetable waste into value added products. Bioresource Technology 87: 167–198
Mamma D, Kourtoglou E, Christakopoulos P (2008) Fungal multienzyme production on industrial by-products of the citrus-processing industry. Bioresource Technology 99: 2373–2383
Mantzavinos D, Kalogerakis N (2005) Treatment of olive mill effluents Part I. Organic matter degradation by chemical and biological processes- an overview. Environment International 31: 289–295
Martin AM (Ed.) (1998) Bioconversion of Waste Materials to Industrial Products, second ed. Blackie Academic and Professional, London
Mazutti, M., Bender, J.P., Treichel, H., Di Luccio, M., 2006. Optimization of inulinase production by solid-state fermentation using sugarcane bagasse as substrate. Enzyme and Microbial Technology 39: 56–59
Misra AK, Mishra AS, Tripathi MK, Prasad R, Vaithiyanathan S, Jakhmola RC (2007) Optimization of solid state fermentation of mustard (*Brassica campestris*) straw for production of animal feed by white rot fungi (*Ganoderma lucidum*) Asian-Australasian. Journal of Animal Sciences 20: 208–213
Mitchell DA, Berovic M, Krieger N (2000) Biochemical engineering of solid state bioprocessing. Advances in Biochemical Engineering/Biotechnology 68: 61–138
Mukherjee R, Nandi B (2004) Improvement of in vitro digestibility through biological treatment of water hyacinth biomass by two *Pleurotus* species. International Biodeterioration and Biodegradation 53: 7–12
Murugesan GS Sathishkumar M, Swaminathan K (2005) Supplementation of waste tea fungal biomass as a dietary ingredient for broiler chicks. Bioresource Technology 96: 1743–1748
Najafpour GD, Klasson KT, Ackerson MD, Clausen EC, Gaddy JL (1994) Biological conversion of poultry-processing waste to single-cell protein. Bioresource Technology 48: 65–70

Nancib N, Nancib A, Boudrant J (1997) Use of waste date products in the fermentative formation of baker's yeast biomass by *Saccharomyces cerevisiae.* Bioresource Technology 60: 67–71

Ngoan LD, An LV, Ogle B, Lindberg JE (2000) Ensiling techniques for shrimp by-products and their nutritive value for pigs. Asian-Australasian Journal of Animal Sciences 13: 1278–1284

Nguyen TTT, Guyot J-P, Icard-Verniere C, Rochette I, Loiseau G (2007a) Effect of high pressure homogenisation on the capacity of Lactobacillus plantarum A6 to ferment rice/soybean slurries to prepare high energy density complementary food. Food Chemistry 102: 1288–1295

Nguyen TTT, Loiseau G, Icard-Verniere C, Rochette I, Treche Sc, Guyot J-P (2007b) Effect of fermentation by amylolytic lactic acid bacteria, in process combinations, on characteristics of rice/soybean slurries: A new method for preparing high energy density complementary foods for young children. Food Chemistry 100: 623–631

Nguyen V, Bednarski W, Kowalewskapiontas J, Tomasik J (1992) The growth-conditions of *Cephalosporium eichhorniae* for the conversion of starch substrates to protein. Acta Biotechnologica 12: 325–332

Nigam JN (1998) Single cell protein from pineapple cannery effluent. World Journal of Microbiology and Biotechnology 14: 693–696

Nigam P, Singh D (1994) Solid-state (substrate) fermentation systems and their applications in biotechnology. Journal of Basic Microbiology 34: 405–423

Nigam P, Singh D (1996) Processing of agricultural wastes in solid state fermentation for microbial protein production. Journal of Scientific and Industrial Research 55: 373–380

Noomhorm A, Ilangantileke S, Bautista MB (1992) Factors in the protein enrichment of cassava by solid-state fermentation. Journal of the Science of Food and Agriculture 58: 117–123

Nortey TN, Patience JF, Sands JS, Zijlstra RT (2007) Xylanase supplementation improves energy digestibility of wheat by-products in grower pigs. Livestock Science 109: 96–99

Obadina AO, Oyewole OB, Sanni LO, Abiola SS (2006) Fungal enrichment of cassava peels proteins. African Journal of Biotechnology 5: 302–304

Oboh G (2006) Nutrient enrichment of cassava peels using a mixed culture of *Saccharomyces cerevisae* and *Lactobacillus spp* solid media fermentation Techniques. Electronic Journal of Biotechnology 9: 46–49

Oboh G, Elusiyan CA (2007) Changes in the nutrient and anti-nutrient content of micro-fungi fermented cassava flour produced from low- and medium-cyanide variety of cassava tubers. African Journal of Biotechnology 6: 2150–2157

Oboh G, Akindahunsi AA (2003) Biochemical changes in cassava products (flour and gari) subjected to *Saccharomyces cerevisae* solid media fermentation. Food Chemistry 82: 599–602

Okine A, Aibibua HY, Okamoto M (2005) Ensiling of potato pulp with or without bacterial inoculants and its effect on fermentation quality, nutrient composition and nutritive value. Animal Feed Science and Technology 121: 329–343

Oliveira MA, Rodrigues C, dos Reis EM, Nozaki J (2001) Production of fungal protein by solid substrate fermentation of cactus *Cereus peruvianus* and *Opuntia ficus indica.* Quimica Nova 24: 307–310

Onwueme I (1999) Taro cultivation in Asia and the Pacific, Food and Agricultural Organisation. RAP Publication 1999/16. FAO Regional Office for Asia and the Pacific. Bangkok Thailand

Onwueme IC, Charles WB (1994) Tropical root and tuber crop production, perspectives and future prospects. FAO Plant Production and Protection paper 126, 288

Ooijkaas LP, Weber F, Buitelaar RM, Tramper J, Rinzema A (2000) Defined media an inert supports: their potential as solid-state fermentation production system. Trends in Biotechnology 18: 356–360.

Orozco AL, Perez MI, Guevara O, Rodriguez J Hernandez M, Gonzalez-Vila FJ, Polvillo O, Arias ME (2008) Biotechnological enhancement of coffee pulp residues by solid-state fermentation with *Streptomyces*. Py-GUMS analysis. Journal of Analytical and Applied Pyrolysis 81: 247–252

Ortega GM, Martinez EO, Betancourt D, Gonzalez AE, Otero MA (1992) Bioconversion of sugarcane crop residues with white-rot fungi *Pleurotus* sp. World Journal of Microbiology and Biotechnology 8: 402–405

Ortiz-Tovar G, Lopez-Miranda J, Cerrillo-Soto MA, Juarez-Reyes A, Favela-Torres E, Soto-Cruz O (2007) Effect of solid substrate fermentation on the nutritional quality of agro-industrial residues. Interciencia 32: 339–342

Pandey A, Soccol CR (2000) Economic utilization of crop residues for value addition: A futuristic approach. Journal of Scientific and Industrial Research 59: 12–22

Pandey A, Soccol CR, Nigam P, Brand D, Mohan R, Roussos S (2000a) Biotechnological potential of coffee pulp and coffee husk for bioprocesses. Biochemical Engineering Journal 6: 153–162

Pandey A, Soccol CR, Nigam P, Soccol VT (2000b) Biotechnological potential of agro-industrial residues. I: sugarcane bagasse. Bioresource Technology 74: 69–80

Pandey, A (2003) Solid-state fermentation. Biochemical Engineering Journal 13: 81–84

Patrabansh S, Madan M (1997) Studies on cultivation, biological efficiency and chemical analysis of *Pleurotus sajor-caju* (Fr) Singer on different bio-wastes. Acta Biotechnologica 17: 107–122

Peng X, Chen H (2008) Single cell oil production in solid-state fermentation by Microsphaeropsis sp from steam-exploded wheat straw mixed with wheat bran. Bioresource Technology 99: 3885–3889

Philippoussis A, Diamantopoulou P, Israilides C (2007) Productivity of agricultural residues used for the cultivation of the medicinal fungus *Lentinula edodes*. International Biodeterioration and Biodegradation 59: 216–219

Plessas S, Koliopoulos D, Kourkoutas Y, Psarianos C, Alexopoulos A,Marchant R, Banat IM, Koutinas AA (2008) Upgrading of discarded oranges through fermentation using kefir in food industry. Food Chemistry 106: 40–49

Prakasham RS, Rao CS, Sarma PN (2006) Green gram husk—an inexpensive substrate for alkaline protease production by *Bacillus* sp. in solid-state fermentation. Bioresource Technology 97: 1449–1454

Raghavarao KSMS, Ranganathan TV, Karanth NG (2003) Some engineering aspects of solid-state fermentation. Biochemical Engineering Journal 13: 127–135

Rahmat H, Hodge RA, Manderson GJ, Yu PL (1995) Solid-substrate fermentation of *Kloeckera-apiculata* and *Candida-utilis* on apple pomace to produce an improved stock-feed. World Journal of Microbiology and Biotechnology 11: 168–170

Rajoka MI (2005) Production of single cell protein through fermentation of a perennial grass grown on saline lands with *Cellulomonas biazotea*. World Journal of Microbiology and Biotechnology 21: 207–211

Rajoka MI, Kiani MAT, Khan S, Awan MS, Hashmi AS (2004) Production of single cell protein from rice polishings using *Candida utilis*. World Journal of Microbiology and Biotechnology 20: 297–301

Ravinder R, Rao LV, Ravindra P (2003) Studies on *Aspergillus oryzae* mutants for the production of single cell proteins from deoiled rice bran. Food Technology and Biotechnology 41: 243–246

Robinson T, Nigam P (2003) Bioreactor design for protein enrichment of agricultural residues by solid state fermentation. Biochemical Engineering Journal 13: 197–203

Rodriguez Z, Boucourt R, Elias A, Nunez O (2005) Effect of the inoculation moment on the fermentative process of mixtures of sugarcane and sweet potato. Cuban Journal of Agricultural Science 39: 289–295

Rodriguez-Ramirez HE, Hernandez-Gomez C, Rodriguez-Muela C, Ruiz-Barrera O, Salvador-Torres F (2007) Protein production by solid state fermentation of apple waste and pomace. Journal of Dairy Science 90: 285–285

Roig A, Cayuela ML, Sanchez-Monedero MA (2006) An overview on olive mill wastes and their valorisation methods. Waste Management 26: 960–969

Roopesh, K., Sumitra Ramachandran, K., Nampoothiri, M., Szakacs, G., Pandey, A., 2006. Comparison of phytase production on wheat bran and oilcakes in solid-state fermentation by *Mucor racemosus*. Bioresource Technology 97: 506–511

Salmones D, Mata G, Waliszewski KN (2005) Comparative culturing of *Pleurotus* spp. on coffee pulp and wheat straw: Biomass production and substrate biodegradationm. Bioresource Technology 96: 537–544

Sampedro I, Annibale AD, Ocampo JA, Stazi SR, Garcia-Romera I (2005) Bioconversion of olive-mill dry residue by *Fusarium lateritium* and subsequent impact on its phytotoxicity. Chemosphere 60: 1393–1400

Sampedro I, D'Annibale A, Ocampo JA, Stazi SR, García-Romera I (2007) Solid-state cultures of *Fusarium oxysporum* transform aromatic components of olive-mill dry residue and reduce its phytotoxicity. Bioresource Technology 98: 3547–3554

Sanchez A, Ysunza F, Beltran-Garcia MJ, Esqueda M (2002) Biodegradation of viticulture wastes by *Pleurotus*: A source of microbial and human food and its potential use in animal feeding. Journal of Agricultural and Food Chemistry 50: 2537–2542

Sands JS, Kay RM (2007) Phyzyme XP phytase improves growth performance and nutrient utilization in wheat-based diets fed to weaned pigs. Livestock Science 109: 264–267

Santana-Delgado H, Avila E, Sotelo A (2008) Preparation of silage from spanish mackerel *comberomoru maculates* and its evaluation in broiler diets. Animal Feed Science and Technology 141: 129–140

Santos MM, Rosa AS, Dal'boit S, Mitchell DA, Kriger N (2004) Thermal denaturation: is solid state fermentation really a good technology for the production of enzymes? Bioresource Technology 93: 261–268

Scerra V, Caparraa P, Fotia F, Lanzab M, Priolob A (2001) Citrus pulp and wheat straw silage as an ingredient in lamb diets: effects on growth and carcass and meat quality. Small Ruminant Research 40: 51–56

Scerra V, Caridi A, Foti F, Sinatra MC, Caparra P (2000) Changes in chemical composition during the colonisation of citrus pulps by a dairy *Penicillium roqueforti* strain. Bioresource Technology 72: 197–198

Schimidt J, Szakacs G, Cenkvari E, Sipocz J, Urbanszki K, Tengerdy RP (2001) Enzyme assisted ensiling of alfalfa with enzymes by solid substrate fermentation. Bioresource Technology 76: 207–212

Schneider O, Sereti V, Machiels MAM, Eding EH, Verreth JAJ (2006) The potential of producing heterotrophic bacteria biomass on aquaculture waste. Water Research 40: 2684–2694

Selle PH, Ravindran V, Ravindran G, Pittolo PH, Bryden WL (2003) Influence of phytase and xylanase supplementation on growth performance and nutrient utilisation of broilers offered wheat-based diets. Asian-Australian Journal Animal Science 16: 394–402

Shaw DM, Narasimha RD, Mahendrakar NS (1998) Effect of different levels of molasses, salt and antimycotic agents on microbial profiles during fermentation of poultry intestine. Bioresource Technology 63: 237–241

Shojaosadati SA, Babaripour V (2002) Citric acid production from apple pomace in multi layer packed bed solid state bioreactor. Process Biochemistry 37: 909–914

Shojaosadati SA, Faraidouni R, Madadi-Nouei A, Mohamadpour I (1999) Protein enrichment of lignocellulosic substrates by solid state fermentation using *Neurospora sitophila*. Resource Conservation and Recycling 27: 73–87

Singh GP, Singh K, Gupta BN (1995) Bioconversion of ammonia to biological protein by *Coprinus-fimetarius* during solid substrate fermentation of ammoniated (urea) wheat-straw. Indian Journal of Animal Sciences 65: 602–605

Singh K (2004) Biotechnological approaches on conversion of agrocellulosic residues and dairy wastes into useful end-products. Indian Journal of Animal Sciences 74: 414–423

Singh K, Puniya AK, Singh S (1996) Biotransformation of crop residues into animal feed by solid state fermentation. Journal of Scientific and Industrial Research 55: 472–478

Skrede A, Nes IF (1988) Slaughterhouse by-products preserved by *Lactobacillus plantamm* fermentation as feed for mink and foxes. Animal Feed Science Technogy 20: 187–198

Slottner D, Bertilsson J (2006) Effect of ensiling technology on protein degradation during ensiling. Animal Feed Science and Technology 127: 101–111

Soccol CR (1996) Biotechnology products from cassava root by solid state fermentation. Journal of Scientific and Industrial Research 55: 358–364

Soler-Rivas C, Garcia-Rosado A, Poloniab I, Junca-Blanch G, Marin FR, Wichers HJ (2006) Microbiological effects of olive mill waste addition to substrates for *Pleurotus pulmonarius* cultivation. International Biodeterioration and Biodegradation 57: 37–44

Srirotha K, Chollakup R, Chotineeranatb S, Piyachomkwan K, Oates CG (2000) Processing of cassava waste for improved biomass utilization. Bioresource Technology 71: 63–69

Stabnikova O, Wang, J-Y, Ding H.O, Jay J-H (2005) Biotransformation of vegetable and fruit processing wastes into yeast biomass enriched with selenium. Bioresource Technology 96: 747–751

Stamford TLM, Decamargo R (1992) Fungal protein enrichment of residual liquor from a sugarcane waste *Saccharum officinarum*. Archivos Latinoamericanos de Nutricion 42: 351–359

Sunita M, Rao DG (2003) Bioconversion of mango processing waste to fish-feed by microalgae isolated from fruit processing industrial effluents. Journal of Scientific and Industrial Research 62: 344–347

Tengerdy RP, Szakacs G (2003) Bioconversion of lignocellulose in solid substrate fermentation. Biochemical Engineering Journal 13: 169–179

Teniola OD, Odunfa SA (2001) The effects of processing methods on the levels of lysine, methionine and the general acceptability of *ogi* processed using starter cultures. International Journal of Food Microbiology 63: 1–9

Titi HH, Tabbaa MJ (2004) Efficacy of exogenous cellulase on digestibility in lambs and growth of dairy calves. Livestock Production Science 87: 207– 214

Tripathi JP, Yadav JS (1992) Optimization of solid substrate fermentation of wheat straw into animal feed by *Pleurotus ostreatus* – a pilot effort. Animal Feed Science and Technology 37: 59–72

Tripathi MK, Mishra AS, Misra AK, Vaithiyanathan S, Prasad R, Jakhmola RC (2008) Selection of white-rot basidiomycetes for bioconversion of mustard (*Brassica compestris*) straw under solid-state fermentation into energy substrate for rumen micro-organism. Letters in Applied Microbiology 46: 364–370

Tsioulpas A, Dimou D, Iconomou D, Aggelis G (2002) Phenolic removal in olive oil mill wastewater by strains of *Pleurotus* spp in respect to their phenol oxidase (laccase) ctivity. Bioresource Technology 84: 251–257

Ubalua AO (2007) Cassava wastes: Treatment options and value addition alternatives. African Journal of Biotechnology 6: 2065–2073

Ugwuanyi JO (2008a) Yield and protein quality of thermophilic *Bacillus* spp. Biomass related to thermophilic aerobic digestion of agricultural wastes for animal feed supplementation. Bioresource Technology 99: 3279–3290

Ugwuanyi JO (2008b) Moisture sorption isotherm and xerophilic moulds associated with dried cocoyam chips in storage in Nigeria. International Journal of Food Science and Technology 43: 846–852

Ugwuanyi JO, Harvey LM, McNeil B (2006) Application of thermophilic aerobic digestion in protein enrichment of high strength agricultural waste slurry for animal feed supplementation. Journal of Chemical Technology and Biotechnology 81: 1641–1651

Ugwuanyi JO, Harvey LM, McNeil B (2008) Protein enrichment of corn cob heteroxylan waste slurry by thermophilic aerobic digestion using *Bacillus stearothermophilus*. Bioresource Technology 99: 6974–6985

Urbaniak M, Sakson G (1999) Preserving sludge from meat industry waste waters through lactic fermentation. Process Biochemistry 34: 127–132

Vadiveloo J (2003) Solid-state fermentation of fibrous residues. Journal of Animal and Feed Sciences 12: 665–676

Valino E, Elias A, Carrasco T, Albelo N (2003) Effect of inoculation of the strain *Trichoderma viride* 137 on auto-femented sugar cane bagasse. Cuban Journal of Agricultural Science 37: 43–49

Valino E, Elias A, Torres V, Albelo N (2002) Study of the microbial content on fresh sugar cane bagasse as substrate for animal feeding by solid state fermentation. Cuban Journal of Agricultural Science 36: 359–364

Varnayte RN, Raudonene VZ (2004) Bioconversion of straw waste by *Micromycetes*. Mikologiya I Fitopatologiya 38: 80–83

Vázquez JA, Docasal SF, Prieto MA, Gonzalez MP, Murado MA (2008). Growth and metabolic features of lactic acid bacteria in media with hydrolysed fish viscera. An approach to bio-silage of fishing by-products. Bioresource Technology 99: 6246–6257

Vendruscolo F, Albuquerque PM, Streit F, Esposito E, Ninow JL (2008) Apple pomace: A versatile substrate for biotechnological applications. Critical Reviews in Biotechnology 28: 1–12

Venugopal V, Shahidi F (1995) Value-added products from underutilized fish species. Critical Reviews in Food Science and Nutrition 35: 431–53

Vidotti RM, Viegas EMM, Carneiro DJ (2003) Amino acid composition of processed fish silage using different raw materials. Animal Feed Science and Technology 105: 199–204

Viera MP, Pinchetti JLG, de Vicose GC, Bilbao A, Suarez S, Haroun RJ, Izquierdo MS (2005) Suitability of three red macroalgae as a feed for the abalone *Haliotis tuberculata coccinea* Reeve. Aquaculture 248: 75–82

Villas-Boas SG, Esposito E, de Mendonca MM (2003) Bioconversion of apple pomace into a nutritionally enriched substrate by *Candida utilis* and *Pleurotus ostreatus*. World Journal of Microbiology and Biotechnology 19: 461–467

Virupakshi S, Gireesh Babu K, Gaikwad SR, Naik GR (2005) Production of a xylanolytic enzyme by a thermoalkaliphilic *Bacillus* sp. JB-99 in solid state fermentation. Process Biochemistry 40: 431–435

Vladimirova EG, Saubenova MG, Ilyaletdinov AN, Karpova GV (1995) Bacterial conversion of plant waste products into feed. Applied Biochemistry and Microbiology 31: 367–369

Volanis M, Zoiopoulos P, Panagou E, Tzerakis C (2006) Utilization of an ensiled citrus pulp mixture in the feeding of lactating dairy ewes. Small Ruminant Research 64: 190–195

Woomer PL, Muzira R, Bwamiki D, Mutetikka D, Amoding A, Bekunda MA (2000) Biological management of water hyacinth waste in Uganda. Biological Agriculture and Horticulture 17: 181–196

Woyengo TA, Guenter W, Sands JSb, Nyachoti CM, Mirza MA (2008) Nutrient utilisation and performance responses of broilers fed a wheat-based diet supplemented with phytase and xylanase alone or in combination. Animal Feed Science Technology, doi:10.1016/j.anifeedsci.2007.11.013

Yang HY, Wang XF, Liu JB, Gao LJ, Ishii M, Igarashi Y, Cui ZJ (2006) Effect of water-soluble carbohydrate content on silage fermentation of wheat straw. Journal of Bioscience and Bioengineering 101: 232–237

Yang SS, Jang HD, Liew, CM, Dupreez JC (1993) protein enrichment of sweet-potato residue by solid-state cultivation with mono-cultures and cocultures of amylolytic fungi. World Journal of Microbiology and Biotechnology 9: 258–264

Yang X, Chen I, Gao HL, Li Z (2001) Bioconversion of corn straw by coupling ensiling with solid state fermentation. Bioresource Technology 71: 277–280

Yang YH, Wang BC, Xiang LJ, Duan CR, Wang QH, Lian J (2003) Construction of a microbial consortium used for solid-state fermentation on rice chaff. Colloids and Surfaces B-Biointerfaces 32: 51–56

Yang SS (1993) protein enrichment of sweet-potato residue with coculture of amylolytic fungi by solid-state fermentation. Biotechnology Advances 11: 495–505

Youssef BM, Aziz NH (1999) Influence of gamma-irradiation on the bioconversion of rice straw by *Trichoderma viride* into single cell protein. Cytobios 97: 171–183

Zadrazil F (1997) Changes in in vitro digestibility of wheat straw during fungal growth and after harvest of oyster mushrooms (*Pleurotus* spp.) on laboratory and industrial scale. Journal of Applied Animal Research 11: 37–48

Zadrazil F, Puniya AK (1995) Studies on the effect of particle size on solid-state fermentation of sugarcane bagasse into animal feed using white-rot fungi. Bioresource Technology 54: 85–87

Zervakis G, Yiatras P, Balis C (1996) Edible mushrooms from olive oil mill wastes. International Biodeterioration and Biodegradation 38: 237–243

Zhang ZY, Jin B, Bai ZH, Wang XY (2008) Production of fungal biomass protein using microfungi from winery wastewater treatment. Bioresource Technology 99: 3871–3876

Zvauya R, Muzondo MI (1994) Some factors affecting protein enrichment of cassava flour by solid-state fermentation. Food Science and Technology-Lebensmittel-Wissenschaft and Technologie 27: 590–591

Chapter 6
Aroma Compounds

Syed G. Dastager

Contents

Abstract The increasing demand for natural products in the food industry has encouraged remarkable efforts towards the development of biotechnological processes for the production of aroma compounds. This chapter deals with major achievements reported in this field, with a special emphasis on the potential lying in plant cell, microbial cultures and enzyme technology for the production of a wide range of flavours. The use of solid-state fermentation as a means to improve economical feasibility of these processes and application of aroma compounds. In order to understand the flavour of (traditional) foods a multitude of scientific investigations were carried out and a number of appropriate analytical tools for flavour research were developed in the past few decades.

Keywords Aroma-production · SSF · Aroma-compounds · Microorganisms · Aroma-application · Terpenes · Alcohols · Vanillin · Methyl ketones · Diacetyles · Pyrizines

---

S.G. Dastager (✉)
National Institute of Interdisciplinary Science and Technology (Formerly RRL), CSIR, Industrial Estate, Thiruvananthapuram-695019, Kerala, India
e-mail: syedmicro@gmail.com

P. Singh nee' Nigam, A. Pandey (eds.), *Biotechnology for Agro-Industrial Residues Utilisation*, DOI 10.1007/978-1-4020-9942-7_6,

6.1 Aroma Compounds

An aroma compound, also known as odorant, aroma, fragrance or flavor, is a chemical compound that has a smell or odor. Aroma compounds can be found in food, wine, spices, perfumes, fragrance oils, and essential oils. The world of aroma is very attractive especially because it concerns the taste of what we eat (Aguedo et al. 2004). From a scientific and technological point of view, this field is also highly exciting since it brings together several different branches of science. Aroma is usually the result of the presence, within complex matrices, of many volatile components of various chemical and physicochemical properties. The processing of mixtures of raw food materials can have various sensorial impacts depending on the properties of each compound. Processing modifies the equilibrium between the different components and, as a result, the original aroma will be perceived as being weaker and as artificial or chemical. The work of the aroma formulator consists in constructing a flavour recalling a true and original aroma in a processed food product with a specific texture and composition. Aroma compounds can be extracted from fruits or vegetables but, as they are required in the product in concentrations comparable to those in the source material, this utilizes high amounts of materials and is generally not economically realistic. Most of them can also be synthesized in a chemical way resulting in chemical compounds that are not well perceived by consumers whose demand, especially in Western Europe, is in favour of natural products. As an alternative, biotechnology proposes to use enzymes or whole cells to produce aroma compounds. Flavors and fragrances constitute a world-wide market of US$ 7 billion a year, with a share of 25% of the food additives market (Armstrong and Yamazaki 1986). The consumer's preference for natural food additives is more important than ever.

6.1.1 Types of Aroma Compounds

There are various families of aroma compounds and the differences used to classify these families can be based not only on chemical structures, physicochemical properties or sensorial properties of the compounds but also, and in fact more commonly, on the chemical family of the substrate. On this latter basis, lipid-derived aroma compounds constitute one of the most important families, which include volatile fatty acids or esters, lactones, aldehydes, alcohols, ketones and some groups such as carotenoid-derived aroma compounds (Table 6.1). Although there are many investigations into the natural generation of these compounds in food products, only a few aroma components are produced by biotechnological routes.

6.2 Production of Aroma Compounds

Aroma production constitutes an important sector in the chemical industry. The aromas are compounds utilized in the manufacture of cosmetics, perfumes, cleaning products and food processing. Traditionally, aromas have been extracted from

Table 6.1 Classification of food aroma compounds based on their chemical structure

Alcohols	Aldehydes	Esters	Fatty acids	Ketones	Lactones	Aromatic compounds	Pyrazines
1,2-butanediol	Acetaldehyde	methyl acetate	Acetate	acetophenone	δ-decalactone	vanillin	2,3-diethyl-5-methyl pyrazine
2-butanol	Decanal	ethyl acetate	Butyrate	acetone	γ-decalactone	benzaldehyde	2-ethyl-3,5-dimethlypyrazine
2,3-butanediol	Heptanal	ethyl butyrate	Caproate	2,3-butanedione	γ -butyrolactone	β-phenethyl alcohol	2-methoxy-3-isopropylpyrazine
Ethanol	(Z)-4-heptenal	ethyl hexanoate	Decanoate	2,3-pentandione	δ -dodecalactone	trimethylbenzene	2-methoxy-3-isopropylpyrazine
2-ethylbutanol	Hexanal	ethyl isobutanoate	Isobutyrate	2-butanone	δ-octalactone		
2-ethylhexanol	2-hexanal	ethyl octonate	2-methylbutyric acid	3-hydroxy-2-butanone	(z)-6-dodecan-δ-lactone		
2-heptanol	Isohexanal	ethyl butanoate	3-methylbutyric acid	2-heptanone			
Hexanol	2-methylbutanal	isobutyl butanoate	Octanoate	2-hexanone			
Isobutanol	3-methylbutanal	2-methyl-1-butyl acetate	phenylacetate	3-methyl-2-butanone			
2-methylbutanol	2-methylpropanal	3-methyl-1-butyl acetate	propionate	4-methyl-2-pentanone			
3-methylbutanol	Nonanal	3-octyl acetate	Valerate	2-nonanone			
2-methylpropanol	(E,E)-2,4-nonadienal	pentyl acetate		2-octanone			

Table 6.1 (Continued)

Alcohols	Aldehydes	Esters	Fatty acids	Ketones	Lactones	Aromatic compounds	Pyrazines
2-nonanol	(Z)-2-nonenal	phenethyl acetate		1-octen-3-one			
(Z)-1,5-octadien-3-ol	(E)-2-nonenal	ethyl butyrate		2-pentanone			
2-octanol	Octanal	propyl butyrate		3-pentanone			
1-octen-3-ol	Butanal	2-hidroxyethyl		2-tridecanone			
1-pentanol	Pentanal	propionate		2-undecanone			
Phenylethanol	Propanal	2-methyl-2-ethyl-3-hydroxyhexyl propionate					
2-phenylethanol	Propenal	ethyl 2-methylbutanoate					
1-nonanol	thiophen-2-aldehyde	ethyl 3-methylbutanoate					

plants, but in general these procedures are low yield processes. Aromas can also be produced by chemical synthesis, however there is a clear consumer preference for products of natural origin. For these reasons, there is an increasing scientific interest in searching for aroma production alternatives, different from processes based on extractive or chemical synthesis. Therefore, several biotechnological approaches have been considered as real options for aroma production (Berger 1995).

The use of biotechnology for the production of natural aroma compounds by fermentation or bioconversion using micro-organisms is an economic alternative to the difficult and expensive extraction from raw materials like plants (Harlander 1994; Janssens et al. 1992). A fungus with aromatic properties and often referred to as a yeast, *Geotrichum candidum*, has been used for commercial cheese ripening (Jollivet et al. 1994). Some strains may produce fatty acids esters, often related to specific fruit aroma (Koizumi et al. 1982; Latrasse et al. 1987). *G. candidum* is highly lipolytic with a whole range of substrate specificity (Jacobsen et al. 1990; Sidebottom et al. 1991). Its proteolytic activity may also form aroma compounds and has been partly characterized by Gueguen and Lenoir (Gueguen and Lenoir 1975, 1976). A vast array of compounds may be responsible for the aroma of the food products, such as alcohols, aldehydes, esters, dicarbonyls, short to medium-chain free fatty acids, methyl ketones, lactones, phenolic compounds and sulphur compounds (Gatfield 1988; Urbach 1997). Since early times, aroma compounds ranging from single to complex substances have been extracted from plant sources. Eventually, after elucidation of their structure, synthetic aroma was produced by chemical synthesis.

Nowadays, aroma represent over a quarter of the world market for food additives and most of the aroma compounds are produced via chemical synthesis or by extraction from natural materials. However, recent market surveys have shown that consumers prefer foodstuff that can be labelled as natural. Although aroma may be produced by chemical transformation of natural substances, the resulting products cannot legally be labelled as natural. Furthermore, chemical synthesis often results in environmentally unfriendly production processes and lacks substrate selectivity, which may cause the formation of undesirable racemic mixtures, thus reducing process efficiency and increasing downstream costs. On the other hand, the production of natural aroma by direct extraction from plants is also subject to various problems. These raw materials often contain low concentrations of the desired compounds, making the extraction expensive. Moreover, their use depends on factors difficult to control such as weather conditions and plant diseases. The disadvantages of both methods and the increasing interest in natural products have directed many investigations towards the search for other strategies to produce natural aroma.

An alternative route for flavour synthesis is based on microbial biosynthesis or bioconversion (Aguedo et al. 2004; Janssens et al. 1992; Krings and Berger 1998; Vandamme and Soetaert 2002). The most popular approaches involve the use of microbial cultures or enzyme preparations, although plant cell cultures have also been reported as suitable production systems (Fig. 6.1). Microorganisms can synthesize aroma as secondary metabolites during fermentation on nutrients such as sugars and amino acids. This capability may be used in two different ways:

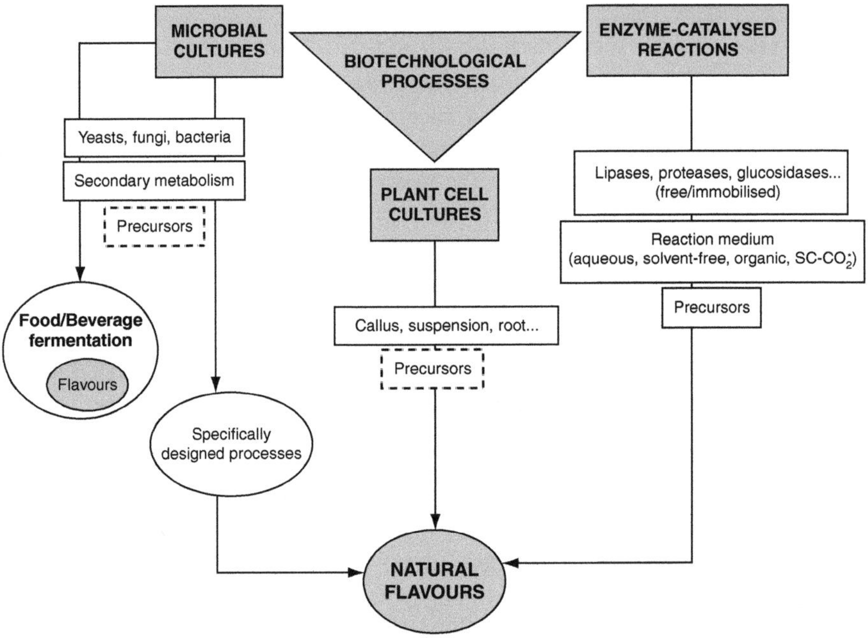

Fig. 6.1 Biotechnological production of Aroma compounds

- In situ flavour generation, as an integral part of food or beverage production processes (i.e. cheese, yogurt, beer, wine) which determines the organoleptic characteristics of the final product
- Microbial cultures specifically designed to obtain aroma compounds that can be isolated and used later as additives in food manufacture. This strategy allows the obtained aroma to be labelled as natural.

In both cases, precursors or intermediates can be added to the culture medium in order to promote the biosynthesis of specific aroma. Also, the information obtained through the investigation of microbial metabolism in food fermentation processes could be utilized to develop suitable production systems for particular aroma additives. On the other hand, enzyme technology offers a very promising option for natural flavour biosynthesis. A number of enzymes (i.e. lipases, proteases, glucosidases) catalyse the production of aroma-related compounds from precursor molecules (Adinarayana et al. 2004; Asther et al. 2002; Kamini et al. 1998; Macris et al. 1987; Miranda et al. 1999). The use of enzyme-catalysed reactions has the notable advantage of providing higher stereo selectivity than chemical routes. Besides, the products thus obtained may possess the legal status of natural substances. Although a considerable amount of current research focuses on the production of aroma compounds, at the moment only a few are obtained by biotechnological routes. The challenge is to put a naturally rich source of substrate in contact with highly active enzymes. In adequate conditions, this can result in the production of aroma compounds in mass fractions of the order of several g/kg, instead of mg/kg encountered in raw materials. The resulting aroma compounds are called natural since they are produced from agro-products through natural biological activities. The ratio of isomers or isotopes is thus comparable to what can be found in extracted products and not to what results from chemical synthesis. However, although the productivity of some of these processes is good, the resulting products are usually more expensive than those from chemical synthesis (Benjamin and Pandey 1997; Besson et al. 1997; Beuchat 1982).

6.2.1 Production of Aroma Compounds from Plant Cell Cultures

Plant cell cultures appear as a viable method to produce a wide range of aromas characteristic of their plant origin (Table 6.2; Dornenburg and Knorr 1996; Kim et al. 2001; Suvarnalatha et al. 1994; Townsley 1972). This approach is based on the unique biochemical and genetic capacity, and the totipotency of plant cells (Scragg 1997). Every cell of a plant culture contains the genetic information necessary to produce numerous chemical components that constitute natural aromas. Feeding intermediates of the biosynthetic pathway can enhance the production of aroma metabolites by precursor biotransformation. Some authors (Mulabagal and Tsay 2004; Rao and Ravishankar 2002) summarised the advantages of plant cell culture technology over conventional agricultural production.

As for specific efforts related to aroma production by plant cell cultures, several researchers have investigated the synthesis of vanillin, a much sought-after flavour compound (Rao and Ravishankar 2000). Plant cell cultures of *Vanilla planifolia* have been initiated from various plant cells and tissues (Davidonis and Knorr 1991), and the convenience of using elicitors to induce vanillic acid synthesis assessed

Table 6.2 Aroma compounds from plant cell cultures

Plant species	Aroma compounds	Literature
Agastache rugosa	2,3-butanedione, (E,Z)-2,6-nonadienal and (E,Z)-2,6-nonadien-1-ol	(Kim et al. 2001)
Malus silvestris	Apple aroma	(Drawert et al. 1984)
Nicotiana tabacum	Cinnamic acid	
Lindera strychnifolia	Caryophyllen	
Oryza sativa	Basmati flavour	(Suvarnalatha et al. 1994)
Theobromo cacao	Cocoa flavour	(Townsley 1972)
Polygonum hydropiper	Flavanol	(Nakao et al. 1999)
Allium sativum	Garlic	(Ohsumi et al. 1993)
Perilla frutescens	Monoterpenes	(Nabeta et al. 1983)
Allium cepa	Onion	(Prince et al. 1997)
Glycyrrhiza glabra glandulifera	Triterpenoid	(Ayabe et al. 1990)
Vanilla planifolia	Vanillin	(Dornenburg and Knorr 1996)

(Funk and Brodelius 1990) and also feeding of the precursor ferulic acid resulted in increase in vanillin accumulation (Romagnoli and Knorr 1988). Furthermore, the production of vanillin from ferulic acid with vanilla aerial roots on charcoal as a product reservoir has been described (Westcott et al. 1994). *Capsicum frutescens* root cultures have also been used for the bioconversion of ferulic acid to vanillin (Suresh et al. 2003). Some other works involve the production of monoterpenes (i.e. limonene, linalool, etc.) in callus tissues and cell suspensions of *Perilla frutescens* (Nabeta et al. 1983; Sahai 1994), and basmati rice volatile flavour components in callus cultures of *Oryza sativa* (Suvarnalatha et al. 1994). In some cases, the flavour profiles obtained in plant cell cultures differ from those encountered in the parent plants. Such was the case in suspension cultures of *Agastache rugosa* Kuntze (Korean mint), which had a marked cucumber/wine-like aroma, and produced some interesting flavour-related alcohols (i.e. 2-phenylethanol) (Kim et al. 2001). This alteration of the original flavour profiles can be deliberately induced by the addition of precursors, as demonstrated in root cultures of *Allium cepa* L. (onion) (Prince et al. 1997).

6.2.2 Aroma Production from Microbial Cultures

Microorganisms have historically played an integral role in the elaboration of the aroma components of many different foods. Products such as wine, vinegar, beer, fermented vegetables, milk, soya and meat have been preserved, modified and flavoured by means of microbial strains. As previously indicated, microbial cultures can be used to produce aroma compounds (Chandrasekaran 1997), either specifically for application as food additives or in situ as a part of food fermentation processes. Detailed information on the production of some commonly used food aroma compounds by microorganisms is presented below.

6.2.2.1 Diacetyl

Diacetyl is mainly related to butter flavour, and therefore extensively used in the imitation of butter and other dairy flavours, as well as whenever butter notes are desirable in food or beverages. This compound is produced by lactic acid bacteria and other microorganisms in several foods (e.g.*Lactococcus lactis, Lactobacillus* sp., *Streptococcus thermophilus, Leuconostoc mesenteroides*) (Escamilla-Hurtado et al. 2005). The studies done by Ibragimova et al. (1980) showed that milk cultures of *Streptococcus lactis, S. cremoris* and *S. diacetilactis* produced high amounts of 2,3-butanedione and acetaldehyde in 24 h at 30°C. Cultures with the best aroma contained 2–5 parts acetaldehyde to 1 part 2,3-butanedione. A number of researchers have investigated the behaviour and/or metabolism of food processing microorganisms, and the enzymes involved in the production of diacetyl and related compounds. The formation of diacetyl by lactic acid bacteria through acetoin dehydrogenase-catalysed dehydrogenation of acetoin has been investigated in the dough products (Bratovanova 2001). Bassit et al. (1995) studied the effect of temperature on diacetyl and acetoin production by a particular strain of *Lactococcus lactis*, with special reference to lactic dehydrogenase, acetolactate synthase, NADH oxidase and diacetyl reductase (Escamilla-Hurtado et al. 2000), the main enzymes involved in pyruvate metabolism. Medina de Figueroa et al. (1998) investigated the effect of citrate in the repression of diacetyl/acetoin reductase, resulting in the accumulation of diacetyl and acetoin in batch cultures of *Lactobacillus rhamnosus*. Genetic manipulation of the gene encoding enzymes involved in diacetyl metabolism, such as diacetyl-acetoin reductase from *Lactococcus lactis*, has been attempted to increase the diacetyl production capacity of lactic acid bacteria (Aungpraphapornchai et al. 1999).

Carroll et al. (1999) cloned and expressed in *E. coli* acetolactate synthase, a key enzyme for the production of the diacetyl precursor acetolactate, with the final objective of increasing diacetyl production in lactococcal strains. In some cases, diacetyl can contribute to off-flavours (i.e. beer production) and strategies should be designed to avoid their formation. Kronlof and Linko (1992) proposed the use of genetically modified brewer's yeast encoding a-acetolactate decarboxylase in immobilised yeast bioreactors for the main fermentation of beer, promoting the direct conversion of a-acetolactate to acetoin without the formation of diacetyl. Sandine et al. (1965) assayed the addition of a crude diacetyl reductase from *Aerobacter aerogenes* as a means to remove diacetyl and 2,3-pentadione from beer by conversion to flavourless acetoin.

6.2.2.2 Lactones

Lactones are cyclic esters of primarily g- and d-hydroxy acids, and they are ubiquitously found in food, contributing to taste and flavour nuances such as fruity, coconut-like, buttery, creamy, sweet or nutty. The possibility of producing a lactone using a biotechnological route was discovered in the 1960s by the group of Okui et al. (1963a, b) during the investigation of hydroxyacid catabolism by several organisms. Dimick et al. (1969) stated in their review that raw milk does not contain free lactones, which only appear after heating. The milky, buttery and coconut-like flavour notes provided by these compounds are generally considered as desirable in dairy products. However, the presence of lactones may contribute to the stale flavour of heated milk, although to a lesser extent than ketones. The compound 6-pentyl-2-pyrone provides a coconut aroma, highly desired by flavourists. It was found by Collin and Halim (1972) to be the major volatile constituent in cultures of the fungus *Trichoderma viride*. Other fungi such as *Tyromyces sambuceus* and *Cladosporium suaveolens* efficiently generate the coconut-flavoured lactones g-decalactone and d-dodecalactone from ricinoleic acid and linoleic acid, respectively (Allegrone et al. 1991; Kapfer et al. 1989). Yeasts such as *Candida tropicalis* or *Yarrowia lipolytica* degraded ricinoleic acid to C16, C14 and C12 acids and, interestingly, accumulated d-decalactone, which exhibits fruity and oily notes important in the formulation of peach, apricot or strawberry aromas. However, the yields of this biotransformation are commonly poor, and they rarely reach concentrations over 4 to 5 g/L in the fermentation broth (Gatfield 1999). Wache et al. (2001) investigated the enzymes involved in g-decalactone production by *Yarrowia lipolytica*, and encountered the reasons for low yields.

6.2.2.3 Esters

Esters are commonly used flavouring agents, very appreciated for the fruity aromas they provide. They are employed in fruit-flavoured products (i.e. beverages, candies, jellies, and jams), baked goods, wines, and dairy products (i.e. cultured butter, sour cream, yogurt, and cheese). Acetate esters, such as ethyl acetate, hexyl acetate, isoamyl acetate and 2-phenylethyl acetate are recognized as important flavour compounds in wine and other grape-derived alcoholic beverages (Geusz and Anderson 1992). Rojas et al. (2001) studied several so-called non-*Saccharomyces* wine yeasts as the producers of acetate ester. Among them, the yeasts *Hanseniaspora guilliermondii* and *Pichia anomala* were found to be potent 2-phenylethyl acetate and isoamyl acetate producers, respectively. In cheese production, ethyl or methyl esters of short-chain fatty acids generally bring about fruity flavours, while thioesters derived from thiols are associated with cabbage or sulphur aromas (Castanares et al. 1992; Liu et al. 2004). The capacity of lactic acid bacteria to synthesize both ethyl esters and thioesters has been reported. The role of a unique esterase from *Lactococcus lactis* in the formation of these aroma compounds has been investigated, and ascertained as at least partially responsible for the esterification reactions leading to the production of aroma ester compounds. This was undertaken by using an esterase negative mutant of *L. lactis* (Nardi et al. 2002).

6.2.2.4 Pyrazines

Pyrazines are heterocyclic, nitrogen-containing compounds which possess a nutty and roasted flavour. They are normally formed during conventional cooking/roasting of food through the Maillard reaction (Seitz 1994). Nowadays, the use of cooking processes that do not favour pyrazine formation (i.e. microwave cooking) has caused the need to supply natural pyrazines with a roasty flavour as food additives. A few microorganisms are also able to synthesize pyrazines. For instance, bacteria such as *Corynebacterium glutamicum* produce important quantities of tetramethylpyrazine from amino acids (Demain et al. 1967).

6.2.2.5 Terpenes

Terpenes are widespread in nature, mainly in plants as constituents of essential oils. They are composed of isoprene units, and can be cyclic, open-chained, saturated, unsaturated, oxidized, etc. The biotransformation of these compounds is potentially of considerable interest for application in the food flavour industry. Among the terpenes, linalool, nerol, geraniol and citronellol are the most flavour-active due to their low sensory threshold. Most of the terpenes obtained in microbial cultures are produced by fungi that belong to the ascomycetes and basidiomycetes species. Schindeler and Bruns (1980) have demonstrated that terpene yields in *Ceratocystis variospora* cultures could be improved when toxic end products were removed using ion exchange resins. The fungus *Ceratocystis moniliformis* produces several aroma products such as ethyl acetate, propyl acetate, isobutyl acetate, isoamyl acetate, citronellol and geraniol. In order to avoid the inhibitory effects detected in these cultures, it is necessary to decrease product concentrations in the bioreactor. Bluemke and Schrader (2001) developed an integrated bioprocess to enhance the production of natural flavours by *C. moniliformis*. The total yield of aroma compounds produced in the integrated bioprocess, with in situ product removal using pervaporation, is higher than in conventional batch cultivation. In addition, permeates obtained from pervaporation consist of highly enriched mixtures of flavours and fragrances.

On the other hand, microbial transformation of terpenes has received considerable attention. Many microorganisms are able to break down terpenes or to carry out specific conversions, creating products with an added value. Dhavlikar and Albroscheit (1973) demonstrated that the inexpensive sesquiterpene valencene can be converted by some bacteria to the important aroma compound nootkatone. Recently, significant research effort has focused on the enzymes related to terpene biosynthesis. The nucleic acid sequence of a monoterpene synthase from sweet basil, a key enzyme for the production of geraniol, has been determined in order to allow the production of recombinant geraniol synthase (Pichersky et al. 2005). Also, a geraniol synthase from the evergreen camphor tree *Cinnamomum tenuipilum* was cloned and expressed in *E. coli* (Yang et al. 2005). Functional genomics has also been applied to identify the genes for monoterpene synthases from *Vitis vinifera* grapes in order to characterize the enzymes by expression in *E. coli* and subsequent analysis (Martin and Bohlmann 2004).

6.2.2.6 Alcohols

In alcoholic fermentations, apart from ethanol, yeast produces long-chain and complex alcohols. These compounds and their derived esters have interesting organoleptic properties. Some authors have proposed strategies for promoting this kind of flavour compounds during alcoholic beverage production. Mallouchos et al. (2002) utilized *Saccharomyces cerevisiae* immobilised on delignified cellulosic material and gluten pellets. The former produced higher amounts of esters, whereas the latter gave higher amounts of alcohols. Kana et al. (1992) evaluated yeast immobilisation on g-alumina and kissiris, and found in the former case an increase in the concentration of amyl alcohols, total volatiles, and ethyl acetate, which led to a fine aroma. One of the most relevant aroma-related alcohols is 2-phenylethanol, which possesses a rose-like smell. It is still predominantly synthesized by petrochemical routes from toluene, benzene, styrene, or methylphenylacetate (Nomura et al. 2001), while the natural 2-phenylethanol is mainly extracted from rose petals through a high-cost process (Fabre et al. 1998; Zheng and Shetty 2000).

Different yeast strains such as *Hansenula anomala*, *Kluyveromyces marxianus* or *Saccharomyces cerevisiae* have shown a high potential for industrial production of aroma compounds, such as 2-phenylethanol, which is derived from 2-phenylalanine by bioconversion (Fabre et al. 1998; Stark et al. 2002). Stark et al. (2002, 2003) reported that the presence of ethanol and 2-phenylethanol in the medium resulted in a synergistic inhibition, which reduced the tolerance of *Saccharomyces cerevisiae* to 2-phenylethanol and thus its final concentration. As a result, the feed rate had to be reduced in fed-batch cultures to avoid ethanol production. Thus, a maximal 2-phenylethanol concentration of 2.35 g/L could be attained in batch cultures, whereas 3.8 g/L were obtained in a fed-batch culture with the limitation of ethanol production (Stark et al. 2002; Topakas et al. 2004).

To enhance the productivity of the bioconversion of 2-phenylalanine by *Saccharomyces cerevisiae*, a novel in situ product recovery strategy was proposed by Serp et al. (2003). An organic solvent (dibutyl sebacate) was entrapped within a polyethylene matrix, in order to form a highly absorbent, chemically and mechanically stable composite resin. The use of this technique increased twofold the volumetric productivity of 2-phenylethanol and significantly facilitated downstream processing. Fabre et al. (1997) screened yeast strains for 2-phenylethanol production. Amongst the different 2-phenylethanol producers, *Kluyveromyces marxianus* was outstanding, which makes this strain a promising candidate to be applied in an industrial process. Moreover, *K. marxianus* has several advantages such as (Wittman et al. 2002)

- It shows optimal production characteristics (Fabre et al. 1997).
- 2-phenylethanol production depends on the medium and temperature used (Etschmann et al. 2004).
- *K. marxianus* is Crabtree-negative, which is an advantage for scale production processes, because the production of toxic by-products (i.e. ethanol) under aerobic conditions can be avoided (Etschmann et al. 2002).

6.2.2.7 Vanillin

Vanillin (4-hydroxy-3-methoxybenzaldehyde) is a universally appreciated flavour chemical that occurs in *Vanilla planifolia* beans. It is widely used in foods, beverages, perfumes, pharmaceuticals, and in various medical industries (Priefert et al. 2001). Chemically synthesized vanillin accounts nowadays for more than 99 % of the total market share (Walton et al. 2000), but there is an increasing demand for natural vanillin. Direct extraction from vanilla beans is expensive and limited by plant supply, which makes this compound a promising target for biotechnological flavour production. Vanillin is an intermediate in the microbial degradation of several substrates, such as ferulic acid, phenolic stilbenes, lignin, eugenol and isoeugenol (Funk and Brodelius 1992). The conversion of natural eugenol and isoeugenol from essential oils into vanillin has been investigated using microbial and enzymatic biotransformations (Overhage et al. 1999; Rao and Ravishankar 1999). Strains including *Pseudomonas putida, Aspergillus niger, Corynebacterium glutamicum, Corynebacterium* sp., *Arthrobacter globiformis* and *Serratia marcescens* (Priefert et al. 2001; Shimoni et al. 2000, 2003) can also convert eugenol or isoeugenol to vanillin (Washisu et al. 1993).

A two-step bioconversion process using filamentous fungi was developed by Lesage-Meessen et al. (1996, 2002) to transform ferulic acid into vanillin. First, *Aspergillus niger* transformed ferulic acid to vanillic acid, and then vanillic acid was reduced to vanillin by *Pycnoporus cinnabarinus*. Bonnin et al. (2000) showed that the yield of vanillin may be significantly increased by adding cellobiose to *P. cinnabarinus* culture medium, due to the decrease in oxidative decarboxylation of vanillic acid. The importance of ferulic acid as precursor of vanillin has brought about a number of efforts in the investigation of its production.

Feruloyl esterase has been identified as the key enzyme in the biosynthesis of ferulic acid, and some researchers have studied the production of this enzyme in microbial cultures of several fungi grown on different pretreated cereal brans, such as wheat, maize, rice bran and sugar cane bagasse (Mathew and Abraham 2005). The metabolism of ferulic acid in some microorganisms has also been investigated (Falconnier et al. 1994; Narbad and Gasson 1998). *Benzaldehyde* It is the second most important molecule after vanillin for its use in cherry and other natural fruit flavours. The world consumption of benzaldehyde amounts to approximately 7000 tonnes per year (Clark 1995). Natural benzaldehyde is generally extracted from fruit kernels such as apricots, leading to the undesirable formation of the toxic hydrocyanic acid.

Nowadays, the fermentation of natural substrates is an alternative route to the production of benzaldehyde without harmful by-products (Lomascolo et al. 1999). However, benzaldehyde is toxic towards microbial metabolism and its accumulation in the culture medium may strongly inhibit cell growth (Lomascolo et al. 2001). For this reason, only a few microorganisms have been reported as benzaldehyde producers. Amongst them, the bacterium *Pseudomonas putida* (Wilcocks et al. 1992; Wilcocks et al. 1992) and the white rot fungi *Trametes suaveolens* (Lomascolo et al. 2001), *Polyporus tuberaster* (Kawabe and Morita 1994), *Bjerkandera adusta* (Lapadatescu et al. 1999) and *Phanerochaete chrysosporium* (Jensen et al. 1994) are mentioned as biocatalysts in the biosynthesis of benzaldehyde from phenylalanine. Park and Jung (2002) proposed the use of calcium alginate-encapsulated whole-cell enzymes from *P. putida* for the production of benzaldehyde from benzoylformate. This allowed the accumulation of benzaldehyde in the capsule core, minimising its subsequent transformation to benzyl alcohol by the action of alcohol dehydrogenase, and thus providing continuous production of benzaldehyde until reactant exhaustion.

6.2.2.8 Methyl Ketones

The methyl ketones, 2-heptanone, 2-nonanone, and 2-undecanone, are the largest contributors to stale flavour in UHT milk (Badings et al. 1981). Moio et al. (1994) similarly report that 2-heptanone and 2-nonanone are the most powerful odorants in UHT milk. These methyl ketones are aromas employed in a wide range of flavouring applications, especially those related to blue cheese and fruit flavours (Hagedorn and Kaphammer 1994). There is not much information on microbial production of these compounds, although Janssens et al. (1992) mention in their review the methyl ketone-producing ability of *Agaricus bisporus*, *Aspergillus niger*, *Penicillium roqueforti* and *Trichoderma viride* TS.

6.3 Using Agro Wastes as Substrates

The tropical agro-industrial residues such as coffee pulp and coffee husk, cassava bagasse, sugar cane bagasse are generated in large amounts during the processing and their disposal rather causes serious environmental problems. In recent years, there has been constant increase in the efforts to utilize these residues as substrates (carbon source) in bioprocessing (Asther et al. 2002; Pandey et al. 1999). Microorganisms play an important role in the generation of natural compounds, particularly in the field of food aromas (Janssens et al. 1992; Jiang 1995). Solid state fermentation (SSF) has been used for the production of aroma compounds by cultivating yeasts and fungi. Numerous microorganisms are capable of synthesizing potentially valuable aroma compounds and enzymes used in flavour manufacturing. However, yields are often disappointingly low, which hampers extensive industrial application.

In the last decades there has been an increasing trend towards the utilization of the solid-state fermentation (SSF) technique to produce several bulk chemicals and enzymes (Adinarayana et al. 2004; Cordova et al. 1998; Gombert et al. 1999; Muniswaran et al. 1994). SSF has been known from ancient times (approximately 2600 BC), and typical examples of this technique are traditional fermentations such as Japanese koji, Indonesian tempeh and French blue cheese. In recent years, SSF has received more and more interest from researchers, since several studies on enzymes (Domínguez et al. 2003; Pandey et al. 1999), flavours (Beuchat 1982; Feron et al. 1996), colourants (Johns and Stuart 1991) and other substances of interest to the food industry have shown that SSF may lead to higher yields or better product characteristics than submerged fermentation (SmF). In addition, costs are much lower due to the efficient utilization and value-addition of wastes (Robinson and Nigam 2003). The main drawback of this type of cultivation concerns the scaling-up of the process, largely due to heat transfer and culture homogeneity problems (Di Luccio et al. 2004; Mitchell et al. 2000). However, research attention has been directed towards the development of designs such as mixed solid-state bioreactor (Nagel et al. 2001), rotating drum bioreactor (Stuart et al. 1999) and immersion bioreactor (Rivela et al. 2000), which overcome these difficulties. SSF could be potentially useful for the production of flavour compounds (Berger 1995; Soccol and Vandenberghe 2003). Feron et al. (1996) reviewed the prospects of microbial production of food flavours and the recommended SSF processes for their manufacture.

Several researchers have studied SSF production of aroma compounds by several microorganisms (Table 6.3; Pastore et al. 1994; Besson et al. 1997; Bramorski et al. 1998a; Medeiros et al. 2001), such as *Neurospora* sp, *Zygosaccharomyces rouxii* and *Aspergillus* sp., using pre-gelatinized rice, miso and cellulose fibres, respectively. Bramorski et al. (1998b) compared fruity aroma production by *Ceratocystis fimbriata* in solid-state cultures using several agro-industrial wastes (cassava bagasse, apple pomace, amaranth and soybean), and found that the medium with cassava bagasse, apple pomace or soybean produced a strong fruity aroma. Soares et al. (2000) also reported the production of strong pineapple aroma when SSF was carried out using coffee husk as a substrate by this strain. Compounds such as acetaldehyde, ethanol, ethyl acetate (the major compound produced), ethyl isobutyrate, isobutyl acetate, isoamyl acetate and ethyl-3-hexanoate were identified in the headspace of the cultures. The addition of leucine increased ethyl acetate and isoamyl acetate production, and then a strong odour of banana was detected.

Bramorski et al. (1998a) and Christen et al. (2000) described the production of volatile compounds such as acetaldehyde and 3-methylbutanol by the edible fun-

Table 6.3 Agro-industrial wastes used for the aroma production by solid state fermentation

Microorganisms in SSF	Substrates	Aroma compounds	Literature
Pediococcus pentosaceus *Lactobacillus acidophilus*	Semisolid maize	Butter flavour	(Escamilla-Hurtado et al. 2005)
Kluyveromyces marxianus	Cassava bagasse and giant palm bran	Fruity aroma	
Ceratocystis fimbriata	Cassava bagasse, apple pomace, amaranth and soybean	Fruity aroma	(Medeiros et al. 2001)
Neurospora sp.	Pre-gelatinized rice	Fruity aroma	(Bramorski et al. 1998a)
Zygosaccharomyces rouxii	Miso	HEMF	(Pastore et al. 1994)
Ceratocystis fimbriata	Coffee husk	Pineapple aroma	(Sugawara et al. 1994)
Bacillus subtilis	Soybeans	Pyrazine	(Soares et al. 2000)
Aspergillus oryzae	Rice koji	Volatile compounds	(Besson et al. 1997)
Rhizopus oryzae	Tropical agro-industrial substrates	Volatile compounds	(Ito et al. 1990) (Christen et al. 2000)

gus *Rhizopus oryzae* during SSF on tropical agro-industrial substrates. The production of 6-pentyl-a-pyrone (6-PP), an unsaturated lactone with a strong coconut-like aroma, was studied using liquid and solid substrates by De Araujo et al. (2002). Sugarcane bagasse was adequate for growth and aroma production; it has been demonstrated that, by solid-state fermentation process, it is possible to produce 6-PP at higher concentration than that reported in literature for submerged process. *Kluyveromyces marxianus* produced fruity aroma compounds in SSF using cassava bagasse or giant palm bran (*Opuntia ficus indica*) as a substrate (Medeiros et al. 2000). SSF was found to be very suitable for the production of pyrazines. Besson et al. (1997) and Larroche et al. (1999) studied the biosynthesis of 2,5-dimethylpyrazine (2,5-DMP) and tetramethylpyrazine (TMP) using SSF cultures of *Bacillus subtilis* on soybeans.

Production of dairy flavour compounds, such as butyric acid, lactic acid and diacetyl in mixed cultures of *Lactobacillus acidophilus* and *Pediococcus pentosaceus* growing on a semisolid maize-based culture, has been reported (Escamilla-Hurtado et al. 2005). Soccol et al. (1994) studied the synthesis of lactic acid by *Rhizopus oryzae* in SSF with sugarcane bagasse as a support. They obtained a slightly higher productivity than in submerged cultivation. Moreover, lactic acid production by lactic acid bacteria *Lactobacillus paracasei* and *Lactobacillus amylophilus* GV6 under SSF conditions using sweet sorghum and wheat bran as both support and substrate, respectively, have been investigated (Naveena et al. 2005a, b; Richter and Träger 1994). It is known that several methyl ketones such as 2-undecanone, 2-nonanone and 2-heptanone are produced at commercial scale by SSF from *Aspergillus niger* using coconut fat as substrate with a yield of 40 % (Allegrone et al. 1991). Several methods have been developed in order to enable vanillin and furanone or pyranone derivatives of natural origin to be produced from agricultural wastes.

6.4 Recovery

Aroma compounds are typically organic compounds that are extremely volatile. Consequently, during thermal processing such as, concentration or pasteurization, change and/or loss of aroma compounds are likely to occur. In many cases these changes are unwanted. In the pasteurization step, for instance, chemical changes can occur due to thermal degradation. The major problem lies through in concentration step by conventional process such as multiple-effect evaporation (Suvarnalatha et al. 1994). It is a problem that may deteriorate the quality and acceptance of the final product. A possible way of minimizing the changes is to use separation techniques for recovery of the aromas. Techniques suitable for this task, both commercially available and developing, are distillation, partial condensation, gas injection, adsorption, super critical fluid extraction and pervaporation (Jørgensen et al. 2004; Yanniotis et al. 2007; Aroujalian and Raisi 2007).

6.5 Application of Aroma Compounds

Aroma compounds have been of high importance for folk medicine, classical medicine, food, perfumery and cosmetics since ancient times. The renaissance of the use of natural products in the last years also led to an increasing interest in aroma components. Especially because of their high biological activity and low toxicity aroma compounds are often used in pharmaceutical products. Although there is a discussion about their usage between researchers in the field of traditional and classical medicine, the number of scientific papers including analytical and biological data on aroma components is at present higher than ever. Additionally, the flavouring and conservation of food stuff by odorous volatiles as well as the search for pleasant smelling raw materials for perfume and cosmetic products in the nature is not only supported by the food, but also by the perfume and cosmetic industry with great commercial significance. And it is also used as defoaming agents for ophthalmic solutions with high concentrations of surfactants Natural aroma compounds are used to improve the shelf life and safety of minimally processed fruits (Anese et al. 1997; Lanciotti et al. 2004).

References

Adinarayana K, Raju KVV et al. (2004) Optimization of process parameters for production of lipase in solid-state fermentation by newly isolated *Aspergillus* species. Indian J Biotechnol 3: 65–69

Aguedo M, Ly MH et al. (2004) The use of enzymes and microorganisms for the production of aroma compounds from lipids. Food Technol Biotechnol 42: 327–336

Allegrone G, Barbeni M, Cardillo R et al. (1991) On the steric course of the microbial generation of (Z6)-gamma-dodecenolactone from (10R,S) 10-hydroxyoctadeca-(E8,Z12)-dienoic acid. Biotechnol Lett 13: 765–768

Armstrong DW, Yamazaki H (1986) Natural flavours production: A biotechnological approach. Trends Biotechnol 4(10): 264–268

Anese M, Manzano M, et al. (1997) Quality of minimally processed apple slices using different modified atmosphere conditions. J Food Quality 20: 359–370

Aroujalian, Raisi A (2007) Recovery of volatile aroma components from orange juice by pervaporation. J Membrane Sci 303: 154–161

Asther M, Haon M, Roussos S et al. (2002) Feruloyl esterase from *Aspergillus niger*: A comparison of the production in solid state and submerged fermentation. Process Biochem 38: 685–691

Aungpraphapornchai P, Griffin HG et al. (1999) Cloning, DNA sequence analysis, and deletion of a gene encoding diacetyl-acetoin reductase from *Lactococcus lactis*. DNA Sequence 10: 163–172

Ayabe S, Takano H, Fujita T et al. (1990) Triterpenoid biosynthesis in tissue cultures of *Glycyrrhiza glabra* var. *glandulifera*. Plant Cell Rep 9: 181–184

Badings HT, Van der Pol JJG et al. (1981) Aroma Compounds which Contribute to the Difference in Flavour Between Pasteurized Milk and UHT Milk. In: Schreier P (ed.) Flavour' 81. Walter de Gruyter, New York

Bassit N, Boquien CY, Picque D et al. (1995) Effect of temperature on diacetyl and acetoin production by *Lactococcus lactis* subsp. *lactis* biovar. *diacetylactis* CNRZ 483. J Dairy Res 62: 123–129

Benjamin A, Pandey A (1997) Coconut cake – A potent substrate for the production of lipase by *Candida rugosa* in solid-state fermentation. Acta Biotechnol 17: 241–251

Berger RG (1995) Aroma Biotechnology. Springer-Verlag. Berlin

Besson I, Creuly C, Gros JB et al. (1997) Pyrazine production by *Bacillus subtilis* in solid state fermentation on soybeans. Appl Microbiol Biotechnol 47: 489–495

Beuchat LR (1982) Flavor chemistry of fermented peanuts. Ind Eng Chem Prod Res Dev 21: 533–536

Bluemke W, Schrader J (2001) Integrated bioprocess for enhanced production of natural flavors and fragrances by *Ceratocystis moniliformis*. Biomol Eng 17: 137–142

Bonnin E, Grange H, Lesage-Meessen L et al. (2000) Enzymic release of cellobiose from sugar beet pulp, and its use to flavour vanillin production in *Pycnoporus cinnabarinus* from vanillic acid. Carbohydr Polym 41: 143–151

Bramorski A, Christen P, Ramirez M et al. (1998a) Production of volatile compounds by the edible fungus *Rhizopus oryzae* during solid-state cultivation on tropical agro-industrial substrates. Biotechnol Lett 20: 359–362

Bramorski A, Soccol CR, Christen P et al. (1998b) Fruit aroma production by *Ceratocystis fimbriata* in static cultures from solid agro-industrial wastes. Rev Microbiol 28: 208–212

Bratovanova P (2001) To the issue of formation of acetoin and diacetyl in dough part manufactured products. Biotechnol Biotec Eq 15: 124–127

Carroll NM, Ross RP et al. (1999) Characterization of recombinant acetolactate synthase from *Leuconostoc lactis* NCW1. Enzyme Microbiol Technol 25: 61–67

Castanares A, McCrae SI et al. (1992) Purification and properties of a feruloyl/p-coumaroyl esterase from the fungus *Penicillium pinophilum*. Enzyme Microbiol Technol 14: 875–884

Chandrasekaran MJ (1997) Industrial enzymes from marine microorganisms: The Indian scenario. J Mar Biotechnol 5: 86–89

Christen P, Bramorski A, Revah S et al. (2000) Characterization of volatile compounds produced by *Rhizopus* strains grown on agro-industrial solid wastes. Bioresour Technol 71: 211–215

Clark GS (1995) Benzaldehyde. An aroma chemical profile. Perfum Flavor 20: 53–60

Collin RP, Halim AF (1972) Characterization of the major aroma constituent of the fungus *Trichoderma viride* (Pers.). J Agric Food Chem 20: 437–438

Cordova J, Nemmaoui M, Ismaili-Alaoui M et al. (1998) Lipase production by solid state fermentation of olive cake and sugar cane bagasse. J Mol Catal B Enzym 5: 75–78

Davidonis G, Knorr D (1991) Callus formation and shoot regeneration in *Vanilla planifolia.* Food Biotechnol 5: 59–66

De Araujo AA , Pastore GM et al. (2002) Production of coconut aroma by fungi cultivation in solid-state fermentation. Appl Biochem Biotechnol 98–100: 747–751

Demain AL, Jackson M et al. (1967) Thiamine-dependent accumulation of tetramethylpyrazine accompanying a mutation in the isoleucine-valine pathway. J Bacteriol 94: 323–326

Dhavlikar RS, Albroscheit G (1973) Microbial transformation of terpenoids: Valencene. Dragoco Rep 20: 251–258

Di Luccio M, Capra F, Ribeiro NP et al. (2004) Effect of temperature, moisture, and carbon supplementation on lipase production by solid-state fermentation of soy cake by *Penicillium simplicissimum.* Appl Biochem Biotechnol 113: 173–180

Dimick PS, Walker NJ et al. (1969) Occurrence and biochemical origin of aliphatic lactones in milk fat – A review. J Agric Food Chem 17: 649–655

Domínguez A, Costas M, Longo MA et al. (2003) A novel application of solid state culture: Production of lipases by *Yarrowia lipolytica*. Biotechnol Lett 25: 1225–1229

Dornenburg H, Knorr D (1996) Production of the phenolic flavour compounds with cultured cells and tissues of *Vanilla planifolia* species. Food Biotechnol 10: 75–92

Drawert F, Berger RG et al. (1984) Regioselective biotransformation of valencene in cell suspension cultures of *Citrus* spp. Plant Cell Rep 3: 37–40

Escamilla-Hurtado ML, Valdes-Martinez S et al. (2000) Effect of some nutritional and environmental parameters on the production of diacetyl and on starch consumption by *Pediococcus pentosaceus* and *Lactobacillus acidophilus* in submerged cultures. J Appl Microbiol 88: 142–153

Escamilla-Hurtado ML, Valdes-Martinez SE et al. (2005) Effect of culture conditionson production of butter flavor compounds by *Pediococcuspentosaceus* and *Lactobacillus acidophilus* in semisolid maize-based cultures. Int J Food Microbiol 105: 305–316

Etschmann MMW, Bluemke W et al. (2002) Biotechnological production of 2-phenylethanol. Appl Microbiol Biotechnol 99: 1–8

Etschmann MMW, Sell D et al. (2004) Medium optimization for the production of the aroma compound 2-phenylethanol using a genetic algorithm. J Mol Catal B Enzyme 29: 187–193

Fabre CE, Blanc PJ et al. (1997) Screening of yeasts producing 2-phenylethanol. Biotechnol Tech 11: 523–525

Fabre CE, Blanc PJ et al. (1998) Production of 2-phenylethyl alcohol by *Kluyveromyces marxianus.* Biotechnol Progr 14: 270–274

Falconnier B, Lapierre C, Lesage-Meessen L et al. (1994) Vanillin as a product of ferulic acid biotransformation by the white-rot fungus *Pycnoporus cinnabarinus* I-937: Identification of metabolic pathways. J Biotechnol 37: 123–132

Feron G, Bonnarme P et al. (1996) Prospects of the microbial production of food flavours. Trends Food Sci Technol 7: 285–293

Funk C, Brodelius P (1990) Phenylpropanoid metabolism in suspension cultures of *Vanilla planifolia* Andr. II. Effects of precursor feeding and metabolic inhibitors. Plant Physiol 94: 95–101

Funk C, Brodelius P (1992) Phenylpropanoid metabolism in suspension cultures of *Vanilla planifolia* Andr. IV. Induction of vanillic acid formation. Plant Physiol 99: 256–262

Gatfield IL (1988) Production of flavour and aroma compounds by biotechnology. Food Technol 10: 110–122

Gatfield IL (1999) Biotechnological Production of Natural Flavor Materials. In: Teranishi R, Wick EL, Hornstein I (eds.) Flavor Chemistry, Thirty Years of Progress. Plenum Press, New York

Geusz SD, Anderson DM (1992) Process of using bacteria that metabolize phenylacetate through mandelate. US patent 5151 353

Gombert AK, Pinto AL, Castilho LR et al. (1999) Lipase production by *Penicillium restrictum* in solid-state fermentation using babassu oil cake as substrate. Process Biochem 35: 85–90

Gueguen M, Lenoir J (1975) Aptitude del'espêce Geotrichum candidum á la production d'enzymes protêolytiques. Note complêmentaire. Lait 549–550, 621–629

Gueguen M, Lenoir J, (1976) Caractères du sys-tème proteèolytique de Geotrichum candidum. Lait 557: 439–448

Hagedorn S, Kaphammer B (1994)Microbial biocatalysis in the generation of flavor and fragrance chemicals. Annu Rev Microbiol 48: 773–800

Harlander S (1994) Biotechnology for the Production of Flavouring Materials. In: Reineccius G (ed.) Source Book of Flavours. Chapman and Hall, New York

Ibragimova AZ, Yakovlev DA et al. (1980) Determination of the ratio of diacetyl and acetaldehyde in *streptococcic* cultures by gas-liquid chromatography. Molochnaya Promyshlennost 2: 43–45

Ito K, Yoshida K, Ishikawa T et al. (1990) Volatile compounds produced by fungus *Aspergillus oryzae* in rice koji and their changes during cultivation. J Ferment Bioeng 70: 169–172

Jacobsen T, Olsen J et al. (1990) Sub-strate speci¢city of Geotrichum candidum lipase preparation. Biotechnol Lett 12: 121–126

Janssens L, De Pooter HL, Schamp NM et al. (1992) Production of flavours by microorganisms. Process Biochem 27: 195–215

Jensen KA, Evans KMC, Kirk TK et al. (1994) Biosynthetic pathway for veratryl alcohol in the ligninolytic fungus *Phanerochaete chrysosporium* Appl Environ Microbiol 60: 709–714

Jiang J (1995) Changes in volatile composition of *Kluyveromyces lactis* broth during fermentation. In: Charalambous G (ed.) Food Flavours, Analysis and Process Influence. Elsevier Science. B.V. 1073–1086

Johns MR, Stuart DM (1991) Production of pigments by *Monascus purpureus* in solid culture. J Ind Microbiol 8: 23–28

Jollivet N, Chataud J, Vayssier Y et al. (1994) Production of volatile compounds in model milk and cheese media by eight strains of Geotrichum candidum Link. J Dairy Res 61: 241–248

Jørgensen RB, Meyer AS, Varming C et al. (2004) Recovery of volatile aroma compounds from black currant juice by vacuum membrane distillation. J Food Eng 64: 23–31

Kamini NR, Mala JGS et al. (1998) Lipase production from *Aspergillus niger* by solid-state fermentation using gingelly oil cake. Process Biochem 33: 505–511

Kana K, Kanellaki M et al. (1992) Volatile by-products formed in batch alcoholic fermentations: Effect of g-alumina and kissiris supported biocatalysts. Food Biotechnol 6: 65–74

Kapfer GF, Berger RG et al. (1989) Production of 4-decanolide by semicontinuous fermentation of *Tyromyces sambuceus*. Biotechnol Lett 11: 561–566

Kawabe G, Morita H (1994) Production of benzaldehyde and benzyl alcohol by the mushroom *Polyporus tuberaster* K2606. J Agric Food Chem 42: 2556–2560

Kim TH, Shin JH, Baek HH, Lee HJ (2001) Volatile flavour compounds in suspension culture of *Agastache rugosa* Kuntze (Korean mint). J Sci Food Agric 81: 569–575

Koizumi T, Kakuta K, Ohsawa R et al. (1982) On the aroma substances produced by tree exudate yeast (On the aroma produced by yeast. Part XII). NipponNâgeikagaku Kaishi 56: 757–763

Krings U, Berger RG (1998) Biotechnological production of flavours and fragrances. Appl Microbiol Biotechnol 49: 1–8

Kronlof J, Linko M (1992) Production of beer using immobilized yeast encoding a-acetolactate decarboxylase. J Inst Brew 98: 479–491

Lanciotti R, Gianotti A, Patrignani F et al. (2004) Use of natural aroma compounds to improve shelflife and safety of minimally processed fruits. Tren Food Sci Technol 15: 201–208

Lapadatescu C, Ginies C, Djian A et al. (1999) Regulation of the synthesis of aryl metabolites by phospholipids sources in the white-rot fungus *Bjerkandera adusta*. Arch Microbiol 171: 151–158

Larroche C, Besson I et al. (1999) High pyrazine production by *Bacillus subtilis* in solid substrate fermentation on ground soy-beans. Process Biochem 34: 67–74

Latrasse A, Dameron P, Hassani M et al. (1987) Production d'un arôme fruité par Geotri-chum candidum (Staron). Sci Aliments 7: 637–645

Lesage-Meessen L, Delattre M, Haon M et al. (1996) A two-step bioconversion process for vanillin production from ferulic acid combining *Aspergillus niger* and *Pycnoporus cinnabarinus*. J Biotechnol 50: 107–113

Lesage-Meessen L, Lomascolo A, Bonnin E et al. (2002) A biotechnological process involving filamentous fungi to produce natural crystalline vanillin from maize bran. Appl Biochem Biotechnol 102: 141–153

Liu SQ, Holland R et al. (2004) Esters and their biosynthesis in fermented dairy products: A review. Int Dairy J 14: 923–945

Lomascolo A, Asther M et al. (2001) Shifting the biotransformation pathways of L-phenylalanine into benzaldehyde by *Trametes suaveolens* CBS 334.85 using HP20 resin. Lett Appl Microbiol 32: 262–267

Lomascolo A, Lesage-Meessen L et al. (1999) Enhanced benzaldehyde formation by a monokaryotic strain of *Pycnoporus cinnabarinus* using a selective solid adsorbent in the culture medium. Can J Microbiol 45: 653–657

Macris BJ, Kekos D, Evangelidou X et al. (1987) Solid-state fermentation of straw with*Neurospora crassa* for CMCase and b-glucosidase production. Biotechnol Lett 9: 661–664

Mallouchos A, Komaitis M, Koutinas A et al. (2002) Wine fermentations by immobilized and free cells at different temperatures. Effect of immobilization and temperature on volatile by-products. Food Chem 80: 109–113

Martin DM, Bohlmann J (2004) Identification of *Vitis vinifera* (–)-a-terpineol synthase by in silico screening of full-length cDNA ESTs and functional characterization of recombinant terpene synthase. Phytochemistry 65: 1223–1229

Mathew S, Abraham TE (2005) Studies on the production of feruloyl esterase from cereal brans and sugar cane bagasse by microbial fermentation. Enzyme Microbiol Technol 36: 565–570

Medeiros A, Pandey A, Christen P et al. (2001) Aroma compounds produced by *Kluyveromyces marxianus* in solid-state fermentation on packed bed column bioreactor. World J Microbiol Biotechnol 17: 767–771

Medeiros ABP, Pandey A, Freitas RJS et al. (2000) Optimization of the production of aroma compounds by *Kluyveromyces marxianus* in solid-state fermentation using factorial design and response surface methodology. Biochem Eng J 6: 33–39

Medina de Figueroa R, Cerutti de Guglielmone G et al. (1998) Flavor compound production and citrate metabolism in *Lactobacillus rhamnosus* ATCC 7469. Milchwissenschaft 53: 617–619

Miranda OA, Salgueiro AA, Pimentel MCB et al. (1999) Lipase production by a Brazilian strain of *Penicillium citrinum* using an industrial residue. Bioresour Technol 69: 145–147

Mitchell DA, Krieger N, Stuart D et al. (2000) New developments in solid-state fermentation II. Rational approaches to the design, operation and scale-up of bioreactors. Process Biochem 35: 1211–1225

Moio L, Etievant P, Langlois D et al. (1994) Detection of powerful odorants in heated milk by use of extract dilution sniffing analysis. J Dairy Res 61: 385–394

Mulabagal V, Tsay HS (2004) Plant cell cultures – An alternative and efficient source for the production of biologically important secondary metabolites. Int J Appl Sci Eng 2: 29–48

Muniswaran PKA, Selvakumar P et al. (1994) Production of cellulases from coconut coir pith in solid state fermentation. J Chem Technol Biotechnol 60: 147–151

Nabeta K, Ohnishi Y, Hirose T et al. (1983) Monoterpene biosynthesis by callus tissues and suspension cells from *Perilla* species. Phytochemistry 22: 423–425

Nagel FG, Tramper J, Bakker MS et al. (2001) Temperature control in a continuously mixed bioreactor for solid-state fermentation. Biotechnol Bioeng 72: 219–230

Nakao M, Ono K, Takio S (1999) The effect of calcium on flavanol production in cell suspension culture of *Polygonumhydropiper*. Plant Cell Rep 18: 759–763

Narbad A, Gasson MJ (1998) Metabolism of ferulic acid via vanillin using a novel CoA-dependent pathway in a newly isolated strain of *Pseudomonas fluorescens.* Microbiology 144: 1397–1405

Nardi M, Fiez-Vandal C, Tailliez P et al. (2002) The EstA esterase is responsible for the main capacity of *Lactococcuslactis* to synthesize short chain fatty acid esters in vitro. J Appl Microbiol 93: 994–1002

Naveena BJ, Altaf M, Bhadrayya K et al. (2005) Direct fermentation of starch to L(+) lactic acid in SSF by *Lactobacillus amylophilus* GV6 using wheat bran as support and substrate: Medium optimization using RSM. Process Biochem 40: 681–690

Naveena BJ, Altaf M, Bhadriah K et al. (2005) Selection of medium components by Plackett–Burman design for production of L(+)lactic acid by *Lactobacillus amylophilus* GV6 in SSF using wheat bran. Bioresour Technol 96: 485–490

Nomura K, Ogura H et al. (2001) Direct synthesis of 2-phenylethanol by hydrogenation of methyl phenylacetate using homogeneous ruthenium-phosphine catalysis under low hydrogen pressure. J Mol Catal A Chem 166: 345–349

Ohnishi K, Yoshida Y et al. (1994) Lipase production of *Aspergillus oryzae.* J Ferment Bioeng 77: 490–495

Ohsumi C, Hayashi T et al (1993) Formation of allin in the culture tissues of *Allium sativum* oxidation of S-allyl-L-cysteine. Phytochemistry 33: 107–111

Okui S, Uchiyama M et al. (1963) Metabolism of hydroxyl fatty acids. I. Metabolic conversion of ricinoleic acid by a certain microorganism to (+)-D-8-hydroxy-*cis*-5-tetradecenoic acid. J Biochem 53: 265–270

Okui S, Uchiyama M et al. (1963) Metabolism of hydroxyl fatty acids. II. Intermediates of the oxidative breakdown of ricinoleic acid by genus *Candida.* J Biochem 54: 536–540

Ortiz-Vazquez E, Granados-Baeza M et al. (1993) Effect of culture conditions on lipolytic enzyme production by *Penicillium candidum* in a solid state fermentation. Biotechnol Adv 11: 409–416

Overhage J, Priefert H, Rabenhorst J et al. (1999) Biotransformation of eugenol to vanillin by a mutant of *Pseudomonas* sp. strain HR199 constructed by disruption of the vanillin dehydrogenase (vdh) gene. Appl Microbiol Biotechnol 52: 820–828

Pandey A, Selvakumar P, Soccol CR et al. (1999) Solid state fermentation for the production of industrial enzymes. Curr Sci 77: 149–162

Park JK, Jung JY (2002) Production of benzaldehyde by encapsulated whole-cell benzoylformate decarboxylase. Enzyme Microbiol Technol 30: 726–733

Pastore GM, Park YK et al. (1994) Production of a fruity aroma by *Neurospora* from beiju. Mycol Res 98: 25–35

Pichersky E, Iijima Y, Lewinsohn E et al. (2005)cloning and sequence of geraniol synthase from sweet basil and use in the production of geraniol and metabolites. International patent PCT WO 2005060553

Prakasham RS, Rao Ch S, Rao RS et al. (2005) Alkaline protease production by an isolated *Bacillus circulans* under solid-state fermentation using agro-industrial waste: Process parameters optimization. Biotechnol Progr 21: 1380–1388

Priefert H, Babenhorst J et al. (2001) Biotechnological production of vanillin. Appl Microbiol Biotechnol 56: 296–314

Prince CL, Shuler ML et al. (1997) Altering flavour profiles in onion (*Allium cepa* L.) root cultures through directedbiosynthesis. Biotechnol Progr 13: 506–510

Rao SR, Ravishankar GA (1999) Biotransformation of isoeugenol to vanilla flavour metabolites and capsaicin in suspended and immobilized cell cultures of *Capsicum frutescens*: Study of the influence of b-cyclodextrin and fungal elicitor. Process Biochem 35: 341–348

Rao SR, Ravishankar GA (2000) Vanilla flavour: Production by conventional and biotechnological routes. J Sci Food Agric 80: 289–304

Rao SR, Ravishankar GA (2002) Plant cell cultures: Chemical factories of secondary metabolites. Biotechnol Adv 20: 101–153

Richter K, Träger A (1994) L(+) lactic acid from sweet sorghum by submerged and solid-state fermentations. Acta Biotechnol 14: 367–378

Rivela I, Rodríguez Couto S et al. (2000) Extracellular ligninolytic enzyme production by *Phanerochaete chrysosporium* in a new solid-state bioreactor. Biotechnol Lett 22: 1443–1447

Robinson T, Nigam P (2003) Bioreactor design for protein enrichment of agricultural residues by solid state fermentation. Biochem Eng J 13: 197–203

Rojas V, Gil JV, Pinaga F et al. (2001) Studies on acetate ester production by non-*Saccharomyces* wine yeasts. Int J Food Microbiol 70: 283–289

Romagnoli LG, Knorr D (1988) Effects of ferulic acid treatment on growth and flavour development of cultured *Vanilla planifolia* cells. Food Biotechnol 2: 93–104.

Sahai OM (1994) Plant tissue culture. In: Gabelman A (ed.) Bioprocess Production of Flavour, Fragrances and Color Ingredients. Wiley, New York

Sandine WE, Elliker PR, Pack MY et al. (1965) Use of diacetyl reductase to remove diacetyl from beer. MBAA TQ 2: 155–159

Schindeler J, Bruns K (1980) Process for Producing Monoterpenes Containing Aroma by Fermentation. German patent 2840143

Scragg AH (1997) The production of aromas by plant cell cultures. In: Scheper P (ed.) Advances in Biochemical Engineering Biotechnology. Vol. 55: 239–263. Springer, Berlin, Germany

Seitz EW (1994) Fermentation production of pyrazines and terpenoids for flavour and fragrances. In: Gabeliman A (ed.) Bioprocess Production of Flavour, Fragrance, and Colour. Wiley, New York

Serp D, von Stockar U et al. (2003) Enhancement of 2-phenylethanol productivity by *Saccharomyces cerevisiae* in two-phase fed-batch fermentations using solvent immobilization. Biotechnol Bioeng 82: 103–110

Shimoni E, Baasov T, Ravid U et al. (2003) Biotransformations of propenylbenzenes by an *Arthrobacter* sp. and its *t*-anethole blocked mutants. J Biotechnol 105: 61–70

Shimoni E, Ravid U et al. (2000) Isolation of a *Bacillus* sp. capable of transforming isoeugenol to vanillin. J Biotechnol 78: 1–9

Sidebottom CM, Charton E, Dunn PPJ et al. (1991) *Geotrichum candidum* pro -duces several lipases with markedly different substrate specificities. Eur J Biochem 202: 487–491

Soares M, Christen P, Pandey A et al. (2000) Fruity flavour production by *Ceratocystis fimbriata* grown on coffee husk in solid state fermentation. Process Biochem 35: 857–861.

Soccol CR, Marin B, Rimbault M et al. (1994) Potential of solid state fermentation for production of L(+) lactic acid by *Rhizopus oryzae*. Appl Microbiol Biotechnol 41: 286–290

Soccol CR, Vandenberghe LPS (2003) Overview of applied solid-state fermentation in Brazil. Biochem Eng J 13: 205–218

Stark D, Münch T, Sonnleitner B et al. (2002) Extractive bioconversion of 2-phenylethanol from L-phenylalanine by *Saccharomyces cerevisiae*. Biotechnol Progr 18: 514–523

Stark D, Zala D, Münch T et al. (2003) Inhibition aspects of the bioconversion of L-phenylalanine to 2-phenylethanol by *Saccharomyces cerevisiae*. Enzyme Microbiol Technol 32: 212–223

Stuart DM, Mitchell DA, Johns MR et al. (1999) Solid-state fermentation in rotating drum bioreactors: Operating variables affect performance through their effects on transport phenomena. Biotechnol Bioeng 63: 383–391

Sugawara E, Hashimoto S, Sakurai Y et al. (1994) Formation by yeast of the HEMF (4-hydroxy-2 (or 5)-ethyl-5(or 2)-methyl-3 (2H)-furanone) aroma components in Miso with aging. Biosci Biotechnol Biochem 58: 1134–1135

Suresh B, Ritu T et al. (2003) Vanilla flavour production through biotransformation using *Capsicum frutescens* root cultures. Biocatal Biotransform 21: 333–340

Suvarnalatha G, Narayan MS, Ravishankar GA et al. (1994) Flavour production in plant cell cultures of Basmati rice (*Oryza sativa* L.). J Sci Food Agric 66: 439–442

Topakas E, Kalogeris E, Kekos D et al. (2004) Production of phenolics from corn cobs by coupling enzymic treatment and solid state fermentation. Eng Life Sci 4: 283–286

Townsley PM (1972) Chocolate from plant cells. J Inst Can Sci Technol Aliment 7: 76–78

Urbach G (1997) The flavour of milk and dairy products: II. Cheese: Contribution of volatile compounds. Int J Dairy Technol 50: 79–89

Vandamme EJ, Soetaert W (2002) Bioflavours and fragrances via fermentation and biocatalysis. J Chem Technol Biotechnol 77: 1323–1332

Wache Y, Aguedo M, Choquet A et al. (2001) Role of b-oxidation enzymes in g–decalactone production by the yeast *Yarrowia lipolytica.* Appl Environ Microbiol 67: 5700–5704

Walton NJ, Narbad A, Faulds CB et al. (2000) Novel approaches to biosynthesis of vanillin. Curr Opin Biotechnol 11: 490–496

Washisu Y, Tetsushi A, Hashimoto N et al. (1993) Manufacture of Vanillin and Related Compounds with *Pseudomonas.* Japanese patent JP52279

Westcott RJ, Cheethan PSJ et al. (1994) Use of organized viable vanilla plant aerial roots for the production of natural vanillin. Phytochemistry 35: 135–138

Wilcocks R, Ward OP (1992) Factors affecting 2-hydroxypropiophenone formation by benzoylformate decarboxylase from *Pseudomonas putida.* Biotechnol Bioeng 39: 1058–1063

Wilcocks R, Ward OP, Scott C et al. (1992) Acyloin formation by benzoylformate decarboxylase from *Pseudomonas putida.* Appl Environ Microbiol 58: 1699–1704

Wittman C, Hans M et al. (2002) Metabolic physiology of aroma-producing *Kluyveromyces marxianus.* Yeast 19: 1341–1363

Yang T, Li J, Wang HX et al. (2005) A geraniol synthase gene from *Cinnamomum tenuipilum.* Phytochemistry 66: 285–293

Yanniotis S,Tsitziloni K, Dendrinos G et al. (2007) Aroma recovery by combining distillation with absorption. J Food Eng 78: 882–887

Zheng Z, Shetty K (2000) Solid-state bioconversion of phenolics from cranberry pomace and role of *Lentinus edodes* b-glucosidase. J Agric Food Chem 48: 895–900

Chapter 7
Production of Bioactive Secondary Metabolites

Poonam Singh nee' Nigam

Contents

Abstract This chapter includes information based on published literature on utilization of agro industrial residues for the production of bioactive compounds. Various approaches using microbial fermentation technology have been explored for the production of bioactive compounds which as secondary metabolites could be produced by selected microorganisms. Certain factors have been found to affect the

P. Singh nee' Nigam (✉)
Faculty of Life and Health Sciences, School of Biomedical Sciences, University of Ulster, Coleraine BT52 1SA, Northern Ireland, UK
e-mail: P.Singh@ulster.ac.uk

P. Singh nee' Nigam, A. Pandey (eds.), *Biotechnology for Agro-Industrial Residues Utilisation*, DOI 10.1007/978-1-4020-9942-7_7,

productivity of these compounds, hence the yield of secondary metabolites may be manipulated by controlling these factors in fermentation system.

Keywords Secondary metabolite · Stationary phase · Idiophase · Biologically-active compounds · Antibiotics · Mycotoxins · Ergot-alkaloids

7.1 Reasons to Use Agro Residues as Starting Material

Despite the obvious problems that agricultural waste can create, the vast quantities of agricultural and agro-industrial residues that are generated as a result of diverse agricultural and industrial practices represent one of the most energy-rich resources on the planet. Accumulation of this biomass in large quantities every year results not only in the deterioration of the environment, but also in the loss of potentially valuable material which can be processed to yield a number of value added products, such as food, fuel, feed and a variety of chemicals. The agro-industrial residues are generated globally and a major portion is left unutilized and left over as wastes in surrounding environment. Such wastes produced annually can be used as a natural bioresource for the production of bioactive compounds such as secondary metabolites from various selected microorganisms.

Secondary metabolites are excreted by microbial cultures at the end of primary growth and during the stationary phase of growth. Secondary metabolites represent some of the most economically important industrial products and are of huge interest. The best known and most extensively studied secondary metabolites are the antibiotics, steroids and alkaloids. A variety of agricultural residues (Table 7.1) such as wheat straw, rice hulls, spent cereal grains, various brans such as wheat and rice bran, and corncobs, are available globally which can be considered the cheaper and often free of cost substrates for the commercial production of secondary metabolites.

7.2 What are Bioactive Compounds

Bioactive compounds are mostly secondary metabolites produced by microorganisms in an active culture cultivation process. Secondary metabolites usually accumulate during the later stage of microbial growth in process of fermentation, known as the "Idiophase". This later stage of microbial growth follows the active growth phase called "Trophophase". Compounds produced in the idiophase have no direct relationship to the synthesis of cell material and normal growth of the microoganisms. Secondary metabolites are formed in a fermentation medium after the microbial growth is completed. Filamentous fungi synthesize many secondary metabolites and are rich in genes encoding proteins involved in their biosynthesis. Genes from the same pathway are often clustered and co-expressed in particular conditions (Khaldi et al. 2008). Most common secondary metabolites are antibiotics

Table 7.1 Use of solid-state fermentation for the production of secondary metabolites and their applications (Adapted from Pandey et al. 2001)

Substrate used	Microorganism employed	Secondary metabolite produced	Application of metabolite
Wheat, oat, rice, maize, peanuts	*Aspergillus oryzae, A. panasitus*	Aflatoxin	Mycotoxin
Impregnated loam based compost	*B. subtillis*	Antifungal volatiles	Antifungal compounds
Coconut waste	*Bacillus thuringenisis*	Bacterial endotoxins	Insecticide
Barley	*Cephalosporium armonium*	Cephalosporin	Antibiotic
Wheat straw with cotton seed cake and sunflower cake	*Streptomyces clavuligerus*	Cephalosporin C	Antibiotic
Wheat straw with cotton seed cake and sunflower cake	*S. clavulingerus*	Clavulanic acid	β-Lactamase inhibitor, antibacterial
Wheat bran	*Tolypocladium infautum*	Cyclosporin A	Immuno suppressive drug
Rice, rice bran, rice husk	*Metarhizum anisopliae*	Destrucxins A and B	Cyclodesipeptides
Sugarcane bagasse	*Claviceps purpurea, C. fusiformis*	Ergot alkaloids	Disease treatment
Wheat bran, corn cob, cassava flour, sugarcane baggase	*Gibberella fujikuroi, Fusarium moniliforme*	Gibberellic acid	Plant growth harmone
Okara, wheat bran	*Bacillus subtilis*	Iturin	Antibiotic
Wheat bran	*P. brevicompactum*	Mycophenolic	–
Wheat, oat, rice, maize, peanuts	*oryzae A. Panasitus*	Ochratoxin	Mycotoxin
Corn cob	*S. rimosus*	Oxytetracycline	Antibiotic
Sugarcane bagasse	*Penicillium chrysogenum*	Penicillin	Antibiotic
Soyabean residue Okara	*B. Subtillis*	Surfactin	Antibiotic
Sweet potato residue	*S. viridifaciens*	Tetracycline, chlortetracycline	Antibiotic
Rice panicles	*Ustilaginoidea virens*	Ustiloxins	Antimitotic cyclic peptides
Corn	*Fusarium monoliforme*	Zearalenone	Growth promoter

and others include mycotoxins, ergot-alkaloids, the widely used immunosuppresant cyclosporin, and fumigillin, an inhibitor of angiogenesis and a suppresser of tumour growth.

7.2.1 Properties of Bioactive Secondary Metabolites

The desired product is released in fermentation medium as secondary metabolite of a particular microorganism grown for the purpose. These metabolites are usually not derived from the primary growth substrate, but a product formed from the primary growth substrate acts as a substrate for the production of a secondary metabolite. Secondary metabolites have the following characteristics:

a. Secondary metabolites of choice can be produced only by few selected microorganisms
b. These compounds are not essential for the organism's own growth and reproduction
c. Growth conditions, especially the composition of medium within a fermentation system, control the formation of secondary metabolites.
d. These compounds are produced as a group of closely related structures.
e. Secondary metabolic compounds can be overproduced.

There are several hypotheses about the role of secondary metabolites. Besides the five phases of the cell's own metabolism i.e. intermediary metabolism, regulation, transport, differentiation and morphogenesis, secondary metabolism is the activity centre for the evolution of further biochemical development. This development can proceed without damaging primary metabolite production. Genetic changes leading to the modification of secondary metabolites would not be expected to have any major effect on normal cell function. If a genetic change leads to the formation of a compound that may be beneficial, and then this genetic change would be fixed in the cell's genome, and becomes an essential one. Now this secondary metabolite would be converted into a primary metabolite.

7.3 Biotechnology Used for Production of Secondary Metabolites from Agro-Industrial Residues

The most commonly used technology is microbial-biotechnology practiced in industry, where secondary metabolites are mostly produced in a microbial fermentation process. This fermentation is performed in liquid state growing culture under submerged (SmF) conditions. This is mainly because the processes associated with scale up are much simplified and easy to manipulate for control of factors associated with whole production process. Liquid state fermentation allows greater control of parameters, such as pH, heat, nutrient conditions etc.

But for using agro-industrial residues for the production of bioactive compounds, another type of technology of microbial culture cultivation process would be ideal. In this technology process is performed under solid state condition rather in liquid state. This biotechnology is solid state (or substrate) fermentation (SSF) and characterized by a fermentation process on a solid support, which has a low moisture content (lower limit $\approx$ 12%), and occurs in a non-septic and natural state. Such technology describes the microbial transformation of biological materials in their natural state. The process is carried out in absence or near absence of free flowing water in the system, and it mainly utilizes fungal species.

Fungal species are ideal for this type of cultivation as these are capable of growth at lower water activity while bacteria require the presence of free water in fermentation system. SSF presents a low-cost system, it utilizes naturally occurring substances such as agricultural residues and forestry as substrates. Fermenters are easily to construct often only incorporating a tray and microorganism can be natural in some processes. The low water volume in SSF has a large impact on the economy of the process mainly due to smaller fermenter-size, reduced downstream processing, obviated or reduced stirring and lower sterilization costs. SSF produces a high product concentration with a relatively low energy requirement. Due to all these advantages over submerged fermentations, SSF has been exploited for the production of primary metabolites as well as secondary metabolite production.

7.3.1 Reason for Selecting Solid State Technology for Bioactive Secondary Metabolites

The use of SSF technology for the production of secondary metabolites should not be discounted. The mycelial morphology associated with the microorganisms that are predominately used for secondary metabolite production is well suited to growth on a solid support. This can also have a detrimental effect on product formation in liquid media, as highly viscous liquid media is required for successful metabolite production and can interfere with oxygen transfer. The filamentous morphology of these microorganisms and the secretion of these metabolites into the growth media can increase viscosity further. Therefore, SSF, technology can be exploited as an alternative, allowing better oxygen circulation (Hesseltine 1977).

Solid-state fermentation (SSF) is an important area of biotechnology, since the last decade has witnessed an unprecedented increase in interest in this technology (Nigam and Singh 1996a,b). This culture-technique is increasingly being used in the development of various bioprocesses in pharmaceutical, industrial and environmental sectors. Solid State (substrate) Fermentation (SSF) can be used successfully for the production of secondary metabolites (Robinson et al. 2001). These products (Nigam and Singh 2000) associate with the stationary phase of microbial growth and can be produced on an industrial scale for use in agriculture and the treatment of disease. Many of these secondary metabolites are still produced by submerged liquid fermentation (Nigam and Singh 1999), although production by this method has been shown less efficient in comparison to SSF. As large-scale production increases further, so does the cost and the growing energy demands. SSF has been shown to produce a more stable product, requiring less energy, in smaller fermenters with easier downstream processing measures. SSF technology has several advantages over submerged fermentation; primarily it represents a low-cost and easy to operate user-friendly system (Nigam and Singh 1994).

Concerning the production of secondary metabolites, SSF have ability to produce higher yields and productivities in certain cases. If the quality of the products could be guaranteed and the process-variables such as temperature, and pH could be controlled, SSF production of secondary metabolites would be very attractive

(Robinson et al. 2003). If these problems could be overcome, SSF technology could reduce the production-cost. It would enable third world countries cheaper access to secondary metabolites. SSF has found important applications particularly in the production of value-added products, such as biologically active secondary metabolites. Some of the important secondary metabolites produced in SSF are antibiotics, alkaloids, and plant growth factors (Balakrishnan and Pandey 1996). SSF systems, which during the previous two decades were termed as "low-technology" systems, appear to be a promising one for the production of value-added "low-volume and high-cost" products such as bio-pharmaceuticals (Pandey et al. 2000a,b). The recent evidence indicates that bacteria and fungi, growing under SSF conditions, are more than capable of supplying the growing global demand for secondary metabolites.

Though there are certain advantages of SSF-production process over conventional SmF systems but many practical advantages have been attributed to the production of biologically-active secondary metabolites through SSF route. Due to the lack of free water smaller fermenters are required for SSF and therefore less effort is required for downstream processing of secondary metabolites. Wild type strains of bacteria and fungi tend to perform better in SSF conditions than genetically modified microorganisms reducing energy and cost requirements further.

Different strategies and processes have been developed utilizing SSF technology for the production of biopharmaceuticals. Potential applications of SSF systems have been realized to produce high value bioactive secondary metabolites. Various secondary metabolites such as mycotoxins, bacterial endotoxins, plant growth factors, antibiotics, immuno-suppressive drugs and alkaloids etc. are among the important group of bioactive compound which can be produced by SSF technology (Table 7.1).

7.4 Biosynthesis of Secondary-Metabolites

Secondary metabolites comprise a diverse range of compounds synthesised by various fungal cultures (Nigam and Singh 2000) and some bacteria such as *Streptomyces*. Fungal secondary metabolites are an important source of bioactive compounds for agro chemistry and pharmacology. Over the past decade, many studies have been undertaken to characterize the biosynthetic pathways of fungal secondary metabolites. This effort has led to the discovery of new compounds, gene clusters, and key enzymes, and has been greatly supported by the recent releases of fungal genome sequences (Collemare et al. 2008). These secondary metabolites are of great commercial importance. Some are beneficial to life such as antibiotics and growth-promoters and some metabolites are mycotoxins, a threat to human and animal life. Various fermentation systems (Nigam and Singh 1999) such as surface-liquid, submerged, batch or fed-batch processes have been used for the production of different secondary metabolites. The use of certain liquid fermentation processes is established industrial practice, there are following reasons for this practice: the relative ease of scaling up liquid culture process; the greater homogeneity of liquid systems and the use of soluble starch; the superior monitoring for the precise regulation of process-parameters which particularly control the biosynthesis of secondary metabolites. As described above due to many other advantages solid state fermentation have been considered for certain secondary metabolite production.

7.4.1 Utilization of Agro-Industrial Residues as Substrate

Following points are worth consideration for the application and suitability of solid agricultural residues in the biosynthesis of secondary metabolites:

1. In several productions, the product formation has been found superior using solid insoluble substrates.
2. The most commonly used microorganisms in the production of secondary metabolites are fungi and *Actinomycetes*; and the mycelial morphology of such organisms is ideal for their invasive growth on solid and insoluble substrates.
3. The fungal morphology is responsible for considerable difficulties in large-scale-submerged processes. These include highly viscous, non-Newtonian broths and foam production. This results in very high power requirements for mixing and oxygen transfer. The presence of chemical antifoam in fermentation broths reduces oxygen transfer efficiency and can lead to problems in the product recovery.
4. In some processes, the final product is required in form of solid consistency, such as antibiotics present in animal feed.
5. The capital cost of overall production process using solid substrates is claimed to be significantly less.
6. The yields of certain secondary metabolites as aflatoxin B_1 and ochratoxin A obtained from liquid culture were found to be very poor. This led to the use of solid substrates and subsequently, a higher yield of 100 g. Similarly the production of the cyclic pentapeptide mycotoxin, malformin C was performed using *Aspergillus niger* in solid culture and a higher yield of 369 mg/kg was obtained compared to the yield of 15–200 mg/kg from liquid fermentations (Kobbe et al. 1977).

The production of extremely toxic mycotoxins by fungi has attracted attention, due to their importance in human and animal food chain. The aflatoxins have considerable economic impact, the poultry industry in U.S. lost US$100 million per year from aflotoxin poisoning in 1970s. Solid state cultivation has been used to produce sufficient quantities of these compounds for toxicity studies and these cultivations have been performed to study the conditions that promote toxin formation on cereal grains (Greenhalgh et al. 1983). The production of gibberellic acid in SSF has been adopted to eliminate the need of cell-removal in downstream-processing after submerged culture process, which contributes a significant part in the production cost. Another concept is the growth of antibiotic producing microorganisms on animal-feed for two purposes, firstly to enrich the protein-content in nutrient-poor feed and secondly to produce antibiotics, such as cephalosporins, tylosin and monensin.

7.4.2 Process Operation for Secondary Metabolites

The biotechnology based process for the production of secondary metabolites can be performed in batch, fed-batch, continuous or plug-flow bioreactor operation. These modes of operation are well suited to solid state process, but a well-mixed fed-batch bioreactor is difficult to operate on a large scale. A plug-flow mode process operating in a continuous system could be more straightforward on industrial-scale compared to a submerged fermentation process. The productivity in a continuous fermentation is higher in such process compared to batch system. Mostly the secondary metabolites are produced in batch reactor system. The productivity of compounds is low in batch process because the time required to achieve the phase for secondary metabolite synthesis is longer, occurring after the primary metabolite production stage and after culture-growth has happened. While a continuous process runs for a longer time with a continuous yield of secondary metabolites once the phase of synthesis has started. Genes responsible for biosynthesis of fungal secondary metabolites are usually tightly clustered in the genome and co-regulated with metabolite production (Patron et al. 2007).

The substrate addition in the fermentation-medium also affects the process-operation. The addition of soluble starch or glucose in the initial wheat-bran medium of *Gibberella fugikuroi* affected the synthesis and a reduced yield of gibberellic acid was obtained. A fed-batch process with intermittent feeding of soluble starch instead of including starch in the beginning increased the yield of gibberellic acid by 18% compared to the lower yield obtained in batch process. As catabolite regulation by glucose is common to the synthesis of many secondary metabolites, solid-state fed-batch operation is likely to be superior to batch systems for most. For example, a 47% of increase in product yield has been obtained using feeds of cornstarch intermittently. A study of the hydrolytic enzymes secreted by *G. fugikuroi* during batch and fed-batch solid-substrate fermentation demonstrated that the quantity and rate of production of enzymes were higher in batch cultures. This was suggested that the glucose levels in the medium would be increased over fed-batch cultures leading to catabolite regulation of gibberellic production by glucose and which resulted in lower yields of the product in batch process.

7.5 Process Control in Synthesis of Desired Metabolite

The production-times for many secondary metabolites have been found similar time-periods required in many submerged state and solid state fermentation using solid substrates. SSF processes performed for the synthesis of secondary metabolites show similar process-kinetic patterns such as microbial-growth, substrate-utilisation and bioconversion, and product-synthesis, similar to the characteristics of a submerged fermentation. Similar patterns have been observed for the production of aflatoxin B1, ochratoxin A (Lindenfelser and Ciegler 1975), trichothecene mycotoxins (Greenhalgh et al. 1983), polyketide pigments (Lin and Lizuka 1982) and gibberellic acid (Kumar and Lonsane 1987a,b,c). The kinetics of spore formation by *Penicillium roqueforti* is similar to those of secondary metabolite production. Process using agro-residues as starting substrates is controlled by a number of process – regulating factors such as: the initial moisture content of the substrate; rate of aeration; mixing of the fermenting solid medium; substrate-type, composition and structure of the substrate, and the constitution of fermenting medium; temperature; and the choice of microorganism.

7.5.1 Preparation of Agro Residues for SM-Production

A variety of solid substrates for secondary metabolites production have been tried in solid state fermentation (Table 7.1). These substrates derived from various sources vary in their nature, structure, and composition. Substrates coming from different origins have ability to provide a range of easily to poorly metabolizable nutritional sources and therefore, various substrates have been utilized as single carbon source or in combination with others, and also some substrates have been used as inert solid-supports for fungal colonization required for secondary metabolites production. Ultimate choice of a substrate for a particular metabolite is made after extensive trials with various types of substrates. Spent cereal grains such as wheat and rice have predominated as substrates for secondary metabolites. Aflatoxin has been produced using corn (Silman et al. 1979), rice (Shotwell et al. 1966), peanuts, corn meal and crushed wheat (Chang et al. 1963). Many brans such as wheat and rice brans are used singly or in combination with grains.

Mostly the production yields of secondary metabolites can be improved with a right choice of substrate or mixture of substrates with appropriate nutrients. A single-selected substrate performs in a different way changing the overall fermentation efficiency of the process; if the substrate is used in its different forms; for example pieces, fibres, particles or flour of a same substrate are metabolized in a different way. Though smaller particle size of a solid substrate has larger surface area for microbial action, but at the same time small particles have tendency of the increasing packing density. Densely packed fermentation system results in higher heat output per unit area of fermenter or output per unit space of fermenting solid medium. A column and other large size bioreactors would have a problem of poor aeration if used with smaller particle size substrates. In a process of gibberellic acid production using *Gibberella fuzikuroi*, higher yields have been achieved with larger particle size of wheat bran such as 0.3 to 0.4 cm (Kumar and Lonsane 1987a,b,c) compared to smaller particles of wheat bran.

7.5.2 Control of SM Production by Temperature

Normally the incubation temperature in a cultivation process is the optimum growth temperature of the particular microorganism used for secondary metabolite production, and this optimal temperature for any secondary metabolite biosynthesis is similar in solid substrate fermentation to that for liquid fermentation. Each process has its range of temperature over which secondary metabolite production occurs, since temperature can not be precisely controlled at all times in all layers of solid substrates and within same system the temperature can vary by few degrees due to metabolic heat generation. Excess temperature due to poor heat transfer in fermenting system may adversely affect the yields of secondary metabolite. Therefore, the temperature regulation is achieved with the mixing of fermenting solid substrates by rotation or agitation and aeration. In a production process of aflatoxin production, the effect of temperature was studied over the range of 27°C to 40°C using flasks and small column fermenters (Silman et al. 1979). Aflatoxin formation was achieved over this range, but it ceased above 40°C; the production of toxin could be restored if the temperature was lowered and controlled within the range of 27°C to 40°C. Hence, it is possible to control the production of secondary metabolites by simply controlling the temperature of fermenting agro-industrial substrates in fermentation system.

7.5.3 Control of SM Production by Agitation of Substrates

The mixing of contents such as solid substrate, nutrient medium and seed-culture is very important to start any process in effective way, but also the mixing is required during fermentation to aid the aeration and to facilitate the heat-transfer in some processes. The extent of mixing or the rate of agitation in system depends primarily on the type and design of the bioreactor used in that particular process of secondary metabolite production. Mixing of fermenting contents in a production process where solid substrates are being used, has generally been found to increase the productivity of the secondary metabolites, whereas lower yields were obtained in a similar process run without agitation or mixing. The only disadvantages noticed of mixing in fermentation are shear damage to the growing microorganism and the extra power requirements to run the agitator. In some processes agitation has increased the products-yield considerably. A possible physiological explanation for the increase or improvement in yields through agitation is that the mixing of fermenting solid insoluble substrates suppresses the process of sporulation. Sporulation occurs simultaneously in system with the production of secondary metabolites and it may compete for common intermediates and substrates. Suppression of this competition has been suggested as the only reason for the increased yield of the desired product.

In some processes a clear advantage of mixing has been noticed such as the rotation at the speed of 16 rpm was found necessary for the production of high yields of ochratoxin A by *A. ochraceus* (Lindenfelser and Ciegler 1975). This culture was grown on wheat in a small-scale rotary drum bioreactor. This was confirmed with the low yields obtained if there was no rotation of the drum-bioreactor or the rotation was just brief and intermittent. Therefore, it was concluded that the superior performance of the rotary-bioreactor was due to the mixing. Similarly, in a utilization process of pearled barley cultivating a culture of *A. clavatus* NRRL 5890 (Demain et al. 1977), the production of crude toxins was increased by 50% in agitated culture over static culture and the crude toxin was enriched in cyochalasin E. In contrast, the cephalosporin production was adversely affected by agitation of solid fermenting mass (Jermini and Demain 1989).

The production of most mycotoxins has been found to be improved in shaken cultures compared to stationary fermentations. Some of such improved-yield fermentations are: aflatoxin production by *A. flavus* NRRL 2999 using rice as substrate (Shotwell et al. 1966), ochratoxin A production by *A. ochraceus*, cytochalasin E and tremorgens production from culture of *A. clavatus*, cyclochorotine and simatoxin from *Penicillium islandicum* (Ghose et al. 1978), and cyclopiozonic acid by *A. flavus* (Luk et al. 1977). The yield of cyclopiozonic acid was obtained almost ten times (10×) in agitated fermentation of white wheat compared to the lower yield (1×) in static culture process (Luk et al. 1977).

7.5.4 Control of SM Production by Aeration of Substrates

Since the production of secondary metabolites has started utilizing agro industrial residues as the starting materials in fermentation process, the effect of aeration on the synthesis of various secondary metabolites has been investigated. The most comprehensive study has been performed on the production of aflatoxin B_1 in a corn storage bin with a capacity of 1266 bushels (Silman et al. 1979). The effect of aeration was studied by passing humidified air of 80–85% relative humidity through the bed of corn at flow rates between 0.001 to 0.04 l/kg corn per min and a recirculation rate of 1.5 l/kg per min. It was noticed that the rate of aflatoxin production and yields were directly proportional to the aeration rate. However, the direction of air-flow through the corn-bed had no apparent effect on aflatoxin yield or on the rate of production.

But the aeration may not be necessarily required in some cases, this needs to be confirmed before the running of process. In one case of secondary-metabolite production aeration proved to be the unnecessary where the production of ochratoxin A performed in a system using a rotating drum bioreactor was adversely affected by aeration (Lindenfelser and Ciegler 1975).

7.5.5 Control of SM Production by Moisture Content

The optimal moisture content of solid fermenting substrates varies for various metabolites production. The control of the water content present in solid substrate fermenting – medium or the maintenance of the initial moisture content in the system is very important factor. The moisture content of solid substrates greatly affects process of any secondary metabolite production. The optimal initial moisture contents have been found to be different according to the reactor-type used in the process. The initial moisture content may vary for the same fermentation process i.e. using same substrate and same microorganism for same metabolite production but during the scaling up of the process it may vary using different designs of bioreactor, for example from flasks to tray type.

Production of another metabolite such as toxin, using substrate grains or corn is greatly favoured at low initial moisture contents. The most favourable moisture content for optimal toxin production has been found to be between 20–40%. The optimal range is characteristically broad, however outside this range the yield of the product and the rate of toxin-formation are severely affected. Aflatoxin B_1 production from corn by *Aspergillus flavus* is negligible at initial moisture content below 17% though the fungal growth takes place (Silman et al. 1979); at moisture contents between 18–20% toxin production occurs at a reduced rate and lower yield is obtained. There is a rapid fungal growth and aflatoxin production at the moisture contents of 20–30% with the optimal moisture content being 22.4% at 33°C in flask culture. Similarly, ochratoxin A production has been found affected due to the variation in initial moisture content using laboratory-scale rotary drum bioreactors. In these culturing processes, the most significant parameter identified has been the effect of initial moisture content that determines the toxin-yields. The optimal moisture content of rice for the production of cephalosporin C was found to be 49–51% using *Acremonium chrysogenum*. In such cultivation-systems, yields were severely affected (inhibited) below the optimal value, whereas at higher levels of moisture content the bacterial contamination became an unavoidable problem.

7.6 Recovery of Secondary Metabolites in Downstream Processing

The solid fermented mass obtained after the completion of the process is extracted using various solvents to recover the product. The extraction is usually performed using the aqueous or other solvents mixing in proper ratio with the solid fermented mass. The extracts obtained from the fermentation of solid substrates usually contain higher concentration of secondary metabolites compared to the concentrations present in the submerged culture medium (Kumar and Lonsane 1988). One advantage of using agro industrial residues as solid substrates in fermentation is that there is no need of cell removal prior to the extraction of metabolites. The fermented liquid medium from a submerged fermentation is subjected to downstream processing for the removal of microbial-cells or mycelial-biomass to obtain a clear supernatant for the extraction of secondary metabolites. The cost of cell-removal from the submerged culture broths prior to the extraction process is estimated to be between 48% to 76% of the total production cost of the final product (Datar 1986).

There could be certain problems associated with the recovery of metabolites from system. The recovery of secondary metabolites from the solid fermented mass offers less flexibility in the choice of the initial unit operation than submerged fermentation. The metabolites diffuse throughout the solid mass during the culturing, which requires longer extraction-times for complete recovery of metabolites. The extraction of larger amount of solids may increase the concentration of impurities in the extract. The cost of purification depends on the quality of extract. The presence and concentration of inert compounds in the extract increase the cost of purification and therefore the cost of recovery is increased. Particularly those secondary metabolites which are used in bulk in the pharmaceutical and health industry and whose purity is governed by stringent regulations need to go through specific purification strategy. Now in antibiotic industry, the problem of culture-biomass in submerged process, is solved to some extent by the application of whole-broth processing. The extraction process uses the whole fermented broth including the cell-biomass. In such extraction the process is carried out using solvent extraction in multi-stage, centrifugal decanters.

The important variables in the extraction of an important secondary metabolite gibberellic acid from the fermented mouldy-bran (Kumar and Lonsane 1987) are the type of solvent to be used for an efficient extraction, concentration of solvent, the ratio of solvent to the solid, and pH. An 87% recovery of gibberellic acid in the extract of 0.9 mg/l was obtained using 2%, v/v ethanol as solvent in a multi-stage, counter-current extraction system. A considerable loss of solvent is common, with up to 68% of the initial solvent added to the bran, remaining absorbed into the solid on separation. In a particular production process the product is left on solid residue for use. Production of antibiotic monensin on various materials is carried out by *Streptomyces cinnamonensis* for use in poultry feeds to control coccidiosis. Tin such application no extraction cost would be involved.

7.7 Scaling-Up of Process For Secondary Metabolites

The elucidation of important scale-up criteria for SSF is very important to guarantee the successful large-scale operation of the processes. The scale-up of the each process results in mixed success for various processes employing different microorganisms and different solid substrates for a range of metabolites production. Since a particular individual process has its own characteristics due to the nature of organism and contents of the system such as type of agro residues used as substrate, each process behaves differently in scaling-up process. On the other hand the synthesis of secondary metabolites is very sensitive to the environmental factors and therefore the control of such factors becomes difficult in a similar process on large-scale, which is naturally expected.

Any process can be operated successfully regulating the various parameters on small-scale but the yield of the secondary-metabolites has been found adversely affected during large-scale cultivation. The synthesis of toxin has been found severely affected in a large-scale system, though the growth of employed microorganisms were always very obvious. In another process for the synthesis of aflatoxin B_1, the scaling-up using 75 g to 1266 bushels of corn have been successful producing good yields of aflatoxin even at larger scale (Silman et al. 1979).

Many of the important scale-up problems are generic to all solid state fermentation processes. The criteria of scale-up of systems utilising solid substrates are different to those established for submerged processes, such as volume-ratio, mixing time, k_1a, and foam-control etc.

7.7.1 Control of Temperature in Process Scale-Up

The precise control of incubation temperature in any bioreactor is a difficult problem, and this problem becomes a difficult engineering challenge with increase in scale of process for a large-scale production. Unlike submerged fermentation, where the dominant mechanism of heat-transfer is convection, the heat-transfer in a fermentation system using solid substrates occurs predominantly via two mechanisms i.e. convection and conduction. The convective medium in solid fermenting biomass is air and therefore the rate of heat-exchange is much smaller than in liquid submerged cultivations.

The microbial-activity on agro residue substrates for the production of secondary-metabolites is similar and not greatly less than in liquid cultures; consequently the heat generation is also considerable. The increase in temperature due to heat generation during the production of amyloglucosidase has been found up to 56–57°C. The rise in temperature affects the rate of microbial-growth and the rate of moisture loss in that process and as a result the production yield of the secondary metabolite is also affected. Excessive heat production may result in the progressive drying of the solid substrate with time in a longer process which requires 10 or more days for the optimum synthesis of secondary metabolites. This is a serious problem, which may affect the yield of the products.

This is natural to face difficulty in maintaining the temperature control on large scale such as if the fermentation is carried in large 1266-bushel storage bins. In such fermenters the temperature up to 47°C have been measured, despite the continual circulation of cool air to regulate the temperature-rise. Even similar difficulties may be experienced on a small-scale process for the production of secondary metabolites. Using a small-scale rotary-drum bioreactor, the temperature increase may be normally noticed to be 1–2°C per day (Lindenfelser and Ciegler 1975).

7.7.2 Control of Factors Related to Substrates in Scale-Up

The yields of secondary metabolites are affected in scale-up due to few more factors other than temperature-rise, which are directly related to solid substrates. The importance of the moisture content of the fermenting solid substrates is very significant in determining the yield of metabolites. However, the control of the moisture content of the solid substrate can be difficult. There are certain reasons for this difficulty such as the combined effects of heat production and moisture release due to microbial-respiration. The excessive wetness may result in bacterial and yeast contamination and also in rotary bioreactors wetness promotes the clumping of the material. Therefore a purposely-built reactor-design is essential to prevent such problems. During the typical process using wheat grains as substrate for ochratoxin A production cycle of 10–12 days, the moisture content has been found to rise from 30–40% and as a result of this the volume of substrate used (wheat) increased by 60–70%. Such problems lead to the reduced yield of the secondary metabolite.

Another factor in scale-up of SSF is the amount of solid substrates used in bioreactors. The yields of secondary metabolites are superior in SSF to those obtained in submerged process based on weight of product obtained per gram substrate and per litre, respectively. Frequently considerable amounts of carbon remain unutilized in fermentation. The yields of secondary metabolites calculated on the basis of the weight of substrate consumed, needs to be ascertained, this is important to adequately compare the productivity in two types of processes. This measure of the productivity of secondary metabolites is important for a commercial operation.

7.8 Prospects of Agro Residues for Secondary Metabolites Production

The use of agro industrial residues for the production of commercially valuable metabolites is at present under-utilized, with a strong preference towards conventional and familiar liquid fermentations. This lack of adoption into industry seems strange, since research in this area clearly shows that SSF produces higher yields in a shorter time period. It is easy to see why a liquid state fermentation using simple sugars as substrate is still prevalent for the production of secondary metabolites; it is a familiar technique, scale-up from lab to industrial fermenter level is much simplified in comparison to SSF with parameters being easier to monitor and control. There are also problems associated with secondary metabolite production in liquid fermentation, such as shear forces, increasing viscosity due to metabolite secretion, and reduction in metabolite stability. There is an increasing demand on science in regards to antibiotic production, with a growing global demand and cases of antibiotic resistance. Therefore, the use of residual agro industrial wastes should be considered by industry, especially when large quantities of secondary metabolites are required in a shorter fermentation period, with minimal expenditure on media and downstream processing.

There is clear evidence that high concentrations of secondary metabolites can be achieved using a variety of cheaper and freely available agro industrial residual substrates, employing best performing microorganisms in suitable properly designed and purposely-built bioreactors for large-scale production. To achieve the high standard of this technology, a number of studies have been performed to elucidate the optimal conditions for the synthesis of various secondary metabolites. Some areas of further interest are the use of fed-batch and continuous plug-flow modes of operation; the study of broader range of secondary-metabolites; and analysis of bioreactors to identify criteria for successful scale-up and therefore to permit effective process control. This last criterion is particularly important for the lengthy nature of these culture-systems. The residual substrate fermentation process for the biosynthesis of secondary metabolites is more likely to be seriously considered for the industrial-scale production of some important secondary metabolites.

References

Balakrishnan K, Pandey A (1996) Production of biologically active secondary metabolites in solid state fermentation. *J Sci Ind Res* **55**: 365–372

Chang SB, Abdel-Kadar M.M, Wick EL, Wogan GN (1963) Aflatoxin B2: chemical identity and biological activity. *Science* **142**: 1191–1192

Collemare J, Billard A, Bohnert HU, Lebrun MH (2008) Biosynthesis of secondary metabolites in the rice blast fungus Magnaporthe grisea: the role of hybrid PKS-NRPS in pathogenicity. *Mycol Res Feb* **112**(Pt 2): 207–215

Datar R (1986) Economics of primary separation steps in relation to fermentation and genetic engineering. *Process Biochem* **21**: 19–26

Demain AL, Hunt AN, Malik V, Kobbe B, Hawkins H, Matsuo K, Wogan GN (1977) Improved procedure for production of cytochalasin E and tremorgenic toxins by *Aspergillus clavatus. Appl Environ. Microbiol* **31**: 138–140

Ghose AC, Manmade A, Townsend JM, Bosquet A, Howes JF, Demain AL (1978) Production of cyclochlorotine and a new metabolite, simatoxin by *Penicillium islandicum. Appl Environ Microbiol* **35**: 1074–1078

Greenhalgh R, Neish GA, Miller D (1983) Deoxynivalenol, acetyl deoxynivalenol, and zearalenone formation by Canadian isolates of *Fusarium graminarium* on solid substrates. *Appl Environ Microbiol* **46**: 625–629

Hesseltine CW (1977) Solid state fermentation-Part I. *Process Biochem* **12**(6): 24–27

Jermini MFG, Demain AL (1989) Solid state fermentation for cephalosporin production by *Streptomyces clauvligerus* and *Cephalosporin acremonium. Experienta* **45**: 1061–1065

Khaldi N, Collemare J, Lebrun MH, Wolfe KH (2008, Jan 24) Evidence for horizontal transfer of a secondary metabolite gene cluster between fungi. *Genome Biol* **9**(1): R18

Kobbe B, Cushman M, Wogan GN, Demain AL (1977) Production and antibacterial activity of malformin C, a toxic metabolite of *Aspergillus niger. Appl Environ Microbiol* **33**: 996–997

Kumar PKR, Lonsane BK (1987a) Gibberellic acid by SSF: consistent and improved yields. *Biotechnol Bioeng* **30**: 267–271

Kumar PKR, Lonsane BK (1987b) Extraction of gibberellic acid from dry mouldy bran produced under solid-state fermentation. *Process Biochem* **22**: 139–143

Kumar PKR, Lonsane BK (1987c) Potential of fed-batch culture in solid-state fermentation for the production of gibberellic acid. *Biotechnol Lett* **9**: 179–182

Kumar PKR, Lonsane BK (1988)]kumar1988 Kumar PKR, Lonsane BK (1988) Batch and fed-batch solid-state fermentations: kinetics of cell growth, hydrolytic enzymes production, and gibberellic acid production. *Process Biochem* **23**(2): 43–47

Lin CF, Lizuka H (1982) Production of pigment by a mutant of *Monascus kaoliang* sp. nov. *Appl Environ Microbiol* **43**: 671–676

Lindenfelser LA, Ciegler A (1975) Solid state fermenter for ochratoxin A production. *Appl Microbiol* **29**: 323–327

Luk KC, Kobbe B, Townsend JM (1977) Production of cyclopiazonic acid by *Aspergillus flavus. Appl Environ Microbiol* **33**: 211–212

Nigam P, Singh D (2000) Secondary Metabolites. In *Encyclopedea of Food Microbiology* (RK Robinson et al. eds.) Academic Publisher, London

Nigam P, Singh D (1999) Characteristics and techniques of fermentation systems in *Biotechnology: Food Fermentation*, Vol. II, (VK Joshi, A Pandey eds) Educational Publishers, N Delhi, pp. 427–466

Nigam P, Singh D (1996a) Processing of agricultural wastes in solid state fermentation for microbial protein production. *J Sci Ind Res* **55**: 373–380

Nigam P, Singh D (1996b) Processing of agricultural wastes in solid state fermentation for cellulolytic enzyme production. *J Sci Ind Res* **55**: 457–467

Nigam P, Singh D (1994) Solid-state (substrate) fermentation systems and their applications in biotechnology. *J Basic Microbiol* **34**: 405–423

Pandey A, Soccol CR, Nigam P, Soccol VT, Vandenberghe LPS, Mohan R (2000a) Biotechnological potential of agro-industrial residues: II cassava bagasse. *Bioresource Technol* **74**(1): 81–87
Pandey A, Soccol CR, Nigam P, Soccol VT (2000b) Biotechnological potential of agro-industrial residues: I sugarcane bagasse. *Bioresource Technol* **74**(1): 69–80
Pandey A, Soccol CR, Rodriguez-Leon JA and Nigam P (2001) Production of organic acids by solid state fermentation. In *Solid state fermentation in Biotechnology-Fundamentals and Applications*, Asitech Publishers N. Delhi, pp. 132–158
Patron NJ, Walker RF, Cozijnsen Aj et al. (2007) Origin and distribution of epipolythiodioxopiperazine (ETP) gene clusters in filamentous ascomycetes. *BMS Evol Biol Sep* **26**(7): 174–179
Robinson T, Nigam P, Singh D (2004) Secondary metabolites. In *Handbook of Fungal Biotechnology*, (DK Arora et al. eds.) Marcel Dekker Inc, NY pp. 267–274 ISBN 08247-4018-1
Robinson T, Singh D, Nigam P (2001) Solid-state fermentation: A promising microbial technology for secondary metabolite production. *Appl Microbio Biotechnol* **55**: 284–289
Shotwell OL, Hesseltine CW, Stubblefield RD, Sorenson WG (1966) *Appl Microbiol* **14**: 425–428
Silman RW, Conway HF, Anderson RA, Bagley EB (1979) Production of aflatoxin in corn by a large-scale solid-substrate fermentation process. *Biotechno Bioeng* **21**: 1799–1808

Chapter 8
Microbial Pigments

Sumathy Babitha

Contents

Abstract Recent increasing concern regarding the use of edible coloring agents has banned various synthetic coloring agents, which have a potential of carcinogenicity and terratogenicity. This circumstance has inevitably increased the demands for safe and naturally occurring natural (edible) coloring agents, one of which is pigment from the fungus *Monascus purpureus*. It has long been known that the microorganisms of the genus *Monascus* produce red pigments, which can be used for coloring the foods. *Monascus* pigments are a group of fungal secondary metabolites, called azaphilones, which have similar molecular structures as well as similar chemical properties. The pigments can easily react with the amino group containing the compounds in the medium to form water-soluble pigments. Due to the high cost of the currently used technology for the microbial pigment production on an industrial scale, there is a need for developing low cost process for the production of the

S. Babitha (✉)
Skin Bioactive Material Laboratory, Inha University, Yong-hyunong, Nam-gu, Incheon 402-751, Republic of Korea
e-mail: babisp2003@yahoo.com

P. Singh nee' Nigam, A. Pandey (eds.), *Biotechnology for Agro-Industrial Residues Utilisation*, DOI 10.1007/978-1-4020-9942-7_8,

pigments that could replace the synthetic ones. Utilizing cheaply available agro-industrial residues as substrate through solid-state fermentation can attain such an objective.

Keywords Pigments · *Monascus* · Azhaphilones · Polyketide · Solid-state fermentation

8.1 Pigments

The word pigment has a Latin origin and initially denoted a color (in the sense of colored matter), but it was later extended to indicate the colored objects such as makeup. In the beginning of the middle ages, the word was also used to describe the diverse plant and vegetable extracts, especially those used for the food coloring. The word pigment is still used in this sense in the biological terminology: the colored matter present in the animals or the plants, occurring in the granules inside the cells or cell membranes as the deposits on the tissues, or suspended in the body fluids (Ullmann 1985).

The modern meaning associated to the word pigment has its origin in the twentieth century, meaning a substance constituted of small particles which is practically insoluble in the applied medium, and is used due to its colorant, protective or magnetic properties (Ullmann 1985). This definition applies well to the pigments of the mineral origin, such as titanium dioxide or carbon black, but for the soluble dyestuffs, usually the organic compounds, the expressions *dye, colorant* or simply *color* (as in the food colors) is more adequate. The terms pigment and color are usually applied for the food coloring matters, sometimes indistinctly (Timberlake and Henry 1986).

8.2 Food Technology and Color

Numerous sociological, technical and economic factors have influenced the food industry over the years in past and the food market has changed rapidly with a much larger proportion of the food being 'processed' before the sale and ready prepared to meet the needs of new consumers such as the working mothers, single parent families and the increasing number of older people in the world. The challenge to the food industry is to provide visually appealing foods that taste good and meet the consumer's demands on the quality and price.

The color production industry aims to meet the food and drink manufactures needs by providing a full range of the colors to suit all the applications, within the legislative constraints. There is, however, a constant ongoing development to improve the stability and handling properties of the colors using the formulation technology, new processing methods and to a much lesser extent (mainly restricted by the legislative controls), the development of totally new pigments.

8.3 Market Trends

The market of the natural products on an international level is in a phase of extraordinary expansion. Among the causes, the boom in the markets, projected beyond US$ 4 milliards in 2000 alone for US, and the emergence of new categories of the natural substances which are rapidly and fundamentally changing the idea of health and diseases (Downham and Collins 2000). There are no reliable published statistics on the size of the color market; however, on a global scale a reasonable estimate would be $940 m which can be segmented as in Fig. 8.1.

In terms of individual sector size, it is estimated as below:

- synthetic colors – $400 m;
- natural colors – $250 m
- nature identical colors – $189 m;
- Caramel colors – $100 m.

Consumer pressure, sociological changes, and technological advances leading to more advances in the food processing industry have increased the overall color market. The most significant growth has been in the naturally derived colors owing to the improvements in the stability as well as the food industries aim to meet the increasing consumer perception that 'natural is best'. Currently, the cost of the natural colors in most cases is higher than the synthetic colors of similar shades but this hurdle can be overcome by the mass production of the natural colors which would bring the cost down, thereby increasing the demand also.

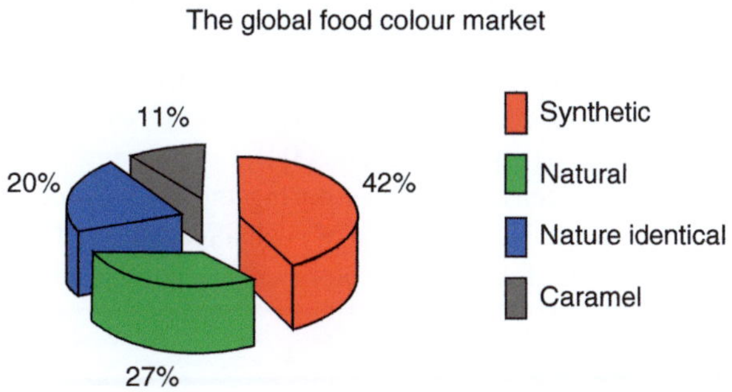

Fig. 8.1 Percentage market share of food colors

8.4 Microbial Pigments

An alternative route for the production of the natural food colorants is through the application of biotechnological tools employing the microorganisms. When the microbial cells are used to produce the color the term refers to 'Microbial pigments'. Inspite of the availability of the variety of the pigments from the fruits and vegetables, there is an ever growing interest in the microbial pigments due to several reasons such as their natural character and safety to use; production being independent of the seasons and geographical conditions, controllable and predictable yield and are not subject to vagaries of natures, for example, the production of cochineal (Francis 1987). Some provide nutrients like vitamins, and have medicinal properties also. The microbial pigment such as ß-carotene gradually shifts its color towards orange-red which is more attractive and this unique property is absent in the plant derived pigments (Pandey and Babitha 2005). Some of the microbial pigments can be produced from industrial residues (starch and juice industry), hence reducing water and environmental pollution (Babitha et al. 2004)

8.5 Pigment Producing Microorganisms

A large number of different species of the bacteria, molds, yeasts and algae can produce pigments but only a few are considered suitable for this purpose. It must satisfy several criteria such as the capability to use a wide range of carbon and nitrogen sources, should have tolerance to pH, temperature, mineral concentration and possess moderate growth conditions, reasonable color yield, should be non-toxic and non-pathogenic and must be easily separable from the cell mass. Table 8.1 gives a list of microbial sources and color shades of pigments produced by them.

Microalgae and several classes of fungi are known to produce a wide range of excreted water-soluble pigments, but the low productivity of algal cultures is a significant bottleneck for their commercialization (Hejazi and Wijffels 2004). Pigments of the basidiomycetous fungi have been used in the past for dying the wool and silk but such fungi are difficult to grow under industrial large-scale conditions. Hence, attention was drawn on the ascomycetous fungi. The use of such fungi to color foodstuffs is not a novel practice; the use of *Monascus* pigments in the food has been carried out traditionally in the Orient for hundreds of years (Teng and Feldheim 2001; Babitha et al. 2004). Considering the apparent heat and pH stability of the *Monascus* derivatives during the food processing, taken together with the socio-climatic independence of such a readily available raw material, fungi seem to be well-worth for further exploration as an alternative source of natural colorants.

8.6 Biodiversity of Filamentous Fungi: A Promising Source of Colorants

Like plants, the filamentous fungi synthesize the natural products because they have an ecological function and are of value to the producer (Firn and Jones 2003, Babitha et al. 2004). Depending on the type of the compound, they serve different functions—varying from a protective action against the lethal photo-oxidations (carotenoids) to protection against the environmental stress (melanins) and acting as cofactors in the enzyme catalysis (flavins). Besides providing the functional diversity to the host, these pigments exhibit a unique structural and chemical diversity with an extraordinary range of the colors. Several characteristic

Table 8.1 Some microbial sources and color shades of pigments produced by them

Microorganism	Pigment color shade
Bacteria	
Janthinobacteriumlividum	Bluish purple
Achromobacter	Creamy
Bacillus sp	Brown
Brevibacterium sp	Orange, Yellow
Corynebacterium michigannise	Greyish to creamish
Pseudomonas sp	Yellow
Rhodococcus maris	Bluish red
Streptomyces sp	Yellow, red, blue
Serratia sp	Red
Fungi	
Aspergillus sp	Orange, red
Blakeslea trispora	Cream
Monascus purpureus	Yellow, orange, red
Helminthosporium catenarium	Red
H. gramineum	Red
H. cynodontis	Bronze
H. avenae	Bronze
Penicilliumcyclopium	Orange
P. nalgeovnsis	Yellow
Yeast	
Rdodotorula sp	Red
Yarrowialipolytica	Brown
Cryptococcus sp	Red
Phaffia rhodozyma	Red
Algae	
Dunaliella salina	Red

non-carotenoid pigments are produced by the filamentous fungi, including quinones such as anthraquinones and naphthaquinones (Baker and Tatum 1998, Medenstev and Akimenko 1998), dihydroxy naphthalene melanin (a complex aggregate of polyketides) (Butler and Day 1998), and flavin compounds such as riboflavin. Anthraquinone (octaketide) pigments such as catenarin, chrysophanol, cynodontin, helminthosporin, tritisporin and erythroglaucin are produced by *Eurotium* spp., *Fusarium* spp., *Curvularia lunata* and *Drechslera* spp. (Duran et al. 2002). Yellow pigments epurpurins A to C were isolated from *Emericella purpurea* (Mapari et al. 2005) and azaphilone derivatives (hexaketides), falconensins A–H and falconensones A1 and B2, have been produced both by *Emericella falconensis* and *Emericella fructiculosa* (Ogasawara et al. 1997). Moreover, known for centuries, *Monascus* spp. produce azaphilone pigments like monascorubrin, rubropunctatin (Juzlova et al. 1996) and, more recently, monascusones from a yellow *Monascus* mutant have been identified (Jongrungruangchok et al. 2004). Another new natural food colorant of fungal origin has been patented by Sardaryan (2002), a red colorant which is an extracellular metabolite of the anthraquinone class produced by a variety of *Penicillium oxalicum*. It is also said to confer the anticancer effects when used in food supplements.

8.7 Microbial Pigments – Production and Market

The development of the products with an attractive appearance has always been an important goal in the food industry. Microbial pigments are advantageous, in terms of production, when compared to similar pigments extracted from the vegetables or animals. The development of superior vegetable or animal organisms is slower than that of microorganisms and algae; therefore, the production of the pigments by bioprocesses involving microorganisms, whose growth velocity is relatively high, is expected to give such a productivity for the processes that rends them industrially competitive. Furthermore, the isolation and development of new strains may provide new, different pigments. Currently, the pigments produced by the microorganisms and used commercially are riboflavin (vitamin B2, a yellow pigment permitted in most countries), by *Eremothecium ashbyii* and *Ashbya gossypi*; the pigments from *Monascus*, by *M. purpureus* and *M. ruber;* carotenoids (yellow pigments produced by several microorganisms, but to this moment produced commercially only from micro algae) such as β-carotene (by *Dunaliella salina* and *D. bardawil*) and astaxanthin (by *Haematococcus pluvialis*); and ficobiliproteins such as phycocianin (a blue pigment used in food and cosmetics), produced by *Spirulina* sp.; the pigments with potential use in the future could be indigoids, anthraquinones and naphtoquinones.

The market for the microbial pigments produced by bioprocesses is hard to estimate, due either for the lack of the statistics of the regional, low-technology products such as annatto extracts, or the fact that the production is pulverized over many small companies worldwide (Carvalho et al. 2003, Babitha et al. 2004). At one side, there is a growing preference for the natural additives in food and cosmetics; at the other side, in some cases, natural substances may be several times more expensive than synthetic analogs. A unique example is the β-carotene produced by the micro-algae, which has an approximate cost of U$1000/kg against U$ 500/kg by synthetic means; although more costly, β-carotene produced by the microbial means competes in a market segments where it is important that all the pigments be "natural"; besides, the microbial pigment is a mixture of cis- and trans- isomers, with therapeutical effects against the cancer that synthetic β-carotene, predominantly cis-, has not (Ruijter 1998).

The world market for the pigments of the natural sources (excluding nature-identical and caramels) was estimated in 1987, as U$ 35 million; in 2000, this market was around U$ 250 million (Downham and Collins 2000). Based on this growth tendency (600% in thirteen years, against 200% for the whole color market in the same period), currently the market for the natural pigments (which excludes nature identical and caramels) is probably on the order of U$ 350 to 600 million (Babitha et al. 2004). The biggest markets for food pigments are Europe and United States. The utilization distribution is not proportional to the food consumption (or to the population), because pigments are used in the processed foods: there is a potential demand for other countries, in which an improvement on the economic profile possibly will cause an improvement on the consumption of processed foods. In the specific case of *Monascus* pigments, the consumption of these pigments in Japan rose from 100 ton in 1981 to 600 ton in 1992, and was estimated as U$12 million, according to a study published in the same year (Lee et al. 1995, Hajjaj et al. 1997).

8.8 Role of Biotechnology

A lot of attention is now paid to the biotechnological synthesis of the colors through the microorganisms. Plant cell and tissue culture, microbial fermentation and gene manipulation have been investigated with respect to the production of pigment. However, extensive safety testing of such products is required before they are given clearance as safe food additives. Single cell algae and fungi appear better options for new biotechnologically derived colorants. The biotechnological production of the natural colors has two fundamental approaches; first is to find new sources of colors and then enhance their color production capacity. The other approach is to obtain enhanced and consistent yields from the already recognized good sources of the colorants either through the strain improvement or through optimizing the process parameters to maximize the yield. There is also the obstacle of research and development investment and manufacturing facilities. The appropriate use of the fermentation physiology together with the metabolic engineering (Nielsen and Olsson 2002) could allow the efficient mass production of the colorants from the fungi. With the advances in the gene technology, attempts have been made to create cell factories for the production of food colorants through the heterologous expression of biosynthetic pathways from either already known or novel pigment producers (Kim et al. 2003, Lakrod et al. 2003).

8.9 *Monascus* – A Potent Source of Food Colorant

Exactly when *Monascus* was discovered is very difficult to investigate, but Chinese have been using it for one thousand years (Su 1970). In the book, Tsu Shuei Ji, of Hsu Jian, Tang Dynasty, he quoted the poem of Wang Tsan of Han Dynasty, Chi Shih, describing the foods he was served when he stayed in Gua Chou. It was "White Steamed Rice blended with Red *Aspergillus* in one to one ratio, it was soft, rich and gliding and spread everywhere after stuffing into mouth". This was the earliest record of *Monascus* in China. In 1884, the French botanist Philippe van Tieghem isolated a purple mold on the potato and linseed cakes and named it *Monascus ruber*. This ascomycete was so named as it has only one polyspored ascus. Then Went in 1895 isolated one from the red mold rice obtained from the market in Java, Indonesia. This fungus was named *Monascus purpureus* (Fig. 8.2a). Then several others species were isolated around the world. *Monascus* is encountered often in the oriental foods, especially in Southern China, Japan and Southeastern Asia. Currently, more than 50 patents have been issued in Japan, United States, France and Germany, concerning the use of *Monascus* pigments for the food.

Historically, *Monascus* has been wildly used in preparing the traditional Chinese medicine and food. Anka (*Monascus*-grown rice) (Fig. 8.2b) is widely used in Japan

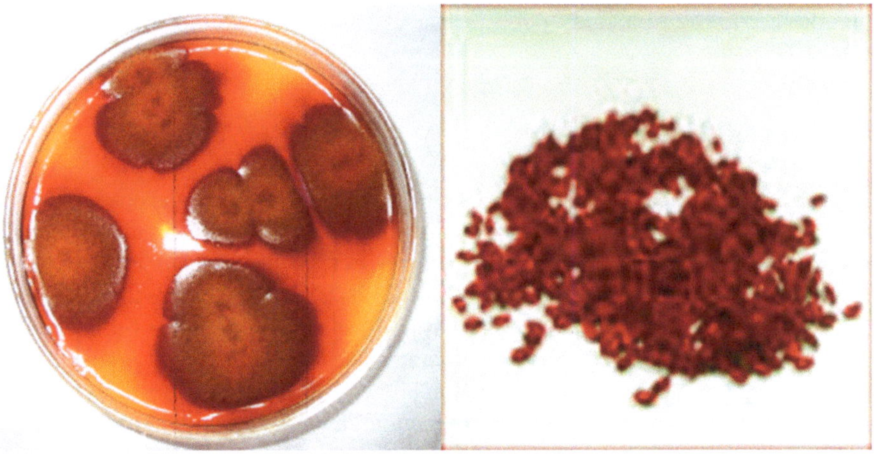

Fig. 8.2 (**a**) *Monascus purpureus* (**b**) Anka

for coloring the protein foodstuff. It has been widely used in China since ancient time up to present for manufacturing the roasted red pork, Chinese-style red sausage, and the like. *Monascus* also produces the enzyme that can decompose the starch, produce alcohol and decompose the protein. It the oriental countries, it also used to produce the red rice wine and red fermented bean curd and the like. At present, several industries produce the red, grinded rice as a natural food supplement capable of lowering the blood cholesterol, and as the dry product or purified extracts as the food colorants. In Europe, it is sold as a natural product. Its use in the rice wine manufacture is due to its high content of alpha-amylase, which promotes the conversion of the starch into glucose (the attractive red color of the rice wine is caused by *Monascus* pigments).

The scientific investigations have confirmed the pharmacological effects of *Monascus* fermentate. *Monascus* extract is marketed in Japan as a dietetic product. The preservative effect of *Monascu*s fermentate has also been confirmed by the scientific investigation. The observation of the bacteriostatic effects has lead to the consideration to use *Monascus* fermentate, at least partially, as a substitute for nitrite in the meat preservation (Fink-Gremmels et al. 1991, Fabre et al. 1993). Sodium nitrite is added to most packaged meat products for imparting the red color to meat products that would otherwise appear to be a putrid gray color is extremely carcinogenic. When combined with saliva and digestive enzymes, sodium nitrite creates cancer-causing compounds known as nitrosamines. The nitrosamines are highly toxic to biological systems (they are actually used to give lab rats cancer in laboratory tests). In the humans, the consumption of sodium nitrite has been strongly correlated with the brain tumors, leukemia, and cancers of the digestive tract (Pszczola 1998). A scientific proof of the flavor-enhancing properties of *Monascus* fermentate is difficult to obtain. However, in a tasting panel, tasters called *Monascus* containing noodles “more salty” than the normal noodles, although there was actually no difference in the salt content. It can be speculated that the relishing effect of *Monascus* could be caused by the flavor-enhancing oligopeptides produced by a partial hydrolysis of rice proteins by *Monascus* enzymes (Jacobson and Wasileski 1994). Lin et al. (2004) have reported a method that used *Monascus* to ferment the alcohol slowly to produce a beer-like and alcohol-free fermented beverage.

8.10 Monascus Pigments

The pigments are secondary metabolites of *Monascus* fermentation; they belong chemically to the group of Azaphilones, which are typical fungus metabolites. Depending on whether the yellow or red pigments predominate or are absent, the color of *M. purpureus* varies from orange-yellow to scarlet to purple-red. The color can be influenced by the culture conditions, in particular by the pH value (Fabre et al. 1993) and by the phosphorus and nitrogen source in the substrate (Wong et al. 1981). *Monascus* produces at least six major related pigments which could be categorized as (1) orange pigments: rubropunctatin ($C_{21}H_{22}0_5$) and monascorubin ($C_{23}H_{26}0_5$); (2) yellow pigments: monascin ($C_{21}H_{26}0_5$) and ankaflavin ($C_{23}H_{30}0_5$); and (3) red pigments: rubropunctamine ($C_{21}H_{23}N0_4$) and monoscorubramine ($C_{23}H_{27}$ $N0_4$) (Sweeny et al. 1981) (Fig. 8.3). The same colour exists in two molecular structures differing in the length of the aliphatic chain. These pigments are produced mainly in the cell- bound state.

Several authors have described the presence of the pigments different from the six classical azaphilones in the cultures of *Monascus*. Blanc et al. (1994) reported the presence of N-glutarylmonascorubramine and N-glutarylrubropunctamine by NMR in *M. purpureus* and *M rub*er grown in defined medium containing glutamate.

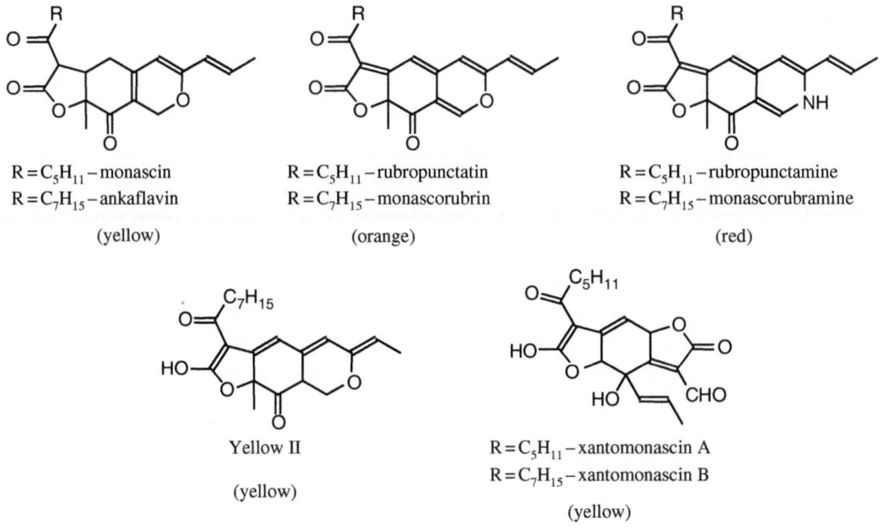

Fig. 8.3 Structure of *Monascus* pigments

These two compounds probably derive from the incorporation of a full glutamic acid molecule instead of the amino group occurring before the cyclization of ring c in the R1 and R2 normal compounds. The addition of a high glucose concentration to the culture induced the formation of glucosyl derivatives of the pigments, such as N-glucosylrubropunctamine and N-glucosylmonascorubramine (Hajjaj et al. 1997), confirming the influence of the media composition in the pigmentation of the cultures. Two yellow pigments, xanthomonascin and yellow-III, were isolated and characterized by Sato et al. (1992) and Yongsmith et al. (1993); these two pigments and the yellow ankalactone molecule described by Nozaki et al. (1991) are azaphilone-derived molecules.

From the previous works (Blanc et al. 1994, Hajjaj et al. 1997), a scheme of the hypothetical routes for the biosynthesis of these various pigments in the filamentous fungi is depicted in Fig. 8.4. The condensation of 1 mol of acetate with 5 mol of malonate leads to the formation of a hexaketide chromophore by the polyketide synthase. Then a medium-chain fatty acid such as octanoic acid, likely produced by the fatty acid biosynthetic pathway, is bound to the chromophore structure by a transesterification reaction to generate the orange pigment monascorubrin (or rubropunctatin upon trans-esterification with hexanoic acid). The reduction of the orange pigment gives rise to the yellow pigment ankaflavin from monascorubrin (or monascin from rubropunctatin), whereas red pigments (monascorubramine and rubropunctamine) are produced by the amination of orange pigments with NH_3 units (Lin et al. 1992). All these pigments remain essentially intracellular because of their high hydrophobicity. They are eventually excreted in the medium after reacting with an NH_2 unit of amino acids (Kyoko et al. 1997, Lin et al. 1992).

The biosynthetic reactions leading to each product are completely open, and it is unclear if they are formed from a common early intermediate by a branched pathway or not. The structure of these compounds indicates that at least one, and possibly two, polyketide synthases must be involved in their biosynthesis. Octanoic or hexanoic acid have been suggested to be precursors of the azaphilone compounds (Hajjaj et al. 2000). Possibly, these fatty acids are formed by a separate polyketide synthase as in aflatoxin biosynthesis (Brown et al. 1996).

The orange pigments, monascorubrin and rubropunctatin, are synthesized on the cytosol from acetyl coenzyme A. These pigments have a structure responsible by their high affinity to compounds containing primary amino groups (thus called aminophiles). Reactions with amino acids lead to formations of hydro soluble red pigments, monascorubramine and rubropunctamine. The mechanism of yellow pigment formations is not yet clear; some authors consider that these are product of the alteration of orange pigments, as others believe it to be pigments with their own metabolic pathway (Lin and Demain 1991, Juzlova et al. 1996).

Since the pigments generally show low water solubility, attempts have been made to make water-soluble pigments. The principle is the substitution of the replaceable oxygen in monascorubrine or rubropunctatine by nitrogen of the amino group of various compounds such as aminoacids, peptides and proteins, changing the color from orange to purple. *Monascus* pigments can be reduced, oxidized and react with other products, especially amino acids, to form various derivative products sometimes

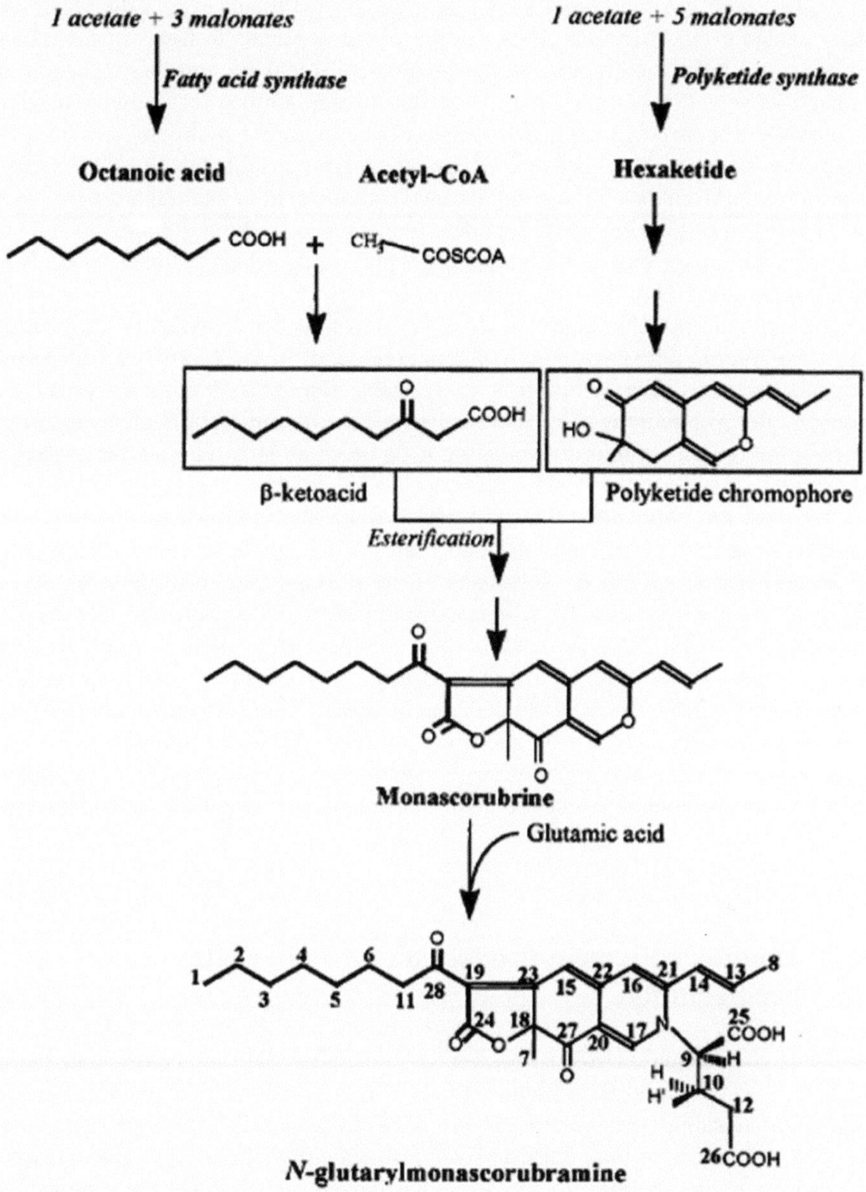

Fig. 8.4 Scheme of the hypothetic metabolic routes leading to the final structure of the water-soluble red pigment N-glutarylmonascorubramine in Monascus. Source: Hajjaj et al., 1997

called the complexed pigments. Glutamyl-monascorubrine and glutamylrubropunctatine were isolated from the broth of a submerged culture. Because of their affinity to amino groups, *Monascus* pigments are frequently associated to proteins (Wong and Koehler 1983), amino acids, nucleic acids to form water-soluble pigments or to the cell wall, forming a complex that may be of difficult extraction. Other authors consider that there may be a fixation of the pigments to lipids of the fungal biomass, so that the extraction would involve cell breaking and dissolution in an organic solvent (St. Martin 1990). Also due to this affinity for amino groups, it is possible to convert orange lipossoluble pigments in red hydro soluble ones by reaction with amino acids and analog compounds in vitro (St. Martin 1990). In that case, nitrogen from the amino group (from the amino acid or analog) takes the place of the oxygen of the ring on rubropunctatin or monascorubrin, yielding analogs of rubropunctamine or monascorubramine, but presenting a radical linked to the N in substitution to the H of the natural red pigments.

Amongst the possible amino acids to be used in order to induce the formation of hydro soluble pigments, *in vivo* is glutamic acid, in the form of monosodium glutamate (MSG) (Lin 1992). In a study using glucose as a carbon source in a liquid medium containing MSG, it was observed the formation of N-glucosil derivatives of the red pigments, corresponding to as much as 10% total pigments (Hajjaj et al. 1997).

Amongst the pigments produced by *Monascus*, the red ones are the most important, since they may be possible substitutes of the synthetic colors such as erythrosine (FD & C red No. 3) (Johns and Stuart 1991); these are stable pigments on the range of pH from 2 to 10, with good stability to temperature and that may be autoclaved (Lin 1992). Some studies show that *Monascus* pigments may be used as substitutes for the traditional food additives, such as nitrites and cochineal, in sausages and other meat products (Fabre et al. 1993). There are other studies on the sensorial response, allergenicity and toxicity, although eastern countries like Japan make extensive industrial use of these pigments since several decades – as examples, yellow hydro soluble pigments for candies, (Watanabe et al. 1997), or red pigments in red rice wine.

8.11 *Monascus* Pigment Production by Solid-State Fermentation

Solid-state fermentation (SSF), however, gives a higher yield and productivity of the pigment than the liquid fermentation. Lin (1973) has demonstrated that the pigment production in submerged culture was only 1/10 of that in solid-state culture. In order to reduce the costs of the industrial-scale fermentations, the use of agro-industrial byproducts as the sources for the microbial biosynthesis has been the subject of much research in recent years, which highlights the biotechnological potential of SSF (Pandey et al. 2000, Carvalho et al. 2007). One another advantage, which SSF offers is the application of fermented solids directly as a colorant without isolating the product. Fungi are the most adequate microorganisms for the SSF as solid substrate presents a more adequate habitat for the fungus (Pandey 1994). The solid-state fermentation of rice by *Monascus* has a long tradition in the East Asian countries, which dates back at least to the first century.

A comparison was made between the liquid and solid media of similar composition, the solid media obtained from the liquid by addition of a gelling agent, followed by the extrusion in rice-sized particles. The solid media thus prepared supported the production of up to three times more pigment than the corresponding liquid media, but the cultivations over rice were still superior (Johns and Stuart 1991). The mycelia of *Monascus* species penetrate into the surface of the solid nutritional medium and grow during the period of fermentation. The pigments produced are absorbed so that the color of the medium turns purple. In addition, the mycelia also penetrate inside the grain particles. This phenomenon facilitates the production of pigments.

8.12 Agro-Industrial Residues as Substrates for Monascus Pigment Production

The selection of substrate for the SSF process depends upon several factors, mainly related with cost and availability. In the SSF process, the solid substrate not only supplies the nutrients to the microbial culture growing in it, but also serves as an anchorage for the cells (Pandey et al. 2000). Traditionally, *Monascus* has been cultivated on the rice (forming ang-kak or red rice), although several other media, have been tested for the pigment production. Rice is the natural substrate which gives the best production, compared to other typical cereals, tubers and leguminous plants (Carvalho et al. 2003). However, some of the other substrates used also presented good biopigment production, especially corn, wheat and cassava. Cassava bagasse gave a low pigment yield, but being an agro-industrial residue whose low price might compensate for its low yield. Very recently jackfruit seed powder has been identified as a potent substrate for pigment production (Babitha et al. 2007).

8.13 Other Applications of *Monascus* Pigments

Tsuyoshi et al. (2004) has invented a decoloring ink for ink jet printing containing a *Monascus* pigment. Inclusion of *Monascus* pigment enables the printed characters and/or images to be preserved as long as they are kept in the dark, but they are quickly decolored on exposure to the visible light and/or ultraviolet light when they are no longer needed. Another landmark application of *Monascus* pigment has been patented by Yamamoto et al. (2006) which relates to a histostain composition for endoscope containing one or more members selected from colors derived from *Monascus*. The stain composition is a staining agent which sharpens the shapes of digestive tract lumen surfaces and the like with a light in the visible wavelength range, having a function of being excited by a light of specific wavelength to emit fluorescence, and being biologically safe and suitable for endoscopy. They unexpectedly found that colors derived from *Monascus* are characterized by being excellent in staining property under a visible light, having fluorescence whose wavelength being different from its excitation wavelength, being useful not only as a staining agent in usual endoscopy but also as a fluorescent dye for interstitial staining in confocal endoscopy, to give a vivid stained image useful in detection of a small affected region, and staining only the cytoplasm without staining cell nuclei, thus indicating that these colors have reduced cellular mutagenicity. Very recently, inhibitory activity of *Monascus* pigments against diet related lipase has been reported (Kim et al. 2007). The lipase produced by pancreatic acinar cells is considered to be essential for the digestion of dietary fats in the intestinal lumen. Therefore, if reactions using the pancreatic lipase in the human body are inhibited, fat absorption and obesity can be controlled.

References

Babitha S, Sandhya C, Pandey A (2004) Natural food colorants. Appl Bot Abstracts 23(4): 258–266

Babitha S, Soccol CR, Pandey A (2007) Solid-state fermentation for the production of *Monascus* pigments from jackfruit seed. Biores Technol 98(8): 1554–1560

Baker RA, Tatum JH (1998) Novel anthraquinones from stationary cultures of *Fusarium oxysporum*. J Ferment Bioeng 85: 359–361

Blanc PJ, Loret MO, Santerre AL, Pareilleux A, Prome D, Laussac JP, Goma G (1994) Pigments of *Monascus*. J Food Sci 59:862–865

Brown DW, Adams TH, Keller NP (1996) *Aspergillus* has distinct fatty acid synthases for primary and secondary metabolism. Proc Natl Acad Sci USA 93: 14873–14877

Butler MJ, Day AW (1998) Fungal melanins: A review. Can J Microbiol 44: 1115–1136

Carvalho JC, Soccol CR, Pandey A, Babitha S, Woiciechowski AL (2007) Production of pigments by SSF, Current Developments in Solid State Fermentation, In- Current Developments in Solid-state Fermentation, A Pandey, C R Soccol CR and C Larroche (Eds), Asiatech Publishers, Inc, New Delhi, pp 335–353

Carvalho JC, Pandey A, Babitha S Soccol CR (2003) Production of *Monascus* biopigments: an overview. Agro Food Ind Hi-tech 14: 37–42

Downham A, Collins P (2000) Coloring our foods in the last and next millennium. Int. J Food Sci Technol 35: 5–22.

Duran N, Tixeira MFS, Conti R, Esposito E (2002) Ecological-friendly pigments from fungi. Crit Rev Food Sci Nutr 42: 53–66

Fabre CE, Santerre AL, Loret MO, Baberian R, Pareilleux A, Goma G, Blank PJ (1993) Production and food application of the red pigments of *Monascus rubber*. J Food Sci 58(5): 1099–1102

Fink-Gremmels J, Dresel J, Leistner L (1991) Use of *Monascus* extracts as an alternative to nitrate in meat products. Fleischwirtsch 71: 1184–1186

Firn RD, Jones CG (2003) Natural products—a simple model to explain chemical diversity. Nat Prod Rep 20: 382–391

Francis F J (1987) Lesser known food colorants. Food Technol 41: 62–68.

Hajjaj H, Klaébé A, Goma G, Blanc PJ, Barbier E, Francois J (2000) Medium-chain fatty acids affect citrinin production in the filamentous fungus *Monascus ruber*. Appl Environ Microbiol 66: 1120–1125

Hajjaj H, Klaébé A, Loret MO, Tzedakis T, Goma G, Blanc PJ (1997) Production and identification of N-glucosylrubropunctamine and N-glucosylmonascorubramine from *Monascus rubber* and the occurrence of electron-donor-acceptor complexes in these red pigments. Appl Environ Microbiol 63: 2671–2678

Hejazi MA, Wijffels RH (2004) Milking of microalgae. Trends Biotechnol 22: 189–194

Jacobson G, Wasileski J (1994) In-Bioprocess Production of Flavor, Fragrance, and Color Ingredients, 1st edn, John Wiley & Sons: New York, pp. 205–237

Johns MR, Stuart DM (1991) Production of pigments by *Monascus purpureus* in solid culture. J Ind Microbiol 8: 23–38

Jongrungruangchok S, Kittakoop P, Yongsmith B, Bavovada R, Tanasupawat S, Lartpornmatulee N, Thebtaranonth Y (2004) Azaphilone pigments from a yellow mutant of the fungus *Monascus kaoliang*. Phytochemistry 65: 2569–2575

Juzlova P, Martinkova L, Kren V (1996) Secondary metabolites of the fungus *Monascus*: A review. J Ind Microbiol 16: 163–170

Kim SJ, Park DH, Ryu YW (2003) High-level production of astaxanthin by fed-batch culture of mutant strain *Phaffia rhodozyma* AJ-6-1. J. Microbiol. Biotechnol 13: 175–181

Kim JH, Kim HJ, Park HW, Youn SH, Choi DY, Shin CS (2007) Development of inhibitors against lipase and a-glucosidase from derivatives of *Monascus* pigment. FEMS Microbiol Lett 276: 93–98

Kyoko S, Yukihiro G, Shiho SS (1997) Identification of major pigments containing d-amino acid units in commercial monascus pigments. Chem Pharm Bull **45**: 227–229

Lakrod K, Chaisrisook C, Skinner DZ (2003) Expression of pigmentation genes following electroporation of albino *Monascus purpureus*. J. Ind. Microbiol 30: 369–374

Lee YK, Chen DC, Chauvatcharin S, Seki T, Yoshida T (1995) Production of *Monascus* pigments by a solid-liquid state culture method. Ferment. Bioeng 79(5): 516–518

Lin TF, Yakushijin K, Büchi GH, Demain AL (1992) Formation of water-soluble *Monascus* red pigments by biological and semi-synthetic processes. J Ind Microbiol 9: 173–179

Lin CF (1973) Isolation and cultural conditions of *Monascus* sp for the production of pigments in a submerged culture. J Ferment Technol 51: 407–14

Lin CF (1992) New application of *Monascus* species to health food. J Food Ind 24: 41–45.

Lin CF, Demain AL (1991) Effects of nutrition of *Monascus* sp on formation of red pigments. Appl Microbiol Biotechnol 36: 70–75

Lin TF, Chen YL, Wang TH (2004) Method for Producing Beer-Like and Alcohol-Free Fermented Beverage. US Patent 20050266120

Mapari SAS, Nielsen KFN, Larsen TO, Frisvad JC, Meyer AS, Thrane U (2005) Exploring fungal biodiversity for the production of water-soluble pigments as potential natural food colorants. Curr Opin Biotechnol 16(2): 231–238

Medenstev AG, Akimenko VK (1998) Naphthoquinone metabolites of the fungi. Phytochemistry 47: 935–959

Nielsen J, Olsson L (2002) An expanded role for microbial physiology in metabolic engineering and functional genomics: Moving towards systems biology. FEMS Yeast Res 2: 175–181

Nozaki H, Date S, Kondo H, Kiyohara H, Takaoda D, Tada T, Nakayama M (1991) Ankalactone, a new $\alpha\beta$-unsaturated γ-lactone from *Monascus anka*. Agric Biol Chem 55: 899–900

Ogasawara N, Mizuno R, Kawai K (1997) Structures of a new type of yellow pigments, falconensones A and B, from *Emericella falconensis*. J Chem Soc Perkins Trans 11: 2527–2530

Pandey A, Babitha S (2005) Microbial Pigments. Advance Biotech 3(8): 17–19

Pandey A (1994) Solid state fermentation—an overview, Solid State Fermentation, A Pandey (Ed), Wiley Eastern Limited, New Delhi, pp. 3–10

Pandey A, Soccol CR, Mitchell D (2000) New developments in solid state fermentation: I-bioprocess and products. Process Biochem 35: 1153–1169

Pszczola D (1998) The ABC's of nutraceutical ingredients. Food Technol 52: 30–37

Ruijter H (1998) New innovative b-carotene product. Food Marketing Technol 98: 16–18

Sardaryan E (2002) Strain of the microorganism *Penicillium oxalicum* var. *Armeniaca* and its application. US Patent 6,340,586 B1.

Sato K, Ywakami S, Goda Y, Okuyama E, Yoshihira K, Ichi T, Odake Y, Guchi H, Sankawa U (1992) Novel colorante from *Monascus anka* U-1. Heterocycles 34: 2057–2060

St. Martin, Edward J (1990) Production of Crystalline Pigments from *Monascus* During Fermentation. US Patent 4927760

Su YC, Chen WL, Fang HY, Wong HC, Wang WH (1970). Mycological study of *Monascus anka*. J Chin Agric Chem Soc. 8: 46–54

Sweeny JG, Estrada-Valdes MC, Iacobucci GA, Sato H, Sakamura S (1981) Photoprotection of the red pigments of *Monascus anka* in aqueous media by 1,4,6-trihydroxynaphalene. J Agric Food Chem 29(6): 1189–1193

Teng SS, Feldheim W (2001) Anka and anka pigment production. J Ind Microbiol Biotechnol 26: 280–282

Timberlake CF, Henry BS (1986) Plant pigments as natural food colors. Endeavour 10(1): 31–36

Tsuyoshi N, Fudou R, Yamanaka S, Furunaga T, Sato K, Kondo Y (2004) Decoloring ink for ink jet printing and ink jet printing method using it. European Patent EP1391486

Ullmann F (1985) Ullmann Encyclopedia of Industrial Chemistry 2nd ed, Edit.,VCH Weinheim, p. 245

Watanabe T, Yamomoto A, Nagai S, Terabe S (1997) Separation and determination of Monascus yellow pigments for food by micellar electrokinetic chromatography. Anal Sci 13: 571–575

Wong H, Koehler PE (1983) Production of water-soluble *Monascus* pigments. J Food Sci 48: 1200–1203

Wong HC, Lin YC, Koehler PE (1981) Regulation of growth and pigmentation of *Monascus purpureus* by carbon and nitrogen concentrations. Mycologia 73: 649–54

Yamamoto A, Iimori Y, Koyama M, Ishiguro M, Saze M (2006) Histostain Composition for Endoscope. US Patent 20070077202

Yongsmith B, Tabloka W, Yongmanitchai W, Bavavoda R (1993) Culture conditions for yellow pigments formation by *Monascus* sp. KB10 grown on cassava medium. World J Microbiol Biotechnol 9: 85–90.

Chapter 9
Production of Mushrooms Using Agro-Industrial Residues as Substrates

Antonios N. Philippoussis

Contents

Abstract Mushroom cultivation as a prominent biotechnological process for the valorization of agro-industrial residues generated as a result of agro-forestry and agro-industrial production. A huge amount of lignocellulosic agricultural crop residues and agro-industrial by-products are annually generated, rich in organic compounds that are worthy of being recovered and transformed. A number of these residues have been employed as feedstocks in solid state fermentation (SSF) processes using higher basidiomycetus fungi for the production of mushroom food, animal feed, enzymes and medicinal compounds. Likewise, the above-mentioned microorganisms have been successfully employed in processes related with the bioremediation of hazardous compounds and waste detoxification. Mushroom cultivation presents a worldwide expanded and economically important biotechnological industry that uses efficient solid-state-fermentation process of food protein recovery from lignocellulosic materials. Several aspects of mushroom physiology along with

A.N. Philippoussis (✉)
National Agricultural Research Foundation, I.A.A.C., Laboratory of Edible and Medicinal Fungi, 13561 Ag, Anargyri, Athens, Greece
e-mail: iamc@ath.forthnet.gr

P. Singh nee' Nigam, A. Pandey (eds.), *Biotechnology for Agro-Industrial Residues Utilisation*, DOI 10.1007/978-1-4020-9942-7_9,

impacts of different environmental and nutritional conditions on mycelium growth and fruiting bodies production are highlined. Moreover, cultivation technologies of *Agaricus bisporus*, *Pleurotus* spp and *Lentinula edodes*, comprising spawn (inoculum) production, substrate preparation and mushroom growing process i.e. inoculation, substrate colonization by the cultivated fungus, fruiting, harvesting and processing of the fruiting bodies, are outlined. Finally, the efficiency of residues conversion into fruiting bodies are outlined in two medicinal mushroom genera, *Pleurotus* and *Lentinula*, widely cultivated for their nutritional value and extensively researched for their biodegradation capabilities. Experimental data concerning residue-substrates used, as well as biological efficiencies obtained during their cultivation were considered and discussed.

Keywords Fungi · Mushroom cultivation · Biotechnology · Agricultural residues · By-products · Solid state fermentation · Fruiting bodies · Yield · Biological efficiency · *Agaricus* spp. · *Pleurotus* spp. · *Lentinula edodes*

9.1 Introduction

On the surface of our planet, around 200 billion tons per year of organic matter are produced through the photosynthetic process (Zhang 2008). However, the majority of this organic matter is not directly edible by humans and animals and, in many cases, becomes a source of environmental problem. Moreover, today's society, in which there is a great demand for appropriate nutritional standards, is characterized by rising costs and often decreasing availability of raw materials together with much concern about environmental pollution (Laufenberg et al. 2003). Consequently, there is a considerable emphasis on recovery, recycling and upgrading of wastes. This is particularly valid for the agro-food industry, which furnishes large volumes of solid wastes, residues and by-products, produced either in the primary agro-forestry sector or by secondary processing industries, posing serious and continuously increasing environmental pollution problems (Boucqué and Fiems 1988, Koopmans and Koppejan 1997, Lal 2005). It is worth mentioning that only crop residues production is estimated to be about 4 billion tons per year, 75% originating from cereals (Lal 2008).

Nevertheless, residues such us cereals straw, corn cobs, cotton stalks, various grasses and reed stems, maize and sorghum stover, vine prunings, sugarcane and tequila bagasse, coconut and banana residues, corn husks, coffee pulp and coffee husk, cottonseed and sunflower seed hulls, peanut shells, rice husks, sunflower seed hulls, waste paper, wood sawdust and chips, are some examples of residues and by-products that can be recovered and upgraded to higher value and useful products by chemical or biological processes (Wang 1999, Fan et al. 2000a, Pandey et al. 2000b, c, Webb et al. 2004). In fact, the chemical properties of such lignocellulosic agricultural residues make them a substrate of enormous biotechnological value. They can be converted by solid state fermentation (SSF) into various different value-added products including mushrooms, animal feed enriched with microbial biomass, compost to be used as biofertilizer or biopesticide, enzymes, organic acids, ethanol, flavours, biologically active secondary metabolites and also for bioremediation of hazardous compounds, biological detoxification of agro-industrial residues, biopulping etc. (Pandey et al. 2000a, Bennet et al. 2002, Sánchez et al. 2002, Tengerdy and Szakacs 2003, Howard et al. 2003, Kim and Dale 2004, Nigam et al. 2004, Zervakis et al. 2005, Manpreet et al. 2005, Krishna 2005).

Among applications of SSF, mushroom cultivation has proved its economic strength and ecological importance for efficient utilization, value-addition and biotransformation of agro-industrial residues (Chang 1999, 2001, 2006, Chiu et al. 2000, Zervakis and Philippoussis 2000). Current literature shows that lignocellulose degrading mushroom species are used in various SSF applications such as bioremediation and biodegradation of hazardous compounds (Pérez et al. 2007), biological detoxification of toxic agro-industrial residues (Pandey et al. 2000d, Fan et al. 2000b, Soccol and Vandenberghe 2003), biotransformation of agro-industrial residues to mushroom food and animal feed (Moore and Chiu 2001, Alborés et al. 2006, Okano et al. 2006), compost and product developments such as biologically active metabolites, enzymes, and food flavour compounds (Ooi and Liu 2000, Cohen et al. 2002, Silva et al. 2007, Nikitina et al. 2007). Moreover, recent research work indicates medicinal attributes in several species, such as antiviral, antibacterial, antiparasitic, antitumor, antihypertension, antiatherosclerosis, hepatoprotective, antidiabetic, anti-inflammatory, and immune modulating effects (Wasser and Weis 1999, Wasser 2002, Daba and Ezeronye 2003, Paterson 2006).

Commercial mushroom production, carried out in a large or small scale, is an efficient and relatively short biological process of food protein recovery from negative-value lignocellulosic materials, utilizing the degrading capabilities of mushroom fungi (Martínez-Carrera et al. 2000, Chiu and Moore 2001). Among mushroom fungi, *L. edodes* and *Pleurotus* species reveal high efficiency in degradation of a wide range of lignocellulosic residues, such as wheat straw, cotton wastes, coffee pulp, corn cobs, sunflower seed hulls wood chips and sawdust, peanut shells, vine prunings and others into mushroom protein, (Ragunathan et al. 1996, Campbell and Racjan 1999, Stamets 2000, Poppe, 2000, Philippoussis et al., 2000, 2001a, b), the productivity of the conversion being expressed by biological efficiency (Chang et al. 1981). Their mycelium can produce significant quantities of a plethora of enzymes, which can degrade lignocellulosic residues and use them as nutrients for their growth and fructification (Bushwell et al. 1996, Elisashvili et al. 2008). However, the nature and the nutrient composition of the substrate affect mycelium growth, mushroom quality and crop yield of this value-added biotransformation process (Kües and Liu 2000, Philippoussis et al. 2001c, 2003, Baldrian and Valášková 2008).

The focus of this work is to highlight significant aspects of utilization of low- or negative-value agro-industrial residues in mushroom biotechnology, emphasizing on their biotransformation to fruiting bodies that are nutritious human foodstuff regarded also as functional food. Aspects to be reviewed in this article include: an overview of availability, sources and types as well as chemical composition of solid lignocellulosic agro-residues suitable for mushroom cultivation, some background on mushroom degrading abilities and of their nutritional and environmental demands, an outline of commercial production technologies of *A. bisporus*, *Pleurotus* spp. and *L. edodes* mushrooms, and finally a consideration and discussion of experimental data regarding productivity (biological efficiency) on various agro-industrial residues during cultivation of*Pleurotus* spp. and *L. edodes*.

9.2 Residue-Based Substrates and their Solid-State Fermentation by Mushroom Fungi

As a result of agro-forestry and agro-industrial production, a huge amount of livestock waste, agricultural crop residues and agro-industrial by-products are annually generated, the major part being lignocellulosic biomass (Kuhad et al. 1997). Although agro-industrial residues contain beneficial materials, their apparent value is smaller than the cost of collection, transportation and processing for beneficial use. However, if residues are utilized, such as to enhance food production, they are not considered as wastes but new resources. A number of agro-industrial residues have been employed as feedstocks in SSF processes, using high basidiomycetus fungi for the production of valuable metabolites (Rajarathnam et al. 1998, Howard et al. 2003, Hölker et al. 2004). However, mushroom production is one of the areas with great potential for exploitation of forest and agricultural residues (Moore and Chiu 2001, Chang 2006, Gregori et al. 2007, Silva et al. 2007).

9.2.1 Types, Availability and Chemical Composition of Raw Materials

Reddy and Yang (2005) and very recently, Zhang (2008), reviewing the global world information about lignocellulose availability, estimated the production of lignocellulosic biomass to be more than 200×10^9 tons per year. Especially, the amount of crop residues produced annually in the world from 27 food crops is estimated at about 4×10^9 tons, from which 3 billion tons account per annum for lignocellulosic residues of cereals (Lal 2005, 2008). Cereals, accounting 75% of global world production (FAO 2004), furnish these outstanding amounts of waste products as wheat residues, rice straw and hulls, barley residue, maize stalks and leaves, millet and sorghum stalks. Sugar cane provides the next sizeable residue with two major crop wastes, leaves and stalk, and bagasse, which is the crop processing residue. The cotton crop also provides significant residue in the form of stalks and husks, while no negligible are the residues furnished by minor crops as sunflower, oil palm, coconut, banana, vines, groundnut and coffee. In fact, this generation of residues is a result of the limited portions of the crops that are actually used. To give the order of magnitude, 95% of the total biomass produced in palm and coconut oil plantations is discarded as a waste material; the respective values for sisal plant and sugar cane biomass are 98% and 83% (Chang 1998). Moreover, in the flax industry only 2% of the produced biomass is effectively used, less than 9% in the palm oil industry and only 8% in the brewing industry (López et al. 2004).

Biomass availability is a primary factor for bio-based industrial production. Indeed, the available resource potential (the amount of residues used for various purposes) is smaller than the one generated. The quantities of crop residues that can be available for bioprocesses are estimated using total grain production, residue to product ratio (RPR), moisture content, and taking into consideration the amount of residue left on the field to maintain soil quality (i.e. maintain organic matter and prevent erosion), grazing and other agricultural activities (Koopmans and Koppejan 1997, Giljum et al. 2005). Concerning cereal straw, the RPRs for rice, barley, wheat and corn are 1.4, 1.2, 1.3 and 1 respectively (Mulkey et al. 2008). Assuming that one quarter of the residues can be harvested and that roughly one third of the harvested straw is used in animal husbandry, 0.22 tons straw per ton cereal grain and 0.25 ton residues per ton maize are available biomass for other uses, as energy, enzyme production or mushroom growing.

As agro-industrial residues accumulate in fields and factories, availability issue tends to become a regional and local matter. Geographical distribution of crop residues is skewed by large crop productions in India and China, where increased quantities of crop residues and agro-industrial by-products are generated because of expanding agricultural production. Furthermore, Asia along with Europe, North America and Australia are world leader mushroom producing regions (Chang 2006) and consequently the major residue demanding for this bio-based industrial activity. Among countries in the Asian and Pacific Region, China produces the largest quantities of agricultural and forest residues, mainly by-products of rice, corn and wheat (Zhang 2008). China's quantities, estimated to reach about 1 billion tons/year (Qu et al. 2006), are followed by India's yielding at least 200 million tons/year of agricultural residues according to Das and Singh (2004), while according to Mande (2005) India's total amount of agro-industrial residues reaches 600 million tons. This quantity comprises 480 tons of crop residues (rice, wheat, millet, sorghum, pulses, oilseed crops, maize stalks and cobs, cotton stalks, sugarcane trash etc.) and 120 tons of processing–based residues (mainly groundnut shells, rice husk, sugarcane bagasse, cotton waste, coconut shell and coir pith). Rice and sugar are Asia's rest southeast countries dominant crops.

Moghtaderi et al. (2006) report that Australian agro-industrial biomass reaches 100 million tons/year, including bagasse, cane trash, wood residues, energy crops etc. As far as Africa is concerned, wheat and barley predominate in the north, millet and sorghum are the main crops in sub-Saharan Africa, while farther south maize is the dominant crop. Kim and Dale (2004), estimated Africa's annual lignocellulosic biomass from rice straw, wheat straw and sugar cane bagasse to be about 40 million tons, indicating that the fraction of most crop residues collectable is less than 30% because of low yields. In the same work, Central and South America's lignocellulosic residues were estimated to be about 140 million tons from rice and wheat straw, corn stover and sugar cane bagasse, not taking into account coffee, banana and other agricultural residues. Concerning North America, according to USDA-US DOE report (2005), USA is able to produce 1.3 billion tons of dry residues per year, including agricultural (933 million tons) and forest resources (368 million tons). Main lignocellulosic by-products in considerable quantity are corn stover, the most abundant agricultural residue in USA, wheat, rice, barley straw, sorghum stalks, coconut husks, sugarcane bagasse, pineapple and banana leaves. Canada, the second largest supplier of wood lignocellulosic biomass, supplies more than 200 million m^3 of lignocellulose annually through commercial operations (Mabee et al. 2005). Finally, Europe is not only a great wheat straw producer, but also outstanding quantities of lignocellulosic residues from barley, maize, sunflower, rapeseed, cotton, olive trees and vines, summarized as 120 million tons/ year (Nikolaou et al. 2003).

Regarding the types of wastes, according to Mande (2005), agricultural residues can be divided into two groups: crop-based residues (generated in the field) and processing-based residues (generated during wood and industrial processing). Crop-based residues, which are plant materials left behind in the field or farm after removal of the main crop produce, are consisted of different sizes, shapes, forms, and densities like straw, stalks sticks, leaves, haulms, fibrous materials, roots, branches, and twigs. Crop-based residues are produced from various sources such as field and seed crops (including straw or stubble from barley, beans, oats, rice, rye, and wheat, stalks or stovers from corn, cotton, sorghum, grasses and reeds, soybeans and alfalfa), fruit, nut, vegetable or energy crops (brushes and orchard prunings, e.g. vine shoots or leaves that remain on the ground after harvesting), and livestock manure. Processing-based agro-industrial residues are by-products of the post-harvest processes of crops such as cleaning, threshing, linting, sieving, and crushing. They are in the form of husk, dust, stalks etc. Food processing wastes that come from plant materials are culls, rinds, seeds, pits, pulp, press cakes, marc, malts, hops and a variety of other by-products from mass food production processes. Some examples of these materials are coffee processing by-products, sugarcane bagasse, hulls and husks, wheat middlings, corncobs, seed meals etc. Moreover, this category comprises wood residues produced either from the primary processing or from secondary manufacturers (producing bark, chips, sawdust, coarse residues, and planer shavings). During the sawing of a log at a typical sawmill, approximately 50% of the initial log volume is converted into wood products and 50% is converted into wood residues (Alderman 1998).

In general, solid agro-industrial residues are heterogeneous water insoluble materials having a common feature, their basic macromolecular structure being cellulose, hemicellulose and lignin and to a lesser extend pectin, starch and other polysaccharides (Thomsen 2005). Cellulose, the most abundant renewable organic resource comprising about 45% of dry wood weight, is a linear homopolymer of glucose units linked with $\beta - 1,4$-glucosidic bonds (Baldrian and Valášková 2008). Hemicelluloses, heteropolysaccharides containing two to four different types of sugars, are divided in three major groups: xylans, mannans and galactans. They consist of short-branched chains of hexoses, e.g. mannose units in mannans and pentoses such as xylose units in xylans (Kuhad et al. 1997). After cellulose, lignin is the second most abundant renewable biopolymer in nature. Lignin, representing between 26 to 29% of lignocellulose, is strongly bounded to cellulose and hemicellulose, imparting rigidity and protecting the easily degradable cellulose from the hydrolase attack (Raimbault 1998). Lignin is an aromatic polyphenol macromolecule, 3-dimensional and amorphous (Pérez et al. 2002). As the proportions of these three structural components characterize residue biomass, their percentages in mushroom substrate ingredients, along with nitrogen content and carbon to nitrogen ratios, are shown in Table 9.1. Additionally, crop residues contain, on a dry weight basis, approximately 0.5–1.5% N, 0.15–0.2% P, 1% K, 1% Ca, 0.5% Mg, 0.2% S, 30 mg Kg^{-1} Mn, 100 mg Kg^{-1} Fe, 30 mg Kg^{-1} Zn, 5 mg Kg^{-1} Cu, 20 mg Kg^{-1} B and about 1 mg Kg^{-1} Mo (Mills and Jones 1996). However, these values differ with crop, plant part, season, soil moisture as well as other factors that affect plant growth.

Substrates used in mushroom cultivation include both field-based residues and processing based-residues (Table 9.1). However, as the nutrient composition of the substrate is one of the factors limiting colonization as well as quantitative and qualitative yield of cultivated mushrooms (Philippoussis et al. 2000, 2002), supplements containing sugars and starch (easily available carbohydrates) and fats (slower degraded and time-lasting nutrient sources) are added to the basal ingredient. Supplements are used to increase nutritional content, speed-up growth and increase mushroom yield, especially in the cultivation of the white-rot mushroom fungi *L. edodes* (Royse et al. 1990, Royse 1996) and *Pleurotus* spp. (Naraian et al. 2008). The various organic supplements used in mushroom cultivation comprise molasses, brewer's grain, grasses and waste paper, cotton and coffee wastes etc. However, soybeans and cereal grains or their milling by-products are the most commonly used supplements, as they are generated in considerable amounts and contain increased levels of protein, fats and easily metabolized carbohydrates: soybeans (carbohydrates 21.5%, N 6.3%), wheat bran (carbohydrates 49.8%, N 2.4%), rice bran (carbohydrates 37.0%, N 2.0%) and millet (carbohydrates 57.3%, N 1.9%), (Przybylowicz and Donoghue 1990).

9.2.2 Nutritional and Environmental Aspects of Mushroom Growing

From about 14000 mushroom-forming fungal species, at least 2000 are edible, of which 80 species are grown experimentally and around 20 are cultivated commercially (Chang 1999, Silva et al. 2007). The most cultivated worldwide species are *A. bisporus*, *P. ostreatus* and *L. edodes*, followed by *Auricularia auricula*, *Flammulina velutipes* and *Volvariella volvacea*. Other mushroom species produced successfully on various substrates include *Agrocybe aegerita*, *Ganoderma* spp., *Grifola frondosa*, *Hericium erinaceus*, *Hypsizygus marmoreus*, *Lepista nuda*, *Coprinus comatus*, *Pholiota nameko* and *Stropharia* spp. (Stamets 2000, Royse 2004). Although the mentioned mushroom species have the ability to degrade lignocellulosic residues in their original or composted form (Rajarathnam et al. 1998), they exhibit differences regarding production of enzymes necessary to degrade lignocellulosic substrates and thus different abilities to grow and fruit on residue-substrates (Bushwell et al. 1996, Chen et al. 2003, Baldrian and Valášková 2008).

Table 9.1 Chemical properties (based on dry matter) of agro-industrial residues used as substrate ingredients in mushroom cultivation

Residue-substrates	Cellulose	Hemicellulose	Lignin (%)	Cellulose/ lignin	Ash (%)	N (%)	C/N	References[b]
Field-based residues								
Corn (maize) stover/husk	36.4–40.0	25.0–29.0	13.0–21.0	2.1–2.3	3.6–7.0	0.6–0.9	55.8–77.3	1, 15, 17, 20
Grass residues	25.0–40.0	13.0–38.0	6.4–17.6	2.4–3.9	4.2–6.2	1.3–2.5	28.0–42.0	1, 8, 20, 22
Reed stems/ residues	34.4–42.6	28.4–30.6	17.1–19.7	1.7–2.0	4.3–4.9	0.3–0.5	150.0–170.0	12, 33
Rice straw	22.8–38.4	17.7–28.5	6.4–18.0	3.6–5.9	8.3–17.8	0.5–1.1	51.4–57.8	8, 14, 17, 29
Vine shoots	34.0–60.8	17.0–21.0	20–22.9	2.0–2.8	NA	NA	NA	7, 10, 31
Wheat straw	31.5–39.5	21.2–29.0	5.6–15.0	2.2–5.3	5.6–8.0	0.4–0.8	48.8–59.6	1, 20, 21, 22, 30, 31
Processing-based residues								
Brewers grains	16.0–18.0	26.4–30.4	27.5–28.1	0.6–0.8	4.6–5.0	4.1–4.5	11.6–12.2	5, 9, 16
Coconut husk/coir	21.0–36.0	12.0–22.7	41.0–48.0	0.6–1.3	2.7–10.2	0.4–1.1	77.6–124.2	1, 24, 25, 29
Coffee pulp/husk	23.0–29.1	15.1–17.1	13.0–26.0	0.89–0.94	4.5–6.3/1.0–6.0	1.4–1.9/0.9–1.0	53.5–59.4	2, 26, 27
Corncob	28.0–45.0	35.0–43.0	11.0–17.0	2.5–2.7	4.4–4.8	0.4–1.1	64.2–71.6	6, 8, 11, 21, 22, 31
Cotton wastes (gin trash /hull)	52.0–90.0	5.0–20.0	4.0–12.0	5.0–11.2	2.6–8.4	0.3–1.4	40.0–59.0	8, 21, 22, 23
Hazelnut husk	24.5–37.5	20.6–24.9	29.6–35.1	0.7–1.2	8.2–8.7	0.8–0.9	50.6–58.6	3, 19, 28
Paper (waste)	54.3–70.0	12.4–25.0	11.3–29.7	3.0–6.0	NA[a]	NA	NA	8, 15, 34
Rice husk	28.0–43.0	17.5–20.6	21.5–22.5	1.3–1.9	16.7–21.4	0.3–0.4	100.0–136.0	9, 13, 31
Sugarcane bagasse	26.6–40.0	19.0–30.0	19.0–23.3	1.4–2.2	1.5–5.0	0.2–0.8	120.0–190.0	8, 18, 24, 26, 31
Sunflower seed hull	31.3–42.7	24.0–25.2	23.2–28.7	1.1–1.8	3.0–3.3	0.6–0.9	60.0–72.4	4, 28
Wood chips/ sawdust (softwood)	37.7–49.5	10.7–25.0	26.1–29.5	1.4–1.7	0.4–0.5	0.1–0.1	310.0–520.0	20, 32, 34
Wood chips/ sawdust (hardwood)	42.9–45.1	22.0–33.0	24.0–26.0	1.7–2.0	0.2–0.3	0.1–0.2	150.0–450.0	6, 20, 21, 32

[a]NA: Data not available

[b]References: [1] USDA-US DOE 2005: http://www.eere.energy.gov/biomass/progs, [2] Brand et al. 2000, [3] Çöpür et al. 2007, [4] Curvetto et al. 2005, [5] Demeke 2007, [6] Gabriel 2004, [7] Gañan et al. 2006, [8] Howard et al. 2003, [9] Anonymous 2006: http://nkk.naro.affrc.go.jp/eng/topics/reseach/2006/9.pdf, [10] Jiménez and González 1991, [11] Laufenberg et al. 2003, [12] Lin 2005, [13] Liou et al. 1997, [14] Mata and Savoie 2004, [15] Mosier et al. 2005, [16] Mussato et al. 2008, [17] Obodai et al. 2003, [18] Ortega et al. 1992, [19] Özçelik and Pekşen 2007, [20] Palonen 2004, [21] Philippoussis et al. 2001a, [22] Poppe 2004, [23] Quian 2004, [24] Ragunathan et al. 1996, [25] Reddy and Yang 2005, [26] Salmones et al. 1999, [27] Salmones et al. 2005, [28] Saura-Calixto et al. 1983, [29] Shashirekha and Rajarathnam 2007, [30] Singh 2000, [31] Thomsen 2005, [32] Tisdale et al. 2006, [33] Ververis et al. 2004, [34] Ward et al. 2000.

In plant residues, cellulose and hemicellulose are the main sources of carbohydrates, often incrusted with lignin, which forms a physical seal around these two components. Lignocellulose is physically hard, dense and recalcitrant, the degradation of which is a complex process requiring a battery of hydrolytic or oxidative enzymes. Taking into consideration that the substrates are insoluble, degradation occurs extracellularly, by two types of extracellular enzymatic systems: the hydrolytic system, which produces hydrolases and is responsible for cellulose and hemicellulose degradation; and a unique oxidative lignolytic system, which depolymerizes lignin (Pérez et al. 2002, Baldrian 2005). The hydrolytic breakdown of cellulose by fungi is catalyzed by extracellular cellobiohydrolases, endoglucanases and β-glucosidases, which hydrolyze the long chains of cellulose, liberating cellobiose and finally glucose, while the major hemicellulose-degrading enzymes are endoxylanases and endomannanases (Tengerdy and Szakacs, 2003). Most of these enzymes have been detected in both wood-degrading mushroom fungi (WDF), like *P. ostreatus* and *L. edodes* (Elisashvili et al. 2008) and litter-decomposing mushroom fungi (LDF), such as *A. bisporus* or *V. volvacea* (Steffen et al. 2007). Due to its complicated structure, lignin is more difficult to break down than cellulose or hemicellulose. The main extracellular enzymes participating in lignin degradation are lignin peroxidase (LiP), manganese peroxidase (MnP) and laccase (Hatakka 1994), with MnP, prooving to be the most common lignin-modifying peroxidase produced by almost all wood-degrading basidiomycetes (Steffen et al. 2007). In addition, litter-decomposing basidiomycetes can degrade lignin e.g. *A. bisporus* produces at least two lignolytic enzymes, laccase and MnP, however, the overall lignin degradation rate by these fungi is lower compared to that of white-rot fungi (Lankinen et al. 2005, Steffen et al. 2007). Besides the lignocellulosic enzyme complex, lignocellulolytic fungi also produce other enzymes, such as pectinases, proteases, lipases and phytases on lignocellulosic substrates (Tengerdy and Szakacs, 2003).

Basidiomycetous fungi comprise diverse ecological groups, i.e. WDF (white rots, brown rots) and LDF, which may insure their nutrition in different ways. White-rot fungi are able of a simultaneous degradation of all wood components (cellulose, hemicellulose and lignin), while brown-rot fungi, a relatively small group of Basidiomycetes, degrade only cellulose and hemicellulose. Given that the majority of cultivated higher basidiomycetes is WDF, while few of them are LDF, emphasis is given here to the nutritional behaviour and degradation potentials of these two groups, represented by the most cultivated species *A. bisporus*, *Pleurotus* spp. and *L. edodes*.

White rot mushroom-forming fungi, comprising cultivated species like *Pleurotus* spp., *L. edodes*, *Ganoderma* spp. etc., are the most efficient degraders, due to their capability to synthesize relevant hydrolytic (cellulases and hemicellulases) and unique oxidative (lignolytic) extracellular enzymes. Their general strategy is to decompose the lignin in wood, so that they can gain access to the cellulose and hemicelluloses embedded in the lignin matrix (Hatakka 1994). However, laccase expression in fungi is influenced by culture conditions, such as nature and concentration of carbon and nitrogen sources, media composition, pH, temperature, presence of inducers and lignocellulosic materials, etc. (Revankar et al. 2007).

A wide variety of lignin degradation efficiency and selectivity abilities, enzyme patterns and substrates enhancing lignin degradation are reported from white-rot fungi (Hatakka 2001, Baldrian and Valášková 2008). An interesting category of white-rot fungi are selective degraders that degrade lignin rather than cellulose, like *Pleurotus* spp., which are used in a wide range of biotechnological applications (Cohen et al. 2002). Lignin degradation by these fungi is thought to occur during secondary metabolism and typically under nitrogen starvation (Hammel 1997). Non-composted, chopped and water-soaked straw is sufficient for the cultivation of *Pleurotus* spp., while *L. edodes* is cultivated on logs or in bags on moisturized sawdust supplemented with cereal bran (Philippoussis et al. 2000, 2004). Although not necessarily optimal, since they are low in readily accessible nutrients, these commercially used substrates satisfy the needs of the fungi for growth and fruiting, and most importantly, help to withstand microbial competitors (Kües and Liu 2000).

In basidiomycetous LDF, comprising cultivated mushroom species like *Agaricus* spp., *Agrocybe* spp., *Coprinus* spp., *Stropharia* spp. and *V. volvacea,* degradation involves a succession of biodegradative activities that precede attack by lignocellulose degraders. However, the ability to break down lignin and cellulose enables some of the LDF to function as typical "white-rot fungi" in soil (Hofrichter 2002). Well known mushroom forming LDF are *A. bisporus* and *V. volvacea*, both grown commercially on composted lignocellulose. As *A. bisporus* contains lignolytic enzymes, degrades both cellulose and lignin, the former more rapidly (Cai et al. 1999, Lankinen et al. 2005). Compost prepared from straw, horse or chicken manure, calcium sulphate (gypsum), water and some nutritional supplements is a cheap cultural substrate for *A. bisporus* and some other saprophytic basidiomycetes. Manure in the compost serves as N source, straw as C source. It must be pointed out that after the initial medium preparation stage, little control can be exerted over the composition of the solid substrate medium. In composted substrates this is particularly crucial since the nutrient composition of the initial medium ingredients has to allow both a successful composting process and good fungal colonization and fruiting (Wood and Smith 1987).

Mushrooms have a two-phase life cycle, the mycelium (vegetative or colonization phase) and the fruiting body (reproductive phase that bears the spores). The mycelium grows through the substrate, biodegrades its components and supports the formation of fruiting bodies. Mushroom growers call the switch from mycelial extension to the production of mushroom primordia "pinning", the successive development of primordia into mushrooms "fruiting". While growth of mycelium lasts for several days, weeks or months, production of fruiting bodies is short lived, and the phenomenon is called 'fructification'. However, both vegetative and reproductive phases are very much influenced by the physiological condition and nutritional state of the mycelium (Wood and Smith 1987).

Since the carbon sources utilized by basidiomycetes are usually of a lignocellulosic character, fungi during vegetative growth produce a wide range of enzymes to degrade the lignocellulosic substrates. Data obtained in various studies demonstrate that the type and composition of lignocellulosic substrate appear to determine the type and amount of enzyme produced by basidiomycetous fungi during vegetative growth (Baldrian 2005, Baldrian and Valášková 2008, Elisashvili et al. 2008).

Moreover, cellulose/lignin ratios of wheat straw and cotton waste substrates were positively correlated to mycelial growth rates and mushroom yields of *P. ostreatus* and *P. pulmonarius* and with the yield of *V. volvacea* (Philippoussis et al. 2001a). According to Kües and Liu (2000), considerable changes in enzyme activities occur during fruiting, indicating a connection to the regulation of fruiting body development. For example, in *A. bisporus* and *L. edodes*, laccase activities are highest just before fruiting body initiation and decline rapidly with primordia formation. Cellulase activities are highest when fruiting body develops (Ohga et al. 2001). Regarding the influence of nitrogen availability, recent studies revealed a positive correlation between the C/N ratio and *P. eryngii* mushroom yield (Philippoussis et al. 2000). They also demonstrated that mycelium growth rates of *Pleurotus* spp. and *L. edodes* were positively correlated to C/N ratio (Philippoussis et al. 2001a, 2003). Similar conclusion was drawn by Silva et al. (2005), indicating that *L. edodes* extension rate is related to bioavailability of nitrogen and is enhanced by supplementation with cereal bran. Moreover, both nature and concentration of nitrogen sources are factors regulating enzyme production by wood rotting basidiomycetes, e.g. in *L. edodes* cultivation on wheat straw, nitrogen supplementation represses MnP and enhances laccase activity (Kachlishvili et al. 2005). According to Kües and Liu (2000), for fruiting body induction it is of importance to keep a balance between C and N sources, e.g. in *A. bisporus* compost, the optimal C/N ratio for fruiting has been determined to lie between 80:1 and 10:1. In addition, substrate supplementation with protein-rich materials proved to enhance yield of *Agaricus*, *Pleurotus* and *Lentinula* strains (Rodriguez-Estrada and Royse 2007, Naraian et al. 2008).

Apart from nutrition, mycelial growth and fruiting of basidiomycetous fungi are also regulated by temperature, gaseous environment, water activity and in certain cases by light. During substrate colonization, the effect of environmental parameters plays an essential role on mycelium growth, and hence confers significantly to the success of the entire cultivation process. In addition, the duration of the substrate colonization phase is of direct economic importance, since media that are non-thoroughly impregnated with the hyphae, are sensitive to fungal and bacterial infections resulting in reduced yields (Philippoussis et al. 2001a, Diamantopoulou et al. 2006). Production of the vegetative mycelium usually occurs over a wide range of temperatures. Zervakis et al. (2001) examined the influence of temperature on mycelium linear growth of *P. ostreatus, P. eryngii, P. pulmonarius, A. aegerita, L. edodes, V. volvacea* and *A. auricula-judae*. Their temperature optima were found to be 35°C for *V. volvacea* strains, while *P. eryngii* grew faster at 25°C, *P. ostreatus* and *P. pulmonarius* at 30°C. Moreover, *A. aegerita* grew faster at 25°C or 30°C and *A. auricula-judae* at 20°C or 25°C depending on the nutrient medium used, and *L. edodes* at 20°C or 30°C depending on the strain examined. It is generally believed that basidiomycetes tolerate relatively high levels of salts for growth, but fruiting body development can be more sensitive. Likewise, mycelial growth is less affected by pH but fruiting body development of several species occurs best at neutral or slightly acidic pH values around 6–7 (Wood and Smith 1987) or, in *L. edodes*, at a pH 4.0 (Ohga 1999). On lignocellulosic substrates, *Pleurotus* and *Lentinula* species are growing with a linear rate (Philippoussis et al. 2001a, Diamantopoulou and Philippoussis 2001), which is influenced by substrate salinity and porosity

(Philippoussis et al. 2002). Measurements of electric conductivity through the entire colonization process of three residue-substrates by *L. edodes* strains revealed an increase of salinity values until mycelium colonized 60 to 75% of the substrate, and then it slightly declined or remained constant until the end of incubation, presenting the highest and lowest values in the wheat straw and oak sawdust media respectively. In addition, a negative correlation was established between final salt content of the substrates and mycelium extension rates. Furthermore, monitoring of CO_2 concentrations in pilot-scale cultivation of *L. edodes* on synthetic blocks, revealed higher respiration rates on oak sawdust and corncobs than on wheat straw, which are further correlated with substrate colonization rates (Philippoussis et al. 2003).

Following colonization of the substrate, fruiting is induced by environmental and/or cultural manipulation. The optimal environmental parameters for mycelial growth and the subsequent fruiting are usually very distinct. Depending on the species and the degree of investment in environmental control technology, temperature is normally manipulated by heating or cooling systems to maintain the optima for vegetative growth or fruiting. However, fruiting body development is often induced after drastically altering the environmental parameters, usually favoured by reducing the temperature by at least five °C compared to mycelium growth. In fact, fruiting is typically induced, after vegetative growth, e.g. in *A. bisporus* to 16–18°C (Kües and Liu 2000), in *P. ostreatus* to 15°C (Zadrazil et al. 2004), and in *L. edodes* to 10–16°C for the cold temperature strains and 16–21°C for the warm temperature strains (Chen et al. 2000). Other parameters of fruiting body initiation and maturation include CO_2 concentration, humidity, salinity and pH. High humidity (90–95%) is favorable for pinning and fruiting but the moisture content of the substrate might be even more critical. The optimal water content for wooden substrates is 35–60% and for other substrates 60–80%. The lower values reflect the oxygen demand of the fungi in the substratum, balanced against their requirement for water (Kües and Liu 2000). Carbon dioxide (CO_2) level is also critical for efficient mycelial growth, fruit body initiation and fruit body development. Higher CO_2 concentrations (e.g. 1% v/v in air) may stimulate mycelial growth and inhibit fruiting. Increased aeration is used to reduce CO_2 levels, which otherwise produces increased elongation of stipe growth and abnormality of cap development (Wood and Smith 1987). Light has been implicated in the fruiting of several mushroom genera e.g. *Lentinula* and especially *Pleurotus* species have an obligate requirement for light for fruiting induction. Brief exposure of the culture to daylight or suitable artificial light is sufficient. Usually, light positively influences hyphal aggregation and fruiting body maturation (Kües and Liu 2000). However, light is not needed for the fruiting of *A. bisporus* (Wood and Smith 1987).

9.2.3 Output and Stages of Mushroom Cultivation

Mushroom industry presents a worldwide expanded and economically important biotechnological application, which can be divided into three main categories: cultivated edible mushrooms, medicinal mushroom products and wild mushrooms, with an annual global market value in excess of $45 billion (Chang 2006). The global annual mushroom output (including production and wild mushroom collection) surpass nowadays 10 million metric tons, with China being the top world producer (about 8.000.000 tons), followed by Europe and USA (Desrumaux 2007, Huang 2007). Commercial mushroom production is an efficient solid state fermentation process of food protein recovery from lignocellulosic materials carried out on a large or small scale (Martínez-Carrera et al. 2000, Chiu et al. 2000). Taking into account the value and volume of the product, the number of people involved in the industry, or the geographical area over which the industry is practiced, mushroom cultivation is the greatest application of exploitation of filamentous fungi using SSF and the biggest (non-yeast) biotechnology industry in the world (Moore and Chiu 2001). The economic strength of mushroom cultivation derives from the successful use as feedstocks of a variety of low- or negative-value residues from agriculture, forestry or industry. These wastes are processed using relatively cheap microbial technology to produce human foodstuff, which could also be regarded as a functional food or as a source of drugs and pharmaceuticals (Wood and Smith 1987). Moreover, the effective exploitation of resources from agricultural solid wastes and by-products, rich in organic compounds that are worthy of being recovered and transformed, is a sound environmental protection strategy (Zervakis and Philippoussis 2000).

There are three major stages involved in mushroom cultivation: (1) inoculum (spawn) production, (2) substrate preparation, and (3) mushroom growing i.e. inoculation of the substrate with propagules of the fungus, growth of the fungal mycelium to colonise the substrate, followed by fruiting, harvesting and processing of the fruiting bodies (Wang 1999, Martínez-Carrera et al. 2000).

Inoculum (spawn) production. In order to achieve reliable and vigorous fungal growth and fruiting bodies production of good quality, inoculum fungal cultures are necessary. Inoculum is produced by inoculation of sterilized cereal grains (usually wheat, rye or millet) from high quality stock mycelial cultures (Stamets 2000, Mata and Savoie 2005b). Essential prerequisite is the selection and breeding work to acquire suitable biological material for commercial cultivation, which ensures good yield and quality. The various mushroom inocula are often the only microbiologically pure part of the whole technology (Wood 1989). Spawn-making is a rather complex task, not feasible for the common mushroom grower, and is produced by specialist companies (spawn-makers) using large scale bulk autoclaving, clean air and other microbiological sterile techniques for vegetative mycelia cultures onto cereal grains, wood chips and plugs or other materials. The colonized cereal grain/mycelium mixture is called spawn and is grown under axenic conditions in autoclavable polyethylene bags, ensuring gas exchange, or rarely in jars. Finally, after quality control to assure biological purity and vigor, spawn is distributed from the manufacturer to individual mushroom farms in the same aseptic containers used for spawn production (Wood and Smith 1987, Royse 2002).

Substrate preparation. Fermentation process involves cultivation on specific substrates imitating the natural way of life of mushroom fungi (Tengerdy and Szakacs 2003). Regarding the litter-decomposer *A. bisporus*, this has come to mean cultivation of a mushroom crop on composted plant litter (Moore and Chiu 2001). On the other hand, the white-rot mushroom fungi *Pleurotus* spp and *L. edodes* are cultivated on non-composted lignocellulosic substrates, using methodologies that exploit their ability to produce enzymes capable of degrading all wood components (Chen et al. 2000, Zadrazil et al. 2004). The first stage of mushroom production has to do with assembly and treatment of the substrate to prepare a growth medium. The substrates used for mushroom production, varying according to cultivated species, are prepared from waste agricultural or forest product materials using ingredients such as manures, cereal straws or other crop residues, sawdusts etc. (Wood and Smith 1987). In certain cases the substrate is be directly inoculated and require very little pre-treatment, e.g. *L. edodes* production using logs. In other cases, the substrate is microbiologically or physically pretreated. Microbiological pre-treatment normally comprises some form of controlled bulk composting process (Wood 1989). Physical pre-treatment could include steam treatment or sterilization by autoclaving. Substrates for fungal growth can be prepared as sterile materials, to produce an axenic growth medium, e.g. bottle cultures of *F. velutipes,* or be non-sterile, e.g. compost substrates to produce *A. bisporus.* One of the aims of substrate preparation is to introduce sufficient water into the substrate to ensure that the water activity of the final medium is optimal for fungal growth. The scale of substrate preparation varies according to the type of species to be cultivated and the size of the production unit. For *A. bisporus* production, many tons of straw are processed per day to produce compost. Thus, large-scale bulk handling machinery is used for this process. After the initial preparation stage, little control can be exerted over the composition of the solid substrate medium. In composted substrates, this is particularly crucial since the nutrient composition of the initial medium ingredients has to allow both a successful composting process and good fungal colonization and fruiting. Although nutrient balances and status, e.g. for carbon, nitrogen, pH and other components, can be measured on the initial ingredients, little can be done to regulate the quantity or feed rate of these once the production processes are under way (Wood and Smith 1987).

Mushroom growing. This stage deals with the two phases of mushrooms life cycle i.e. the mycelium (vegetative phase) and the fruiting body formation (reproductive phase). Following inoculation, the mycelium, grows through the substrate, biodegrades its ingredients and supports the formation of fruiting bodies. Mycelial growth and fruiting during this stage are regulated by temperature, gaseous environment, nutrient status, water activity and in certain cases by light e.g. *Pleurotus* spp. has an obligate requirement for light for fruiting induction, *Agaricus* spp. have no light requirement (Wood 1989, Zadrazil et al. 2004). The level of environment and cultural control used is determined by the type of production technology. In controlled environment growing system, temperature is manipulated by heating or cooling systems to maintain the optima for vegetative growth or fruiting. Carbon dioxide (CO_2) level and humidity are also controlled. Basidiomata production on the culture medium surface occurs as a series of cycles (flushes). Depending on the fate of the harvested product as fresh or preserved material, the fruit bodies are harvested either by hand or mechanically and processed accordingly. After harvesting, mushrooms are normally cooled down to retard fruiting body metabolism, packed and sent to the fresh market, or processed further through freezing, canning, drying etc., depending on marketing strategies (Martínez-Carrera et al. 2000).

9.3 Bioconversion of Solid Residue-Substrates Through Mushroom Cultivation

9.3.1 Commercial Mushroom Production Processes

In the suite, the principles of production of the litter decomposing*Agaricus* spp. and. of two wood-degrading mushroom species (*Pleurotus* spp. and *L. edodes*) are presented.

Agaricus bisporus cultivation. It is the most commonly cultivated mushroom worldwide, mainly cultivated in Europe, North America, China and Australia (Chang 1999). *A. bisporus*, belonging to the Agaricaceae family, is a litter-decomposing basidiomycete that in nature usually grows on grasslands and forests (Kirk et al. 2001). Western countries, focused on *Agaricus* for consumption, led to progress in cultivation technology including farm design, quality control in compost production, microprocessor control and records for growing, mechanical harvesting and processing. The Netherlands, practising high-technology cultivation systems, has the highest mushroom yield per unit area worldwide (Chiu et al. 2000). The substrate used for *A. bisporus* cultivation is a complex culture medium made from straw- and manure-based compost. Its preparation is a two-stage process in which the first stage includes composting of the raw material consisting of straw, horse or poultry manure and gypsum (Sánchez 2004). During composting that lasts about 3 weeks, the lignocellulose waste is modified by various bacteria and fungi to a better-digested form suitable for *A. bisporus*. In the second week stage, the compost is pasteurized before inoculation with *A. bisporus* spawn. The final mushroom compost is a selective growth medium for this organism. Natural drop in temperature and lack of free ammonia are signs that the composting process has been completed (Moore and Chiu 2001).

Cultivation begins with inoculation (spawning; the process that introduces the mushroom mycelium into the compost) and growth of the mushroom mycelia into the compost under high humidity and temperature 25°C. At complete colonization, after 2–3 weeks, a casing layer containing peat moss and limestone is spread on the top of the compost. 'Casing' is needed only by *Agaricus,* the procedure is not necessary when cultivating other species such as *Pleurotus* spp. and *L. edodes*. After allowing 7 to 9 days for the *Agaricus* mycelium to grow into the casing layer, a machine with rotating tines is run across the mushroom bed to mix the casing layer thoroughly. The above process is called 'ruffling' and it serves in breaking up the mycelial strands and encourages the mushroom mycelia to grow and colonize the surface of the casing layer. The mushroom mycelium grows into the casing layer in similar conditions to those of compost colonization, and when it reaches the upper surface of the casing layer the fruiting process starts comprising environmental manipulation. The growing room is ventilated to decrease the concentration of carbon dioxide (usually to $< 0.1\%$) and to help reduce the temperature to 16–18°C. The temperature, humidity and CO2 level are then adjusted to trigger fructification and to favor the development of mushrooms. The first pin initials begin to appear about 2 weeks after casing. One layer of compost produces 2–4 crops called flushes. In general, the production of *A. bisporus* is time-consuming due to the long composting stage (Sánchez 2004).

In modern mushroom growing process, a specialist compost producer may complete the outdoor stages of composting and the ready-to-use compost (spawn run or not) can be delivered in bulk to a mushroom farm. Therefore, the mushroom production industry comprises spawn makers, phase I, phase II and phase III compost suppliers. Phase III compost is completely colonized by the mushroom mycelia, which if placed in a suitable environment in a mushroom farm will produce the fruit body crop readily. For a commercial mushroom farmer, the use of phase I compost gives the most flexibility to optimize farm conditions for cultivation of any mushroom strain. Purchase of phase II compost enables a farmer to choose which mushroom strain to spawn. The use of phase III compost, though it is obviously more costly, it guarantees the production of a crop in a short time and requires the least investment in facilities (Moore and Chiu 2001). Moreover, there are different growing systems, while the process can be separated into specialized stages. In the shelf-bed growing system, shelving is usually made of metal and arranged to give four to six layers of 1.4 m wide fixed shelves in a cropping room with centre and peripheral access gangways. Special machinery for compost filling, emptying, spawning, casing and other cultivation operations is necessary. In the bag growing system, growing bags of about 25 kg are usually supplied to the farm already spawned and may be arranged on the floor of the cropping house or on tiered shelving. Each arrangement makes its own demands on techniques and equipment.

Pleurotus spp. cultivation. Pleurotus species (like *P. ostreatus*, *P. sajor-caju*, *P. pulmonarius*, *P. eryngii*, *P. cornucopiae*, *P. tuber-regium*, *P. citrinopileatus* and *P. flabellatus*) are commercially very important edible mushrooms, found all over the world. These mushrooms present several advantages related with rapid mycelial growth, high ability for saprophytic colonization, simple, inexpensive cultivation techniques and several kinds of species available for cultivation under different climatic conditions. The production of *Pleurotus* mushrooms is a sharp contrast with the technology used for *Agaricus* production. Both pasteurized and sterilized substrate of a wide range of residues can be used (Fig. 9.1) and no casing is required. The primary ingredients used for *Pleurotus* spp. production is chopped wheat straw (*Triticum aestivum* L.) or cottonseed hulls (*Gossypium hirsutum* L.) or mixtures of them. For production on wheat straw, the material is chopped from 2 to 6 cm, water is added and pH of the material is adjusted with limestone to about 7.5 or higher to provide selectivity against *Trichoderma* green mold (Royse 2004). The substrate is then pasteurized in tunnels with aerated steam at 60–70°C for 12 hr by passing the air-steam mixture through the substrate. After pasteurization is complete (a proximate two-day process), filtered air is passed through the substrate for cooling to 25°C. Then grain spawn (inoculum) is added at about 3–5% of the fresh weight and the substrate, packed into 15–20 kg plastic bags or blocks, is placed in a dark room at

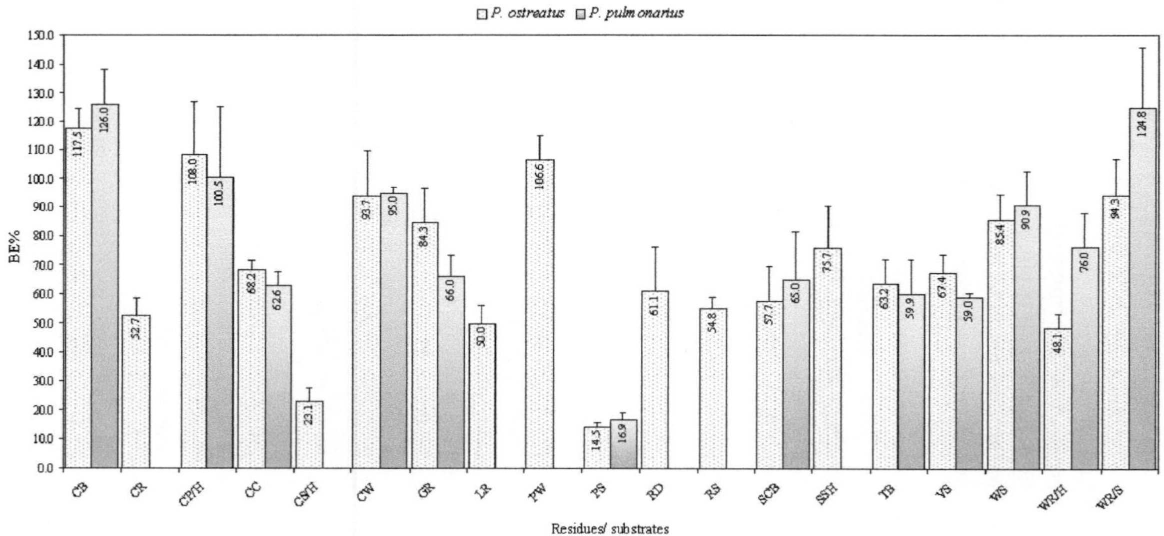

Residues	References
Cardboard (CB)	10
Coconut residues (CR/CH)	28
Coffee pulp/ husk (CP/H)	12, 13, 21, 25, 27
Corn cobs (CC)	1, 31
Corn stover/ husk (CS/H)	15
Cotton wastes (CW)	18
Grasses residues (GR)	1, 5, 7
Leaves residues (LR)	15, 30
Paperwaste (PW)	10
Peanut shells (PS)	18
Reed stems/ residues (RD)	6, 28
Rice straw (RS)	15, 29
Sugarcane bagasse (SCB)	16, 19
Sunflower seed hulls (SSH)	2, 4, 9
Tequila bagasse (TB)	9
Vine shoots/ prunings (VS)	22, 23
Wheat straw (WS)	16, 18, 20, 21, 24, 26
Wood residues/ hard (WR/H)	17, 24, 29
Wood residues/ soft (WR/S)	3, 11, 15, 23, 24, 28

Fig. 9.1 Biological efficiency (BE; % ratio of fruiting body fresh weight over the dry weight of substrate) of *P. ostreatus* and *P. pulmonarius* cultivated on residue-based substrates. References: [1] Coutiño et al. 2004, [2] Curvetto et al. 2002, [3] Croan 2003, [4] Darjania et al. 1997, [5] Das and Mukherjee 2007, [6] Gezer et al. 2007, [7] Hernández et al. 2003, [8] Kalmiş and Sargin 2004, [9] Lara et al. 2002, [10] Mandeel et al. 2005, [11] Marino et al. 2003, [12] Martínez-Carrera et al. 1985, [13] Martínez-Carrera et al. 2000, [14] Moda et al. 2005, [15] Obodai et al. 2003, [16] Pant et al. 2006, [17] Philippoussis et al. 2000, [18] Philippoussis et al. 2001a, [19] Ragunathan et al. 1996, [20] Royse and Schisler 1987, [21] Salmones et al. 2005, [22] Sánchez et al. 2002, [23] Serra and Kirby 1999, [24] Shah et al. 2004, [25] Soto et al. 1987, [26] Upadhyay et al. 2002, [27] Velázquez-Cedeño et al. 2002, [28] Vetayasuporn 2006, [29] Yildiz et al. 2002, [30] Zhang et al. 2002, [31] Zervakis and Balis 1992

25–30°C with 80 per cent humidity. Depending on the strain, complete colonization of the substrate is achieved in 2–3 weeks. In Asia, small containers are used, e.g. in Japan, bottle production of *Pleurotus* mushrooms is common. Substrate is filled into bottles, sterilized and inoculated with *Pleurotus* spawn. Upon completion of spawn run, bottle lids are removed and mushroom emerge from the surface of the substrate (Wood and Smith 1987, Royse 2003a, 2004).

Fructification of *P. ostreatus* is triggered by lowering the air temperature to 12–15°C (cold-shock treatment) although no such treatment is required for other *Pleurotus* spp. The fruiting is light-dependent, requiring a 8–12 hour light cycle (solar or fluorescent lamp light) and adequate ventilation is given to keep CO_2 levels lower than 500 ppm. Insufficient ventilation generally leads to mass primordial development with little differentiation into fruit bodies (Zadrazil et al. 1996). Three to four weeks after spawning depending on strain, amount of supplement used and temperature of spawn run, mushrooms begin to form around the edges of bag perforations and they are harvested from the substrate (Royse 2004). Throughout cropping, mushroom houses are kept at 12–17°C, the substrate usually being a few degrees higher. Under ideal growing conditions, 1 kg of well-colonized substrate will yield about 1 kg of marketable mushrooms (after two flushes) the completely growing cycle being completed in about 70 days. Due to the absence of a velum covering the gills of *Pleurotus* fruitbodies, spore discharge begins at a very early stage. Very large spore deposits within mushroom houses can cause allergy problems (Wood and Smith 1987).

Lentinula edodes cultivation. L. edodes (Berk.) Pegler is the second most popular edible mushroom in the world because of its flavour, taste, nutritional and medicinal properties. (Smith et al. 2002). This fungus can grow on synthetic logs as well as natural logs. The most traditional but laborious cultivation is carried out in wood logs, mainly oak. The wood logs are holed and the mycelia plugs are inserted in these holes. After inoculation the logs are stored several months for mycelium colonization and finally for the formation of fruit bodies. This method is still used because of its high quality mushroom product (Royse 2001, Silva et al. 2007), although leads to a severe threat to natural forests (Chiu et al. 2000). The last decades, new methods for *L. edodes* cultivation on residue-based substrates have been developed using milled wood residues (e.g. oak, hornbean, sweetgum, poplar, alder, ironwood, beech, willow, pine, maple and birch sawdust) supplemented with nitrogen sources (e.g. rice bran). The main advantages of using synthetic medium over natural logs are time and efficiency (Royse 2004). Some formulations used consist of 80% sawdust and 20% bran; 80% sawdust, 10% bran and 10% wheat or millet; and 84% sawdust, 5% rice bran, 5% wheat bran, 3% soybean and 3% lime (Kalberer 1987).

Other agricultural wastes that can be used as substrates (alone or in combination with other supplements) in *L. edodes* cultivation are cereal straw, corn cobs, sugarcane bagasse, tea waste, sunflower seed hulls, peanut shells, vineyard prunings, cotton straw and seed hulls etc. (Fig. 9.2; Curvetto et al. 2002; Philippoussis et al. 2003; Rossi et al. 2003, Gaitán-Hernández and Mata 2004, Mata and Savoie 2005a, b, Fan and Soccol 2005, Özçelik and Pekşen 2007, Royse

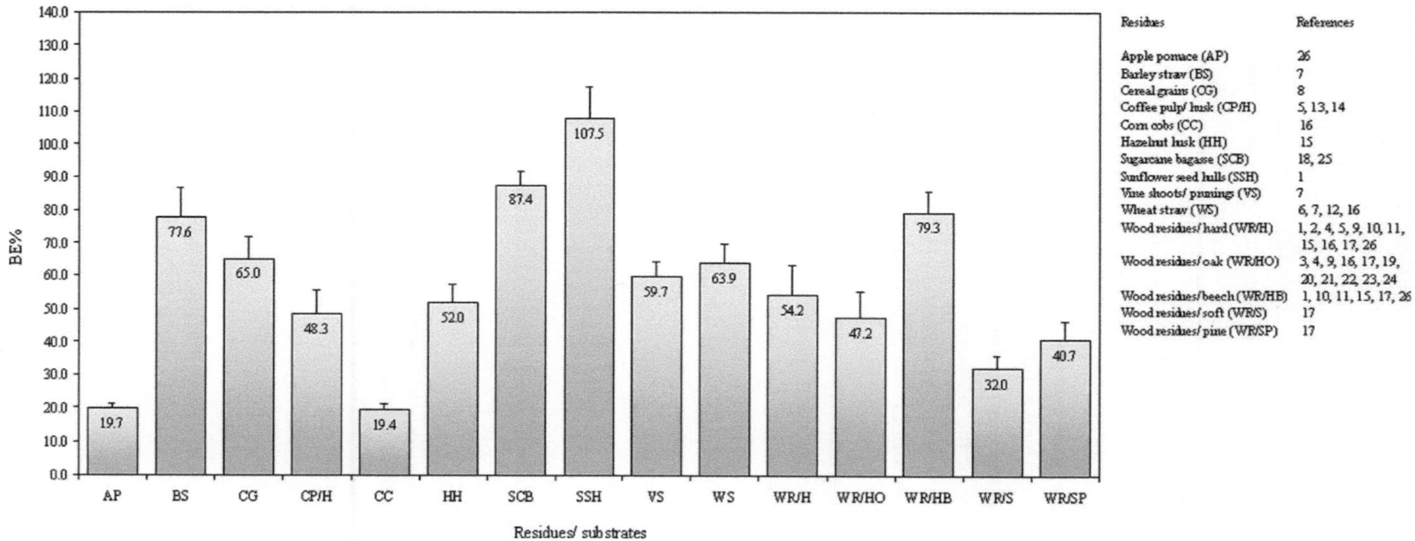

Fig. 9.2 Biological efficiency (BE; % ratio of fruiting body fresh weight over the dry weight of substrate) of *L. edodes* mushroom cultivated on residue-based substrates. References: [1] Curvetto et al. 2005, [2] Diehle and Royse 1986, [3] Donoghue and Denison 1995, [4] Donoghue and Denison 1996, [5] Fan and Soccol 2005, [6] Gaitán-Hernández and Mata 2004, [7] Gaitán-Hernández et al. 2006, [8] Hiromoto 1991, [9] Kalberer 2000, [10], Kawai et al. 1996, [11] Kirchhoff and Lelley 1991, [12] Levanon et al. 1993, [13] López et al. 2004, [14] Martínez-Carrera et al. 2000, [15] Özçelik and Pekşen 2007, [16] Philippoussis et al. 2003, [17] Pire et al. 2001, [18] Rossi et al. 2003, [19], Royse and Bahler 1986, [20] Royse and Sanchez 2007, [21] Royse and Sanchez-Vazquez 2001, [22] Royse and Sanchez-Vazquez 2001, [23] Royse et al. 1990, [24] Royse 2002, [25] Salmones et al. 1999, [26] Worrall and Yang 1992

and Sanchez 2007). The principle of the cultivation method comprises mixing and compacting ingredients into plastic bags, followed by sterilization, inoculation with fungal mycelia, incubation in dark rooms with controlled temperature and humidity for 30 to 80 days and finally fruiting induced by temperature reduction. Regardless of the main ingredients used, starch-based supplements such as wheat bran, rice bran, millet, rye, corn, etc. are added to the mixture in a 10 to 40% ratio to the main ingredient. These supplements serve as nutrients to provide an optimum growing medium (Royse et al. 1990, Royse 1996, 2003b). Substrate's ingredients are mixed, watered to gain a moisture content around 60% and filled into polypropylene bags 1–3 kg/ bag. The filled bags are stacked on racks, loaded into an industrial-sized autoclave, sterilized for 2 hours at 121°C, cooled and inoculated with spawn. After a 20 to 25 days spawn run, the bags are removed and the substrate blocks are exposed to an environment conducive for browning of the exterior log surfaces. As the browning process reaches completion (4 weeks), primordia begin to form about 2 mm under the surface of the bag-log indicating that it is ready to produce mushrooms (Royse 2004). Primordia maturation is stimulated by soaking the substrate in water (12°C) for 3 to 4 hours (or 3 to 4 min if vacuum soaking is used; Royse 2002). Soaking allows water rapidly to displace carbon dioxide contained in air spaces, providing enough moisture for one flush of mushrooms. Approximately 9 to 11 days after soaking, mushrooms are ready to harvest (Royse 2001). This method decreases the production time and increases productivity. While in the traditional cultivation the logs need 8 months to 1 year of cultivation to produce 10–15 kg/100 kg of substrate, the cultivation on agro-forestry residues can furnish a yield of 60–80 kg/100 kg of substrate in 80 days harvest period (Israilides and Philippoussis 2003, Royse 2004, Silva et al. 2007).

9.3.2 Efficiency of Residue Conversion to **Pleurotus sp.** *and* **L. edodes** *Fruiting Bodies*

Two particular basidiomycetus mushroom genera that have received considerable attention for their nutritional value, medicinal properties and biodegradation abilities are *Pleurotus* and *Lentinula* (Elisashvili et al. 2008). These widely cultivated edible mushrooms are efficient colonizers and bioconverters of lignocellulosic agro-industrial residues into palatable human food with medicinal properties (Zervakis and Philippoussis 2000, Philippoussis et al. 2004, Zadrazil et al. 2004, Silva et al. 2007, Gregori et al. 2007). The efficacy of this value-added bioconversion process and the productivity of the mushroom crop are assessed by the biological efficiency (Chang et al. 1981). Biological efficiency (BE) expresses the bioconversion of dry substrate to fresh fruiting bodies and indicates the fructification ability of the fungus utilizing the substrate (Fan et al. 2000a). BE is calculated as the percentage ratio of the fresh weight of harvested mushrooms over the weight of dry substrate at inoculation (Chang and Chiu 1992, Philippoussis et al. 2001b, Diamantopoulou et al. 2006). Yet, in a scarce number of reports, biological efficiency has been defined in terms of dry fruit bodies yield over the dry weight of the substrate used (Bisaria et al. 1987, Wang et al. 2001). Nevertheless, it should be made clear that apart from the type of substrate and stain used, yield response is determined by the duration of the cropping period and cultivation practice applied e.g. high spawn levels enhance mushroom yields (Obodai et al. 2003). Nevertheless, for considering the *Pleurotus* cultivation profitable, BE value must be over 50% (Patra and Pani 1995).

The genus *Pleurotus* comprises some the most popular edible mushrooms due to their favourable organoleptic and medicinal properties, fast mycelial growth and undemanding cultivation conditions. These mushrooms are commercially grown on pasteurized straw-based substrates or hardwood sawdust, fermented or not, with added supplements. However, as these fast-growing mushrooms display a complete lignocellulolytic enzyme system (Bushwell et al. 1996, Elisashvili et al. 2007), they can use a wide spectrum of agricultural and industrial wastes that contain lignin and cellulose for growth and fruiting (Poppe 2000). A significant number of agro-industrial lignocellulosic materials are used as substrates for the production of *Pleurotus* spp., like corn cobs, various grasses and leaves, reed stems, maize and sorghum stover, rice and wheat straw, vine shoots, cardboard and paper, wood sawdust and chips, coffee pulp, cottonseed hulls, peanut shells, sunflower seed hulls, sugarcane and tequila bagasse etc. Average experimental BE values varying from 14.5–126.0% are presented in Fig. 9.1. Further evaluation of the overall BE values obtained on these residue-substrates for *P. ostreatus* and *P. pulmonarius* strains indicated that among all residues cardboard, coffee pulp, paper wastes and softwood residues, presented the highest ($\geq$ 100%) biological efficiencies (Martínez-Carrera et al. 2000, Croan 2003, Mandeel et al. 2005). BEs between 75% and 100% were recorded on cotton wastes and wheat straw (Upadhyay et al. 2002, Philippoussis et al. 2001a). Regarding straw pre-treatment, data demonstrated an approximate 20% reduction of overall BE when *P. ostreatus* is cultivated on non fermented wheat straw, as compared to fermented substrate (mean values 70.5% and 85.5% respectively). On pretreated wheat straw, supplementation with cotton seed cake and soybean cake proved to enhance productivity of *P. ostreatus* (Upadhyay et al. 2002, Shah et al. 2004). Finally, satisfactory productivity (BEs 50–75%) is demonstrated by most of agro-industrial residues, namely corncobs, various grasses and reed stems, vine shoots, cottonseed hulls and sugarcane bagasse. From the considered data it can be assumed that *P. pulmonarius* furnished significantly better yields than *P. ostreatus* on cardboard and soft-wood residues (respective BEs: 126.0 and 124.3%). Moreover, yield of *P. pulmonarius* is favored on coffee and cotton wastes as well as on wheat straw (Pant et al. 2006, Zervakis and Balis 1992, Philippoussis et al. 2001a, Velázquez-Cedeño et al. 2002).

Our previous studies, concerning evaluation of a wide range of residues available in the Mediterranean region as *P. ostreatus*, *P. eryngii* and *P. pulmonarius* cultivation substrates, demonstrated significantly higher colonization rates of these mushrooms on wheat straw and cotton waste (Philippoussis et al. 2000). Moreover, faster colonization was achieved on non-composted than on composted wheat straw and cotton waste substrates. Cellulose/lignin ratios in substrates were positively correlated to mycelial growth rates and mushroom yields of *P. ostreatus* and *P. pulmonarius*. In addition, there was a positive correlation between the C/N ratio and *P. eryngii* mushroom yield (Philippoussis et al. 2001a). Additional data furnished by the 'race-tube' method experiments provided an estimate of the potential of wheat straw, cotton gin-trash, peanut shells, poplar sawdust and corn cobs to serve as alternative mushroom cultivation substrates (Zervakis et al. 2001). Furthermore, in recent studies conducted to evaluate different grass and reed stalks as cultivation substrates of *Pleurotus* species (Philippoussis et al. 2007, Diamantopoulou et al. 2007), bean plant residues (BRP) and a mixture of reed-grass maces (TCP) supported fast colonization rates for both genera strains, while BRP enhanced laccase and endoglucanase activities (Diamantopoulou et al. 2007). Moreover, fructification assay by the 'tube fruiting technique' provided a quick estimate of the potential of these wastes to support basidiomata formation. Basidiomata produced by both fungi, on all tested residues, with TCP supporting remarkably better fruiting results compared to wheat straw (Philippoussis et al. 2007).

Lentinula edodes (Berk.) Pegler is one of the best-known species among cultivated mushrooms, grown on natural or artificial logs, composed of either sawdust or of locally available agricultural wastes. *L. edodes* produces hydrolytic and oxidative enzymes responsible for lignocellulose degradation (Ohga and Royse 2001, Mata and Savoie 2005c). The production of enzymes is specifically related to and dependent on substrate composition and environmental factors such as temperature and moisture (Bushwell et al. 1996, Silva et al. 2005, Elisashvili et al. 2008). Since *L. edodes* is an efficient wood degrader, it can be grown on a variety of agro-industrial residues such as oak, ash, poplar, alder, eucalypt, beech, pine, maple and birch sawdust, cereal straws (mainly barley and wheat), corn cobs, sugarcane bagasse, sunflower seed hulls, peanut shells, cotton straw and seed hulls, vine shoots, coffee husk and pulp etc. Figure 9.2 presents comparatively the biological efficiencies obtained on these substrates during productivity evaluation experiments. Data indicate that the nature of the substrate affects remarkably *L. edodes* basidiomata yield. The highest average biological efficiencies were achieved with sunflower seed hulls (BE: 107.5%; Curvetto et al. 2005), followed by sugarcane bagasse (BE: 87.4%; Salmones et al. 1999). More or less similar interesting results (BE $\approx$ 80%) appeared to be obtained with hard-wood residues (beech) and barley straw (Kirchhoff and Lelley 1991, Kawai et al. 1996, Gaitán-Hernández et al. 2006). Among other substrates, progressively lower BEs, in the range of 65–50% and in descending order, were furnished by cereal grains, wheat straw, vine shoots, hard-wood residues of various trees and hazelnut husk (Hiromoto 1991, Philippoussis et al. 2003, Gaitán-Hernández et al. 2006, Özçelik and Pekşen 2007). Oak sawdust (comprising all types of oak), pine sawdust and coffee residues (all types) exhibited BEs in the range of 40–50% (Donoghue and Denison 1996, Royse and Sanchez-Vazquez 2001, Royse and Sanchez 2007), while the overall lower BE values (around 20%) were detected on apple pomace and corncobs (Worrall and Yang 1992, Philippoussis et al. 2003). Nevertheless, reliable estimations of the residues impact on *L. edodes* yield cannot be withdrawn from the presented average BE values, which only as indicative could be regarded. The main problem is that data from different substrate formulas, co-substrates and supplements, media treatments, strains used, cultivation periods, experimental conditions etc. are compared. It is well known that apart from substrate nature and composition, strain and length of incubation (to mention only a few factors), are important parameters for *L. edodes* production on artificial substrates (Royse and Bahler 1986, Zadrazil 1993, Kalberer 1995, Sabota 1996, Chen et al. 2000, Philippoussis et al. 2002, 2003). Comparing Figs. 9.1 and 9.2, BEs obtained on residue-substrates used for both *L. edodes* and *Pleurotus* spp. cultivation, the overall assumption is that wood chips and wheat straw, followed by sugarcane bagasse and vine prunings can support good basidiomata yield of *L. edodes*, as well as of *P. ostreatus* and *P. pulmonarius* strains. Coffee pulp and corncobs supported significantly higher yields for *Pleurotus* strains, while sunflower seed hulls favoured higher productivity for *L. edodes*.

Our previous studies, evaluating six commercial and wild *L. edodes* strains as regards their efficacy of mycelium growth on wheat straw, cotton wastes, oak-wood sawdust and corncobs, have demonstrated that oak-wood sawdust and wheat straw supported faster growth than corncobs and cotton wastes (Philippoussis et al. 2001c). In addition, a strain-dependent behaviour was detected since three strains performed much better on oak-wood sawdust and wheat straw, while one commercial strain performed satisfactorily on the other two substrates. In general, significantly lower linear growth rates were recorded for corncobs and cotton wastes (Philippoussis et al. 2003). Results were verified by the fruiting technique conducted in glass tubes, that furnishes a remarkable reduction in the time necessary for the first fructification ($\geq$ 2 months), conducing to a quick evaluation of the production potential of tested substrates. This, along with the use of growth rate measurement for valorization of the substrate incubation efficacy, renders the 'glass-tube' method a dependable technique for screening-selecting purposes. Additionally, measurements of mycelium respiration during the incubation phase in bag-log cultivation demonstrated very low respiration rates on cotton wastes, irrespective of the strain used (Philippoussis et al. 2002), while further experiments demonstrated the suitability of wheat straw and to a lesser extend supplemented corncobs for the cultivation of *L. edodes* (Philippoussis et al. 2003, 2004). Nevertheless, cotton wastes generated in large quantities in many countries, have proved to be a very good substrate for the cultivation of *Pleurotus* species (Philippoussis et al. 2001a, Cohen et al. 2002).

9.4 Closing Remarks

Current mushroom industry is based on both application of techniques for the production of mushroom fruiting bodies and the application of modern biotechnological techniques to produce medicinally beneficial compounds and nutraceutical products (Chang 2006, 2007). The medicinal properties of bioactive substances, like polysaccharides with antitumor and immunostimulating properties occurring in higher basidiomycetes have become a subject of numerous recent reviews (Mizuno 1999, Wasser and Weis 1999, Kidd 2000, Ooi and Liu 2000, Wasser 2002, Daba and Ezeronye 2003, Paterson 2006). Among them, the commercial polysaccharide of *L. edodes*, lentinan, has been researched extensively as it offers the most clinical evidence for antitumor activity (Wasser 2002, Nikitina et al. 2007). Recently, some mushroom polysaccharides have shown to exert a direct cytotoxic effect on cancer cells in vitro (Jiang et al. 2004, Wong et al. 2007, Israilides et al. 2008).

Moreover, solid-state fermentations other than fruiting body production are suggested for upgrading and valorizing lignocellulosic residues using basidiomycetous cultures, either through protein enhancement and transformation of residues into animal feed (Zadrazil et al. 1996, Zadrazil 2000), or for enzyme production (Revankar et al. 2007, Elisashvili et al. 2008). In the first case, agro-industrial residues such as rice straw, coffee pulp, sugarcane bagasse, banana leaves etc. have been fermented by white-rot basidiomycetes to improve the digestibility of the residues for use as ruminate feed supplement (Vega et al. 2005, Alborés et al. 2006, Okano et al. 2006). In the second case, lignocellulose degrading mushroom species like *Pleurotus* sp, *Lentinula edodes*, *Trametes versicolor*, *Flammulina velutipes* are used for the production of enzymes of industrial importance, such as cellulases, xylanases and laccases, using as substrates agro-industrial residues, from which wheat straw and bagasse are the most commonly used substrates (Krishna 2005, Silva et al. 2007). Most results, however, come from laboratory, or semi-pilot-scale experiments (Pandey et al. 2000a, Cohen et al. 2002). Additionally, lignocellulolytic mushroom fungi like*Pleurotus ostreatus* and *Trametes versicolor* have been investigated for bioremediation and biodegradation of toxic and hazardous compounds like caffeinated residues (Fan et al. 2000a, b) as well as toxic chemicals such as pesticides, polycyclic aromatic hydrocarbons (PAHs) and polychlorinated biphenyls (PCBs), chlorinated ethenes (CIUs) etc., in polluted soils or contaminated ground water (Pointing 2001, Pérez et al. 2007, Rigas et al. 2007).

In terms of food production process, the aim of mushroom growing should be to follow the holistic concept of production, according to Laufenberg et al. (2003). This approach tries to connect differing goals, such as highest product quality and safety, highest production efficiency and integration of environmental aspects into product development and food production. An outstanding example of integrated crop management practice of mushroom cultivation is the use of spent substrate, that is the residual growth medium after cropping (Rinker 2002) (1) as animal feed, since the mushroom mycelium boosts its protein content (Zhang et al. 1995), (2) as soil conditioner and fertilizer as it is still rich in nutrients and with polymeric components that enhance soil structure (Castro et al. 2008), (3) as a source of enzymes (Ko et al. 2005), (4) for the biological control of plant pathogens (Philippoussis et al. 2004, Davis et al. 2005) and even (5) used for bioremediation purposes as to digest pollutants on land-fill waste sites because it contains populations of microorganisms able to digest the natural phenolic components of lignin (Eggen 1999, Fermor et al. 2000). In this concept, solid state fermentation processes are not only the methods of mushroom production for food and nutraceutical purposes but also examples of an organic system integrated with waste treatment that contributes to sustainability and benefits the human population, health and environment.

Acknowledgments This work was partially financial supported by the General Secretariat of Research and Technology through the 05 PAV 105 project. The author wish to express his gratitude to Diamantopoulou Panagiota MSc and to Kontogiorgi Ioanna, members of Laboratory of Edible and Medicinal Fungi of NAGREF, for their great assistance in the preparation of the manuscript. In addition, Dr. Serapheim Papanikolaou is thanked for his comments and improvements made to the document.

References

Alborés S, Pianzzola MJ, Soubes M et al. (2006) Biodegradation of agr-oindustrial wastes by *Pleurotus spp* for its use as ruminant feed. Electron J Biotechnol 9(3): 215–220

Alderman DR Jr (1998) Assessing the availability of wood residues and residue markets in Virginia. Master thesis in: Wood Science and Forest products, Blacksburg, Virginia

Anonymous (2006) Listing of the properties of biomass materials for the selection of additives to adjust the components of compost produced from livestock manure. http://nkk.naro.affrc.go.jp/eng/topics/research/2006/9.pdf. Accessed 5 June 2008

Baldrian P (2005) Fungal laccases-occurrence and properties. FEMS Microb Rev 30: 215–242

Baldrian P, Valášková V (2008) Degradation of cellulose by basidiomycetous fungi. FEMS Microbiol Rev 32: 501–521

Bennet JW, Wunch KG, Faison BD (2002) Use of fungi Biodegradation. In: Hurst CC (ed) Environmental Microbiology. 2nd edition, ASM Press Washington DC

Bisaria R, Madan M, Bisaria VS (1987) Biological efficiency and nutritive value of *Pleurotus sajor-caju* cultivated on agro-wastes. Biol Wastes 19(4): 239–255

Boucqué CV, Fiems LO (1988) Vegetable by-products of agro-industrial origin. Livest Prod Sci 19: 97–135

Brand D, Pandey A, Roussos S et al. (2000) Biological detoxification of coffee husk by filamentous fungi using a solid state fermentation system. Enzyme Microb Technol 27(1–2): 127–13

Bushwell JA, Cai YJ, Chang ST (1996) Ligninolytic enzyme production and secretion in edible mushroom fungi. In: Royse DJ (ed) Mushroom Biology and Mushroom Products, Penn State University, Pensnylvania

Cai YJ, Chapman SJ, Buswell JA et al. (1999) Production and distribution of endoglucanase, cellobiohydrolase, and β-glucosidase components of the cellulolytic system of *Volvariella volvacea*, the edible straw mushroom. Appl Environ Microbiol 65(2): 553–559

Campbell AC, Racjan M (1999) The commercial exploitation of the white-rot fungus *Lentinula edodes* (Shiitake). Int Biodeteriorat Biodegrad 43: 101–107

Castro RIL, Delmastro S, Curvetto NR (2008) Spent Oyster Mushroom substrate in a mix with organic soil for plant pot cultivation. Micol Appl Int 20(1): 17–26

Chang ST (1998) A global strategy for mushroom cultivation – a challenge of a "non-green revolution". In: Lu M, Gao K, Si HF et al. (eds) Proceedings of the '98 Nanjing International Symposium, Science and Cultivation of Mushrooms, Nanjing

Chang ST (1999) World production of cultivated edible and medicinal mushrooms in 1997 with emphasis on *Lentinus edodes* (Berk.) Sing. in China. Int J Med Mush 1: 291–300

Chang ST (2001) A 40-year journey through bioconversion of lignocellulosic wastes to mushrooms and dietary supplements. Int J Med Mush 3: 80

Chang ST (2006) The World Mushroom Industry: Trends and technological development. Int J Med Mush 8(4): 297–314

Chang ST (2007) Products of Medicinal Mushrooms as a good source of dietary supplements for HIV/AIDS patiens. Int J Med Mush 9(3–4): 189–190

Chang ST, Chiu EW (1992) Mushroom production – An economic measure in maintenance of food security. In: DaSilva EJ, Ratledge C, Sasson A (eds) Microbial Technology: Economic and Social Aspects, Cambridge University Press, Cambridge

Chang ST, Lau OW, Cho KY (1981) The cultivation and nutritional value of *Pleurotus sajor caju*. Europ J Appl Microbiol 12: 58–62

Chen AW, Arrold N, Stamets P (2000) Shiitake cultivation systems. In: Van Griensven JLJD (ed) Science and Cultivation of Edible Fungi, Balkema, Rotterdam

Chen DM, Bastias BA, Taylor AFS et al. (2003) Identification of laccase-like genes in ectomycorrhizal basidiomycetes and transcriptional regulation by nitrogen in *Piloderma byssinum*. New Phytolog 157: 547–554

Chiu SW, Law SC, Ching ML (2000) Themes for mushroom exploitation in the 21st century: Sustainability, waste management, and conservation. J Gen Appl Microbiol 46: 269–282

Chiu SW, Moore D (2001) Threats to biodiversity caused by the traditional mushroom cultivation in China. In: Moore D, Nauta M, Rotheroe M (eds) Fungal Conservation: The 21st Century issue, Cambridge

Cohen R, Persky L, Hadar Y (2002) Biotechnological applications and potential of wood-degrading mushrooms of the genus *Pleurotus*. Appl Microbiol Biotechnol 58: 582–594

Çöpür Y, Güler, C, Akgül. M et al. (2007) Some chemical properties of hazelnut husk and its suitability for particleboard production. Build Environ 42: 2568–2572

Coutiño F, Jiménez L, Sánchez JE et al. (2004) *Digitaria decumbens* Grass Substrate prepared by Alkaline Immersion for culture of *Pleurotus spp*. In: Romaine CP, Keil CB, Rinker DL et al. (eds) Science and Cultivation of Edible and Medicinal Fungi. Penn State University, Pennsylvania

Croan SC (2003) Utilization of treated conifer wood chips by *Pleurotus* (Fr.) P. Karst. species for cultivating mushrooms. Mush Int 91: 4–7

Curvetto NR, Figlas D, Devalis et al. (2002) Growth and productivity of different Pleurotus ostreatus strains on sunflower seed hulls supplemented with N-NH4+ and/or Mn(II). Bioresour Technol 84(2): 171–176

Curvetto NR, Figlas D, Gonzalez Matute R et al. (2005) Sunflower seed hulls. In: Gush R (ed) Mushroom's Grower's Handbook 2, Mush World, Seoul

Daba AS, Ezeronye OU (2003) Anti-cancer effect of polysaccharides isolated from higher basidiomycetes mushrooms. African Journal of Biotechnology 2(12): 672–678

Darjania L, Curvetto N, Schapiro M et al. (1997) Sunflower seed hulls as a substrate for cultivation of *Pleurotus ostreatus*. Mush News 45(2): 6–10

Das N, Mukherjee M (2007) Cultivation of *Pleurotus ostreatus* on weed plants. Bioresour Technol 98: 2723–2726

Das N, Singh SK (2004) Useful byproducts from cellulosic wastes of agriculture and food industry – a critical appraisal. Crit Rev Food Sci Nutr 44: 77–89

Davis DD, Kuhns LJ, Harpster TL (2005) Use of Mushroom compost to suppress artillery fungi. J Environ Hort 23(4): 212–215

Demeke S (2007) Comparative nutritive value of *Atella* and industrial brewers grains in chicken starter ration in Ethiopia. Livest Res Rural Develop 19: http://www.cipav.org.co/lrrd/lrrd19/1/deme19008.htm

Desrumaux B (2007) European market 2006. Mush Bus 22: 4

Diamantopoulou P, Philippoussis A (2001) Production attributes of *Agaricus bisporus* white and off-white strains and the effect of $CaCl_2$ irrigation on productivity and quality. Sci Hortic 91 (3–4): 379–391

Diamantopoulou P, Philippoussis A, Kastanias MA et al. (2006) Effect of famoxadone, tebuconazole and trifloxystrobin on *Agaricus bisporus* productivity and quality. Sci Hortic 109: 190–195

Diamantopoulou P, Varzakas T, Papadopoulou K et al. (2007) Evaluating the growth and lignocellulolytic enzymes activity of *Ganoderma* species during Solid State Fermentation (SSF) of agricultural and littoral residues. In: Book of Abstracts II International Conference on Environmental, Industrial and Applied Microbiology (Bio Micro World 2007), Seville

Diehle DA, Royse DJ (1986) Shiitake cultivation on sawdust: Evaluation of selected genotypes for biological efficiency and mushroom size. Mycol 78(6): 929–933

Donoghue JD, Denison WC (1995) Shiitake cultivation: Gas phase during Incubation Influences Productivity. Mycol 87(2): 239–244

Donoghue JD, Denison WC (1996) Commercial production of shiitake (*Lentinula edodes*) using whole-log chip of *Quercus, Lithocarpus*, and *Acer*. In: Royse DJ (ed.), Mushroom Biology and Mushroom Products, Penn State University, Pennsylvania

Eggen T (1999) Application of fungal substrate from commercial mushroom production – *Pleurotus ostreatus* – for bioremediation of creosote contaminated soil. Int Biodeterior Biodegrad 44: 117–126

Elisashvili V, Asatiani M, Kachlishvili ET et al. (2007) Basidiomycetes as a source of food, enzymes, polysaccharides, lectins, and antioxidants. Int J Med Mush 9(3–4): 206

Elisashvili V, Penninckx M, Kachlishvili E et al. (2008) *Lentinus edodes* and *Pleurotus species* lignocellulolytic enzymes activity in submerged and solid-state fermentation of lignocellulosic wastes of different composition. Bioresour Technol 99: 457–462

Fan L, Soccol CR (2005) Coffee residues. In: Gush R (ed) Mushroom Growers' Handbook 2, Mush World, Seoul

Fan L, Pandey A, Mohan R et al. (2000a) Use of various coffee industry residues for the cultivation of *Pleurotus ostreatus* in solid state fermentation. Acta Biotechnol 20(1): 41–52

Fan L, Pandey A, Soccol CR (2000b) Solid state cultivation – an efficient method to use toxic agro-industrial residues. J Basic Microbiol 40(3): 187–197

FAO (2004) Food and Agriculture Organization of the United Nations. www.fao.org. Accessed 5 June 2008

Fermor T, Watts N, Duncombe T et al. (2000) Bioremediation: Use of composts and composting technologies. In: Van Griensven JLJD (ed) Science and Cultivation of Edible Fungi, Balkema, Rotterdam

Gabriel V (2004) Cereal straw and corncobs. In: Gush R (ed) Mushroom Growers' Handbook 1, Mush World, Seoul

Gaitán-Hernández RG, Mata G (2004) Cultivation of the edible mushroom *Lentinula edodes* (Shiitake) in pasteurized wheat straw – Alternative use of geothermal energy on Mexico. Eng Life Sci 4(4): 363–367

Gaitán-Hernández RG, Esqueda M, Gutiérrez A et al. (2006) Bioconversion of agrowastes by *Lentinula edodes*: the high potential of viticulture residues. Appl Microbiol Biotechnol 71: 432–439

Gañan J, Al-Kassir Abdula A, Cuerda Correa EM et al. (2006) Energetic exploitation of vine shoot by gasification processes. A preliminary study. Fuel Proc Technol 87: 891–897

Gezer ED, Yildiz ÜC, Yildiz S et al. (2007) Use of Common Reed (*Phragmites communis* Trin.) as raw material in the cultivation of *Pleurotus ostreatus*. http://www.mushworld.com/tech/view.asp?vid=7744&cata_id=1110 (1 of 7) 22/6/2007 6:02:23. Accessed 10 May 2008

Giljum S, Kovanda J, Niza S et al. (2005) Material Input Data for the GINFORS model. Technical Report: Word Package 3.1. http://www.mosus.net/documents/MOSUS%20MFA_technical%20report.pdf. Accessed 10 May 2008

Gregori A, Svagelj M, Pohleven F (2007) Cultivation techniques and medicinal properties of *Pleurotus* spp. Food Technol Biotechnol 45(3): 236–247

Hammel KE (1997) Fungal Degradation of Lignin. In: Cadisch G, Giller KE (eds) Driven by Nature: Plant Litter Quality and Decomposition. CAB Int, pp. 33–45

Hatakka A (1994) Lignin-modifying enzymes from selected white-rot fungi: production and role in lignin degradation. FEMS Microbiol Rev 13: 125–135

Hatakka A (2001) Biodegradation of lignin. In: Hofrichter M, Steinbüchel A (eds) Biopolymers, Wiley-VCH, Weinheim

Hernández D, Sánchez JE, Yamasaki K (2003) A simple procedure for preparing substrate for *Pleurotus ostreatus* cultivation. Bioresour Technol 90: 145–150
Hiromoto BT (1991) Comparative analysis of Shiitake culture systems. In: Maher MJ (ed) Science and Cultivation of Edible Fungi, Balkema, Rotterdam
Hofrichter M (2002) Review: Lignin conversion by maganese peroxidase (MnP). Enzyme Microb Technol 30: 454–466
Hölker U, Höfer M, Lenz J (2004) Biotechnological advantages of laboratory-scale solid-state fermentation with fungi. Appl Microbiol Biotechnol 64: 175–186
Howard RL, Abotsi E, Jansen van Rensburg EL (2003) Lignocellulose biotechnology: issues of bioconversion and enzyme production. African J Biotechnol 2(12): 602–619
Huang Y (2007) Chinese market trends. Mush Bus 23: 10–11
Israilides C, Philippoussis A (2003) Bio-technologies of recycling agro-industrial wastes for the production of commercially important polysaccharides and mushrooms. Biotechnol Genet Eng Rev 20: 247–259
Israilides C, Kletsas D, Arapoglou D et al. (2008) Cytostatic and immunomodulatory properties of the medicinal mushroom *Lentinula edodes*. Phytomed 15: 512–519
Jiang J, Slivova V, Valachovicova T et al. (2004) Ganoderma lucidum inhibits proliferation and induces apoptosis in human prostate cancer cells PC-3. Int J Oncol 24: 1093–1099
Jiménez L, González F (1991) Study of the physical and chemical properties of lignocellulosic residues with a view to the production of fuels. Fuel 70: 947–950
Kachlishvili E, Penninckx MJ, Tsiklauri N et al. (2005) Effect of nitrogen source on lignocellulolytic enzyme production by white-rot basidiomycetes under solid-state cultivation. W J Microb Biotechnol 22(4): 391–397
Kalberer PP (1987) Water potentials of casing and substrate and osmotic potentials of fruit bodies of*Agaricus bisporus*. Sci Hortic 32:175–182
Kalberer PP (1995) An investigation of the incubation phase of a shiitake (*Lentinus edodes*) culture. Mush Sci 14: 375–383
Kalberer PP (2000) Influence of urea and ammonium chloride on crop yield and fruit body size of Shiitake (*Lentinula edodes*). In: Van Griensven LJLD (ed) Science and Cultivation of Edible Fungi, Balkema Rotterdam
Kalmiş E, Sargin S (2004) Cultivation of two *Pleurotus* species on wheat straw substrates containing olive mill waste water. Int Biodeterior Biodegrad 53: 43–47
Kawai G, Kobayashi H, Fukushima Y et al. (1996) Effect of liquid mycelial culture used as a spawn on sawdust cultivation of shiitake (*Lentinula edodes*). Mycosci 37: 201–207
Kidd PM (2000) The Use of Mushroom Glucans and Proteoglucans in Cancer treatment. Alternative Medicine Review 5(1): 4–27.
Kim S, Dale BE (2004) Global potential bioethanol production from wasted crops and crop residues. Biomass Bioenerg 26: 361–375
Kirchhoff B, Lelley J (1991) Investigations of Shiitake (*Lentinus edodes (Berk.*) Sing.) bag-log cultivation to increase the yield in Germany. In: Maher MJ (ed) Science and Cultivation of Edible Fungi, Balkema, Rotterdam
Kirk PM, Cannon PF, David JC et al. (2001) Ainsworth and Bisby Dictionary of the fungi. 9th edition, CABI Bioscience, Egham
Ko HG, Park SH, Kim SH (2005) Detection and recovery of hydrolytic enzymes from spent compost of four mushroom species. Folia Microbiol 50(2): 103–106
Koopmans A and Koppejan J (1997) Agricultural and forest residues – Generation, utilization and availability. Regional Consultation on Modern Applications of Biomass Energy, Kuala Lumpur, Malaysia
Krishna C (2005) Solid-state fermentation systems-an overview. Critic Rev Biotechnol 25: 1–30
Kuhad RC, Singh A, Eriksson K-EL (1997) Microorganisms and enzymes involved in the degradation of plant fiber cell walls. In: Eriksson K-EL (ed) Advances in Biochemical Engineering Biotechnology, Springer-Verlag, Berlin
Kües U, Liu Y (2000) Fruiting body production in basidiomycetes. Appl Microbiol Biotechnol 54: 141–152

Lankinen P, Hildén K, Aro N et al. (2005) Manganese peroxidase of *Agaricus bisporus:* grain bran-promoted production and gene characterization. Appl Microbiol Biotechnol 66: 401–407

Lal R (2005) World crop residues production and implications of its use as a biofuel. Environ Int 31 (4): 575–584

Lal, R (2008) Crop residues as soil amendments and feedstock for bioethanol production. Waste Manag 28 (4): 747–758

Lara M, Arias A, Villasenor L (2002) Cultivation of *Pleurotus ostreatus* and *P. Pulmonarius* on spent brewer's grain and tequila maguey bagasse. In: Sánchez JE, Huerta G, Montiel E (eds) Mushroom Biology and Mushroom Products, Universidad Autónoma del Estado de Morelos, Mexico

Laufenberg G, Kunz B, Nystroem M (2003) Transformation of vegetable waste into value added products: (A) the upgrading concept; (B) practical implementations. Bioresour Technol 87: 167–198

Levanon D, Rothschild N, Danai O (1993) Bulk treatment of substrate for the cultivation of shiitake mushrooms (*Lentinus edodes*) on straw. Bioresour Technol 45: 63–64

Lin Z (2005) Grass. In: Gush R (ed) Mushroom Growers' Handbook 2, Mush World, Seoul

Liou TH, Chang F-W, Lo JJ (1997) Pyrolysis kinetics of acid-leached rice husk. Ind Eng Chem Res 36: 568–573

López CJ, Valencia NR, Chang ST (2004) Cultivation of Shiitake on Coffee Waste. In: Romaine CP, Keil CB, Rinker DL et al. (eds) Science and Cultivation of Edible and Medicinal Fungi, Penn State University, Pennsylvania

Mabee WE, Gregg DJ, Saddler JN (2005) Assessing the emerging biorefinery sector in Canada. Appl Biochem Biotechnol 121–124: 765–778

Mande S (2005) Biomass gasifier-based power plants: potential, problems, and research needs for decentralized rural electrification. In: Lal B, Reddy MRVP (eds) Wealth from Waste: Trends and Technologies, New Delhi

Mandeel Q, Al-Laith A, Mohamed S (2005) Cultivation of oyster mushrooms (*Pleurotus spp.*) on various lignocellulosic wastes. World J Microbiol Biotechnol 21: 601–607

Manpreet S, Sawraj S, Sachin D et al. (2005) Influence of process parameters on the production of metabolites in solid-state fermentation. Mal J Microbiol 1(2): 1–9

Marino RH, Ferreira da Eira A, Kuramae EE (2003) Morphomolecular characterization of *Pleurotus ostreatus* (Jacq. Fr.) kummer strains in relation to luminosity and temperature of fructification. Sci Agric 60(3):531–535

Martínez-Carrera D, Aguilar A, Martínez W (2000) Commercial production and marketing of edible mushrooms cultivated on coffee pulp in Mexico. In: Sera T, Soccol C, Pandey A et al. (eds) Coffee Biotechnology and Quality, Klewer Academic Publishers, Dordrecht

Martínez-Carrera D, Guzmán G, Soto C (1985) The effect of fermentation of coffee pulp in the cultivation of *Pleurotus ostreatus* in Mexico. Mush Newslet Trop 6: 21–28

Mata G, Savoie JM (2005a) Wheat straw. In: Gush R (ed) Mushroom's Grower's Handbook 2, Mush World, Seoul

Mata G Savoie JM (2005b) Improvement of spawn for cultivation in alternative substrates. In: Gush R (ed) Mushroom's Grower's Handbook 2, Mush World, Seoul

Mata G Savoie JM (2005c) Extracellular enzyme activities in six *Lentinula edodes* strains during cultivation in wheat straw. World J Microb Biotechnol 14: 513–519

Mills HA, Jones JB (1996) Plant analysis handbook II. Micromacro Publishing Inc., Athens, USA

Mizuno T (1999) The extraction and development of antitumour-active polysaccharides from medicinal mushrooms in Japan. Int J Med Mush 1: 9–29.

Moda EM, Horii J, Spoto FMH (2005) Edible mushroom *Pleurotus sajor-caju* production on washed and supplemented sugarcane bagasse. Sci Agric 62(2): 127–132

Moghtaderi B, Sheng C, Wall TF (2006) An overview of the Australian biomass resources and utilization technologies. Bioresour 1(1): 93–115

Moore D, Chiu SW (2001) Filamentous fungi as food. In: Pointing SB, Hyde KD (eds) Exploitation of Filamentous Fungi, Fungal Diversity Press, Hong Kong

Mosier N, Wyman C, Dale B, Blander R (2005) Features of promising technologies for pretreatment of lignocellulosic biomass. Bioresour Technol 96: 673–686

Mulkey S, Alavalapati J, Hodges A, Wilkie AC et al. (2008) Opportunities for Greenhouse Gas Reduction Through Forestry and Agriculture in Florida. University of Florida, School of Natural Resources and Environment. http://snre.ufl.edu

Mussato SI, Rocha GJM, Roberto IC (2008) Hydrogen peroxide bleaching of cellulose pulps obtained from brewer's spent grain. Cellul. doi 10.1007/s10570-008-9198-4

Naraian R, Sahu RK, Kumar S et al. (2008) Influence of different nitrogen rich supplements during cultivation of *Pleurotus florida* on corn cob substrate. Environ. doi 10.1007/s10669-008-9174-4

Nigam P, Robinson T, Singh D (2004) Solid-state fermentation: An overview. In: Arora D (ed) Handbook of Fungal Biotechnology, Mycology 20, CRC Press, London

Nikitina VE, Tsivileva OM, Pankratou AN et al. (2007) *Lentinula edodes* biotechnology – from lentinan to lectins. Food Technol Biotechnol 45(3): 230–237

Nikolaou A, Remrova M, Jeliazkov I (2003) A report of the project—Lot 5: Bioenergy's role in the EU Energy Market, Biomass availability in Europe

Obodai M, Cleland-Okine J, Vowotor KA (2003) Comparative study on the growth and yield of *Pleurotus ostreatus* mushroom on different lignocellulosic by-products. J Industr Microbiol Biotechnol 30: 146–149

Ohga S (1999) Effect of water potential on the fruit body formation of *Lentinula edodes* in sawdust-based substrate. J Wood Sci 45: 337–342

Ohga S, Royse DJ (2001) Transcriptional regulation of laccase and cellulase genes during growth and fruiting of *Lentinula edodes* on supplemented sawdust. FEMS Microbiol Let 201: 111–115

Ohga S, Smith M, Thurston, CF et al. (2001) Transcriptional regulation of laccase and cellulase genes in the mycelium of *Agaricus bisporus* fruit body development on the solid substrate. Mycol Res 103: 1557–1560

Okano K, lida Y, Samsuri M (2006) Comparison of *in vitro* digestibility and chemical composition among sugarcane bagasses treated by four white-rot fungi. Anim Sci J 77: 308–313

Ooi, CVE, Liu F (2000) Immunomodulation and anti-cancert activity of polysaccharide-protein complexes. Cur Med Chem 7(7): 715–729

Ortega GM, Martínez EO, Betancourt, D (1992) Bioconversion of sugar cane crop residues with white-rot fungi *Pleurotus sp.* World J Microbiol Biotechnol 8: 402–405

Özçelik E, Pekşen A (2007) Hazelnut husk as a substrate for the cultivation of shiitake mushroom (*Lentinula edodes*). Bioresour Technol 98: 2652–2658

Palonen H (2004) Role of lignin in the enzymatic hydrolysis of lignocellulose. Dissertation for the degree of Doctor of Technology, Helsinki University of Technology, Helsinki

Pandey A, Soccol CR, Mitchell D (2000a) New developments in solid state fermentation: I – Bioprocesses and products. Process Biochem 35: 1153–1169

Pandey A, Soccol CR, Nigam P et al. (2000b) Biotechnological potential of agro-industrial residues: I – Sugarcane bagasse. Bioresour Technol 74: 69–80

Pandey A, Soccol CR, Nigam P et al. (2000c) Biotechnological potential of agro-industrial residues: II – Cassava bagasse. Bioresour Technol 74: 81–87

Pandey A, Soccol CR, Nigam P et al. (2000d) Biotechnological potential of coffee pulp and coffee husk for bioprocesses. Biochem Eng J 6: 153–162

Pant D, Gangi Reddy U, Adholeya A (2006) Cultivation of oyster mushrooms on wheat straw and bagasse substrate amended with distillery effluent. World J Microbiol Biotechnol 22: 267–275

Paterson RRM (2006) *Ganoderma* – A therapeutic fungal biofactory. Phytochem 67: 1985–2001

Patra AK, Pani BK (1995) Evaluation of banana leaf as a new alternative substrate to paddy straw for oyster mushroom cultivation. J Phytol Res 8:145–148

Pérez J, Munoz-Dorado J, Martínez J (2002) Biodegradation and biological treatments of cellulose, hemicellulose and lignin: An overview. Int Microbiol 5: 53–63

Pérez SR, García Oduardo N, Bermúdez RC et al. (2007) Decolourisation of mushroom farm wastewater by *Pleurotus ostreatus*. Biodegrad. doi 10.1007/s10532-007-9157-z

Philippoussis A, Zervakis G, Diamantopoulou P (2000). Potential for the cultivation of exotic mushroom species by exploitation of Mediterranean agricultural wastes. In: Van Griensven LJLD (ed) Science and Cultivation of Edible Fungi, Balkema, Rotterdam

Philippoussis A, Zervakis G, Diamantopoulou P (2001a) Bioconversion of lignocellulosic wastes through the cultivation of the edible mushrooms *Agrocybe aegerita, Volvariella volvacea* and *Pleurotus* spp. World J Microbiol Biotechnol 17(2): 191–200

Philippoussis A, Diamantopoulou P, Zervakis G (2001b) Calcium chloride irrigation influence on yield, calcium content, quality and shelf-life of the white mushroom *Agaricus bisporus.* J Sci Food Agric 81(15): 1447–1454

Philippoussis A, Diamantopoulou P, Euthimiadou H et al. (2001c) The composition and porosity of lignocellulosic substrates influence mycelium growth and respiration rates of *Lentinus edodes* (Berk.) Sing. Int J Med Mush 3(2–3): 198

Philippoussis A, Diamantopoulou P, Zervakis G (2002) Monitoring of mycelium growth and fructification of *Lentinula edodes* on several agricultural residues. In: Sánchez JE, Huerta G, Montiel E (eds), Mushroom Biology and Mushroom Products, UAEM, Cuernavaca

Philippoussis A, Diamantopoulou P, Zervakis G (2003) Correlation of the properties of several lignocellulosic substrates to the crop performance of the shiitake mushroom *Lentinula edodes.* World J Microbiol Biotechnol 19(6): 551–557

Philippoussis A, Zervakis GI, Diamantopoulou P et al. (2004). Use of spent mushroom compost as substrate for plant growth and against plant infections caused by Phytophora spp. In: Romaine CP, Keil CB, Rinker DL et al. (eds) Science and Cultivation of Edible and Medicinal Fungi, Penn State University, Pennsylvania

Philippoussis A, Diamantopoulou P, Israilides C (2007) Production of functional food from the sporophores of the medicinal mushroom *Lentinula edodes* through exploitation of lingocellulosic agricultural residues. Int Biodeteriorat Biodegrad 59(3): 216–219

Przybylowicz P, Donoghue J (1990) *Shiitake* Growers Handbook, The Art and science of Mushroom cultivation. Kendall/Hunt Publishing Company, Iowa

Pire DG, Wright JE, Albertó E (2001) Cultivation of Shiitake using Sawdust from widely available local woods in Argentina. Micol Appl Int 13(2): 87–91

Pointing SB (2001) Feasibility of bioremediation by white-rot fungi. Appl Microbiol Biotechnol 57: 20–33

Poppe J (2000) Use of Agricultural waste materials in the cultivation of mushrooms. In: Van Griensven LJLD (ed) Science and Cultivation of Edible Fungi, Balkema, Rotterdam

Poppe J (2004) Agricultural wastes as substrates for oyster mushrooms. In: Gush R (ed) Mushroom Growers' Handbook 1, Mush World, Seoul

Quian G (2004) Cottonseed hulls. In: Gush R (ed) Mushroom Growers' Handbook 1, Mush World, Seoul

Qu Y, Zhu M, Liu K (2006) Studies on cellulosic ethanol production for sustainable supply of liquid fuel in China. J Biotechnol 1: 1235–1240

Ragunathan R, Gurusamy R, Palaniswamy M (1996) Cultivation of *Pleurotus* spp. on various agroresidues. Food Chem 55(2): 139–144

Rajarathnam S, Shashireka MNJ, Bano Z (1998) Biodegradative and biosynthetic capacities of mushrooms: Present and future strategies. Critic Rev Biotechnol 18(2–3): 91–236

Raimbault M (1998) General and microbiological aspects of solid substrate fermentation. Electron J Biotechnol 1(3): 1–15

Reddy N, Yang Y (2005) Biofibers from agricultural byproducts for industrial applications. Trends Biotechnol 23(1): 22–27

Revankar MS, Desai KM, Lele SS, (2007) Solid-state fermentation for enhanced production of Laccase using indigenously isolated *Ganoderma sp.* Appl Biochem Biotechnol 143: 16–26

Rigas F, Papadopoulou K, Dritsa V et al. (2007) Bioremediation of a soil contaminated by lindane utilizing the fungus *Ganoderma australe* via response surface methodology. J Haz Mat 140: 325–332

Rinker DL (2002) Handling and using "Spent" Mushroom Substrate around the world. In: Sánchez JE, Huerta G, Montiel E (eds), Mushroom Biology and Mushroom Products, UAEM, Cuernavaca

Rodriguez-Estrada AE, Royse DJ (2007) Yield, size and bacterial blotch resistance of Pleurotus eryngii grown on cottonseed hulls/oak sawdust supplemented with manganese copper and whole soybean. Biores Technol 98: 1898–1906

Rossi IH, Monteiro AC, Machado JO (2003) Shiitake *Lentinula edodes* Production on a sterilized bagasse substrate enriched with rice bran and sugarcane molasses. Braz J Microbiol 34: 66–71

Royse DJ (1996) Yield Stimulation of Shiitake by Millet. Supplementation of Wood Chip Substrate. In: Royse DJ (ed) Mushroom Biology and Mushroom Products, Penn State University, Pensnylvania

Royse DJ (2001) Cultivation of Shiitake on Natural and Synthetic Logs. College of Agricultural Sciences, Penn State University, Pennsylvania

Royse DJ (2002) Influence of Spawn rate and commercial delayed release nutrient levels on *Pleurotus cornucopiae* (oyster mushroom) yield, size, and time to production. Appl Microbiol Biotechnol 58: 527–531

Royse DJ (2003a) Cultivation of Oyster Mushrooms. College of Agricultural Sciences, The Pennsylvania State University, University Park, PA.

Royse DJ (2003b) Influence of precipitated calcium carbonate ($CaCO_3$) on Shiitake (*Lentinula edodes*) yield and mushroom size. Bioresour Technol 90: 225–228

Royse DJ (2004) Specialty mushrooms. In: Mushroom Fact Sheet, Mushroom Spawn Laboratory, Penn State University, Pennsylvania

Royse DJ, Bahler CC (1986) Effects of genotype, spawn run time, and substrate formulation on biological efficiency on shiitake. Appl Environ Microbiol 52(6): 1425–1427

Royse DJ, Schisler LC (1987) Yield and size of *Pleurotus ostreatus* and*Pleurotus sajor-caju* as affected by delayed-release nutrient. Appl Microbiol Biotechnol 26: 191–194

Royse DJ, Sanchez JE (2007) Ground wheat Straw as a substitute for portions of oak wood chips used in Shiitake (*Lentinula edodes*) substrate formulae. Biorsour Technol 98: 2137–2141

Royse DJ, Bahler BD, Bahler CC (1990) Enhanced Yield of shiitake by saccharide amendment of the synthetic substrate. Appl Environ Microbiol 56(2): 479–482

Royse DJ, Sanchez-Vazquez JE (2001) Influence of substrate wood-chip particle size on Shiitake (*Lentinula edodes*) yield. Bioresour Technol 76: 229–233

Royse DJ, Rhodes TW, Sánchez JE (2002) Vacuum-soaking of wood chip Shiitake (*Lentinula edodes*) logs to reduce soak time and log weight variability and to stimulate mushroom yield. Appl Microbiol Biotechnol 58: 58–62.

Sabota C (1996) Strain of Shiitake mushroom [*Lentinula edodes (Berk.) Pegler*] and wood species affect the yield of Shiitake mushrooms. HortTechnol 6(4): 388–393

Salmones D, Mata G, Waliszewski KN (2005) Comparative culturing of *Pleurotus spp.* On coffee pulp and wheat straw: Biomass production and substrate biodegradation. Bioresour Technol 96: 537–544

Salmones D, Mata G, Ramos LM, Waliszewski KN (1999) Cultivation of shiitake mushroom, *Lentinula edodes,* in several lignocellulosic materials originating from the subtropics. Agronomie 19(1): 13–19

Sánchez C (2004) Modern aspects of mushroom culture technology. Appl Microbiol Biotechnol 64: 756–762

Sánchez A, Ysunza F, Beltrán-García MJ (2002). Biodegradation of Viticulture Wastes by *Pleurotus*: a source of microbial and human food and its potential use in animal feeding. J Agric Food Chem 50: 2537–2542

Saura-Calixto F, Cañellas J, García-Raso J (1983) Determination of hemicellulose, cellulose and lignin contents of dietary fibre and crude fibre of several seed hulls. Data Comparison. Z Lebensm Unters Forsch 177: 200–202

Serra M, Kirby R (1999) Development of *Pleurotus* (Oyster) mushroom production in Southern Africa using alien wood species as lignocellulose substrate. Int J Mush Sci 2(3): 49–55

Shah ZA, Ashraf M, Ishtiaq CM (2004) Comparative study on cultivation and yield performance of oyster mushroom (*Pleurotus ostreatus*) on different substrates (Wheat Straw, Leaves, Saw Dust). Pakistan J Nutr 3(3): 158–160

Silva EM, Machuca A, Milagres AMF (2005) Effect of cereal brans on *Lentinula edodes* growth and enzyme activities during cultivation on forestry waste. Lett Appl Microbiol 40: 283–288

Silva ÉS, Cavallazzi JRP, Muller G (2007) Biotechnological applications of *Lentinus edodes*. J Food, Agric Environ 5(3,4): 403–407

Singh MP (2000) Biodegradation of lignocellulosic wastes through cultivation of *Pleurotus* sajor-caju. In: Van Griensven JLJD (ed) Science and Cultivation of Edible Fungi, Balkema, Rotterdam

Smith JE, Rowan NJ, Sullivan R (2002) Medicinal mushrooms: a rapidly developing area of biotechnology for cancer therapy and other bioactivities. Biotechnol Lett 24: 1839–1845

Soccol CR, Vandenberghe LPS (2003) Overview of applied solid-state fermentation in Brazil. Biochem Eng J 13: 205–218

Soto C, Martínez-Carrera D, Morales P, Sobal M (1987) La pulpa de café secada al sol, como una forma de almacenamiento para el cultivo de *Pleurotus ostreatus*. Rev Mex Mic 3: 133–136

Stamets P (2000) Growing Gourmet and Medicinal Mushrooms. Ten Speed Press, Berkeley

Steffen KT, Cajthaml T, Šnajdr J, Baldrian P (2007) Differential degradation of oak (*Quercus petraea*) leaf litter by litter-decomposing basidiomycetes. Res Micobiol 158: 447–455

Tengerdy RP, Szakacs G (2003) Bioconversion of lignocellulose in solid substrate fermentation. Biochem Eng J 13: 169–179

Thomsen MH (2005) Complex media from processing of agricultural crops for microbial fermentation. Appl Microbiol Biotechnol 68: 598–606

Tisdale TE, Miyasaka SC, Hemmes DE (2006) Cultivation of the oyster mushroom (*Pleurotus ostreatus*) on wood substrates in Hawaii. World J Microbiol Biotechnol 22: 201–206

Upadhyay RC, Verma RN, Singh SK et al. (2002) Effect of organic nitrogen supplementation in Pleurotus species In: Sánchez JE, Huerta G, Montiel E (eds) Mushroom Biology and Mushroom Products, Universidad Autónoma del Estado de Morelos, Mexico

USDA-US DOE (2005). Biomass as feedstock for a bioenergy and bioproducts industry: The technical feasibility of a billion-ton annyal supply. http://feedstockreview.ornl.gov/pdf/ billion_ton_vision.pdf. Accessed 10 May 2008

Vega A, Caballero RE, Garcia JR (2005) Bioconversion of agro-industrial residues by *Pleurotus ostreatus* cultivation. Revist Mex Micol 20: 33–38

Velázquez-Cedeño MA, Mata G, Savoie JM (2002) Waste-reducing cultivation of *Pleurotus ostreatus* and *Pleurotus pulmonarius* on coffee pulp: changes in the production of some lignocellulolytic enzymes. World J Microbiol Biotechnol 18: 201–207

Ververis C, Georghiou K, Christodoulakis N (2004) Fiber dimensions, lignin and cellulose content of various plant materials and their suitability for paper production. Ind Crops Prod 19: 245–254.

Vetayasuporn S (2006) Oyster mushroom cultivation on different cellulosic substrates. Res J Agric Biol Sci 2(6): 548–551

Wang HH (1999) Development and/or reclamation of bioresources with solid state fermentation. Proc Nat Sci Counc 23(2): 45–61

Wang D, Sakoda A, Suzuki M (2001) Biological efficiency and nutritional value of *Pleurotus ostreatus* cultivated on spent beer grain. Bioresour Technol 78(3): 293–300

Ward PL, Wohlt JE, Zajac PK et al. (2000) Chemical and physical properties of processed newspaper compared to wheat straw and wood shavings as animal bedding. J Dairy Sci 83: 359–367

Wasser SP (2002) Medicinal mushrooms as a source of antitumor and immunomodulating polysaccharides. Appl Microbiol Biotechnol 60: 258–274

Wasser SP, Weis AL (1999) Medicinal properties of substances occurring in higher basidiomycetes mushrooms: current perspectives. Int J Med Mush 1: 31–62

Webb C, Koutinas AA, Wang R (2004) Developing a sustainable bioprocessing strategy based on a generic feedstock. Adv Biochem Eng Biotechnol 86: 195–268

Wong SM, Wong KK, Chiu L et al. (2007) Non-starch polysaccharides from different developmental stages of *Pleurotus tuber-regium* inhibited the growth of human acute promyelocytic leukemia HL-60 cells by cell-cycle arrest and/or apoptotic induction. Carbohydr Polym 68: 206–217

Wood DA, Smith JF (1987) The cultivation of mushrooms. In: Norris JR, Pettipher GL, (eds) Essays in Agricultural and Food Microbiology, John Willey and Sons Ltd, London

Wood DA (1989) Mushroom Biotechnology. Int Ind Biotechnol 9: 5–9

Worrall JJ, Yang CS (1992) Shiitake and oyster mushroom production on apple pomace and sawdust. HortSci 27(10): 1131–1133

Yildiz S, Yildiz ÜC, Gezer ED (2002) Some lignocellulosic wastes used as raw material in cultivation of the *Pleurotus ostreatus* culture mushroom. Proc Biochem 38: 301–306

Zadrazil F (1993) *Lentinula (=Lentinus) edodes*: physiology and condition of industrial production. Mush Inf 6: 5–27

Zadrazil F (2000) Is conversion of lignocellulosics into feed with white-rot fungi realizable? Practical problems of scale-up and technology. In: Van Griensven LJLD (ed) Science and Cultivation of Edible Fungi, Balkema, Rotterdam

Zadrazil F, Kamra DN, Isikhuemhen OS et al. (1996) Bioconversion of lignocellulose into ruminant feed with white rot fungi – Review of work done at the FAL, Braunschweig. J Appl Anim Res 10: 105–124

Zadrazil F, Compare G, Maziero R (2004) Biology, cultivation and utilization of *Pleurotus* species. In: Romaine CP, Keil CB, Rinker DL et al. (eds) Science and Cultivation of Edible and Medicinal Fungi, Penn State University, Pennsylvania

Zervakis G, Balis C (1992) Comparative study on the cultural characters of *Pleurotus* species under the influence of different substrates and fruiting temperatures. Micol Neotrop Aplicada 5: 39–47

Zervakis G, Philippoussis A (2000) Management of agro-industrial wastes through the cultivation of edible mushrooms. In: Proceedings o f IV European Waste Forum 'Innovation in waste management', C.I.P.A., Milan

Zervakis G, Philippoussis A, Ioannidou S et al. (2001) Mycelium growth kinetics and optimal temperature conditions for the cultivation of edible mushroom species on lignocellulosic substrates. Folia Microbiol 46(3): 231–234

Zervakis G, Papadopoulou K, Ehaliotis C et al. (2005) Use of composts deriving from Mediterranean agro-industrial wastes in vegetable crops: effects on disease suppression and plant growth. In: de Kreij C, Warmenhoven M (eds) Proceedings of the International Symposium on the Use of Composted Organic Wastes in Horticulture, Wageningen

Zhang CK, Gong F, Li DS (1995) A note on the utilization of spent mushroom compost in animal feeds. Biores Technol 52(1):89–91

Zhang R, Li X, Fadel JG (2002) Oyster mushroom cultivation with rice and wheat straw. Bioresour Technol 82: 277–284

Zhang YHP (2008) Reviving the carbohydrate economy via multi-product lignocellulose bioefineries. J Ind Microbiol Biotechnol 35: 367–375

Chapter 10
Solid-State Fermentation Technology for Bioconversion of Biomass and Agricultural Residues

Poonam Singh nee' Nigam and Ashok Pandey

Contents

Abstract Solid-state fermentations (SSF) have attracted a renewed interest and attention from researchers due to recent developments in the field of microbial-biotechnology. Hence, for the practical, economical and environmentally-friendly

P. Singh nee' Nigam (✉)
Faculty of Life and Health Sciences* University of Ulster, Coleraine BT52 1SA, Northern Ireland, UK
e-mail: P.Singh@ulster.ac.uk

P. Singh nee' Nigam, A. Pandey (eds.), *Biotechnology for Agro-Industrial Residues Utilisation*, DOI 10.1007/978-1-4020-9942-7_10,

bioconversion of agro-industrial wastes, solid state or substrate fermentation has been researched globally and proved to be the ideal technology for this purpose. In this chapter some important aspects of solid-state cultivation system have been discussed, including the variety of substrates and microorganisms used in SSF for the production of various end products; and the performance control of system by regulation of important factors.

Keywords Solid substrates · Agricultural residues · Solid state fermentation · Water activity · Moisture · Bioreactor

10.1 Agro-Residue Bioconversion in SSF

Commonly used substrates in SSF are natural agricultural products, as well as agro-industrial waste residues and by-products serve as a source of carbon in SSF (Table 10.1). Lignocellulosic materials of agriculture origin compose more than 60% of plant biomass produced annually through the process of photosynthesis. This vast resource is the potential and renewable source of biofuels, biofertilizers, animal feed and chemical feedstocks. Lignocellulose may be a substrate for the production of value-added products (Nigam and Singh 1996a, b; Nigam 1988, 1989a, b) (Table 10.2), such as biofuels, biochemicals, biopesticides, biopromoters, or may even be a product itself after biotransformation (e.g. compost, biopulp).

In all applications the primary requirement is the hydrolysis of lignocellulose into fermentable sugars by lignocellulolytic enzymes, or appropriate modification of

Table 10.1 Diverse range of agro-residues utilization in SSF technology

Substrates for SSF	Microorganisms used in SSF	Reference
Starchy raw materials	*Aspergillus spp*	Czajkowska and Ilnicka 1988
Bannana waste	*A. niger*	Baldensperger et al. 1985
Barley Husk	*Bjkendra adusta*	Robinson and Nigam 2008
Corn cob	*A. niger*	Singh et al. 1989
Citrus peel	*A. niger*	Rodriquez et al. 1985
Sugarcane by-products	*A. terreus*	Gonzalez-Blanco et al. 1990
Cassava	*Rhizopus oryzae*	Daubresse et al. 1987
Sugarbeet pulp	*Trichoderma viride*	Durand 1998
Cassava	*T. resei & yeast*	Opoku and Adoga 1980
Wheat straw	*T. reesei & Endomycopsis fibuleger*	Laukevics et al. 1984
Wheat straw	*T. reesei, Chaeotominum*	Abdullah et al. 1985
Sugarbeet pulp	*T. reesei and Fusarium oxysporum*	Nigam and Vogel 1988, 1990
Sugarcane bagasse	*Polyporus spp*	Nigam 1990
Saccharum munja-	*Pleurotus spp.*	Gujral et al. 1987
Residues Wheat straw	*Coprinus spp.*	Yadav 1989, 1988
Cassava	*Sporotrichum pulverulentum*	Smith et al. 1986
Straw	*Candida utilis*	Han 1987
Sweet potato	*Pichia bartonii*	Yang 1988
Fodder beets	*Saccharomyces cerevisiae*	Gibbon et al. 1984

Table 10.2 Agro-industrial residues used for addcd-value products

Substrate	Microorganisms	Product	Reference
Oat cereal	*Rhizopus oryzae*	Lactic acid; various value added products	Koutinas et al. (2007)
Agro-residues	*Aspergillus niger*	Citric acid	Prado et al. (2004)
Distillers grain	*Aspergillus niger*	Citric acid	Xie and West (2006)
Wheat bran	*Mucor meihei*	Rennet	Thakur et al. 1990
Wheat bran	*Rhizopus oligosporus, Mucor meihei*	Rennet	Karanth (1988)
Agar	*Trichoderma viride*	Cheese aroma	Gervais (1988)
Agro-residues	*Fungal cultures*	Cheese flavour	Revah and Lebeault (1988)
Polished rice	*Neurospora spp*	Aroma	Yamauchi et al. (1989)
Barley	*Strptomyces Cephcdosportum aermonium*	Cephalosporin	Jermini and Demain (1989)
Sweet potato	*Aspergillus*	Tetracyclines	Yang and Ling (1984)
Rice grains	*Streptomyces*	Cephalosporin	Wang et al. (1984)
Bagasse	*Penicillium chrysogenum*	Penicillin	Barrios-Gonzalez et al. (1988a, b)
Cassava	*Aspergillus niger*	Aflatoxins	Barrios-G et al. (1990)
Corn	*A. flavus*	Mycotoxin	Hesseltine (1972)
Soya	Various moulds	Mycotoxin	Bhumiratna et al. 1980
Oat straw	*Pofyporovs spp*	Lignin degradation	Bone and Munoz (1984)
Birch lignin	*Phanerochaete chrysosporhun*	Lignin conversion	Mudagett and Paradis (1985)
Maple Wood	*Polyporus anceps*	Lignin conversion	Matteau and Bone (1980)
Bagasse	*2 Polyporous spp*	Lignin conversion	Nigam (1990)
Aspen Wood	*Phubia tremelloasa*	Delignification	Reid (1989a, b)
Wheat bran	*Fusarium monoliforme, Gibberrela fujikuroi*	Gibberellic acid	Kumar and Lonsane (1987a, b, c)
Wheat bran	*Fusarium monoliforme, Gibberrela fujikuroi*	Gibberellic acid	Prema et al. (1988)
Wheat straw	*Poms tignium*	Hydroenperoxi de	Maltseva et al. (1989)
Soya bean Cassava	*R. oligsoporus*	Tempeh and Koji	Hesseltine (1972)
Koji-type SSF	*Filamentous fungi*	Fungal spores	Vezina and Singh (1975)
Soya bean	*Filamentous fungi*	Fungal spores	Lotong and Suwarnarit (1983)

the structure of lignocellulose. Economical and effective lignocellulolytic enzyme complexes, containing cellulases, hemicellulases, pectinases and ligninases may be prepared by SSF (Table 10.3). Lignocellulose is also the raw material of the paper industry. To fully utilize the potential of lignocellulose, it has to be converted by chemical and/or biological processes. Solid substrate fermentation (SSF) plays an important role, and has a great perspective for the bioconversion of plant biomass. Lignocellulose may be a good feedstock for the production of biofuels, enzymes and other biochemical products by SSF. Crop residues (straw, corn by-products, bagasse, etc.) are particularly suitable for this purpose, since they are available in large quantities in processing facilities (Pandey et al. 2001).

Lignocellulose in wood may be transformed into good quality paper products with the help of SSF biopulping and biobleaching. Agricultural residues may be

Table 10.3 Agro-residues used in SSF for enzyme production

Substrates	Microorganisms	Enzymes	Reference
Bagasse, sawdust,corn cobs	*A. niger*	cellulose, beta glucosidase	Madamwar et al. (1989)
Corn cobs	*A. niger*	cellulose	Singh et al. (1989)
Wheat bran	*A. niger*	glucoamylase	Pandey (1990)
Wheat bran	*A. niger*	glucoamylase	Ramakrishnana et al. (1982)
Sugarbeet pulp	*A. phoenicis*	beta glucosidase	Deschamps and Huet (1984)
Wheat bran	*A. flavus*	protease	Malathi and Chakrabarty (1991)
Wheat bran	*A. carbonarius*	pectinase	Karanth (1988)
Wheat bran	*A. niveus*	catalase	Karanth (1988)
Sugarbeet pulp	*T. viride and A. niger*	cellulase and amylase	Desgrenges and Durand (1990)
Wheat bran and rice straw	*Trichoderma spp. A. ustus, Botritis spp., S. pulverulentum*	Cellulose, beta-glucosidase, Xylanse	Shamala and Sreekantiah (1986)
Wheat bran	*Pencillium spp. Geotrichwn Candidum, Mucor meihei & 2 Rhizopus spp.*	lipase	Munoz et al. (1991)
Sugarbeet pulp	*P. capsulatum*	enzymes	Considine et al. (1988)
Citrus pulp-pellets	*P. charlesii, Talaromyces flavus, Tubercularia vulgaris*	Pectic enzymes	Siessere and Said (1989)
Citrus pulp	*T. vulgaris*	pectic enzymes	Vieira et al. (1991)
Bagasse	*Polyporous spp.*	Cellulase & ligninase	Nigam et al. (1987b)
Lignocellulo sis	*Lentinula edodus*	enzymes	Mishra and Leatham (1990)
Wheat bran	*Bacillus licheniformis*	alpha amylase	Ramesh and Lonsane (1987a, b, 1990)
Wheat bran	*Bacillus subtilis*	protease	Jermini and Demain (1989)
Straw	*Neurospora crasse*	Carboymethyl cellulase, beta glucosidase	Macris et al. (1987)

converted into animal feed enriched with microbial biomass, enzymes, biopromoters, and made more digestible by SSF. Lignocellulosic waste may be composted to targeted biofertilizer, biopesticide and biopromoter products. Post-harvest residue may be decomposed on site by filamentous fungi and recycled to the soil with improved biofertilizer and bioprotective properties.

10.1.1 Nature of Substrates

The major organic material available in nature are polymeric in nature e.g. polysaccharides (cellulose, hemicellulose, pectins, and starch etc.), lignin and protein,

Table 10.4 Agro-residues used for microbial growth studies

Substrate	Microorganism	Reference
Cassava	*A. niger*	Raimbault and Alazard, 1980, Casteada et al., 1990, Oriol et al. 1988
Barley Husk and Barley straw	*Bjkendra adusta*	Robinson and Nigam 2008
Citrus peel	*A. niger*	Rodriguez et al. 1985,
Cassva	*Rhizosporus oligosporus*	Mitchell et al. (1988, 1990)
Soya bean	*R. oligosporus*	Rathbun and Shuler 1983
Sugarbeet pulp	*T. viride, A. niger*	Desgrenges and Durand 1990
Sugarrbeet pulp	*T. viride, Sporotrichum Pulverulentum and Thermoascus auranticus*	Grajek 1988
Buckwheat seeds	*Penicillium roqueforti*	Desfarges et al. 1987
Wheat straw	*Corinus fimetarius*	Singh et al. 1990

which can be metabolized by different microorganisms as a source of energy. These substrates that are insoluble in water, absorb water onto their matrix, which provides required moisture in SSF system for the growth and metabolic activities of microorganisms. Bacterial and yeast cultures grow on the surface of substrate fibrils and particles while fungal mycelia penetrate into the particles of substrate for nutrition.

The solid phase in SSF provides a rich and complex source of nutrients that may be sufficient or sometimes insufficient and incomplete with respect to the overall nutritional requirements of that particular microorganism that is cultivated on that substrate. The constituents in the agricultural solids are approximately analysed in terms of total carbohydrates, proteins, lipids, various elements and ash content. The solid substrates generally contain some small carbon compounds whereas the bulk of total dry weight is a complex polymer. The polymeric forms require enzymatic hydrolysis for their mineralisation as carbon-energy sources in microbialmetabolism. In comparison with liquid-state fermentation, which generally use less complex carbon energy sources, solid insoluble substrates provide mixed ingredients of high molecular weight carbon compounds. Such complex carbon compounds may contribute inhibition, induction, or repression mechanism in microbial metabolism during solid state cultivation (Table 10.4).

10.2 A Bio-Technology Solid State Fermentation

Solid substrate systems have been defined in several ways:

1. Solid substrate fermentation (SSF) is the microbial transformation of biological materials in their natural state, in contrast with liquid or submerged fermentation that is carried out in dilute solutions or slurries (Pandey et al. 2001, 2004).
2. Solid substrate fermentation is generally defined as the growth of microorganisms on solid substrates or sometimes referred to as solid-state fermentation since the process taking place is in the absence or near-absence of *free* water in the system (Nigam and Singh 1994). The substrate however, must contain

enough moisture, which exists in absorbed form within the solid substrate matrix and simulates the fermentation reaction occurring in nature. These moist solid-substrates are insoluble in water and polymeric in nature, are a source of carbon and energy, vitamins, minerals, nutrients and also provide their absorbed water for microbial growth as well as anchorage.

3. Solid-state or solid-substrate fermentation means that the substrate is moistened, often with a thin layer of water on the surface of the particles, but there is not enough water present to make fluid mixture. Weight ratios of water to substrate in SSF are usually between 1:1 and 1:10.
4. SSF can be defined as a system with solid matrix particles, a liquid phase bound to them and a gaseous phase entrapped within the particles. The physical properties of this system such as the water potential and water holding capacity, (can be used as an index of aeration) and bulk density (which predicates the volume of pore space) help to define the conditions of solid-state fermentation.

10.3 Advantages of SSF Over Conventional Liquid Fermentation

Traditional SSF came about for two primary reasons:

1. The desire for more tasty food, as with Oriental fermented foods and mould-ripened cheese; and
2. The need to dispose of agricultural and farm waste materials (as in composting).

A closer examination of SSF processes in recent years in several research centres throughout the world has led to the realisation of its numerous economical and practical advantages (Lonsane et al. 1985; Steinkraus 1984). The attraction of SSF comes from its simplicity and its closeness to the natural way of life for many microorganisms. Since large amount of water are not added to the biological systems, fermenter volumes remain small, necessary manipulations become less expensive and the cost of water removal at the end of fermentation in minimised. This type of fermentation is especially suitable for growing mixed cultures of microorganisms where symbiosis stimulates better growth and productivity (Bushell and Slater 1981). Solid-state fermentations are clearly distinguished from submerged cultures by the fact that microbial colonisaton occur at or near the surfaces of solid substrate, or in few cases the soluble substrate supported on the solid insoluble-matrix in the environment of low-moisture contents. In contrast to liquid fermentation, the substrates traditionally fermented in the solid-state are renewable agricultural products, such as wheat, rice, millet, barley, corn and soybeans. The non-traditional substrates, which can be used in industrial process development, include an abundant availability of agricultural, forest and food-processing wastes.

From an engineering point of view, SSF offers many attractive features in comparison to conventional stirred tank reactors or aerated liquid medium fermentations because no free water is present, this leads to many benefits.

Solid-state fermentations can be used to provide low-shear environments for the cultivation of shear-sensitive mycelial organisms. Solid state cultivations can be and have been used for mass production of spores, which can than be used for the transformation of organic compounds such as steroids, antibiotics, fatty acids, and carbohydrates. Fungal spores have applications in the production of food-flavours and insecticides. The advantage of solid state fermentation includes simplicity, yields and the homogeneity of spore preparations. The expected advantages of SSF over submerged fermentations are:

a. Smaller fermenter volume, relative to the yield of the product, as there is no excess water taking space in the fermenter,
b. Lower sterilisation energy costs, as less volume of water needs to be heated,
c. Seed tanks are not necessary in all cases, as the spore inocula can be successfully used to inoculate the solid medium.
d. Easier aeration, as air can circulate easily and freely between the substrate particles, and also because the liquid film covering the substrate has a large surface area compared to its volume. Aeration is facilitated by spaces between substrate particles and particle mixing.
e. Reduced or eliminated capital and operating costs for stirring, since occasional stirring is sufficient.
f. Lower costs of product recovery and drying; in many cases the product is concentrated in the substrate and can be used directly e.g. Oriental foods and cheeses, or the products can be directly incorporated into animal feeds.
g. If the product is to be extracted from the substrate e.g. enzymes and other metabolites, then much less solvent is needed. The fermented solids may be extracted immediately by direct addition of solvents or maintained in frozen storage before extraction.
h. Reduced or eliminated capital and operating costs for effluent treatment due to lower water content in the system.

The other benefits are:

1. The media are relatively simple; a natural, as opposed to a synthetic, medium is used;
2. A more natural environment for microorganisms, e.g. agricultural wastes degrading organisms: many of these fungi grow and perform better under SSF than submerged conditions;
3. A less favourable environment for many bacteria, which require a high moisture level to survive, lowering the risk of contamination, therefore many SSF processes need no sterilisation;
4. SSF is adaptable to either continuous or batch process and the complexity of equipment is no greater than that required for submerged reactors.

Above described advantages are so attractive for the biological processing of agricultural by products that most of the work has used SSF process. These advantages can outweigh the disadvantages of SSF, which are the slowness of fermentation and the difficulty of controlling the process precisely.

10.4 Performance Control of SSF Process

The difference in process control between SSF and SmF is mainly due to the use of solid substrates with a very low moisture content in system. The disadvantages of large-scale solid cultures are due to the problems of process-control, process scale-up and the major problem of heat build-up. Despite these drawbacks, large-scale SSF processes have been developed successfully in Japan for the manufacture of a variety of products, including fermented foods and food-products, enzymes, and organic acids. The drawbacks have been overcome by carrying these fermentations in stationary and rotary tray processes, where the temperature and humidity-controlled air is circulated through the stacked beds of fermenting solid substrate particles. These tray methods of cultivation have been used for centuries in the manufacture of traditional food products and the cultures experience the shear-sensitivity in some of these processes. These are main reasons of less frequent use of rotary drum-type fermenters.

Little information is available in the West on the details of modern control systems in large-scale solid-state cultivations. The control of temperature and humidity within practical limits is exercised through water temperatures, which is used to humidify the circulating air. The humidified air is circulated at flow-rates to meet the requirements of heat and mass transfer. The gas environment has been found to significantly affect the rate and extent of culture colonisation and product formation in SSF. In the commercial production of amylase using rice substrate in SSF, oxygen pressures above atmospheric have been found to significantly stimulate the enzyme productivity, suggesting oxygen limitation at normal atmospheric pressure. The DNA measurements revealed that this only caused a little effect on biomass formation, but the carbon dioxide pressures above 0.01 atm severely affected the process through the inhibition in amylase productivity.

In a protein production process by *Aspergillus* species using alfalfa residues, cellulase and pectinase activities have been found stimulated by oxygen and carbon dioxide pressures above atmospheric levels, and with no effect on biomass formation. These studies have been conducted in controlled gas environments at constant partial pressures, which is maintained by admitting pure oxygen on demand at pressures below a set point and purging carbon dioxide in 30% KOH at pressures above a set point in a closed aeration system. In another type of SSF performed for the degradation of natural birch lignin employing *Phanerochaete chrysosporium*, high oxygen pressures have been found to be stimulating, whereas the high carbon dioxide pressures have been found inhibiting the process. The stimulatory effect of oxygen on breakdown of lignins has been confirmed in laboratory studies by using labeled synthetic lignins and natural wood lignins.

Given the present state of the art, the most promising approach in solid state fermentation processes development happens to be the measurements and control of various parameters and process variables, similarly as in any liquid fermentation. In SSF processes, various methods are selected to analyse the temperature, pH, humidity, oxygen and carbon dioxide concentrations in gas phases, biochemical analysis of fermented and unfermented solids and their extracts. The manufacturing productivities of some industrial scale submerged liquid fermentations have increased significantly over years, e.g. antibiotic production. This development has been possible due to applied and basic research in microbial-biochemistry, microbial-physiology, and genetics. To some extent the contribution also goes to engineering research based on concepts of stoichiometry, kinetics, thermodynamics, and heat and mass transfer in control of the microbial fermentation process and its environment.

Direct economic comparisons of solid-state and liquid-state fermentations are not possible, it is apparent that the large-scale solid-state fermentations (known as Koji in Orient) have been developed in Japan on an economic basis. Potential economic advantages of such processes to employ suitable microbe-substrate system include:

1. reduced thermal processing requirements, since many processes are not aseptic;
2. reduced energy requirements for agitation, since surface-to-volume ratios for gas transfer are high and many processes do not require agitation due to their shear-sensitivity;
3. high extracellular product concentrations, that can be efficiently recovered by superficial-extraction or leaching methods.

10.4.1 Performance Control by Particle Size of Agro Residues

SSF processes performance can be varied and controlled by changing physical and chemical factors. It has been reported that substrates with finer particles showed improved degradation due to an increase in surface area for enzymatic action (Moloney et al. 1984). The greater growth of fungal cultures has been found stimulated by smaller particle size substrates. Higher enzyme productivity in SSF has been achieved with substrates, which contained particles of mixed sizes from 180 μm to 1.4 mm.

Particles and kernels of grain must be of suitable size, but not be too small in order to avoid particle agglomeration. The particle size must be in a limited size range to be maintained at relatively low moisture content to prevent contamination. The smaller particle size provides a larger surface area which facilitates heat transfer and gas exchange. Smaller particle sizes also distribute equivalent moisture concentrations in thinner films on external surfaces exposed to the gas environments, given the same void volume fraction (porosity) and pore size distribution. Internal pores maintain the same surface-to-volume ratios with respect to solid surfaces, based on geometric considerations of spherical particles. This results in higher surface nutrient concentrations and the diffusion of nutrients takes place via shorter pathways at the surfaces as well as in the pores of those substrates which have same tortuosity.

Too small a particle size may result in closer packing densities of the substrates and the void space between particles becomes considerable reduced. The reduced space between particles tends to reduce the available area for heat transfer and gas-exchange with the surrounding environment. If such condition arises, densely packed particles in a cultivation system have to be sufficiently agitated to provide a better separation of particles for the exchanges of gases and heat transfer. There may be a lower limit in particle size at which the heat transfer or gas exchange becomes rate limiting and there may be an upper limit at which the nutrient transfer becomes limiting. Conclusively under any condition, the particle size of the substrate to be used is one of the major variables in the SSF-process development. Various methods are available to obtain particle sizes such as milling, grinding, chopping and sieving to obtain substrates of particular particle-sizes. In the case of lignocellulosic substrates, smaller particle size substrate is usually obtained through ball-milling.

10.4.2 Performance Control by Medium Preparation of Agro-Residues

Some SSF systems do not require any nutritional supplements as do most of the traditional food fermentations. Medium supplementation is necessary in non-traditional SSF fermentations, as it induces enzyme-synthesis, provides balanced growth conditions for mycelial-colonisation and biomass formation, as well as prolonging the production of secondary metabolites. SSF employing brown-rot fungi, require an additional carbon source for the induction of enzymes for the cellulose-utilisation. Certain fungi including *Lentinus lapidus, Poria monticola*, and *Lezites trabea* can be cultivated on lignin-containing natural wood substrates from aspen, pine and spruce, when the SSF medium is supplemented with glucose or cellobiose in smaller quantities of 0.5%, w/v, and an even smaller amount of peptone, asparagine and yeast extract. In unsupplemented media, growth of these fungi was very slow as negligible. A co-metabolite, such as glucose or cellulose, stimulates the lignin-degrading system in white-rot fungi such as *Phaenerochaete chrysosporium* and *Coriolus versicolor* when these organisms are cultivated on spruce lignin. Other supplementations of cellobiose, mannose, xylose, glycerol or succinate have been found less effective.

Studies for the nutritional requirements for a developmental microbe-substrate system to be used on a large-scale SSF, can be done in preliminary experiments in small-scale liquid or SSF on laboratory scale. There is a procedure for evaluating the effects of nutritional supplements on culture-growth and product formation, in which microbial-cultures and the solid substrate are contained in separate compartments divided by a membrane with a molecular-weight-cut-off. The membrane permits the passage of enzymes and small molecular weight compounds but restricts microbial and substrate solids. One of the major difficulties in the development of solid state fermentations has been the problem in separating microbial biomass from the solid substrate particles after the mycelial growth has covered the substrate surfaces. In solid culture cultivation the microorganism and substrate are intimately associated making the analytical methods of limited value in stoichiometric analysis of SSF. The analysis of biomass yield and growth rate by the measurement of glucosamine, protein, RNA, DNA, oxygen consumption, and carbon dioxide or heat evolution, can not be accurately used in samples of SSF.

Solid cultures for the production of secondary metabolites may have another problem in that the nutrient, whose deficiency triggers the pathway leading to formation of secondary metabolite, may be available in excess when the microbial growth becomes limited by other nutrient. Therefore, the selection of a solid substrate and required-supplements is more critical for a SSF process for antibiotic production that for a SSF designed for enzyme and organic acid biosynthesis.

10.4.3 Performance Control by Moisture Content of Agro Residues

Solid-state or solid-substrate fermentation means that the substrate is moistened, often with a thin layer of water on the surface of the particles, although there is not enough water present to make a fluid mixture. Weight ratios of water to substrate in SSF are usually between 1:1 and 1:10 (Reid 1989a, b). Since biological activity ceases below a moisture content of about 12%, this establishes the lower limit at which SSF can take place. The upper limit is a function of absorbency and hence, moisture content varies with the substrate material type.

Solid substrates may be viewed as gas-liquid-solid mixtures. The aqueous phase in such mixtures is intimately associated with solid surfaces in various states of sorption. The aqueous phase in a cultivation system is in contact with the gas phase continuous with the external gas environment. Different types of solid substrates can absorb different amounts of water. Depending on the moisture content of the solid; some of the water is tightly bound to solid surfaces, some amount of water is less tightly bound and remaining water may exist in a free state inside the capillary regions of the solid substrates. The gas-liquid interface provides a boundary for gaseous exchange between carbon dioxide and oxygen as well as for heat exchanges.

Water in biological materials exists in three states. The moisture isotherm measurements determines that the solids sorb or desorb water vapour in equilibrium with relative humidities in a gas phase (water activities), which can be maintained by saturated salt solutions at a constant temperature. Water is tightly bound to solid surfaces at the surface in a monolayer region. In case of agricultural residues, monolayer binding is generally 5 to 10 g per 100 g of dry solids. Beyond the surface monolayer in a multilayer region, water is less tightly bound in additional layers at progressively decreasing energy levels. Then beyond the multilayer region, free water exists in a region of capillary condensation. In terms of relationships between water activity and moisture content, the distinction between the multilayer and capillary regions is ambiguous. The electric measurements of an agricultural residue containing high starch content has been used to determine the dividing line between multilayer and capillary regions. The dividing line was defined by a moisture content of about 25 to 30% by weight at a water activity of 80 to 85%, which is the lower limit for microbial growth except for some halophilic or osmophilic microbes.

The sorption isotherm may vary from one type of product to another, the hysteresis is seen in sorption and desorption isotherms. Water may exist in free state at moisture levels of interest in solid state fermentation, which is in contrast with general perception about SSF that the free water does not exist in such systems. Moisture is a critical factor in SSF of aflatoxin production on rice; the yields of aflatoxins have been found decreasing rapidly at moistures above 40%. The rice particles become sticky at moistures above 30 to 35%. Moisture content plays an important role on the growth of lactic acid bacteria on feedlot wastes liquids mixed with cracked corn; growth and acid production was limited at moisture level less than 35%, whereas the higher level above 42% in SSF-mixtures caused the contents to become gummy and aggregate. One of the secrets of a successful SSF-process is to keep the fermenting substrate moist enough for fungal-growth and colonisation and to avoid higher moisture level not to promote the unwanted bacterial growth. Therefore, the optimum moisture content for a particular type of SSF for its microbe-substrate system should be determined for a particular end-product and cultivation conditions of that SSF.

The level of moisture content affects the process productivity significantly in any SSF system, when available in lower or higher quantities than the optimum value (Lonsane et al. 1985). Hence, it should be in limited and required amounts in system. The presence of an optimum moisture content in SSF medium has been emphasised also for the cultivation of bacterial cultures (Ramesh and Lonsane 1990). The process productivities are affected by water content because the physiochemical properties of the solids depend and vary with moisture available to them. Therefore, the major key factors determining the outcome of the SSF-process are the moisture content and the relative humidity levels (Lonsane et al. 1985).

Heat removal during fermentation is mostly achieved by evaporative cooling. This leads to an uneven distribution of water in system due to large quantities of water evaporation. Workers have practised various ways to maintain the moisture content of the solids (Lonsane et al. 1985; Ahmed et al. 1987).

10.4.3.1 Control of Water Activity Factor in SSF

Water activity of the substrate has been proposed as the condition of growth and viability of the microbes and hence, the importance of a_w in SSF has widely been studied (Nishio et al. 1979; Raimbault and Alazard 1980; Kim et al. 1985). Water activity is defined as the relative humidity of the gaseous atmosphere in equilibrium with the substrate and the water activity factor, a_w of the substrate quantitatively expresses the water requirement for microbial activity (Smith et al. 1985).

$a_w = -Vm\,\phi/55.5$ where,
V = number of ions formed,
m = Molar concentration of solute
ϕ = Molar osmotic coefficient, and
55.5 = molar concentration of a solution of pure water.

Pure water has an $a_w = 1.00$ and it will decrease with the presence of solutes.

The types of the microorganisms that can grow in SSF systems are determined by the water activity factor, a_w. Bacteria mainly grow at higher a_w values while filamentous fungi and some yeasts can grow at lower a_w values (0.6–0.7). The microorganisms capable of carrying out their metabolic activities at lower a_w values are suitable for SSF processes. High a_w favours sporulation in the course of growth in SSF, but low a_w favours spore germination and mycelial growth.

Numerous experiments have demonstrated the influence of a_w on microbial-metabolism, such as, on growth rate and sporogenesis of filamentous fungi (Gervais et al. 1988), on enzyme biosynthesis by fungi (Grajek and Gervais 1987), and on cheese aroma production (Gervais et al. 1988).

The a_w of the medium is a fundamental parameter for mass transfer of the water and solutes across the cell membrane (Gervais and Sarrette 1990). The control of this parameter could be used to modify the metabolic production or excretion of a microorganism (Gervais 1989, 1990). A theoretical calculation based on the Ross equation showed that a_w decreased towards the end of the SSF-culture (Oriol et al. 1988). A kinetic model which relates the rate constant of the death of the microbial cells to a_w and temperature has been proposed by Moser (1988), using the equation

$$k = k_\alpha \, a_w \, \exp - E_A \, a_w / RT$$

Constants k_α and E_A are calculated from the experimental value of a_w. Regulation of the a_w can be controlled by the relative humidity of the air. Gervais and Bazelin (1986) reported a SSF process allowing the control of a_w and Gervais (1989) developed a new sensor for the continuous a_w measurement in SSF.

10.5 Microorganisms Used for Agro-Residues Bioconversion

Selection of a suitable microorganism is one of the most important criteria in SSF. The vast majority of wild type microorganisms are incapable of producing commercially acceptable yields of the desired products. The unique characteristics of solid-state cultivations are their ability to provide a selective environment at lower concentrations of moisture ideal for mycelial organisms. The mycelial organisms are capable of producing a range of extracellular enzymes required for the hydrolysis of complex, polymeric solid substrates. Such microorganisms are able to colonise at high nutrient concentrations near solid surfaces. The mycelial organisms include a large number of filamentous fungi and a few bacteria of actinomycetes. The importance of microorganisms can be seen from the fact that a culture of *Aspergillus niger* can produce as many as 19 types of enzymes, while enzyme alpha amylase can be produced by some 28 different types of cultures (Fogarty and Kelly 1979; Pandey 1992). SSF processes can be placed in two main classes based on the type of microorganism involved:

1. *Natural (Indigenous) SSF*: Ensiling and composting are SSF processes, that utilise natural microflora. In nature, SSF is often carried out by mixed cultures in which several microorganisms show symbiotic cooperation.

2. *Pure culture SSF*: Known purified microorganisms are used in such processes either singly or in mixed culture. SSF using a pure culture is known since antiquity e.g. the Koji process with *Aspergillus oryzae*. A pure culture is necessary in industrial SSF process for improved rate of substrate utilisation and controlled product formation. A typical example of pure mixed culture SSF is the bioconversion of agricultural residues to fungal biomass (protein) using two pure cultures of *Chaetomium cellulolyticum* and *Candida utilis*.

Several microorganisms have been employed in a wide range of SSF processes for various objectives. The cultivation of filamentous fungi on solid substrates has been widely used for different purposes at laboratory scale e.g. for Koji fermentation, for lignocellulose fermentation (Matteau and Bone 1980), for fungal spores (Lotong and Suwarnarit 1983), and for mycotoxin production (Hesseltine 1972; Bhumiratna et al. 1980). For various purposes, among the filamentous fungi three classes, viz. Phycomycetes (*Mucor* and *Rhizopus*), Ascomycetes (*Aspergillus* and *Penicillium*) and Basidiomycetes (Nigam and Prabhu 1985), have been most widely used.

SSF has been most commonly used employing *Aspergillus niger* for protein enrichment (Rodriquez et al. 1985; Baldensperger et al. 1985; Czajkowska and Ilnicka 1988) as well as for enzymes production, such as, cellulase (Singh et al. 1989; Madamwar et al. 1989), amylase, glucoamylase (Ramakrishna et al. 1982; Pandey 1990), beta glucosidase, and protease (Malathi and Chakrabarty 1991). Production of alcohols, ketones and aldehyde in rice fermentation was achieved by the use of *A. oryzae* (Ito et al. 1990). For protein enrichment and kinetic studies related to SSF process *Rhizopus oligosporus* has been employed (Rathbun and Shuler 1983; Mitchell et al. 1988, 1990).

Fungal rennet has been produced by *R. oligosporus* and *Mucor meihei* (Karanth 1988). For enzyme production and protein enrichment cultures of *Trichoderma* spp. have been employed in pure, single and mixed SSF (Daubresses et al. 1987; Grajek 1988). Lipase enzyme production has been reported (Munoz et al. 1991) using six species of *Penicillium*, two species of *Rhizopus, Geotrichum candidum* and *Mucor meihei*, whereas the maximum lipase activity was obtained with *P. candidum, P. camembertii* and *M. meihei*. For the production of several other enzymes e.g. hydrolases and pectic enzymes (Siesser and Said 1989) several other species of *Penicillium* have been employed in SSF.

Production of the antibiotic penicillin was achieved in a non-sterile SSF process on sugar cane bagasse impregnated with culture medium using *Penicillium chrysogenum*. Protein enrichment of lignocellulosic substrates for animal feed production (Nigam 1990; Nigam and Vogel 1990a, b), lignin degradation (Bone and Munoz 1984), and cellulase and ligninase enzyme production (Nigam et al. 1987a, b) have been obtained by white-rot cultures in SSF.

Production of gibberellic acid has been reported using *Fusarium monoliforme* and *Gibberella fugikuroi* (Kumar and Lonsane 1987a, b). Bacterial alpha amylase production is reported using *Bacillus licheniformis* in SSF (Ramesh and Lonsane 1987a, b, 1990). Several yeasts have been used for protein enrichment and ethanol fermentation in SSF. For protein enrichment of straw (Han 1987) *Candida utilis* was used whereas *Saccharomyces cerevisiae* has most commonly been employed for ethanol production (Gibbons et al. 1984; Kargi et al. 1985).

10.6 Designing and Types of SSF

10.6.1 Fermenter Design for SSF

Several miscellaneous types of fermenters have been used in batch or continuous mode in SSF processes (Hardin 2004). Process parameters are very important factors and they have to be considered in a bioreactor design for any SSF. Design considerations in types of SS-fermenters used by various researchers are described by Aidoo et al. (1982). The engineering aspects, with major types of fermenters describing their advantages and drawbacks has been reviewed by Fernandez et al. (2004). Solid state cultivations are not as well characterised on a fundamental scientific or engineering basis, as are the liquid fermentation systems that are used in the West for the industrial production of microbial-metabolites. Solid-state fermentations are, however, widely used in the Orient and therefore, the old traditional methods of cultivation systems which have been used in food-processing for more than 2,000 years, have now been modernised and well characterised for their extended application to non-traditional products. Mitchell et al. (2004) have described in detail the modelling aspects of SSF.

The physical state of the substrate and the products to be produced in the system characterise the design-type of solid state cultivation process:

a. Low-moisture solids are fermented

   1. without any agitation for the production of Tempeh and Natto;
   2. by occasional stirring for the production of Miso and Soy sauce;
   3. with continuous stirring for the production of Aflatoxin.

b. Suspended solids are fermented in packed bed columns

   1. through which the liquid is circulated, as for the production of rice-wine;
   2. which contain stationary or agitated liquid media, for the production of Kaffir beer.

10.6.2 Types of SSF Systems

There are two types based on process design:

Type one- Fermentation in static reactor
e.g. Tray fermentations (Lonsane et al. 1985; Viesturs et al. 1987)

Type two- Fermentation with occasional or continuous agitation
e.g. Production of aflatoxin, ochtratoxin and enzymes (Lindenfelser and Ciegler 1975; Hesseltine 1977; Silman 1980).

Type two has 4 variations according to the need of process:

1. Occasional agitation, without forced aeration
2. Slow continuous agitation, without forced aeration
3. Occasional agitation with forced aeration
4. Continuous agitation with forced aeration.

10.6.3 SSF Bioreactors

Three basic groups of reactor exist for SSF, and these may be distinguished by the type of mixing and aeration used. In laboratory scale, SSF occurs mainly in flasks whereas following reactors are used for large-scale product-formation.

10.6.3.1 Tray Bioreactors

Tray bioreactors tend to be very simple in design, with no forced aeration or mixing of the solid substrate. Such reactors are restrictive in the amount of substrate that can be fermented, as only thin layers can be used, so as to avoid overheating and maintain aerobic conditions. Tray undersides are perforated to allow aeration of the solid substrate, each arranged above each other. In such reactors, temperature and relative humidity are the only controllable external parameters (Durand 1998). Wooden trays were initially used for soy sauce production in Koji fermentations by *Aspergillus oryzae*. The use of tray fermenters in large-scale production is limited as they require a large operational area and tend to be labour intensive. The lack of adaptability of this type of fermenter makes it an unattractive design for any large-scale production.

10.6.3.2 Drum Bioreactors

Drum bioreactors are designed to allow adequate aeration and mixing of the solid, whilst limiting the damage to the inoculum or product. As previously mentioned, mixing and aeration of the medium has been explored in two ways: by rotating the entire vessel or through the use of various agitation devices. Rotation or the use of agitation can be carried out on a continuous or periodic basis. In contrast to tray reactors, growth of the inoculum in drum bioreactors is considered to be better and more uniform. Increased sheer forces through mixing, can however, have a detrimental affect on the ultimate product yield.

Although the mass heat transfer, aeration and mixing of the substrate is increased, damage to inoculum and heat build up through sheer forces may affect the final product yield. Application of drum reactors for large-scale fermentations also poses handling difficulties.

10.6.3.3 Packed Bed Bioreactors

Columns are usually constructed from glass or plastic with the solid substrate supported on a perforated base through which forced aeration is applied. They have been successfully used for the production of enzymes, organic acids and secondary metabolites. Forced aeration is generally applied at the bottom of the column, with the humidity of the air kept high to avoid desiccation of the substrate. Disadvantages associated with packed bed column bioreactors for SSF include difficulties in retrieving the product, non-uniform growth, poor heat removal and scale-up problems.

10.7 Scale-Up Stages of SSF

Scale-up of SSF has been defined in many ways. There are mainly four stages:

10.7.1 Flask Level

This is smallest scale using 50–1000 g substrate working capacity, and used for the selection of the organism, optimisation of the process and experimental variables in a short time and at low cost. The vessels used are conical flasks and beakers (Mitchell et al. 1986; Nigam et al. 1987a, b), Roux bottles (Gervais et al. 1988; Nigam 1990), jars (Hang et al. 1986), and glass tubes (Raimbault and Alazard 1980).

10.7.2 Laboratory Fermenter Level

This is next to flask scale using a 5–20 kg substrate working capacity. It is used for a selection of procedures such as, inoculum development, medium sterilisation, aeration, agitation and downstream processing. Standardisation of various parameters, selection of control strategies and instruments, evaluation of economics of the process and its commercial feasibility are also examined at this level. The fermenters used are glass incubators (Deschamps and Huet 1984; Oriol et al. 1988; Smith et al. 1986), column fermenters (Oriol et al. 1988); polypropylene bags (Yadav 1988), and miscellaneous types of fermenters (Raimbault and Alazard 1980; Viesturs et al. 1981).

10.7.3 Pilot Fermenter Level

This scale is a stage before the commercial scale using 50–5000 kg of substrate. This level is necessary for the confirmation of laboratory data and selection of optimised procedures. It facilitates market trials of the product, physicochemical characterisation and determination of viability of the process. Most large scale SSFs employ tray type fermenters as in the oldest soy sauce Koji process (Daubresse et al. 1987), rotating drum type (Lindenfelser and Ciegler 1975; Han and Anderson 1975; Hesseltine 1977), horizontal paddle fermenters and mixed layer pilot plant fermenters (Laukevics et al. 1984). Durand and Chereau (1988) reported the use of a pilot reactor having a one ton working capacity.

10.7.4 Production Fermenter Level

The commercial scale fermenter utilises 25–1000 tonnes of substrate and is performed for streamlining of the developed process. Yokotsuka (1985) described deep trough methods and mechanical continuous equipment for Koji production generating 50–100 tonnes of Koji per day.

10.8 Factors Affecting SSF

Each microbe-substrate system is unique and the process variables must be considered in terms of the physical properties and chemical composition of its substrate, growth characteristics and physiological properties of the microorganisms to be cultivated in SSF. The nature of the product, if the process involves the synthesis of primary or secondary metabolite may be based on the synthesis of extracellular enzymes in growth-associated metabolism. The process variables affecting a solid state cultivation include, pretreatment of substrates, particle-size of substrates, medium-ingredients, supplementation of growth medium, sterilisation of SSF-medium, moisture-content, inoculum-density, temperature, pH, agitation and aeration. These variables should be considered in process-development of a SSF to be carried out for different purposes. Some of these variables have been discussed in some sections as above, the rest are discussed below.

10.8.1 Significance of Aeration and Mixing in SSF

In any SSF-process an adequate supply of oxygen is required to maintain the aerobic conditions and for the transfer of excess carbon dioxide produced during metabolism. This requirement can be achieved through the process of aeration and mixing of the fermenting solids. In certain cases, the mixture can not be agitated vigorously or in some cases, at all, if the microorganism used in SSF is shear sensitive. The shear sensitivity is attributed to disruption of mycelial-substrate contact; this is particularly concerned to those organisms which possess mycelial-bound enzymes required for the hydrolysis of solid substrate-polymers. Most Koji processes in Japan performed for the commercial production of enzymes do not involve great agitation. The fermenting substrate is gently turned periodically just to bring the bottom of Koji to the top. These processes have been developed in highly controlled environments, using automated systems for inoculum mixing, and turning of the fermenting substrate.

Most of the traditional food-fermentation in Japan use the rotary-tray method for SSF with the circulation of humidified air to create the conditions suitable for gas-exchange and heat-transfer. In the SSF for the production of certain secondary metabolites such as aflatoxin and ochratoxin, and in some processes for the enzyme production, mixing and particle separation are achieved by agitation on shakers or in rotating vessels with circulating conditioned air. Maximum rotation rates generally decrease with the size of the fermentation-vessel. Therefore, solid-state fermentations are ideal for the cultivation of those microorganisms that are extremely sensitive to the shear rates of the impeller speeds required for stringent oxygen demand rates in liquid fermentaton. Such microorganisms colonise the solid substrates by microbe-substrate attachment and there is no pellet formation in solid-state cultivation, which is added advantage to SSF.

Aeration plays an important role in solid state fermentations as compared to liquid fermentation where it only helps in gas transfer. Aeration facilitates in heat, gas and moisture transfer between the fermenting solid particles and the gas environment of the system. The temperature of the gas phase serves by supplying or removing heat, in maintaining the relative humidity in equilibrium with the liquid phase. In liquid fermentations the substrates are dissolved in at low substrate concentrations in large volumes of fluid, but in solid cultures with respect to moisture transfer, the loss or gain of moisture during SSF is extremely sensitive to the water activity of the gas-phase. Therefore, small changes in the relative humidity of the gas phase in equilibrium with the solids may cause the large changes of moisture content in the solid state, depending on the sorption-desorption characteristics of the solid substrate.

There are two main functions of the gas phase in SSF, the primary function is to supply oxygen and remove the carbon dioxide from the system. The secondary function of aeration is in heat and moisture transfer that is more important, when the rates of oxygen and carbon dioxide are not limiting. The gas phase can facilitate in the control of solid cultures, due to the fact that direct measurements can not be performed to estimate dissolved oxygen or carbon dioxide concentrations in low-moisture solids during the course of the fermentation on either a continuous or sampling basis. The methods of aeration may cause the conditions of gas transfer being relatively stagnant. This condition may be responsible for the oxygen limitation at small penetration depths or may lead to inhibitory carbon dioxide concentrations in normal atmospheric environments. The gas phase in the SSF during the course of microbial metabolism, can be analysed for oxygen, and carbon dioxide pressures using analysers which function on thermal-conductivity, paramagnetism, or infrared absorption. The technique of gas chromatography can also be used for gas-analysis of the gas phase of a SSF.

10.8.2 Significance of Control of Temperature and pH in SSF

Two significant variables affecting any SSF are the incubation temperature and the pH of SSF-medium. Both variables are specific for each SSF process depending on the microoganisms to be cultivated and the product to be formed. Unlike submerged fermentation, these factors are difficult to control in SSF. These variables can not be directly measured in the liquid phase, as these are associated with the solids at lower moisture content without any free liquid in the fermenting medium. The other difficult situation arises when the growth temperature of cultivated microorganism is different than the optimal temperature for the product formation. Such systems require a possible need for temperature profiling or shift in the later stages of fermentation. The thermal gradients may be induced within SSF-mixture due to the rate of heat generation in SSF-system at high levels of biological activity. This gradient may limit the heat transfer and may lead to sub-optimal conditions for microbial-biomass and product formation.

The local pH levels at solid surfaces near which the biological activity occurs, may be considerable different than the bulk pH of the liquid phase. This difference in pH levels happens due to surface charge effects and ionic equilibria modified by solute transport effects. There is no suitable method to measure the precise pH of fermenting solid residues in SSF. A general method used for measuring pH of solid agricultural residues involves mixing one part of fermented solids (dry weight) and three parts of freshly boiled and cooled water, and measuring the pH of the resultant liquid after five minutes using a glass electrode. This procedure can be used to monitor pH changes during fermentation on intervals using minimum one gram of the SSF-mixture.

It is easier to measure temperature of the fermenting SSF-mixture, in comparison to pH measurement. Temperature can be measured using thermistor or thermocouple probes at various depths of the SSF-mixture below the medium-surface. In various SSF-processes for the production of enzymes, mycelial-biomass or organic acids, total heat generation of up to 600 kcal per kilogram of fermenting solids has been observed. A study of composting of animal wastes and agricultural residue has revealed that such heat generations may lead to rapid temperature rise of the fermenting mass in the system limited by heat transfer. The study also revealed that the biological activity was considerably higher near the surface of the compost pile than in the depth of pile that was at lower oxygen pressure. This phenomenon happens due to a decrease in interior oxygen concentrations inside the SSF-mixture pile of compost. Thus the heat generation in such fermentations is coupled to conditions for heat as well as mass transfer.

10.9 Processes and Products of SSF

Various processes and products from bioconversion of agro-residues of industrial, pharmaceutical, and environmental importance have been discussed in detail in further chapters 11–24 under sections II, III.

References

Abdullah AL, Tengerdy RP, and Murphy VG (1985) Optimization of solid-state fermentation of wheat straw. *Biotechnol Bioeng* **27**:20–27

Ahmed SY, Lonsane BK, Ghildyal NP, and Ramakrishna SV, (1987a) Design of solid-state fermenter for production of fungal metabolites on large-scale. *Biotechnol Tech* **1**:97–102

Aidoo K E, Henry R, and Wood BJB (1982) Solid substrate fermentations. *Adv Appl Microbiol* **28**:201–237

Baldensperger J, Le MerJ, Hannibal L, and Quinto PJ (1985) Solid-state fermentation of banana wastes. *Biotechnol Lett* **7**(10):743–748

Barrios-Gonzalez J, Rodriguez GM, and Tomasini A (1990) Environmental and nutritional factors controlling aflotoxin production in cassava SSF. *J Ferment Bioeng* **70**(5):329–333

Barrios-Gonzalez J, Tomasini A, Viniegra-Gonzalez G, and Lopez L (1988a) Penicillin production by solid-state fermentation. *Biotechnol Lett* **10**(11):793–798

Barrios-Gonzaloz J, Tomasini A, Viniegra-Gonzalez G, and Lopez L (1988b) *Solid-State Fermentation in Bioconversion of Agro-Industrial Raw Materials* (M. Raimbault ed). ORSTOM Centre Montpellier, Montpellier, France, pp. 39–51

Bhumiratna A, Flegel TW, Glinsukon T, and Somporan W (1980) Isolation and analysis of moulds from soy sauce koji in Thailand. *Appl Environ Microbiol* **39**: 430–435

Bone DH, and Munoz EL (1984) Solid-state fermentation of oat straw by Poyporus spp. *Biotechnol Lett* **6**(10):657–662

Bushell ME, and Slater JH (1981) Mixed Culture Fermentations. Special publication No 5 *Soc. General Microbiology*, Academic Press, London

Castaneda GS, Rojas M, Bacquet G, Raimbault M, and Gonzalez GV (1990) Heat transfer simulation in solid-state fermentation. *Biotechnol Bioeng* **35**:802–808

Considine PJ, O'Rorke A, Hackett TJ, and Coughlan MP (1988) Hydrolysis of beet pulp polysaccharides by extracts of solid-state cultures of *Penicillium capsulatum. Biotechnol Bioeng* **31**(5):433–438

Czajkowska D, and Ilnicka O (1988) Biosynthesis of protein by microscopic fungi in solid-state fermentation. *Acta Biotechnologica* **8**(5):407–413

Daubresse P, Ntibashirwa S, Gheysen A, and Meyer JA (1987) A process for protein enrichment of cassava by SSF in rural conditions. *Biotechnol Bioeng* **29**:962–968

Deschamps F, and Huet MC (1984) β-glucosidase production in agitated solid-state fermentation. *Biotechnol Lett* **6**:55–60

Desgrenges C, and Durand A (1990) Effect of pCO_2 on growth, conidiation and enzyme production in solid-state culture on *A. niger* and *T. viride. Enzyme Microb Technol* **12**:546–551

Desfarges C, Larroche C, and Gros JB (1987) Spore production of *P. roquefortii* by SSF: stoichiometry, growth and sporulation behaviour. *Biotechnol Bioeng* **29**:1050–58

Durand A (1998) Solid state fermentation. *Biofuture* **181**:41–43

Durand A, and Chereau D (1988) A new pilot reactor for solid-state fermentation: application to protein enrichment of sugar beet pulp. *Biotechnol Bioeng* **31**:476–486

Fernandez MF, Pe'rez-Correa R, and Agosin E (2004) Engineering aspects of SSF. In *Concise Encyclopedia of Bioresource Technology*, (A. Pandey, ed.). The Haworth press Inc. NY, pp.690–699

Fogarty WM, and Kelly CT (1979) *Progress in Industrial Microbiology* (M. J. Bull, ed.). 15 Elsevier

Gervais P, and Bazelin C (1986) Development of a solid substrate fermenter allowing the control of the substrate water activity. *Biotechnol Lett* 8(3):191–196

Gervais P, Belin JM, Grajek W, and Sarrett M (1988) Influence of water activity on aroma production by *Trichoderma viride* growing on solid substrate. *J Ferment Technol* **66**(4): 403–407

Gervais P (1989) New sensor allowing continuous water activity measurement of submerged or solid-state fermentations. *Biotechnol Bioeng* **33**: 266–271

Gervais P (1990) Water activity: A fundamental parameter of aroma production by microorganisms. *Appl Microbiol Biotechnol* **33**(1):72–75

Gervais P, and Sarrette M (1990) Influence of age of mycelium and water activity of the medium on aroma production. *J Ferment Technol* **69**(1):46–50

Gibbon WR, Westby CA, and Dobbs TL (1984) A continuous farm scale solid phase fermentation process for fuel ethanol and protein production from fodder beets. *Biotechnol Bioeng* **26**: 1098–1107

Gonzalez-Blanco P, Sancedo-Castarieda G, and Viniegra-Gonzalez G (1990) Protein enrichment of sugarcane by-products using solid state cultures of *Aspergillus terreus*. *J Ferment Bioeng* 70:351–354

Grajek W (1988) Production of protein by thermophilic fungi from sugar beet pulp in SSF. *Biotechnol Bioeng* **32**(2):255–260

Grajek W, and Gervais P (1987) Influence of water activity on the enzyme biosynthesis and enzyme activities produced by *Trichoderma viride* TS in solid state fermentation. *Enz Microbial Technol* 11:658–668

Gujral GS, Bisaria R, Madan M, and Vasudevan P (1987) SSF of saccharum munja residues into food through *Pleurotus* cultivation. *J Ferment Technol* **65**(1):101–106

Han YW (1987) Oxygen requirements for growth of *Candida utilis* on semi-solid straw substrate. *Biotechnol Bioeng* **30**(5):672–674

Han YW, and Anderson AW (1975) Semisolid fermentation of rye-grass straw. *Appl Microbiol* **30**:930–934

Hang YD, Lee CY, and Woodams EE (1986) Solid-state fermentation of grape pomace for ethanol production. *Biotechnol Lett* **8**(1):53–56

Hardin M (2004) Design of bioreactors in SSF. In *Concise Encyclopedia of Bioresource Technology* (A. Pandey, ed.). The Haworth press Inc. NY, pp. 679–688

Hesseltine CW (1972) Biotechnology report: Solid-state fermentations. *Biotechnol Bioeng* **14**: 517–532

Hesseltine CW (1977) Solid-state fermentation. *Process Biochem* **12**:24–27(a), 30–32(b)

Ito K, Yoshida K, Ishikawa T, and Kobayashi S (1990) Volatile compounds produced by the fungus *Aspergillus oryzae* in rice koji and their changes during cultivation. *J Ferment Bioeng* **70**(3):169–172

Jermini MFG, and Demain AL (1989) SSF for cephalosporin production by *S. clavuligerus* and *Cephalosporium acremonium. Experentia* **45**:1061–1065

Karanth NG (1988) CFTRI work on solid state fermentation. In *International Seminar on SSF*. ORSTOM, Montepellier, France, pp. 25–27

Kargi F, and Curme JA (1985) Solid-state fermentation of sweet sorghum to ethanol in a rotary drum fermenter. *Biotechnol Bioeng* **27**:1122–1125

Kim DH, Hosobuchi M, and Ryu D (1985) Cellulase production by a solid-state culture system. *Biotechnol Bioeng* **27**:1445–1450

Koutinas, AA, Malbranque F, Wang R-H, Campbell GM, and Webb C, (2007). Development of an oat-based biorefinery for the production of lactic acid by *Rhizopus oryzae* and various value-added co-products. *J Agric Food Chem* 55:1755–1761

Kumar PKR, and Lonsane BK (1987a) Extraction of gibberellic acid from dry mouldy bran produced under solid-state fermentation. *Process Biochem* **22**:139–143

Kumar PKR, and Lonsane BK (1987b) Potential of fed-batch culture in solid-state fermentation for production of gibberellic acid. *Biotechnol Lett* **9**:179–182

Kumar PKR, and Lonsane BK (1987c) Gibberellic acid by SSF: consistent and improved yields. *Biotechnol Bioeng* **30**:267–271

Laukevics JJ, Apsite AF, Viesturs HE, and Tengerdy RE (1984) Solid-state fermentation of wheat straw for fungal protein. *Biotechnol Bioeng* **26**:1465–1474

Lindenfelser LA, and Ciegler A (1975) Solid-substrate fermentations for ochratoxin-A production. *Appl Microbiol* **29**:322–327

Lonsane BK, Ghildyal NP, Budiatman S, and Ramakrishna, SV (1985) Engineering aspects of solid-state fermentation. *Enzyme Microb Technol* **7**: 258–265

Lotong N, and Suwarnarit P (1983) Production of soya sauce koji mold spore inoculum in plastic bags. *Appl Environ Microbiol* **46**:1224–1226

Macris BJ, Kekos D, Evangelidou X, Panayotou MG, and Rodis P (1987) SSF of straw for CMCase and B-glucosidase production. *Biotechnol Lett* **9**(9):661–664

Madamwar D, Patel S, and Parik H (1989) Solid-state fermentation for cellulases and betaglucosidase production by *Aspergillus niger. J Ferment Bioeng* **67**:424–426

Malathi S, and Chakrabarty R (1991) Production of alkaline protease by a new *Aspergillus flavus* isolated under solid-substrate fermentation. *Appl Environ Microbiol* **57**(3): 712–716

Maltseva OV, Golovleva LA, Leont'evskii AH, Nerud F, Misurcova Z, and Musilek V (1989) Dynamics of enzymes generating hydrogen peroxide in SSF of *Panus tigrinus* on wheat straw. *Folia Microbiol* **34(3)**:261–266

Matteau PP, and Bone DH (1980) Solid-state fermentation of maple wood by *Polyporus anceps*. *Biotechnol Lett* **2**:127–132

Mishra C, and Leatham GF (1990) Recovery and fractionation of the extracellular degradative enzymes from cultures on a solid lignocellulosic substrate. *J Ferment Bioeng* **69**(1): 8–15

Mitchell DA, Greenfield PF, and Doelle HW (1986) A model substrate for solid state fermentation. *Biotechnol Lett* **8**(11):827–832

Mitchell DA, Doelle HW, and Greenfield PF (1988) Improvement of growth of *Rhizopus oligosporus* on a model solid substrate. *Biotechnol Lett* **10**:497–501

Mitchell DA, Greenfield PF, and Doelle HW (1990) An empirical model of growth of *Rhizopus oligosporus* in SSF. *World J Microbiol Biotechnol* **6**(2): 201–208

Mitchell DA, Von Meien OF, Krieger N (2004) Modelling in SSF. In *Concise Encyclopedia of Bioresource Technology* (A. Pandey, ed.). The Haworth press Inc. NY, pp. 709–717

Moloney AP, O'Rorke A, Considine PJ, and Coughlan MP (1984) Enzymatic saccharification of sugar beet pulp. *Biotechnol Bioeng* **26**:714–718

Moser A (1988) *Bioprocess Technology. Kinetics and Reactors*. Springer-Verlag, Berlin pp. 198–204

Mudgett RE, and Paradis AJ (1985) SSF of natural birch lignin by *P. chrysosporium*. *Enzyme Microb Technol* **7**:150–154

Munoz GR, Valencia JRT, Sanchez S, and Farres A (1991) Production of microbial lipases in a SSF system. *Biotechnol Lett* **13**(4): 277–280

Nigam P, and Singh D (1996a) Processing of agricultural wastes in solid state fermentation for microbial-protein production. *J Sci Ind Res* 55(5–6) pp 373–380

Nigam P, and Singh D (1996b) Processing of agricultural wastes in solid state fermentation for cellulase production. *J Sci Ind Res* **55**(5–6): 457–463

Nigam P, and Singh D (1994) Solid-state (substrate) fermentation systems and their applications in Biotechnology. *J Basic Microbiol* 34(6):405–423

Nigam P (1988) Protein enrichment of bagasse by solid-state fermentation for animal feed. *Proceedings of 5th Convention and Symposium of Bioenergy Society of India,* Baroda Oct 30–31 Dept of non-conventional energy resources, New Delhi

Nigam P (1989a) Mixed culture solid state fermentation of bagasse for animal feed. Production. In *Proceedings of 52nd Annual Convention of Sugar-Technologists' Association of India,* pp. G 53–59

Nigam P (1989b) Studies on dairy-effluent utilization in SSF of bagasse for feed production. In *Symposium Impact of Pollution in and from Food Industries and its Management CFTRI,* Mysore May 4–5, pp. FPM **13**:29

Nigam P (1990) Investigation of some factors important for SSF of bagasse for animal feed production. *Enz Microbial Technol* **12**(10):808–811

Nigam P, and Prabhu KA (1985) Fermentation of bagasse for animal feed. *International Sugar Journal* **87**(1033):17–19

Nigam P, Pandey A, and Prabhu KA (1987a) A note on utilization of bagasse for the production of proteinaceous cattle feed. *Biological Wastes* **19**(4):275–280

Nigam P, Pandey A, and Prabhu KA (1987b) Cellulase and ligninase production by Basidiomycetes culture in solid-state fermentation. *Biological Wastes* **20**(1):1–9

Nigam P, and Vogel M (1988) Selection of preculture conditions for solid-state fermentation of sugar beet pulp. *Biotechnol Lett* **10**(10):755–758

Nigam P, and Vogel M (1990a) Protein-enrichment solid-state fermentation of sugar beet pulp. *American Society of Micribiology Conference on Biotechnology*. ASM Chicago, Illenois, pp. 7–10

Nigam P, and Vogel M (1990b) *Process for the Production of Beet Pulp Feed by Fermentation.* Patent No. DE 3812612 C2 1.3.1 990

Nishio N, Tai K, and Nagai S (1979) Hydrolase production by *A. niger* in solid-state cultivation. *Eur J Appl Micrbiol Biotechnol* **8**:263–270

Opoku AR, and Adoga PA (1980) Two-stage fermentation method for production of protein-enriched feed from cassava. *Enzyme Microb Technol* **2**:241–243

Oriol E, Raimbault M, Roussos S, and Gonzalea GV (1988) Water and water quality in SSF of cassava starch by *A. niger. Appl Microbiol Biotechnol* **27**:498–503

Pandey A (1990) Improvements in solid-state fermentation for gluco-amylase production. *Biological Wastes* **34**(1):11–19

Pandey A (1992) Recent process developments in SSF. *Process Biochem* **27**: 109–117

Pandey A, Francis F, Sabu A, and Soccol CR (2004) General aspects of SSF. In: *Concise Encyclopedia of Bioresource Technology* (A. Pandey, ed.). The Haworth press Inc. NY, pp. 702–708

Pandey A, Soccol CR, Rodreguez-Leon JA, and Nigam P. (2001) *Solid state fermentation in Biotechnology: Fundamentals and Applications.* Asiatech publishers Inc, Delhi

Prado FC, Vandenberghe LPS, Lisboa C, Paca J, Pandey A, and Soccol CR (2004) Relation between citric acid production and respiration rate of *Aspergillus niger* in solid state fermentation. *Eng Life Sci* 4: 179–186

Prema P, Thakur MS, Prapulla SG, Ramakrishna SV, and Lonsane BK (1988) Production of gibberellic acid by solid-state fermentation. *Indian J microbiol* **28**:78–81

Rafmbault M, and Alazard D (1980) Culture method to study fungal growth in solid fermentation. *EurJAppl Microbiol Biotechnol* **9**:199–209

Ramakrishna SV, Suseela T, Ghildyal NP, Jaleel SA, Prema P, Lonsane BK, and Ahmed SY (1982) Recovery of amyloglucosidase from mouldy bran. *Indian J Technol* **20**:476–480

Ramesh MV, and Lonsane BK (1987a) Solid-state fermentation for production of alpha-amylase. *Biotechnol Lett* **9**:323–328

Ramesh MV, and Lonsane BK (1987b) A novel bacterial thermostble alpha-amylase system produced under solid-state fermentation. *Biotechnol Lett* **9**: 501–504

Ramesh MV, and Lonsane BK (1990) Chracteristics and novel features of thermostable alpha-amylase produced under solid-state fermentation. *Starch/Starke* **42**:233–238

Rathbun BL, and Shuler ML (1983) Heat and mass transfer effects in static solid-substrate fermentations. *Biotechnol Bioeng* **25**:929–938

Reid ID (1989a) Solid-state fermentation for biological delignification. *Enz Microbiol Tech* **11**:786–802

Reid ID (1989b) Optimization of SSF for selective delignification of aspen wood. *Enzyme Microb Technol* **11**:804–809

Revah S, and Lebeault JM (1988) In *Solid-State Fermentation in Bioconversion of Agro-industrial Raw Materials* (M. Raimbault, ed.). ORSTOM, Centre Montpellier, France, pp. 53–59

Robinson T, and Nigam P (2008) Remediation of textile dye-waste water using a white rot fungus *Bjerkandera adusta* through solid-state fermentation (SSF). *Appl Biochem Biotechnol* 151(2-3):618–28

Rodriquez JA, Echevania J, Rodriguez FJ, Sierra N, Daniel A, and Martiner O (1985) *Biotechnol Lett* **7**(8):577–580

Shamala TR, and Sreekantiah KR (1986) Production of cellulase and D-xylanase by some selected fungal isolates. *Enzyme Microb Technol* **8**:178–182

Siessere V, and Said S (1989) Pectic enzymes production in SSF using citrus peel pellets. *Biotechnol Lett* **11**(5):343–344

Silman RW (1980) Enzyme formation during solid-state fermentation in rotating vessel. *Biotechnol Bioeng* **22**:411–420

Singh AB, Abidi AB, Darmwal NS, and Agrawal AK (1989) Evaluation of chemical for biodegradation of agricultural lignocellulosic wastes by *A. niger. MIRCEN J Appl Microbiol Biotechnol* **5**(4):451–456

Singh K, Rai SN, Neelkantan S, and Han YW (1990) Biochemical profiles of solid state fermented wheat straw with *Coprinus fimetarius. Indian J Anim Sci* **60**(8):484–490

Smith RE, Osothsilp C, Bicho P, and Gregory KF (1986) Improvement in the protein content of cassava by *S. pulverulentum* in solid state culture. *Biotechnol Lett* **8**(1):31–36

Steinkraus KH (1984) Solid-state (solid substrate) food/beverage fermentations involving fungi. *Acta Biotechnol* **4**:83–88

Thakur MS, Karanth NG, and Krishna N (1990) Production of Fungal rennet using solid state fermentation. *Appl Microbiol Biotechnol* **32**:409–413

Vezina C, and Singh K (1975) Transformation of organic compounds by fungal spores. In *The Filamentous Fungi* vol. 1, (J.E. Smith, J.E. and D.R. Berry, eds.). Edward Arnold, London, pp. 158–192

Vieira MJF, Spadaro ACC, and Said S (1991) Separation of the components of pectolytic complex produced by *T. vulgaris* in solid state culture. *Biotechnol Lett* **13**(1): 39–42

Viesturs UE, Apsite AF, Laukevics JJ, Ose VP, Bekers MJ, and Tengerdy RP (1981) SSF of wheat straw with *C. cellulolyticum* and T. *lignorum. Biotechnol Bioeng Symp* **11**:359–369

Viesturs UE, Strikauska SV, leite MP, Berzincs AJ, and Tengerdy RP (1987) Combined submerged and solid-state fermentation for the bioconversion of cellulose. *Biotechnol Bioeng* **30**:282–288

Wang HH, Chiou JY, Wang JY, Hong CY, and Tein WC (1984) Cephalosporin production by SSF of rice grains. *J Microbiol Immunol* **17**:55–69

Xie G, and West TP (2006) Citric acid production by *Aspergillus niger* on wet corn distillers grains. *Lett Appl Microbiol* **43**(3):269–273

Yadav JS (1988) SSF of wheat straw with *Alcalphilic coprinus. Biotechnol Bioeng* **31**(5):414–417

Yamauchi H, Akita O, Obata T, Amachi T, Hara S, and Yoshizawa K (1989) Production and application of a fruity odor in a solid-state culture of *Neurospora* spp. *Agri Biological Chem* **53**(1):2881–2886

Yang SS, and Ling MY (1989) Tetracycline production with sweet potato residue by SSF. *Biotechnol Bioeng* **33**:1021–1028

Yokotsuka T (1985) Fermented protein foods in the Orient, with emphasis on shoyu and miso in Japan. In *Microbiology of Fermented Foods.* (B.J.B. Wood, ed.). Elsevier Applied Science Publisher, London, pp. 197–247

Part III
Biotechnological Potential of Agro-Industrial Residues for Bioprocesses

Chapter 11
Biotechnological Potentials of Cassava Bagasse

Rojan P. John

Contents

Abstract Crop residues such as cassava bagasse are annually renewable sources of energy. Though they are rich in carbohydrate, their utilization for any direct application is very less due to the low content of protein and poor digestibility. However, the utilization of such agro-industrial residues provides alternative substrate for bioprocesses and will solve the problem of environmental pollution to an extent. Several processes have been developed to utilize cassava bagasse, the fibrous residue of the tropical tuber for the production of value added products such as organic acids, ethanol, aroma, mushroom etc. The chapter focuses on the wide spectrum applications of cassava bagasse in bioprocess technology.

Keywords Agro-industrial residues · Bioprocess technology · Cassava bagasse · Value addition

---

R.P. John (✉)
Biotechnology Division, National Institute for Interdisciplinary Science and Technology, CSIR, Trivandrum–695 019, Kerala, India
e-mail: rojanpj@yahoo.co.in

P. Singh nee' Nigam, A. Pandey (eds.), *Biotechnology for Agro-Industrial Residues Utilisation*, DOI 10.1007/978-1-4020-9942-7_11,

11.1 Introduction

The concept of utilizing excess biomass or waste from agricultural and agro-industrial residues to produce energy, feeds or foods, and other useful products is not necessarily new. Our whole world thinks in the same path to overcome the pollution problems in environmentally sound methods like composting, reuse, recycling, bioconversion etc. Researchers from all over the world adopted many technologies to reduce the pollution and cost of value added products by the use of agricultural wastes. Several processes have been developed to utilize renewable raw materials for the production of value added products such as enzyme, organic acids, ethanol, amino acids, aroma, single cell protein, mushroom etc. Currently, a great deal of attention is being paid on biotechnological potential of agro industrial residues such as cassava bagasse, sugarcane bagasse, sugar beet pulp, coffee husk and pulp, apple pomace, oilcakes, wheat/rice bran etc. for their use as raw material in bioprocesses (John et al. 2007a). Approximately 3.5 billion tons of agricultural residues are produced per annum in the world. Though they are rich in carbohydrate, their utilization is very less due to the low content of protein and poor digestibility (Elkholy and Eltantawy 2000). Recently, fermentation of biomass such as cassava bagasse, wheat bran etc. has gained considerable attention because of the forthcoming scarcity of fossil fuels, and because it is necessary to increase the world food and feed supplies. This chapter focuses on the potentials of cassava bagasse, starchy waste generated from the starch extraction, to generate value added products in eco-friendly manners.

11.2 Cassava- a Global View

Cassava (*Manihot esculenta Cranz*), tropical root crop, is the third most important source of calories in the tropics, after rice and corn. According to FAO, more than 600 million people depend on the cassava in Africa, Asia and Latin America (Tonukari 2004). It gives the highest yield of starch per hectare of any crop ranking it as 4th crop in worldwide production after rice, wheat and maize. Total cassava production is estimated as approximately 166 million tons and is projected to reach approximately 266 million tons by 2020 based on the current production rate. Most of the contribution in the production is from the developing countries and from these the sub-Saharan Africa (Soccol 1996). It is known as tapioca in Asian countries, as aipin, castelinha, and macaxeira in Brazil, as yuca in Spanish-speaking countries of Latin America, and as manioc in French-speaking countries in Africa (Pandey et al. 2000). Cassava is considered to have originated in Venezuela during 2700 BC (Pandey et al. 2000). In a significant research published during May 1999, biologists from the Washington University in St. Louis discovered that cassava originated from the southern border of the Amazon River basin in Brazil. It was introduced in Africa during the 16th century and from there into Asia during the 18th century. It is a bushy plant producing tubers and is made up of an aerial part and an underground part. The aerial part can be as high as 2–4 m with a trunk and branches on it. The underground part consists of two types of roots: the ones responsible for the plant nutrition, and the others with axial disposition surrounding the trunk. These are called tubers and are the edible parts of the plant. Each plant may have 5–20 tubers, and each tuber may attain a length of 20–80 cm and a diameter of 5–10 cm. The fresh weight of each tuber may vary between a few hundred grams and 5 kg.

11.3 Cassava Bagasse

Cassava (*Manihot esculenta* Crantz) belongs to Euphorbiaceae is a short-lived perennial, 1 to 5 meters tall. The origin of cassava is in South America, presumably eastern Brazil. From stem cuttings, the plant produces 5 to 10 very fleshy adventitious roots up to 15 centimeters in diameter. Cassava (*Manihot esculenta* Crantz), tropical root crop, is the third most important source of calories in the tropics, after rice and corn.

According to FAO, more than 600 million people depend on the cassava in Africa, Asia and Latin America (Elkholy and Eltantawy 2000). Total cassava production is estimated as approximately 166 million tons and is projected to reach approximately 266 million tons by 2020 based on the current production rate. Most of the contribution in the production is from the developing countries and from these the sub-Saharan Africa (Tonukari 2004). It is known as tapioca in Asian countries, as aipin, castelinha, and macaxeira in Brazil, as yuca in Spanish-speaking countries of Latin America, and as manioc in French-speaking countries in Africa (Soccol 1996; Pandey et al. 2000). Cassava is considered to have originated in Venezuela during 2700 B.C. (Soccol 1996).

In a significant research published during May 1999, biologists from the Washington University in St. Louis discovered that cassava originated from the southern border of the Amazon River basin in Brazil. It was introduced in Africa during the 16th century and from there into Asia during the 18th century. It is a bushy plant producing tubers and is made up of an aerial part and an underground part. The aerial part can be as high as 2–4 m with a trunk and branches on it. The underground part consists of two types of roots: the ones responsible for the plant nutrition, and the others with axial disposition surrounding the trunk. These are called tubers and are the edible parts of the plant. Each plant may have 5–20 tubers, and each tuber may attain a length of 20–80 cm and a diameter of 5–10 cm. The fresh weight of each tuber may vary between a few hundred grams and 5 kg.

Industrial processing of cassava tubers is mainly done to isolate flour and starch, which generates more liquid and solid residues (processing for flour generates solid residues while for starch generates more liquid residues). Solid residues include brown peel, inner peel, unusable roots, bagasse and flour refuse, among which bagasse is the main residue (Fig. 11.1). Processing about 250–300 t of fresh tubers results about 280 t of wet cassava bagasse. Cassava bagasse is made up fibrous root material and contains starch that physically process could not be extracted. Poor processing conditions may result even higher concentrations of starch in cassava bagasse.

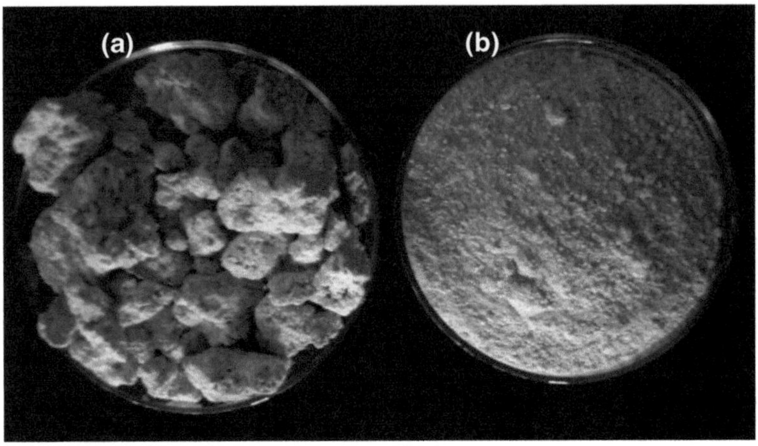

Fig. 11.1 Cassava bagasse (**a**) residue after starch extraction and (**b**) its powder

11.3.1 General Process in Starch Extraction and Generation of Bagasse

Generally, factories only purchase sufficient cassava tubers to fulfill a daily crushing capacity and collected roots are deposited onto a cement floor for short-term storage. Soil and sand are removed as the roots pass through a cylindrical root sieve, rotating at a low rpm. Soil, sand, pieces of broken peels and impurities pass through the sieve. Roots after sieving are transported into a water chamber where they are washed and moved by a paddle blade rotating at low rpm. Water used for washing is re-circulated from later processing stages. Washed roots are chopped with a cutting blade. The small root pieces are gravimetrically fed to the raspers. Fresh rasped root slurry from the rasper is pumped through a series of extractors, from coarse to fine. The extractors are continuous centrifugal perforated baskets. Pulp from the coarse extractor is repeatedly re-extracted using the same screen aperture until minimal starch content of the pulp is achieved. Moisture content of the pulp is then reduced by a screw press and the pulp finally discharged out of the process. Whole processes are generalized in Fig. 11.2. The moisture content of pressed pulp is about 60 to 70% and starch content about 40 to 70% (dry basis). High starch content of the pulp reflects both low efficiency of the rasping stage and composition of the cell membrane and cell walls. The compositions of polysaccharides are in turn influenced by variety and environmental factors.

11.3.2 General Properties of Cassava Bagasse

Bagasse has a large absorption capacity and may contain about 75% moisture. Cassava tuber contains cyanide content but cassava bagasse does not show any cyanide content. However, its poor nitrogen content (2.3%) makes it unattractive as an animal feed. Because of the poor nutrient content other than carbohydrates this waste is

Fig. 11.2 Schematic of cassava starch extraction

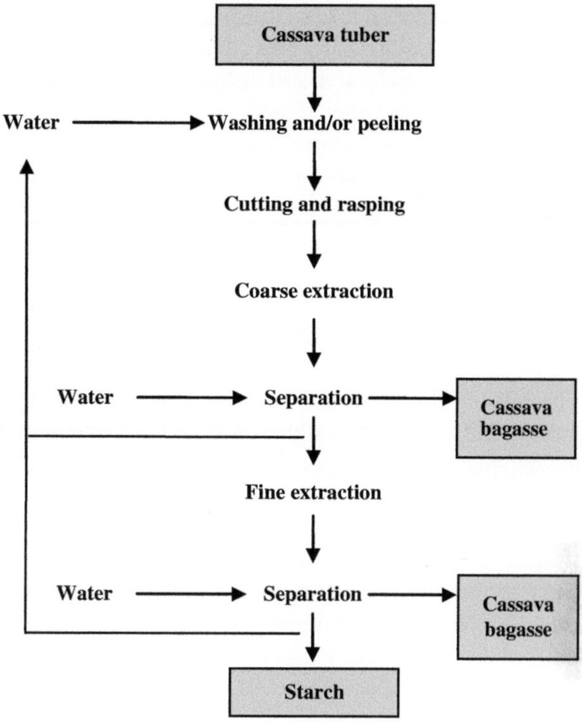

mainly used as the land fill and creates environmental pollution. Application of cassava bagasse as substrate would provide an alternative substrate, and solve pollution problem. Such agro-industrial residues constitute promising alternative substrates for bioprocesses. On the other hand, because of its low ash content (1.44%), it could offer numerous advantages in comparison to other crop residues such as rice straw and wheat straw, which have 7.5 and 11%, respectively, for usage in bioconversion process using microbial cultures. Other components such as calcium, potassium, lipids, phosphorous can be also be found in low concentrations (<1%). Table 11.1 shows the major physico-chemical composition of cassava bagasse reported by different studies.

Table 11.1 Physico-chemical composition of cassava bagasse (g/100 g dry weight)

Composition	Soccol 1994	Cereda 1994	Stertz 1997	Vandenberghe 1998
Moisture	5.02	9.52	10.70	11.20
Protein	1.57	0.32	1.60	1.61
Lipids	1.06	0.83	0.53	0.54
Fibers	50.55	14.88	22.20	21.10
Ash	1.10	0.66	1.50	1.44
Carbohydrates	40.50	63.85	63.40	63.00

Source: Pandey et al. 2000.

Extraction of starch from the tuber is not fully success as a large part of it trapped in the fibrous residue. The only alternatives that can help reduce environmental impact of pulpy waste and add value to the pulp is to recover the starch, either as starch or sugar. In order to accomplish this, physical or biotechnological methods need to be developed. Improved starch recovery from cassava roots and waste can be achieved by using multi-enzyme preparations. Cassava starch entangled with fibers was more efficiently saccharified after treatment with fungal cellulases.

11.3.3 General Cultivation Systems Using Cassava Bagasse

The microbial cultivation on cassava bagasse can broadly be classified into three groups and are processes based on liquid fermentation using hydrolysate prepared by enzymatic or acid hydrolysis, simultaneous saccharification of cassava bagasse and fermentation and processes based on solid-state fermentation (SSF). Due to high water retention capacity makes it an ideal substrate for SSF processes like cultivation of mushrooms and other fungi. Some of the work carried out using cassava bagasse hydrolysate and inert support in SSF processes (Rojan et al. 2005; John et al. 2006a, 2007c). Submerged fermentation (SmF) processes have rarely been utilized due to obvious reasons of cost effectiveness. But simultaneous saccharification and fermentation can be effectively utilized for the production of organic acid like lactic acid using bacterial cultures.

11.3.4 Cassava Bagasse Hydrolysis

Cassava bagasse can be hydrolysed using acid or enzymes. Even though acid hydrolysis may be fast or cost effective, it needs the neutralization of medium and thus increase in salt concentration may occur. Efforts were taken by many researchers for the hydrolysis of cassava bagasse by commercial starch degrading enzymes (Pandey et al. 2000). Woiciechowski et al. 2002 studied the hydrolysis of cassava bagasse starch by acid and enzymatic hydrolysis. They reported both methods were quite efficient when considering one or other parameter like the percentage of hydrolysis, time and cost of the chemicals and energy consumption. Although acid hydrolysis is time saving and cost effective, there will be a neutralizing step after acid hydrolysis and which will create the unnecessary increase of salts in the medium and it will affect the growth and production of lactic acid. So the enzymatic hydrolysis is better as it yields a high percentage of reducing sugars from cassava bagasse. But the enzymatic hydrolysis of 150 kg cassava bagasse required US$ 2470 as power for long time saccharification and cost of enzymes. Simultaneous saccharification reduces the cost of energy consumption for the liquefaction and saccharification and thus it is cost effective and time saving process for bioconversion of cassava bagasse in production of organic acid like lactic acid.

11.3.5 Value Addition of Cassava Bagasse

The molds such as *Rhizopus stolonifer, Neurospora sitophila* and lactic acid bacteria, *Leuconostoc pseudomesenteroides, Leuconostoc mesenteroides, Enterococcus faecium, Weissella cibaria, Lactobacillus plantarum, L. manihotivorans* etc. were identified as the natural microflora in cassava. There is a restriction of the growth of a wide variety microorganism in cassava due to its cyanogenic content. As there is no cyanogenic content in cassava bagasse, it was used as a suitable substrate for many microorganisms to produce the value added products and Table 11.2 shows the microorganism which grow on the bagasse or bagasse hydrolysate. Fibrous starchy cassava waste could utilize to produce many value added products by biotechnological, chemical and physical processes.

11.3.5.1 Xanthan Gum

Xanthan gum is a microbial desiccation-resistant polymer prepared commercially by aerobic submerged fermentation from *Xanthomonas campestris*. Woiciechowski et al. (2004), reported use of cassava bagasse hydrolysate for the production of xanthan gum using a bacterial culture of *X. campestris*. Cassava bagasse hydrolysate with an initial concentration of approx 20 g of glucose/L proved to be the best

Table 11.2 Microorganisms used for the production of various products from cassava bagasse

Organisms	Process	Product	Reference
Aspergillus niger LPB 21	SSF	Citric acid	Kolicheski et al. (1997)
A. niger NRRL 2001	SSF	Citric acid	Vandenberghe et al. (2000)
A. niger CFTRI 30	SSF	Citric acid	Shankaranand and Lonsane (1994)
Candida lipolytica	SmF	Citric acid	Vandenberghe et al. (1998)
C. fimbriata	SSF	Aroma compounds	Bramorski et al. (1998a)
Kluyveromyces marxianus	SSF	Aroma compounds	Medeiros (1998)
Pleurotus sajor-caju	SSF	Mushroom	Barbosa et al. (1995)
Rhizopus arrahizus	SmF	Fumaric acid	Carta et al. (1999)
R. ciricians	SmF	Fumaric acid	Carta et al. (1999)
R. delemer	SmF	Fumaric acid	Carta et al. (1999)
R. formosa	SmF	Fumaric acid	Carta et al. (1998, 1999)
R. oligosporus	SmF	Fumaric acid	Carta et al. (1999)
R. oryzae	SmF	Fumaric acid	Carta et al. (1999)
R. oryzae	SSF	Aroma compounds	Bramorski et al. (1998b)
Lactobacillus casei	SSF	Lactic acid	Rojan et al. 2005
L. delbrueckii	SsF	Lactic acid	Anuradha et al. 1999
L. delbrueckii	SSF	Lactic acid	John et al. 2006a
L. casei and *L. delbrueckii*	SsF	Lactic acid	John et al. 2006b
Xanthomonas campestris	SmF	Xanthan gum	Woiciechowski et al. 2004

SSF: Solid-state Fermentation, SmF: Submerged Fermentation, SsF: Simultaneous Saccharification and Fermentation.

substrate concentration for xanthan gum production. Maximum xanthan gum (about 14 g/L) was produced when fermentation was carried out with a medium containing 19.8 g/L of initial reducing sugars supplemented with potassium nitrate and fermented for 72 h, and it remained almost the same until the end of fermentation (96 h).

11.3.5.2 Aroma Compounds

Bramorski et al. 1998a compared fruity aroma production by *Ceratocystis fimbriata* in solid cultures from several agro-industrial wastes: cassava bagasse, apple pomace, amaranthus and soya bean. Cassava bagasse was used in combination with soya bean or apple pomace. Media containing cassava bagasse with apple pomace or soya bean produced a strong fruity aroma. The components present in the headspace of fermenter were acid, alcohols, aldehyde, ketones and esters. Bramorski et al. 1998b also studied the production of volatile compounds by the edible fungus *Rhizopus oryzae* during solid-state cultivation on tropical agro industrial substrates. When *R. oryzae* was grown on a medium containing cassava bagasse plus soybean meal (5:5, w/w), CO_2 production rate was at its highest (200 mL/L), whereas the highest volatile metabolite production was with amaranth grain as the sole substrate (282.8 mL/L). In the headspace, ethanol was the most abundant compound (more than 80%). A strain of the yeast *Kluyveromyces marxianus* was used for the production of a fruity aroma in SSF and showed the feasibility of cassava bagasse as a substrate (Medeiros 1998).

11.3.5.3 Biopigments

Now, people prefer natural colourants to synthetic colours. Some filamentous fungi, like *Monascus purpureous* can be used to produce the GRAS level pigments. Carvalho et al. (2005) showed the possibility of producing pigments by *Monascus* sp. with cassava bagasse. Good results were obtained with cassava bagasse. In this case, growth and pigment extraction were higher comparing to that commonly obtained with rice. An optimized medium composition was found, showing great perspectives for this process.

11.3.5.4 Mushroom Culturing

Mushrooms are the rich source of nutrients and they can be produced by the utilization of many agro-residues. Barbosa et al. (1995) also compared cassava bagasse and sugarcane bagasse for mushroom production. They used *Pleurotus sajor-caju* and best results were obtained when cassava bagasse was used in a mixture with sugarcane bagasse (8:2, dry weight basis).

11.3.5.5 Citric Acid

Citric acid is a weak organic acid and is used as a natural preservative and is also used to add sour taste to foods and drinks. Citrus fruits are rich source of citric acid and were used in the industrial production of citric acid. In 1917, the American food chemist James Currie discovered that certain strains of the mold *Aspergillus niger* could be efficient citric acid producers and their microbial production replaced the position of citrus fruits for the industrial production of citric acid. Kolicheski et al. 1997 studied citric acid production on three cellulosic supports in SSF. Out of the six strains of *Aspergillus niger* one, LPB 21, was selected for cultivation on cassava bagasse, sugarcane bagasse and vegetable sponge. Cassava bagasse was found to be a good substrate, giving 13.64 g citric acid per 100 g dry substrate. This corresponded to 41.78% yield. Under improved fermentation conditions, the citric acid production increased to 27 g/100 g dry substrate, which corresponded to 70% yield (based on sugars consumed). Shankaranand and Lonsane (1994) presented a comparative profile of citric acid production from various agro industrial residues, such as cassava bagasse, wheat bran, rice bran, sugarcane pressmud, coffee husk, etc., using an indigenous strain of *A. niger* CFTRI 30. Cassava bagasse gave the highest yield of citric acid based on the total starch or sugars present initially in the medium (Shankaranand and Lonsane 1994). Vandenberghe 2000 used three substrates, sugarcane bagasse, coffee husk and cassava bagasse for citric acid production with *A. niger* NRRL 2001. Cassava bagasse best supported the mould's growth, giving the highest yield of citric acid among the tested substrates. The citric acid production reached a maximum (88 g/kg dry matter).

11.3.5.6 Fumaric Acid

Fumarate is an intermediate in the citric acid cycle formed by the oxidation of succinate by the enzyme succinate dehydrogenation. Fumaric acid is a food acidulent and used in beverages. Carta et al. (1998, 1999) studied the prospects of production of fumaric acid from cassava bagasse. Submerged fermentation was carried out using enzymatic hydrolysate of cassava bagassse nourished with different nitrogen sources by *Rhizopus* strains. The strain *Rhizopus formosa* MUCL 28422 was found to be the best fumaric acid producer, yielding 21.28 g/L in a medium containing cassava bagasse hydrolysate as the sole carbon source. Moresi et al. (1992) studied the production of fumaric acid by *Rhizopus arrhizus* from hydrolysates of corn, cassava and potato starch.

11.3.5.7 Bioethanol

Ethanol can be used as a fuel, mainly as a biofuel alternative to gasoline, and is widely used in cars in countries like Brazil. Because it is easy to manufacture and process, and can be made from very common crops, such as sugar cane and maize, it is an increasingly common alternative to gasoline in some parts of the world. Agu et al. (1997) studied the combined effect of heat treatment and acid hydrolysis (various concentrations) on cassava grate waste (CGW) biomass for ethanol production was investigated. Sixty percentage process efficiency was achieved with 0.3 M H_2SO_4 in hydrolysing the cellulose and lignin materials present in the CGW biomass. From three litres of the CGW biomass hydrolysate obtained from hydrolysis with 0.3 M H_2SO_4, ethanol yield was 3.5% (v/v) after yeast fermentation. However, although the process resulted in gainful utilization of CGW biomass, additional costs would be required to effectively dispose new by-products generated from CGW biomass processing. Improved alcohol production from cassava starch residues was achieved by saccharification with multi-enzymatic preparation consisting of cellulase, D-xylanase, α-D-glucosidase, α-amylase, amyloglucosidase, and pectinase. They got 29–36 mL alcohol/100g of sun-dried cassava starch residue (Shamala and Sreekantiah 1986).

11.3.5.8 Natural Insecticide

The long residual action and toxicity of the chemical insecticides have brought about serious environmental problems such as the emergence and spread of insecticide resistance in many species of vectors, mammalian toxicity, and accumulation of pesticide residues in the food chain. Entomo-pathogenic *Bacillus thuringiensis* (*Bt*) and *Bacillus sphaericus* (*Bs*) are two safe biological control agents. In the conventional *Bt* production process, the cost of raw materials varied between 30 and 40% of the total cost depending on the plant production capacity. Therefore, local production of this insecticide in developing countries should depend on the use of production media made of cheap, locally available sources including agro-industrial by-products. The cassava waste was proved as the potent carbon source for the production of natural insecticide *Bt* (Ejiofore 1991).

11.3.5.9 Biotransformation of Cassava Bagasse for Feed or Food

Using *Rhizopus* sp., an edible fungus, the high content starch of cassava bagasse biotransformed to food or feed by the successful improvement in the protein content from 1.67% to 12%. The optimized conditions were incubation temperature 28–32°C, inoculum size 10^5 spores/g dry cassava bagasse, initial moisture 70% and pH 5.7–6.4. Tray fermenter with the substrate thickness of 6–8 cm was proved best bioreactor during the scale up studies using different bioreactors (Soccol et al. 1995a, b, c). The bio-transformed bagasse obtained showed an excellent sanitary condition and is perfectly inside the standards of the sanitary legislation, considering that cassava bagasse was bio-transformed without any thermal process of sterilization, being only dehydrated at 60°C for about 24 h after fermentation. The results showed no growth of undesirable bacteria such as *Staphylococcus aureus*, *Bacillus cereus*, *Salmonella* and faecal coliforms. But there was the presence of some moulds and yeasts in the bio-transformed one and which was very low from the not bio-transformed one. Great possibilities were found for the biotransformation products obtained from cassava bagasse by *Rhizopus* showing some facilities and economical advantages such as very short time of fermentation (Soccol and Vandenberghe 2003).

11.3.5.10 Lactic Acid

Lactic acid (2-hydroxypropionic acid), is an important organic acid having a prime position due to its versatile applications in food, pharmaceutical, textile, leather and chemical industries. The demand for lactic acid has been increasing considerably, owing to the promising applications of its polymer, the polylactic acid an environment-friendly alternative to plastics derived from petrochemicals. There have been numerous investigations on the development of biotechnological processes for lactic acid production, with the ultimate objective to enable the process to be more effective and economical. Cassava bagasse starch was used as the carbon source for the lactic acid fermentation by lactobacilli in submerged, solid-state, and simultaneous saccharification and fermentation (Rojan et al. 2005; John et al. 2006a, b, 2007b, c, d, 2008). Cassava bagasse was saccharified with amylase enzyme and hydrolysate was used for the submerged and solid-state fermentation. It was proven that cassava bagasse under simultaneous saccharification and fermentation is far better than the submerged or solid-state fermentations with respects to its productivity and yield.

11.4 Conclusion

Biological conversion has an important role in waste utilization, and it is likely that various food processing wastes may contain useful substrates, which can be used for value added product production. The biotechnological production of these products offers several advantages compared to chemical synthesis, like low cost of substrates, high product specificity, low production temperature and low energy consumption. Fermentation technologies are widely used in many environmental-friendly and economic industrial sectors. Environmental pollution is no longer accepted as inevitable in technological societies. Over the past century, there has been a tremendous increase in awareness of the effects of pollution, and public pressure has influenced both industry and government. There is increasing demand to replace traditional chemical processes with biotechnological processes involving microorganisms, which not only provide an economically viable alternative but are also more environmentally friendly. The use of nutrient-rich renewable resources such as cassava bagasse opens an avenue in a dual working manner for value addition through an eco-friendly green technology.

Acknowledgments Rojan P. John thanks Council of Scientific and Industrial Research, New Delhi, India for the Senior Research Fellowship.

References

Agu RC, Amadife AE, Ude CM, Onyia A, Ogu EO, Okafor M, Ezejiofor E (1997) Combined heat treatment and acid hydrolysis of cassava grate waste (cgw) biomass for ethanol production. Waste Manage 17: 91–96

Anuradha R, Suresh AK, Venkatesh KV (1999) Simultaneous saccharification and fermentation of starch to lactic acid. Proc Biochem 35: 367–375

Barbosa M, Soccol CR, Marin B, Todeschini ML, Tonial TM, Flores V (1995) Prospect to production of *Pleorotus sajor-caju* from cassava fibrous waste, In: Roussos S, Lonsane BK, Raimbault M, Viniegra-Gonzalez G (Eds.) Advances in Solid State Fermentation: Edible Mushrooms/FungiKluwer Academic Publishers, London

Bramorski A, Soccol CR, Christen P, Revah S (1998a) Fruity aroma production by *Ceratocystis fimbriata* in solid cultures from agro-industrial wastes. Rev Microbiol 29: 208–212

Bramorski A, Christen P, Ramirez M, Soccol CR, Revah S (1998b) Production of volatile compounds by the edible fungus *Rhizopus oryzae* during solid state cultivation on tropical agro-industrial substrates. Biotechnol Lett 20: 359–362

Carta FS, Soccol CR, Machado L, Machado CMM (1998) Prospect of using cassava bagasse waste for producing fumaric acid. J Sci Ind Res 57: 644–649

Carta FS, Soccol CR, Ramos LP, Fontana JD (1999) Production of fumaric acid by fermentation of enzymatic hydrolysates derived from cassava bagasse. Bioresour Technol 68: 23–28

Carvalho JC, Oishi BO, Pandey A, Soccol CR (2005) Biopigments from *Monascus*: strain selection, citrinin production and color stability. Braz Arch Biol Technol 48: 885–894.

Cereda MP (1994) Caratecrizaçãode resíduos da industrialização da mandioca, in Industrialização da mandioca no, Cereda MP, editor. Paulicía, SãoPaulo, Brazil, 11–50.

Ejiofore AO (1991) Production of *Bacillus thuringiensis* serotype H–14 as bioinsecticides using a mixture of spent brewer's yeast and waste cassava starch as the fermentation medium. Disc Innovation 3: 85–88

Elkholy H, Eltantawy A (2000) The world of cassava production: An overview. J Root Crops 26: 1–5

John RP, Dhanya G, Nampoothiri KM (2008) Genome shuffling of *Lactobacillus delbrueckii* mutant and *Bacillus amyloliquefaciens* through protoplasmic fusion for L-lactic acid production from starchy waste. Bioresour Technol doi:10.1016/j.biortech.2008.03.058

John RP, Nampoothiri KM, Pandey A (2006a) Solid-state fermentation for L-lactic acid production from agro wastes using *Lactobacillus delbrueckii*. Proc Biochem 41: 759–763

John RP, Nampoothiri KM, Pandey A (2006b) Simultaneous saccharification and fermentation of cassava bagasse for L(+)-lactic acid production using lactobacilli. Appl Biochem Biotechnol 134: 263–272

John RP, Nampoothiri KM, Pandey A (2007a) Fermentative production of lactic acid from biomass: an overview on process developments and future perspectives. Appl Microbiol Biotechnol 74: 524–534

John RP, Nampoothiri KM, Pandey A (2007b) Production of L(+) lactic acid from cassava starch hydrolysate by immobilized *Lactobacillus delbrueckii*. J Basic Microbiol 47: 25–30

John RP, Nampoothiri KM, Pandey A (2007c) Polyurethane foam as an inert carrier for the production of L(+) lactic acid by *Lactobacillus casei* under solid-state fermentation. Lett Appl Microbiol 44: 582–587

John RP, Sukumaran RK, Nampoothiri KM, Pandey A (2007d) Statistical optimization of simultaneous saccharification and L(+) lactic acid fermentation from cassava bagasse using mixed culture of lactobacilli by response surface methodology. Biochem Eng J 36: 262–267

Kolicheski MB, Soccol CR, Marin B, Medeiros E, Raimbault M (1997) Secondary metabolites, aroma, pigments and bio-pesticides, citric acid production on three cellulosic supports in solid state fermentation, In: Roussos S, Lonsane BK, Raimbault M, Viniegra-Gonzalez G (Eds.) Advances in Solid State Fermentation: Edible Mushrooms/Fungi Kluwer Academic Publishers, London

Medeiros BPA (1998) Production and Composition of Aromatic Volatile Compounds by *Kluyveromyces marxianus* in Solid State Fermentation. Master's Thesis, Federal University of Parana,Curitiba, Brazil

Moresi M, Parente E, Petruccioli M, Federici F (1992) Fumaric acid production from hydrolysates of starch-based substrates. J Chem Technol Biotechnol 54: 283–290

Pandey A, Soccol CR, Nigam P, Soccol VT, Vandenberghe LPS, Mohan R (2000) Biotechnological potential of agro-industrial residues. II: cassava bagasse. Bioresour Technol 74: 81–87

Rojan PJ, Nampoothiri KM, Nair AS, Pandey A (2005) L(+)-Lactic acid production using *Lactobacillus casei* in solid-state fermentation. Biotechnol Lett 27:1685–1688

Shamala TR, Sreekantiah KR (1986) Saccharification of tapioca starch residue with a multienzyme preparation of *Aspergillus ustus*. Starch/Starke 38: 428–432

Shankaranand VS, Lonsane BK (1994) Citric acid by solid state fermentation a case study for commercial exploitation. In: Pandey A (ed) Solid State Fermentation. Wiley Eastern, New Delhi, India

Soccol CR (1994) Contribuição ao Estudo da Fermentação no Estado Sólido em Relacção com a Producção de Aácido Fumárico, Biotransformação de Resíduo Sólido de Mandioca por Rhizopus e Basidiomacromicetos do Género Pleurotus. Curitiba, Tese (Professor Titular), Universidade Federal do Paraná

Soccol CR (1996) Biotechnological products from cassava roots by solid state fermentation. J Sci Ind Res 55: 358–364

Soccol CR, Stertz SC, Raimbault M, Pinheiro LI (1995a) Biotransformation of solid waste from cassava starch production by *Rhizopus* in solid state fermentation. Part I. Screening of strains. Arch Biol Tecnol 38: 1303–1310

Soccol CR, Stertz SC, Raimbault M, Pinheiro LI (1995b) Biotransformation of solid waste from cassava starch production by *Rhizopus* in solid state fermentation. Part II. Optimization of the culture conditions and growth kinetics. Arch Biol Tecnol 38: 1311–1318

Soccol CR, Stertz SC, Raimbault M, Pinheiro LI (1995c) Biotransformation of solid waste from cassava starch Production by *Rhizopus* in solid state fermentation. Part III. Scale-up studies in different bioreactors. Arch Biol Tecnol 38: 1319–1326

Soccol CR, Vandenberghe LPS (2003) Overview of applied solid-state fermentation in Brazil. Biochem Eng J 13: 205–218

Stertz SC (1997) Bioconversao da Farinha de Mandioca Crua (Manihot Esculenta, Crantz) por Fungos do Genero Rhizopus em Fermentacão no Estado Solido. Tese Mestrado, Universidade Federal do Parana, Curitiba

Tonukari NJ (2004) Cassava and the future of starch. Electronic J Biotechnol 7: 5–7

Vandenberghe LPS, Soccol CR, Lebeault JM, Krieger N (1998) Cassava wastes hydrolysate an alternative carbon source for citric acid production by *Candida lipolytica*. Paper presented in Internatinal Congress of Biotechnology'98, Portugal

Vandenberghe LPS, Soccol CR, Pandey A, Lebeault JM (2000) Solid state fermentation for the synthesis of citric acid by *Aspergillus niger*. Bioresour Technol 74: 175–178

Woiciechowski AL, Nistsche S, Pandey A, Soccol CR (2002) Acid and enzyme hydrolysis to recover reducing sugars from cassava bagasse: an economic study. Braz Arch Biol Technol 45: 393–400

Woiciechowski AL, Soccol CR, Rocha SN, Pandey A (2004) Xanthan gum production from cassava bagasse hydrolysate with *Xanthomonas campestris* using alternative sources of nitrogen. Appl Biochem Biotechnol 118: 305–312

Chapter 12
Sugarcane Bagasse

Binod Parameswaran

Contents

Abstract Advances in industrial biotechnology offer potential opportunities for economic utilization of agro-industrial residues. Sugarcane bagasse is the major by-product of the sugar cane industry. It contains about 50% cellulose, 25% hemicellulose and 25% lignin. Due to its abundant availability, it can serve as an ideal substrate for microbial processes for the production of value-added products such as protein enriched animal feed, enzymes, amino acids, organic acids and compounds of pharmaceutical importance etc. Since untreated bagasse is degraded very slowly by micro-organisms, a pre-treatment step may be useful for improved substrate utilization. This chapter reviews the developments on processes and products developed for the value-addition of sugarcane bagasse through the biotechnological means and it also discuss about various pre-treatment methods for efficient utilization of this substrate for the production of fermentable sugars.

Keywords Sugarcane bagasse · Industrial enzymes · Value-added products · Bioethanol · Bioplastics

---

B. Parameswaran (✉)
Bioenergy Research Centre, Korea Institute of Energy Research (KIER), Yusong, Daejon 305-343, Republic of Korea
e-mail: binodkannur@yahoo.com

P. Singh nee' Nigam, A. Pandey (eds.), *Biotechnology for Agro-Industrial Residues Utilisation*, DOI 10.1007/978-1-4020-9942-7_12,

12.1 Introduction

Sugarcane is the common name of a species of herb belonging to the grass family. The official classification of sugarcane is *Saccharum officinarum*, and it belongs to the family Gramineae. It is common in tropical and subtropical countries throughout the world. It can grow from eight to twenty feet tall, and is generally about 2 inches thick. Several different horticultural varieties are known, and they differ by their stem color and length. About 200 countries cultivate this crop and Brazil is the world's largest sugar cane producer, responsible for around 25 percent of total world production, followed by India, Pakistan, China and Thailand. India is second largest producer of sugar in the world. In India there are about 571 sugar mills which produce a total quantity of 19.2 million tones (MT). Uses of sugar cane include the production of sugar, Falernum, molasses, rum, soda, cachaça (the national spirit of Brazil) and ethanol for fuel.

12.2 Processing of Sugarcane

Sugar processing begins when the cane plant arrives at the sugar mill. Rotating knives, shredders, and crushers extract the juice from the cane. Heating the juice evaporates off excess water and condenses the juice into thick syrup. Sugar granules act as seed crystals when they are added to the syrup, making the dissolved sugar in the syrup crystalize. When as much sugar as possible has crystallized in the syrup, the mix is spun in a centrifuge, which separates the remaining syrup (now called molasses) from the raw sugar crystals. The fibrous residue of cane stalk left over after the crushing and extraction of juice from the sugar cane is called bagasse.

12.3 Composition of Bagasse

Bagasse consists of approximately 50% cellulose and 25% each of hemicellulose and lignin. Chemically, bagasse contains about 50% α-cellulose, 30% pentosans and 2.4% ash. Because of its low ash content, bagasse offers numerous advantages for usage in bioconversion processes using microbial cultures. Also, in comparison to other agricultural residues, bagasse can be considered as a rich solar energy reservoir due to its high yields (about 80 tonnes per hectare in comparison to about 1, 2 and 20 tonnes per hectare for wheat, other grasses and trees, respectively) and annual regeneration capacity (Pandey et al. 2000).

12.4 Biotechnological Potential of Sugarcane Bagasse

Sugar cane bagasse is a lignocellulosic material providing an abundant and renewable energy source. It is one of the largest cellulosic agro-industrial by-products. Several processes and products have been reported that utilize sugarcane bagasse raw material. These include electricity generation, pulp and paper production, and products based on fermentation. The various products, which have been obtained from the processes involving bagasse include chemicals and metabolites such as alcohol and alkaloids, mushrooms, protein enriched animal feed (single cell protein), and enzymes etc. One of the significant applications of the bagasse is the production of protein-enriched cattle feed and enzymes. Bagasse could also been used for the production of industrially important enzymes and biofuel.

12.4.1 Sugarcane Bagasse as Animal Feed

Bagasse has most commonly been used for the production of protein enriched animal feed. It can be used as a substrate for solid state fermentation (SSF) for animal feed production. Zadrazil and Puniya (1995) differentiated bagasse into four fractions of particle size (>1 mm, 1–3 mm, 3–5 mm and 5–10 mm) with a view to enhancing its nutritive value as animal feed. They found a varying degree of degradation by white-rot fungi and also variation in in vitro rumen digestions. It was concluded that the mechanical separation of a substrate into different particle size could be useful if it was utilized as a substrate to be fermented by filamentous fungi to produce animal feed. Puniya et al. (1996) subjected bagasse to SSF using a strain of *Pleurotus sajor-caju* in a closed system, with the aim of optimising the gaseous atmosphere and developing a cost-effective and simple technology for animal feed production. A patent was obtained on the application of bagasse, softened with alkali treatment, for feedstuff, fertilizer, and sweetener by cultivating *Enterococcus faecium* in SSF (Iritani et al. 1995).

12.4.2 Sugarcane Bagasse for the Production of Industrially Important Enzymes

Sugarcane bagasse has been used for the production of various industrially important enzymes. Many microorganisms, including filamentous fungi, yeasts and bacteria, have been cultivated on this material in fermentation processes.

12.4.2.1 Cellulase

Amongst the various enzymes produced in SSF of bagasse, cellulases have most extensively been studied. It is well established that the hydrolysis of the lignocellulosic residues using enzyme largely depends upon the cost of the production of cellulases. Application of bagasse in SSF for this purpose appears attractive. Sharma et al. (1991, 1995) reported production of cellulases from different fungal strains. Roussos et al. (1992) used a mixture of bagasse and wheat bran (4:1) for the production of cellulases. They suggested hydraulic pressing as a good technique to leach out the enzymes from the fermented matter. Modi et al. (1994) reported higher yields of cellulase from a strain of *Streptomyces* sp. HM29 when grown on bagasse in comparison to rice straw, rye straw and corncobs. Yields were comparable with those obtained from rice bran but lower than those from wheat straw, wheat bran and newspaper.

Often, cultivation of two different strains as mixed culture and pre-treatment of bagasse showed desirable impact on fermentation. Gupte and Madamwar (1997) reported that production of cellulolytic enzymes under SSF by co-culturing of two fungal strains showed improved hydrolytic and β-glucosidase activities as compared to the occasions when they were used separately. Alkali pre-treatment improved the enzyme production (Gupte and Madamwar 1994). Similarly, Gutierrez-Correa and Tengerdy (1997) also reported higher cellulase productivity in co-culturing of a basidiomycete strain with another filamentous fungi. A mutual synergism was observed between the parent strain (of *Trichoderma reesei* LM-UC4) and the *Aspergillus phoenicis* QM 329, resulting in enhanced combined biomass production and corresponding increase in cellulase, endo-glucanase and β-glucosidase activities. When co-culturing was carried out using a mutant strain of *T. reesei* LM-UC4E1, such synergism was absent, suggesting that in the hypermutation the ability for co-operative interaction with other microbes was lost. Treatment of bagasse with ammonia (80%, w/w moisture content) resulted higher enzyme productivity (Duenas et al. 1995).

When *Trichoderma reesei* LM-UC4 and its hypercellulolytic mutant LM-UC4E1 were co-cultured with *Aspergillus niger* ATCC 10864 in solid substrate fermentation on alkali-treated sugar cane bagasse for cellulolytic enzyme production, the mutant strain was more responsive to mixed culturing than the parent strain (Gutierrez-Correa et al. 1999). Bagasse was supplemented with either soymeal or with ammonium sulfate and urea, and fermented at 80% moisture content and 30 degrees C. Mixed culturing produced better results with the inorganic supplement. Their study shows that mixed culturing is beneficial for the economic production of cellulases on nutritionally poor agricultural residues, without the need for supplementation with expensive organic supplements.

12.4.2.2 Xylanase

Xylanase has been another enzyme produced in SSF of bagasse. Xylanases are typically important enzymes for the degradation of plant materials (hemicellulose, which comprises mainly of xylan). Xylans are formed mainly by heteropolysaccharide of a chain of β-1,4 xylanopyranose units highly substituted by acetyl, arabinosyl and glucopiranosyl residues. Most of the commercially available xylanases are being produced from fungi which are active at neutral or acidic pH and their optimum temperature for activity is below 45°C. Thermophilic xylanases, which are active at alkaline conditions, have great potential for industrial applications. Jain (1995) used a thermophilic fungus for the production of extra-cellular xylanase on various agro-residues, including bagasse. Fungus grew well on untreated bagasse and enzyme titres were lower when fungus was grown on treated (alkali or acid chlorite treatment) bagasse. Acetyl esterase was produced concurrently, maximal activity being with bagasse in comparison to other substrates. Gutierrez-Correa and Tengerdy (1998) also performed xylanase production in SSF using bagasse. They co-cultured *T. reesei* and *A. niger* or *A. phoenicis* and achieved high xylanase titres (2,600–2,800 IU/g dry wt). Sugar cane bagasse was chemically treated to generate different bagasse samples with varying quantities of lignin and hemicellulose, keeping the cellulose fraction intact. These bagasse samples were evaluated for the production of cellulase and xylanases enzymes by *Penicillium janthinellum* NCIM 1171 and *Trichoderma viride* NCIM 1051 in the production medium (Adsul et al. 2004).

12.4.2.3 Amylase

Studies by Rajagopalan and Krishnan (2008) showed that sugar cane bagasse hydrolysate (SBH) can be used for α-amylase production. Utilization of sugar cane bagasse has not been possible for α-amylase production by *Bacillus* sp. and there is no previous report for the production of α-amylase from *Bacillus* sp. in submerged or solid state fermentation. This is due to the fact that hydrolysis of sugar cane bagasse forms simple sugars primarily glucose, xylose and arabinose that repress α-amylase synthesis through catabolic repression. A new isolate of *Bacillus subtilis* KCC103 showed absence of repression by glucose during α-amylase synthesis. The level of α-amylase produced in sugar cane bagasse hydrolysate medium was equivalent to that in starch medium, therefore replacement of starch by SBH in production medium is highly feasible to produce α-amylase at low cost.

12.4.2.4 Inulinase

Optimization of inulinase production by *Kluyveromyces marxianus* NRRL Y-7571 using sugarcane bagasse as substrate was studied by Marcio et al. (2006). The best fermentation conditions found after optimization was 36°C and 20% of corn steep liquor, which yielded about 390 Ug^{-1}. Maximum productivity was 3.34 $Ug^{-1}\,h^{-1}$, the highest reported in literature to date. Sugarcane bagasse seems to present a great nutritional potential for growth of *K. marxianus* NRRL Y-7571 and production of inulinase.

12.4.2.5 Lipase

The use of solid state fermentation for the production of thermostable lipases is an interesting alternative to the valorization of bagasse and olive oil cake. Lipase production could be optimized by adding the appropriate precursors found in olive oil cake. Olive oil cake and sugar cane bagasse were used for lipase production using thermostable fungal cultures of *Rhizomucor pusillus* and *Rhizopus rhizopodiformis* by Cordova et al. (1998). The maximum production of lipase by *Rhizomucor pusillus* and *Rhizopus rhizopodiformis* in solid state fermentation using SCB, was 4.99 Ug^{-1} DM equivalent to 1.73 Uml^{-1} and 2.67 Ug^{-1} DM equivalent to 0.97 Uml^{-1}, respectively. However, the mixture of olive oil cake and sugarcane bagasse, 50% each, increased the lipase activity as high as 79.6 Ug^{-1} DM equivalent to 43.04 Uml^{-1} and 20.24 Ug^{-1} DM equivalent to 10.83 Uml^{-1} obtained by *Rhizopus rhizopodiformis* and *Rhizomucor pusillus*, respectively.

12.4.2.6 Other Value-Added Products

Nampoothiri and Pandey (1996) reported production of L-glutamic acid using sugarcane bagasse in which bagasse was impregnated with a medium containing glucose, urea, mineral salts and vitamins. Maximum yields (80-mg glutamic acid/g dry bagasse) were obtained when bagasse of mixed particle size was fermented with 85–90% moisture and 10% glucose. Impregnated bagasse was also used by Hernandez et al. (1993) to grow a fungus culture for the production of ergot alkaloids. They used a total of sixteen different combinations of liquid media and concluded that there existed the possibilities of achieving tailor-made spectra of ergot alkaloids by changing the liquid nutrient media composition used for impregnation. Barrios-Gonzalez et al. (1993) studied the effect of particle size, packing density and agitation on penicillin production in SSF using bagasse as inert substrate. The use of a large particle size (14-mm) bagasse increased penicillin production by 37%. Christen et al. (1994, 1997) reported production of fruity aroma on bagasse when it was fermented with a nutritive medium containing glucose (200 g/L). Twenty-four compounds were separated and twenty of them were identified from the headspace analysis of the fermenter by GC. Aroma production was dependent on the growth and the maximum aroma intensity was detected at about time of the maximum respirometric activity.

SSF carried out on inert support materials, which differs from the process of microbial growth on or in solid particles floating in a liquid medium has been regarded as one of the future development of SSF systems (Aidoo et al. 1982, Pandey 1991, 1992). The use of solid inert material impregnated with suitable liquid media would provide homogenous aerobic conditions throughout the bioreactor and the purity of the product would also be relatively high. Soccol et al. (1994) evaluated potential of bagasse, impregnated with a liquid medium containing glucose and calcium carbonate, to be used as inert substrate, for lactic acid production from a strain of *Rhizopus oryzae* NRRL 395. Keeping glucose level at 120 and 180 gL^{-1} for liquid and solid state fermentation, an yield of 93.8 and 137.0 gL^{-1} of L(+)-lactic acid was obtained, respectively. The productivity was 1.38 and 1.43 $gL^{-1}h^{-1}$ in liquid and solid fermentation, respectively.

Lactobacillus delbrueckii was able to grow in a solid support of sugarcane bagasse and it effectively utilized the sugar available in the medium used for moistening the substrate (Rojan et al. 2006). Cassava hydrolysate prepared from the bagasse was the sole carbon source used with minimum supplementation of ammonium salt and yeast extract. Under the optimised conditions, the strain produced up to 249.1 mg lactic acid g^{-1}DM with a conversion efficiency of more than 99% of the total sugar available to lactic acid. This process offers a cost effective, eco friendly technology to scale up lactic acid production.

Citric acid was another organic acid, which was produced in SSF using bagasse as inert carrier. Manonmani and Sreekantiah (1987) conducted citric acid production using enzymatic hydrolyzate of alkali treated bagasse by SSF. The studies by Kianoush and Alaleh (2008) showed that among all pre-treatment methods (acid, alkali and urea) which had an important role in increasing citric acid productivity from sugar cane bagasse, urea pre-treatment was the most influential support for the acid production. Tosmani et al. (1997) compared gibberllic acid production in submerged fermentation with SSF when latter showed excellent fungal growth.

Pectinases were produced in SSF using bagasse, impregnated with high glucose concentration solution (Solis-Pereyra et al. 1996). The fermentation was carried out in packed bed column fermenter for SSF. In a similar study, Huerta et al. (1994) concluded that SSF carried out on inert substrates (they referred it as adsorbed substrate fermentation technique) not only allowed the design of culture medium to produce important metabolites, but also the study of fungal metabolism in artificially controlled SSF processes. Acuna-Argulles et al. (1994) studied the effect of water activity on pectinases production using bagasse impregnated with a medium containing pectin and sucrose. Ethylene glycol, sorbitol and glycerol were used as water activity depressors. Results indicated that although polygalacturonase production decreased at low a_w values, this activity was present at a_w values as low as 0.90. The specific activity increased up to 4.5-fold by reducing a_w from 0.98 to 0.9.

Chiu and Chan (1992) described production of pigments using bagasse in roller bottle cultures of *Monascus purpurea*. Fungus was cultivated in wet bagasse containing PGY medium with corn oil in SSF when it produced red and yellow pigments.

12.4.2.7 Bioethanol

Sugarcane bagasse is a potential lignocellulosic feedstock for ethanol production, since it is cheap and readily available as an industrial waste product. Because of its high carbohydrate content and relatively low lignin content, bagasse is particularly appropriate substrate for bioconversion to ethanol. After glucan, xylan is the most abundant carbohydrate in bagasse. Xylose can account for almost one-third of the total sugar content in bagasse hydrolysates. Therefore, micro-organisms able to ferment both glucose and xylose are required for an efficient conversion of bagasse to ethanol. The geometrical ratio and size of sugar cane bagasse fibres strongly influence the profile of lignocellulosic enzyme activity. The hemicellulose fraction of sugar cane bagasse contains up to 35% of the total carbohydrate that can be readily hydrolyzed to monomeric sugars by dilute sulfuric acid. However, the concentration of reducing sugar in hydrolysate is relatively low due to high liquid/solid ratio during the acid hydrolysis (Cheng et al. 2008). Hydrolysis of bagasse by using dil. sulfuric acid or hydrochloric acid at elevated temperature and pressure was studied by Lavarack et al. (2002). In this technique, the elevated temperature softens the lignin protective layer around the hemicellulose fibres and allows the aid to hydrolyse the hemicellulose to form polysaccharides and monosaccharides of mainly xylose and arabinose. The potential of a genetically engineered xylose-utilising *Saccharomyces cerevisiae* strain for fermenting sugarcane bagasse enzymatic hydrolysates was demonstrated by Martin et al. (2002b).

A process have been developed for the continuous production of xylitol from hemicellulosic hydrolysate utilizing *Candida guilliermondii* cells immobilized onto natural sugar cane bagasse fibres. Xylitol is an alternate high added-value sweetner with anti-carcinogenic properties of great concern for both the food industry and the biomedical sector (Diego et al. 2008).

12.4.2.8 Bioplastics

Studies by Jian and Heiko (2008) showed that sugarcane bagasse can be pretreated in dilute acid solution under moderately severe conditions, releasing sugars and other hydrolysates including volatile organic acids, furfurals and acid soluble lignin and utilization of these hydrolysates by an aerobic bacterium, *Ralstonia eutropha* for biosynthesis of value-added bioplastics, polyhydroxyalkanoates (PHAs). PHA biopolyesters were synthesized and accumulated to 57 wt% of cell mass under appropriate C/N ratios. Poly(3-hydroxybutyrate) was the predominant biopolyester formed on the hydrolysates, but the cells could also synthesize co-polyesters that exhibit high ductility. Table 12.1 shows microorganisms cultivated on sugarcane bagasse and the products.

12.5 Pre-treatment Methods for Sugarcane Bagasse

Different pre-treatment methods for preparing bagasse enzymatic hydrolysates have been investigated with focus on obtaining high sugar yields. The pre-treatment methods include steam explosion (Martin et al. 2002a), liquid hot water pre-treatment (Laser et al. 2002) and pre-treatments with peracetic acid (Teixeira et al. 1999) or with ammonia water (Kurakake et al. 2001). Martin et al. (2007) reported wet oxidation as a pre-treatment method for enhancing the enzymatic convertibility of sugarcane bagasse. Wet oxidation (WO) is the process of treating material with water and air or oxygen at temperatures above 120°C (McGinnis et al. 1983). Two types of reactions occur during WO, a low-temperature hydrolytic reaction and a high-temperature oxidative reaction. The advantages of WO as a pre-treatment method for lignocellulose have been presented in the 1980s (McGinnis et al. 1983) and have been confirmed in recent years at Risø National Laboratory (Varga et al. 2003). In a recent work, the enzymatic convertibility and the fermentability of bagasse pretreated by WO at different pH values were investigated (Martin et al. 2006). The studies by Martin et al. (2007) revealed that wet oxidation is an appropriate method for fractionating sugarcane bagasse and for enhancing its enzymatic hydrolysis. Alkaline WO at 195°C during 15 min gave the best results, yielding a solid material with nearly 70% cellulose content, a solubilisation of approximately 93% of hemicelluloses and 50% of lignin, and an enzymatic convertibility of cellulose of around 75%. Although acidic WO at 195°C for 15 min gave good fractionation of bagasse, a significant part of the polysaccharides was lost due to degradation and formation of by-products, mainly carboxylic acids, and the enzymatic convertibility of the pretreated material was poor. From the study they found the tendency of WO

Table 12.1 Micro-organisms cultivated on sugarcane bagasse and the products

Product	Organism	Reference
Ethanol	*Pachysolen tannophilus*	Cheng et al. 2008
	Escherichia coli KO11,	Caroline et al. 2000
	Pichia stipitis	Roberto et al. 1991
	Pichia stipitis	Vanzyl et al. 1991
	Recombinant bacteria	Katzen and Fowler 1994
	Yeast	Gong et al. 1993
	Yeast	Navarro et al. 1982
Xylitol	*Candida guilliermondii*	Diego et al. 2008
	Candida guilliermondii	Carvalho et al. 2005
	Candida guilliermondii	Gurgel et al. 1995
	Candida guilliermondii	Roberto et al. 1995
	Yeast	Dominguez et al. 1996
	Candida guilliermondii	Felipe et al. 1996
	Candida guilliermondii	Alves et al. 1998
SCP/protein enriched feed	Polyporus sp.	Nigam et al. 1987
	Aspergillus niger	Zayed and Mostafa 1992
	Cellulomonas flavigena and *Xanthomonas* sp	Rodriguez-Vazquez and Diazcervantes (1994)
	white-rot fungus	Elsayed et al. 1994
	Candida tropicalis	Pessoa et al. 1996
	Trichoderma reesei	Aiello et al. 1996
Mycoprotein	*Neurospora sitophila*	MooYoung et al. 1993
Aroma	*Ceratocystis fimbriata*	Christen et al. 1994
Cellulases	*Piromyces* sp.	Teunissen et al. 1992
	Aspergillus niger	Ray et al. 1993
	Trichoderma reesei	Aiello et al. 1996
	white-rot fungi	Breccia et al. 1997
	Aspergillus niger	Ray et al. 1993
	Aspergillus ellipticus and *Aspergillus fumigatus*	Gupte and Madamwar 1994, 1997
	Trichoderma harzianum	Roussos et al. 1992
Xylanases	*Cellulomonas flavigena:*	Mayorga-Reyes et al. 2002
	Penicillium janthinellum	Milagres et al. 1993
	Piromyces sp.	Teunissen et al. 1992
	Melanocarpus albomyces	Jain 1995
	Cellulomonas flavigena	Perezavalos et al. 1996
Laccase	*Flammulina velutipes* and *Trametes versicolor*	Pal et al. 1995
Gibberllic acid	*Gibberella fujikuroi*	Tosmani et al. 1997
Pigments	*Monascus purpurea*	Chiu and Chan 1992
Glutamic acid	*Brevibacterium* sp.	Nampoothiri and Pandey 1996
Ergot alkaloids	*Claviceps purpurea*	Hernandez et al. 1993
Lactic acid	*Rhizopus oryzae*	Soccol et al. 1994
	Lactobacillus delbrueckii	Rojan et al. 2006
Citric acid	*Aspergillus niger*	Manonmani and Sreekantiah 1987
Pectinases	*Aspergillus niger*	Huerta et al. 1994
	Aspergillus niger	Acuna-Arguelles et al. 1994
Bioplastics	*Ralstonia eutropha*	Jian and Heiko 2008

to catalyze the transfer of hemicellulose from the solid phase to the liquid phase without a major hydrolysis of the solubilized hemicellulose molecules. It was also reported that more xylose was formed by WO and more glucose by steam explosion (Martin et al. 2008). According to McGinnis et al. (1983) monosaccharide are oxidized to carboxylic acids by WO, while oligosaccharides are more resistant to oxidation due to the stability of the glycosidic linkages. Therefore sugar oligomers will be present in the wet oxidation filtrate.

12.6 Conclusions

It can be concluded that bioconversion of bagasse could be economically advantageous for the production of enzymes, animal feed, bioethanol and bioplastics. Since untreated bagasse is degraded very slowly by micro-organisms, a pre-treatment step may be useful for improved substrate utilization. Evidently, additional research on the pre-treatment of bagasse is required to improve components yield and cellulose digestibility to the extent which would make its use economically viable. Similarly, although many efforts have been made on sugarcane bagasse hydrolysis using pre-treatment methods as well as enzymatic saccharification, its effective conversion into bioethanol is an area which needs further inputs in terms of research and development.

Abbreviations

SSF: Solid State Fermentation
PHA: polyhydroxyalkanoates
WO: Wet Oxidation

References

Acuna-Arguelles M, Gutierrez-Rojas M, Viniegra-Gonzalez G, Favela-Torres E (1994) Effect of water activity on exo-pectinase production by *Aspergillus niger* CH4 on solid state fermentation. Biotechnol Lett 16:23–28

Adsul MG, Ghule JE, Singh R, Shaikh H, Bastawde KB, Gokhale DV, Varma AJ (2004) Polysaccharides from bagasse: applications in cellulase and xylanase production. Carbohydr Polym 57:67–72

Aidoo KE, Hendry R, Wood JB (1982) Solid substrate fermentations. Adv Appl Microbiol 28: 201–237

Aiello C, Ferrer A, Ledesma A (1996) Effect of alkaline pre-treatment at various temperatures on cellulase and biomass production using submerged sugarcane bagasse fermentation with *Trichoderma reesei* QM-9414. Bioresour Technol 57:13–16

Alves LA, Felipe MGA, Silva JBAE, Silva SS, Prata AMR (1998) Pre-treatment of sugarcane bagasse hemicellulose hydrolysate for xylitol production by *Candida guilliermondii.* Appl Biochem Biotechnol 70–72:89–98

Barrios-Gonzalez J, Gonzalez H, Mejia A (1993) Effect of particle size, packing density and agitation on penicillin production in solid state fermentation. Biotechnol Adv 11:539–547

Breccia JD, Bettucci L, Piaggio M, Sineriz F (1997) Degradation of sugarcane bagasse by several white-rot fungi. Acta Biotechnol 17:177–184

Caroline MT, Katia GCL, Debora FT, Flavio A (2000) Fermentation of sugarcane bagasse hemicellulosic hydrolysate and sugar mixtures to ethanol by recombinant *Escherichia coli* KO11. World J Microbiol Biotechnol 16:829–834

Carvalho W, Santos JC, Canilha L, Silva SS, Perego P, Converti A (2005) Xylitol production from sugarcane bagasse hydrolysate: Metabolic behaviour of *Candida guilliermondii* cells entrapped in Ca-alginate. Biochem Eng J 25:25–31

Cheng Ke-Ke, Cai Bai-Yan, Zhang Jian-An , Ling Hong-Zhi, Zhou Yu-Jie, Geb Jing-Ping, Xu Jing-Ming (2008) Sugarcane bagasse hemicellulose hydrolysate for ethanol production by acid recovery process. Biochem Eng J 38:105–109

Chiu SW, Chan SM (1992) Production of pigments by *Monascus purpurea* using sugar cane bagasse in roller bottle cultures. World J Microbiol Biotechnol 8:68–70

Christen P, Meza JC, Revah S (1997) Fruity aroma production in solid state fermentation by *Ceratocystis fimbriata*: influence of the substrate type and the presence of precursors. Mycol Res 101:911–919

Christen P, Villegas E, Revah S (1994) Growth and aroma production by *Ceratocystis fimbriata* in various fermentation media. Biotechnol Lett 16:1183–1188

Cordova J, Nemmaoui M, Ismaili-Alaoui M, Morin A, Roussos A, Raimbault M, Benjilali B (1998) Lipase production by solid state fermentation of olive cake and sugar cane bagasse. J Mol Catl B Enzym 5:75–78

Dominguez JM, Gong CS, Tsao GT (1996) Pre-treatment of sugarcane bagasse hemicellulose hydrolysate for xylitol production by yeast. Appl Biochem Biotechnol 57–58:49–56

Diego TS, Boutros FS, Juan DR, Attilio C, Silvio SS (2008) Use of sugarcane bagasse as biomaterial for cell immobilization for xylitol production. J Food Eng 86:542–548

Duenas R, Tengerdy RP, Guierrez-Correa M (1995) Cellulase production by mixed fungi in solid substrate fermentation of bagasse. World J Microbiol Biotechnol 11:333–337

Elsayed SA, Zaki MT, Abouelkhair AW (1994) Bioconversion of sugarcane bagasse into a protein rich product by white-rot fungus. Resources Conserv Recycl 12:195–200

Felipe MGA, Vitalo M, Mancilha IM (1996) Xylitol formation by *Candida guilliermondii* grown in a sugarcane bagasse hydrolysate – effect of aeration and inoculum adaptation. Acta Biotechnologica 16:73–79

Gong GS, Chen CS, Chen LF (1993) Pre-treatment of sugarcane bagasse hemicellulose hydrolysate for ethanol production by yeast. Appl Biochem Biotechnol 39:83–88

Gupte A, Madamwar D (1994) High strength cellulase and beta-glucosidase formation from *Aspergillus* sp. under solid state fermentation. In: Pandey A (ed) Solid State Fermentation. Wiley Eastern Publishers, New Delhi, pp. 130–133

Gupte A, Madamwar D (1997) Solid state fermentation of ligno-cellulosic wastes for cellulase and beta-glucosidase production by co-culturing of *Aspergillus ellipticus* and *Aspergillus fumigatus*. Biotechnol Progr 13:166–169

Gurgel PV, Mancilha IM, Pecanha RP, Siqueira JFM (1995) Xylitol recovery from fermented sugarcane bagasse hydrolysate. Bioresour Technol 52:219–223

Gutierrez-Correa M, Tengerdy RP (1997) Production of cellulase on sugar cane bagasse by fungal mixed culture solid substrate fermentation. Biotechnol Lett 19:665–667

Gutierrez-Correa M, Tengerdy RP (1998) Xylanase production by fungal mixed culture solid state fermentation on sugar cane bagasse. Biotechnol Lett 20:45–47

Gutierrez-Correa M, Portal L, Moreno P, Tengerdy RP (1999) Mixed culture solid substrate fermentation of *Trichoderma reesei* with *Aspergillus niger* on sugar cane bagasse. Bioresour Technol 68:173–178

Hernandez MRT, Lonsane BK, Raimbault M, Roussos S (1993) Spectra of ergot alkaloids produced by *Claviceps purpurea* in solid state system – influence of the composition of liquid medium used for impregnating sugar cane bagasse. Process Biochem 28:23–27

Huerta S, Favela E, Lopez-Ulibarri R, Fonseca A, Viniegra-Gonzalez G, Gutierrez-Rojas M (1994) Absorbed substrate fermentation for pectinase production with *Aspergillus niger*. Biotechnol Technol 8:837–842

Iritani S, Mitsuhashi M, Chaen H, Miyake T (1995) New Alkali Treated Bagasse Used as Livestock Feed. PN:GB 2285806; 26.07.95

Jain A (1995) Production of xylanase by thermophilic *Melanocarpus albomyces* IIS-68. Process Biochem 30:705–709

Jian Y, Heiko S (2008) Microbial utilization and biopolyester synthesis of bagasse hydrolysates. Biores Technol 99:8042–8048

Katzen R, Fowler DE (1994) Ethanol from lignocellulosic wastes with utilisation of recombinant bacteria. Appl Biochem Biotechnol 45–46:697–707

Kianoush KD, Alaleh Z (2008) Comparison of pre-treatment strategies of sugarcane baggase: Experimental design for citric acid production. Biores Technol 99:6986–6993

Kurakake M, Kisaka W, Ouchi K, Komaki T (2001) Pre-treatment with ammonia water for enzymatic hydrolysis of corn husk, bagasse and switchgrass. Appl Biochem Biotechnol 90:251–259

Laser M, Schulman D, Allen SG, Lichwa J, Antal MJ, Lynd LR (2002) A comparison of liquid hot water and steam pre-treatments of sugar cane bagasse for bioconversion to ethanol. Biores Technol 81:33–44

Lavarack BP, Griffin GJ, Rodman D (2002) The acid hydrolysis of sugarcane bagasse hemicellulose to produce xylose, arabinose, glucose and other products. Biomass Bioenergy 23:367–380

Manonmani HK, Sreekantiah KR (1987) Studies on the conversion of cellulase hydrolysate into citric acid by *Aspergillus niger*. Process Biochem 22:92–94

Marcio M, Joao PB, Helen T, Marco DL (2006) Optimization of inulinase production by solid-state fermentation using sugarcane bagasse as substrate. Enzyme Microb Technol 39:56–59

Martin C, Galbe M, Nilvebrant NO, Jonsson LJ (2002a) Comparison of the fermentability of enzymatic hydrolysates of sugarcane bagasse pretreated by steam explosion using different impregnating agents. Appl Biochem Biotechnol 98:699–716

Martin C, Galbe M, Fredrik WC, Hahn-Hägerdal B, Jönsson LJ (2002b) Ethanol production from enzymatic hydrolysates of sugarcane bagasse using recombinant xylose-utilising *Saccharomyces cerevisiae*. Enzyme Microb Technol 31:274–282

Martin C, Gonzalez Y, Fernandez T, Thomsen AB (2006) Investigation of cellulose convertibility and ethanolic fermentation of sugarcane bagasse pretreated by wet oxidation and steam explosion. J Chem Technol Biotechnol doi:10.1002/jctb.1586

Martin C, Helene BK, Anne BT (2007) Wet oxidation as a pre-treatment method for enhancing the enzymatic convertibility of sugarcane bagasse. Enzyme Microb Technol 40:426–432

Martin C, Marcet M, Anne BT (2008) Comparison between wet oxidation and steam explosion as pre-treatment method for enzymatic hydrolysis of sugarcane bagasse. Bio Resour 3: 670–683

Mayorga-Reyes L, Morales Y, Salgado LM, Ortega A, Ponce-Loyola T (2002) *Cellulomonas flavigena*: characterization of an endo-1,4-xylanase tightly induced by sugarcane bagasse. FEMS Microbiol Lett 214:205–209

McGinnis GD, Wilson WW, Mullen CE (1983) Biomass pretreated with water and high pressure oxygen: The WO process. Ind Eng Chem Prod Res Dev 22:352–357

Milagres AMF, Lacis LS, Prade RA (1993) Characterization of xylanase production by a local isolate of *Penicillium janthinellum*. Enzyme Microb Technol 15:248–253

Modi HA, Patel KC, Ray RM (1994) Solid state fermentation for cellulase production by *Streptomyces* sp HM-29. In: Pandey A (ed) Solid State Fermentation. Wiley Eastern Publishers, New Delhi, pp. 137–141

MooYoung M, Chisti Y, Vlach D (1993) Fermentation of cellulosic materials to mycoprotein foods. Biotechnol Adv 11:469–479

Nampoothiri KM, Pandey A (1996) Solid state fermentation for L-glutamic acid production using *Brevibacterium* sp. Biotechnol Lett 18:199–204

Navarro AR, Lucca ME, Callieri DAS (1982) Continuous production of ethanol by yeast cells immobilized on sugarcane bagasse pith. Acta Cient Venez 33: 214–218

Nigam P, Pandey A, Prabhu KA (1987) Mixed cultures fermentation for bioconversion of whole bagasse into microbial protein. J Basic Microbiol 27:323–327

Pal M, Calvo AM, Terron MC, Gonzalez AE (1995) Solid state fermentation of sugar cane bagasse with *Flammulina velutipes* and *Trametes versicolor*. World J Microbiol Biotechnol 11:541–545

Pandey A (1991) Aspects of design of fermenter in solid state fermentation. Process Biochem 26:355–361

Pandey A (1992) Recent developments in solid state fermentation. Process Biochem 27:109–117

Pandey A, Soccol CR, Nigam P, Soccol VT (2000) Biotechnological potential of agro-industrial residues. I: sugarcane bagasse. Biores Technol 74:69–80

Perezavalos O, Poncenoyola T, Maganaplaza I, Delatorre M (1996) Induction of xylanase and beta xylosidase in *Cellulomonas flavigena* growing on different carbon sources. Appl Microbiol Biotechnol 46:405–409

Pessoa A, Mancilha IM, Sato S (1996) Cultivation of *Candida tropicalis* in sugarcane hemicellulose hydrolysate for microbial protein production. J Biotechnol 51:83–88

Puniya, AK, Shah, KG, Hire, SA, Ahire, RN, Rathod, MP, Mali, RS (1996) Bioreactor for solid state fermentation of agro-industrial wastes. Ind J Microbiol 36:177–78

Rajagopalan G, Krishnan C (2008) α-Amylase production from catabolite derepressed *Bacillus subtilis* KCC103 utilizing sugarcane bagasse hydrolysate. Bioresour Technol 99:3044–3050

Ray L, Pal A, Ghosh AK, Chattopadhyay P (1993) Cellulases and beta-glucosidase from *Aspergillus niger* and saccharification of some cellulosic wastes. J Microb Biotechnol 8:85–94

Roberto IC, Lacis LS, Barbosa MFS, Demancilha IM (1991) Utilization of sugarcane bagasse hemicellulose hydrolysate by *Pichia stipitis* for the production of ethanol. Process Biochem 26:15–21

Roberto IC, Felipe MGA, Demancilha IM, Vitalo M, Sato S, Dasilva SS (1995) Xylitol production by *Candida guilliermondii* as an approach for the utilization of agro-industrial residues. Biores Technol 51:255–257

Rodriguez-Vazquez R, Diazcervantes D (1994) Effect of chemical solutions sprayed on sugarcane bagasse pith to produce single cell protein – physical and chemical analysis of pith. Biores Technol 47:159–164

Rojan PJ, Nampoothiri KM, Pandey A (2006) Solid-state fermentation for L-lactic acid production from agro wastes using *Lactobacillus delbrueckii.* Process Biochem 41: 759–763

Roussos S, Raimbault M, Geoffroy F, Saucedo-Castaneda G, Lonsane BK (1992) Efficient leaching of cellulases produced by *Trichoderma harzianum* in solid state fermentation. Biotechnol Tech 6:429–432

Sharma DK, Niwas S, Behera BK (1991) Solid state fermentation of bagasse for the production of cellulase enzyme from cellulolytic fungi and extent of simultaneous production of reducing sugars in the fermenter. J Microb Biotechnol 6:7–14

Sharma DK, Tiwari M, Behera BK (1995) Solid state fermentation of new substrates for production of cellulase and other biopolymers hydrolysing enzymes – scientific note. Appl Biochem Biotechnol 51–52:495–500

Soccol CR, Marin B, Raimbault M, Lebeault JM (1994) Potential of solid state fermentation for production of L(+)-lactic acid by *Rhizopus oryzae*. Appl Biochem Biotechnol 41:286–290

Solis-Pereyra S, Favela-Torres E, Gutierrez-Rojas M, Roussos S, Saucedo-Castaneda G, Gunasekaran P, Viniegra-Gonzalez G. (1996) Production of pectinases by *Aspergillus niger* in solid state fermentation at high initial glucose concentrations. World J Microbiol Biotechnol 12:257–260

Teixeira LC, Linden JC, Schroeder HA (1999) Optimizing peracetic acid pre-treatment conditions for improved simultaneous saccharification and cofermentation (SSCF) of sugar cane bagasse to ethanol fuel. Renew Energy 16:1070–1073

Teunissen MJ, Dekort GVM, Dencamp HJMO, Tveld JHJHI (1992) Production of cellulolytic and xylanolytic enzymes during growth of the anaerobic fungus *Piromyces* sp. on different substrates. J Gen Microbiol 138:1657–1664

Tosmani A, Fajardo C, Barrios-Gonzalez J (1997) Gibberllic acid production using different solid state fermentation systems. World J Microbiol Biotechnol 13:203–206

Vanzyl C, Prior BA, Dupreez JC (1991) Acetic acid inhibition of D-xylose fermentation by *Pichi stipitis*. Enz Microb Technol 13:81–86

Varga E, Schmidt AS, Reczey K, Thomsen AB (2003) Pre-treatment of corn stover using wet oxidation to enhance enzymatic digestibility. Appl Biochem Biotechnol 104:37–50

Zadrazil F, Puniya AK (1995) Studies on effect of particle size on solid state fermentation of sugar cane bagasse into animal feed using white-rot fungi. Biores Technol 54: 85–87

Zayed G, Mostafa N (1992) Studies on the production and kinetic aspects of single cell protein from sugarcane bagasse saccharified with *Aspergillus niger.* Biomass Bioenergy 3:363–367

Chapter 13
Edible Oil Cakes

Swetha Sivaramakrishnan and Dhanya Gangadharan

Contents

Abstract The very sustainability of the growing bioprocess industry depends on the progressive reduction of expensive nutrient inputs into fermentation media. The use of cheap agricultural and food-processing by-products such as oil cakes, as feedstock is highly favored so as to improve the commercial feasibility of bioprocess technology. Due to stringent nutritional requirements of these edible oil cakes as animal feed, there is considerable interest in using them as substrates in the fermentation industry. This chapter will provide an impetus to further research in this area enabling better utilization of edible oil cakes as sources of protein and carbohydrates for economic viability of the bioprocess industry.

Keywords Edible oil cakes · Agro-residues · Fermentation · Enzyme production · Mushroom production

---

S. Sivaramakrishnan (✉)
Biotechnology Division, National Institute for Interdisciplinary Science and Technology (NIIST), CSIR, Trivandrum 695 019, Kerala, India
e-mail: swsj79@yahoo.co.in

P. Singh nee' Nigam, A. Pandey (eds.), *Biotechnology for Agro-Industrial Residues Utilisation*, DOI 10.1007/978-1-4020-9942-7_13,

13.1 Introduction

Fermentation medium plays a major role in the commercialization of a bioprocess in terms of cost, high product yield and efficient recovery. This necessitates the formulation of a medium developed from cheap agro-industrial by-products which has to provide nutritional conditions necessary for growth and synthesis of the product. Various crop-residues such as bran, straw, bagasse, molasses and oil cakes are being evaluated in terms of efficiency and economics in bioprocesses for the production of chemicals and value added products such as enzymes, amino acids, organic acids, surfactants, pigments, flavors, bioactive products etc. For the past two decades, significant attention is being given to the utilization of oil cakes in bioprocess and related industries as feedstock because of its high nutritional value, year-round availability and competitive pricing.

Oil cakes/meals are solid residues obtained after the extraction of oil from the plant part such as seed, by expelling or solvent extraction. Those cakes resulting from edible oil-bearing seeds which are being used to meet a part of the nutritional requirements of either animal-feed or of human consumption are called as edible oil cakes and those which cannot be used as feed stuff due to the presence of toxic compounds and other impurities are differentiated as non-edible (Mitra and Misra 1967). Most of the non edible oil cakes such as from neem, castor, mahua, karanja are used as manures. The world oil cake market is dominated by eight major edible oil cakes such as soybean cake, rapeseed cake, cottonseed cake, groundnut cake, sunflower cake, groundnut cake, copra cake and linseed cake. Among these soybean cake occupies 54% of the total production volume of the eight cakes followed by rapeseed cake (10%) and cottonseed cake (10%). According to Food and Agricultural Policy Research Institute (FAPRI) the world oil cake production and consumption is expected to have a dynamic growth of 2.3% per annum over the forecast period 2006–2015. The annual production of major oil cakes in the world for the year 2006–2007 according to Food and Agricultural Organisation (FAO) is given in the Table 13.1.

Table 13.1 World production of major oil cakes for the year 2006–2007

Oil cake	Production in 2006/07 (Megatons)	Plant source	Plant part
Soybean oil cake	236	*Glycine max*	seed
Rapeseed/mustard/canola oil cake	47.0	*Brassica napus*	seed
Cotton seed oil cake	44.5	*Gossypium*	seed
Groundnut/peanut oil cake	33.8	*Arachis hypogaea*	seed
Sunflower oil cake	29.7	*Helianthus annuus*	seed
Palm kernel oil cake	10.0	*Elaeis guineensis*	kernel
Copra cake (coconut oil cake)	4.9	*Cocos nucifera*	endocarp
Linseed/flax oil cake	1.1	*Linus usitatissimum*	seed
Safflower oil cake	0.3	*Carthamus tinctorius*	seed
Sesame/gingelly oil cake	1.2	*Sesamum indicum*	seed

13.2 Chemical and Nutrient Composition

The utility of any feed stock depends on its chemical and nutrient composition. As these oil cakes are the preferred protein sources for livestock rations they have been a subject of extensive investigation since the oil-production industry started a century ago. They have been thoroughly examined for use as feedstuff for almost all economically important species of animals. Even though the feed-conversion is high, feeding requirements are critical for acceptable production in case of animal feed (Schumacher 1993). The presence of anti-nutrients is a major constraint to the practical use of several oil cakes as animal feed, making them less demanding. Moreover pre-processing of meals for the removal of toxicity will further increase the operational cost. But these limitations are not encountered in the fermentation industry creating renewed interest in their efficient utilization and value addition.

The composition of an oil cake depends on the processing and mode of oil extraction from the oil-bearing source. All oilseeds require pre-processing such as dehulling before oil extraction. Hulls are fibrous outer covering enclosing the seed which needs to be removed for efficient oil extraction. The extent of dehulling varies with different oilseeds. The extraction of oil from oilseeds is either by mechanical pressing or by solvent extraction. The expeller pressed oil cake contains more residual oil than the desolventized cake. The chemical compositions of major oil cakes (solvent extracted) in the world are given in the Table 13.2. Most of these oil cakes

Table 13.2 Chemical composition of major oil cakes

Oil cake	DM %	CP %	Carbohydrate %	Crude fiber %	Ash %	Fat %	Reference
Soybean oil cake	90.3	51.8	23.6	17.8	7.3	0.9	(Castro et al. 2007)
Rapeseed oil cake	90.75	42.8	32.2	12.1	7	4.1	(Bell 1984)
Cottonseed oil cake	91.53	41.5	27.0	14.67	6.46	5.75	(Briggs and Heller 1942)
Groundnut oil cake	90	45.6	14.1	8.3	5.02	2.47	(Batal et al. 2005)
Sunflower oil cake	93	35.6	23.0	28.41	7.36	1.68	(Villamide and San Juan 1998)
Palm kernel oil cake	93	17.5	45.5	11.9	4.8	7.4	(Carvalho et al. 2006)
Sesame oil cake	93.9	48.2	21.0	6.4	12.6	2.3	(Yamauci et al. 2006)
Linseed oil cake	88.9	33.2	36.0	8.1	5.4	2.8	(Loosli et al. 1960)
Safflower oil cake	93.1	44	20.1	12.1	7.2	5.9	(Lyon et al. 1979)
Copra cake	89.9	20.9	42.4	11.5	5.5	8.0	(Thampan 1975)
Olive oil cake	72.8	4.77	10.1	49.14	2.36	8.72	(Vlyssides et al. 2004)

DM- dry matter, CP- crude protein

are rich in protein content, the highest for soybean cake and the lowest for olive oil cake.

Soybean cake is a rich source of protein and energy with lower fiber content than most other oil cakes. They are widely used as feed ingredients both for animal and human food products owing to their high digestibility and palatability. Methionine was found to be the major limiting amino acid in soybean cake while threonine, valine and lysine were found to be marginal in a study conducted on chick performance (Smith 1986). Carbohydrates in soybeans are largely present as oligosaccharides (15%) such as sucrose, raffinose and stachyose. Low levels of anti-nutrients such as saponins, lectins and phytates were reported. The comparative amino acid compositions of important oilseed meals are presented in Table 13.3. Being highly produced and priced meal in oil-cake market, soybean cake is followed by rapeseed cake in production. The cultivated rapeseed types are represented by two species such as *Brassica napus* and *B. campestris*. Rapeseed oil cake is comparable well with that of soybean cake in amino acid balance and is richer in sulphur amino acids such as methionine and cysteine. Carbohydrates in rapeseed cake are mainly pectins (14.5%), cellulose, fuco-amyloid, arabinan and arabinogalactan. Its use in poultry feeding is limited by the presence of anti-nutrients such as glucosinates, sinapine, tannins, erucic acid and phytates. The cultivars of rapeseed (*Brassica campestris*) which have low content of both erucic acid and glucosinates widely cultivated in Canada are known as canola. Mustard is an oil seed crop as well as a condiment belonging to the same genus of rapeseed. The genetic types *Brassica juncea* and *Brassica nigra* are widely cultivated in India, the former for oil and the latter is used only as a condiment and not pressed for oil. The chemical and amino acid composition of mustard oil cake compares closely with that of rapeseed cake, but contains more glucosinates than rapeseed, although of different kinds (Achaya 1990).

Cotton seed oil cake comprises 45% of the seed and has high protein content. But its utilization as a feed ingredient for non ruminants faces severe constraints due to the presence of toxic metabolite gossypol, high fiber content and low lysine, cysteine and methionine levels. Cotton seed genetically devoid of gossypol-containing glands has been developed and oil cake obtained from this variety is reported to have immense potential as protein diet. Groundnut oil cake, a high protein content solid residue is rich in arginine levels, but low in essential amino acids such as lysine. Aflatoxins, toxic fungal metabolites from *Aspergillus flavus* group are frequent contaminants of these cakes and their presence has severe implications on animal performance. Sunflower oil cake is similar to that of cotton seed cake in composition and is rich in sulphur amino acids, but remarkably low in lysine. It is mainly being used for swine and poultry. Chlorogenic acid, a polyphenolic compound reported to inhibit hydrolytic enzymes has an adverse effect on animal performance. No anti-nutritional factors such as tripsin inhibitors are known in sunflower oil cake and its mineral content is satisfactory. Copra cake or coconut oil cake has a protein content of about 20% with low concentrations of essential amino acids notably tryptophan, lysine, methionine and histidine. Palm kernel cake is dry and gritty with high fiber content. This reduces its suitability for monogastric animals. Like sunflower cake, it is deficient in lysine and rich in sulphur containing amino acids. It has a better

Table 13.3 Comparative amino acid composition of important oil cakes

AA %	SBOC[a]	RSOC[a]	CSOC[b]	GOC[a]	SFOC[a]	PKOC[a]	SOC[a]	LSOC[a]	SfOC[a]	COC[a]	OOC[c]
Gly	2.05	2.58	4.9	–	1.88	0.82	1.04	1.74	0.77	0.89	6.64
Ala	2.15	1.69	4.5	1.76	1.36	–	0.96	1.04	0.6	0.81	8.56
Pro	2.05	1.31	4.3	1.96	1.21	–	0.79	0.81	0.62	0.71	4.8
Val	2.12	2.09	5.1	1.87	1.52	0.93	0.98	1.74	0.72	0.89	10.7
Leu	3.81	2.85	6.6	2.89	2.07	1.11	1.44	2.02	0.92	1.21	11.4
Ile	2.08	1.63	3.4	1.54	1.2	0.62	0.77	1.68	0.61	0.6	4.64
Met	0.69	0.73	1.8	0.52	0.67	0.3	0.6	0.54	0.14	0.37	1.41
Phe	2.44	1.64	5.7	2.27	1.42	0.73	0.94	1.46	0.53	0.81	5.69
Tyr	1.81	0.48	3.4	1.64	0.71	0.38	0.67	1.09	0.35	0.46	1.62
Trp	0.77	–	–	0.45	–	0.17	0.28	0.51	0.14	–	–
Ser	2.57	1.98	4.7	1.71	1.29	0.55	0.99	1.92	0.63	0.96	3.98
Thr	1.98	1.92	3.6	1.17	1.21	0.2	0.76	1.22	0.45	0.66	4.05
Cys	0.69	0.5	1.2	0.64	0.53	0.2	0.38	0.61	0.21	0.24	1.04
Lys	2.99	2.45	4.4	1.54	1.01	0.59	0.58	1.16	0.46	0.4	6.53
His	1.26	1.21	2.8	1.1	0.73	0.29	0.52	0.69	0.35	0.41	3.16
Arg	3.64	2.5	11.5	5.04	2.43	2.18	2.58	2.94	1.28	1.96	7.74
Asp	–	2.1	10.1	4.96	2.88	–	1.75	1.92	1.42	1.62	2.67
Glu	–	2.14	21.7	8.36	5.99	–	4.15	4.1	2.77	3.64	5.54
Reference	(Castro et al. 2007)	(Bell 1984)	(O'Mara et al. 1997)	(Batal et al. 2005)	(Villamide and San Juan 1998)	(Sundu et al. 2006)	(Ravindran and Blair 1992)	(Barbour and Sim 1991)	(Ravindran and Blair 1992)	(Creswell and Brooks 1971)	(Martín et al. 2003)

[a]- % of DM, [b]- g/ 100 g of total amino acids, [c]-g/100 g of total nitrogen, Soybean oil cake (SBOC), Rapeseed oil cake (RSOC), Cotton seed oil cake (CSOC), Groundnut oil cake (GOC), Sunflower oil cake (SFOC), Palm kernel oil cake (PKOC), Sesame oil cake (SOC), Linseed oil cake (LSOC), Safflower oil cake (SfOC), Coconut oil cake (COC), Olive oil cake (OOC)

amino acid index relative to coconut oil cake. The digestibility of the cake is low for poultry and swine due to high fiber content (Ravindran and Blair 1992).

The nutrient composition of sesame oil cake is comparable to soybean cake with an average protein content of 40% and fiber content of 8%. It's an excellent source of methioinine, cysteine and tryptophan but deficient in lysine. Even though sesame cake is a rich source of minerals, its availability is lower due to the presence of oxalates and phytates in the hull fraction of the seed. Removal of hull increases mineral availabity and reduces fiber content. However complete dehulling is not always possible due to small size of sesame seeds. Linseed cake comprises of 60–55% of flax seed with a protein content of 32%. But poor protein quality due to lysine deficiency, presence of antipyridoxine factor linatin, cyanogenic glucoside linamarin and indigestible mucilage adversely limits its nutritive value as animal feed. The crude protein content of safflower oil cake ranges from 20–60%, the composition being dependent on dehulling. Partially dehulled seed cakes comprise of 40% crude protein and 15% crude fiber. Two phenolic glucosides 2-hydroxyarctiin and matairesinol impart a bitter flavour to the cake (Ravindran and Blair 1992). Crude olive oil cake has a very low protein content of 5–10%, high crude fiber content (35–50%), phenolic content of 4.3% with high moisture content ranging 35–30%. The fibers comprise of hemicellulose (1.5%), cellulose (1.72%) and lignin (2.75%) (Vlyssides et al. 2004).

India is known for the production of a wide variety of oil cakes from its native oil seeds and other oleaginous plant sources. About 86 different types of minor oil cakes are reported to be produced in India apart from the major oil cakes described earlier. Production of various non conventional oleaginous sources and oil cakes are given in Table 13.4. Castor cake, a minor oil cake is produced mainly by Brazil followed by India and China. It contains 35% crude protein and 25% fiber. It contains three anti-nutritional compounds such as ricin (toxic protein), ricinine (alkaloid) and castor bean allergen, a potent allergen. It is mainly used as a fertilizer. Kapok cake (*Ceiba pentandra*) is reported to have low feeding value due to its high fiber content and presence of tannins (Ravindran and Blair 1992). Among the minor oil cakes, water melon oil cake and musk melon oil cake are rich in proteins, with negligible fiber content and no anti-nutrients (Achaya 1990).

Table 13.4 Annual production of various non-conventional oil bearing materials and oil cakes in India during 1983–1984 (Achaya 1990)

Oil bearing material	Plant source	Production of cake (Tonnes)
Rice bran	*Oryza sativa*	1,212,522
Neem seed	*Azadirachta indica*	62,600
Karanja seed	*Pongamia pinnata*	7,895
Kusum seed	*Schleichera oleosa*	2,866
Mahua seed	*Madhuca indica*	55,480
Mango kernel	*Mangifera indica*	2,507
Tobacco seed	*Nicotiana tabacum*	1,246
Spent coffee	*Coffea arabica*	224

Table 13.5 Composition of non-conventional oil cakes from native Indian oil-seeds (Achaya 1990)

Oil cake	CP %	Crude fiber %	Ash %	Unusual constituents
Rice-bran	14	10	4	High crude fibre and high sand & silica (8%)
Tapioca seed	50	6.5	3.5	Cyanogenic glucosides
Maize germ	18	10	3	None
Tobacco seed	30	17	10	None
Rubber seed kernel	25	5	9	Cyanogenic glucosides
Jute seed	22	19	7	High crude fiber
Spent coffee	16	22	3.7	High crude fiber
Tea seed kernel	8	30	2.5	Saponins (10%)
Neem kernel	21	28	9.5	High crude fiber, high ash, tannins (1.5%)
Karanja kernel	34	6	3.5	Furanoflavones like karanjin, pongamol & mucilage (13.5%).
Kusum kernel	22	10	5.3	None
Mahua kernel	16	8	6.3	Toxic saponins (6.8%), tannins (1.2%)
Teak kernel	60	4.5	12	High ash
Tamarind kernel	19	1.1	3.4	–
Mango kernel	6	4.5	3.6	Tannins
Water melon kernel (dehulled)	51	4	9	–
Musk melon kernel (dehulled)	66	3	9	–

Presence of anti-nutritional factors, high fiber content and poor protein quality precludes the utilization of these minor cakes as animal feed. The compositions of various non-conventional oil cakes produced in India are given in the Table 13.5.

13.3 Applications of Edible Oil Cakes for Bioprocesses

The major challenge of the bioprocess industry lies in developing cost-and eco-efficient processes using renewable raw materials to deliver high value-added products at costs, acceptable to the general public. Oil seeds are second only to grain crops in the supply of plant proteins for human and animal consumption. The global oilseed production reached 417 million tonnes in 2006–2007 and the oil cake production was brought to 106 million tonnes (FAO). In view of increasing the utility of renewable feedstock in bioprocesses, the exploration of potential usefulness of oil-cakes is desirable.

13.3.1 Production of Enzymes

Enzymes are biological catalysts, the catalysts for cell metabolism. They are widely occurred throughout the biological system, and complex network of reactions brought about by enzymes are the basis for the continuity of living world that evolved over millennia. They offer efficient biocatalytic conversion potentials to technologies ranging from food industry to personal care industry. They are one of the most important products obtained from microbial sources for human requirements. Enzymes are produced industrially either by submerged (SmF) or solid state fermentation (SSF). Both bacteria and fungi are employed for the industrial production of enzymes. The production of various enzymes utilizing oil cakes either as the major substrate or media supplement is enumerated in Table 13.6. The critical analysis of literature shows that enzyme production by fungi is favored to bacteria employing oil cakes under solid state fermentation. This is attributed to the morphology and physiology of these molds which enable them to penetrate and colonize solid substrates (Sivaramakrishnan et al. 2007). *Aspergillus*, *Penicillium* and *Rhizopus* are the widely reported fungal species for enzyme production. Among bacteria *Bacillus* sp has been most commonly used for enzyme production utilizing oil cakes. Oil cakes are used either as a single substrate or mixed with various other substrates in different combinations for solid state fermentation. In the case of submerged fermentation, oil cakes are supplemented either as carbon or nitrogen source for enzyme production. The use of minor oil cakes such as babassu oil cake and almond oil cake for enzyme production point out to the constant search in the bioprocess industry to evaluate and valorize indigenous sustainable agro-resources.

13.3.2 Production of Media Supplements

The high cost of synthetic media has a negative impact on the development of studies in the field of microbiology and fermentation and molecular biology in developing countries. Efforts are being made to develop organic nitrogen supplements from oil cakes equivalent to commercially available yeast extract, beef extract and peptone. Phillipchuk and Jackson, 1979 compared the effect of acidic and enzymic digests of rapeseed cake with commercial media such as Sabouraud dextrose broth and mycological media in supporting the growth of *Candida utilis*. Acidic and trypsin digests of rapeseed cake supplemented with dextrose gave growth responses equivalent to that on commercial media with the exception of the peptic digest of rapeseed cake. Gupta et al. 2005 prepared growth medium from oil cakes such as linseed, mustard and neem to study the growth and morphology of *Catenaria anguillulae*, a nematicidal fungus. The oil cake media were prepared by boiling and filtering 5g of oil cake and 15g of agar to yield 0.5% of oil cake. These three media were compared with commercially available media such as beef extract agar, Emerson agar and YPSS agar. Optimal growth and morphology of the fungus was recorded on linseed oil cake agar. Defatted cotton seed protein was used for the production of protease by *Entomophthora coronata* in 6L fermenter (Jönsson 1968). Enzymic digests and water soluble extracts of various oil cakes such as soybean cake, cottonseed cake were used as nitrogen sources in the commercial production of various bioactive molecules. Soybean cakes are processed to yield an array of products such as soy flour, soybean protein concentrate, soy protein isolate. Various commercial media has been developed for micro-organisms utilizing mainly soybean cake such as trypticase soy agar, soy peptone etc.

Table 13.6 Enzyme production employing oil cakes as nutrient source

Utilization of oil cake	Enzyme	Micro-organism	Method	Magnitude	Reference
Soybean cake	Lipase	*Penicillim simplicissimum*	SSF	Static flask	(Di Luccio et al. 2004)
Soybean cake	Protease	*Penicillium* sp.	SSF	Static flask	(Germano et al. 2003)
Soybean cake (1% w/w) supplemented to spent brewing grain	Alpha amylase	*Aspergillus oryzae* NRRL 6270	SSF	Static flask	(Francis et al. 2003)
Soybean cake (3% w/v)	Carboxy peptidase	*Aspergilus saitoi*	SmF	6L Fermenter	(Ichishima et al. 1973)
Soybean cake as nitrogen source (1.5%w/v)	Alkaline protease	*Bacillus sp. I-312*	SmF	Shake flask	(Joo and Chang 2003)
Soybean cake (4% w/v)	Lipase	*Penicillium camembertii Thom* PG	SmF	Shake flask	(Tan et al. 2004)
Canola cake	Phytase	*Aspergillus ficuum* NRRL 3135	SSF	Static flask	(Ebune et al. 1995)
Groundnut oil cake + Wheat bran (1:1 w/w)	Alpha amylase	*Bacillus amyloliquefaciens* ATCC 23842	SSF	Static flask	(Gangadharan et al. 2005)
Copra cake	Glucoamylase	*Aspergillus niger* NCIM 1245	SSF	Static flask	(Pandey et al. 1995)
Copra cake	Alpha amylase	*Aspergillus oryzae* IFO 30103	SSF	Static flask	(Ramachandran et al. 2004)
Copra cake	Phytase	*Mucor racemosus* ATCC 46129	SSF	Static flask	(Boger et al. 2003)
Coconut oil cake	Lipase	*Candida rugosa*	SSF	Static flask	(Benjamin and Pandey 1997)
Coconut oil cake + Sesame oil cake (1:1 w/w)	Phytase	*Rhizopus oryzae* NRRL 1891	SSF	Static flask	(Ramachandran et al. 2005)
Sesame oil cake + Wheat bran (1:3 w/w)	Lipase	*Aspergillus niger* MTCC 2594	SSF	Static flask	(Mala et al. 2007)
Sesame oil cake	Phytase	*Mucor racemosus* NRRL 1994	SSF	Shake flask	(Roopesh et al. 2006)
Sesame oil cake	L-Glutaminase	*Zygosaccharomyces rouxii* NRRL-Y 2547	SSF	Static flask	(Kashyap et al. 2002)
Sesame oil cake	Lipase	*Aspergillus niger* MTCC 2594	SSF	Static flask	(Kamini et al. 1998)
Sunflower oil cake + Sugar beet oil cake +Wheat bran (3:1:1)	Alpha amylase	*Penicillium chrysogenum*	SSF	Static flask	(Ertan et al. 2006)

Table 13.6 (continued)

Utilization of oil cake	Enzyme	Micro-organism	Method	Magnitude	Reference
Palm kernel oil cake	Tannase	*Aspergillus niger* ATCC 16620	SSF	Static flask	(Sabu et al. 2005)
Palm kernel oil cake + Wheat bran (2:1)	Xylanase	*Aspergillus niger* ATCC 6275	SSF	Static flask	(Prasertsan et al. 1997)
Palm kernel oil cake	Alpha amylase	*Bacillus licheniformis* CUMC305	SmF	Shake flask	(Krishnan and Chandra 1982)
Mustard oil cake (2% w/v)	Lipase	*Rhizopus rhizopodiformis*	SSF	Packed bed column reactor	(Cordova et al. 1998)
Olive oil cake + Sugarcane bagasse (1:1 w/w)	Lipase	*Penicillium simplicissimum*	SSF	Fixed bed reactor	(Cavalcanti et al. 2005)
Babassu oil cake	Lipase	*Penicillium restrictum*	SSF	Static flask	(Gombert et al. 1999)
Almond oil cake	Lipase	*Rhizopus oligosporus* GCBR-3	SSF	Static flask	(Haq et al. 2002)

13.3.3 Production of Secondary Microbial Metabolites

Secondary metabolites of microbial origin such as antibiotics have significant role in the development of humanity. The fermentation industry received its greatest impetus for expansion with the advent of antibiotics as chemotherapeutic agents. The requirement of antibiotics in huge numbers to combat bacterial diseases led to the extensive research on producing them economically at the highest concentration with minimum energy input. Arun and Dharmalingam (1999) have reported the production of daunorubicin, an antitumor antibiotic produced by *Streptomyces peucetius* using 5% sesame oil cake as carbon source thus reducing the production cost by 96%. Soy peptone, water-soluble enzymatic digest of soybean cake is reported to enhance the production of pactamycin, an antitumor antibiotic produced by *Streptomyces pactum* by two fold (Bhuyan 1962). Rifamycin, an antibiotic used against *Mycobacterium tuberculosis* and *M. leprae* inhibits RNA synthesis by binding to the β-subunit of RNA polymerase. Significant improvement (4 fold) of rifamycin production is acheived by optimization of media which comprised of 1% peanut cake and 1% soybean cake as nitrogen source (Krishna et al. 1998).

Cotton seed cake (2%) was used as a nitrogen source for the production of an antimicrobial agent called lomofungin produced by *Streptomyces lomondensis* (Johnson and Dietz 1969). Bacitracin A, the predominating antibiotic belonging to the group polypeptidic antibiotics is produced by *Bacillus licheniformis*. Different processes for industrial production of bacitracin utilized soybean cake and cottonseed cake individually as nitrogen sources under submerged conditions. Matelová and Břečka (1967) studied the influence of peanut cake (6%) as nitrogen source and oxygen transfer in bacitracin biosynthesis to effect maximal production. Utilization of soybean cake extract (water soluble) as nitrogen source was reported to trigger significant production of synnematin B (identical to cephalosporin N) by *Emericellopsis terricola* var. *glabra* (Nara and Johnson 1959). Sarada and Sridhar (1998) reported significant enhancement in the production of cephamycin C produced by *Streptomyces clavuligerus* LC 21 when sunflower oil cake (12.5%) was supplemented to the medium. Cephamycin C belongs to the class of microbially synthesized β-lactam antibiotics, highly resistant to β-lactamases. Melingimycin is a potent broad-spectrum insecticide produced by *Streptomyces nanchangensis*. It belongs to the family of milbemycins, which have a 16-membered macrocyclic lactone. Soybean cake (1%) is supplemented as nitrogen source for the optimal production of this biocontrol agent (Zhuang and Chen 2006).

Biosurfactants are surface active compounds having a wide range of industrial applications such as enhanced oil recovery, lubricants, bioremediation of pollutants, food processing etc. The structures of these complex molecules include lipopeptides, glycolipids, polysaccharide protein complexes, fatty acids and phospholipids. Optimal production of biosurfactant (glycolipid) by *Bacillus megaterium* was obtained in 3L laboratory scale fermenter when peanut oil cake (2%) was used as carbon source (Thavasi et al. 2008). Carotenoids are important natural pigments with a range of applications as colorants, feed supplements and neutraceuticals. Lycopene is a red coloured intermediate of the ß-carotene biosynthetic pathway and is an important dietary carotenoid. It is reported to inhibit the harmful effect of ferric nitrilotriacetate on DNA in rats and prevents liver necrosis. López-Nieto et al. 2004 reported the development of a semi-industrial process (800 L fermentor) for lycopene production by mated fermentation of *Blakeslea trispora* plus (+) and minus (−) strains. This process describes the critical requirement of soybean cake (44g/L) as nitrogen source for optimal lycopene production. Mustard oil cake (6%) in the presence of Mg^{2+} ions is reported to improve lactic acid production ability of agar-gel immobilized *Lactobacillus casei* after 48 hours, when further addition of the substrate (whey lactose) failed to maintain the process efficiency (Tuli et al. 1985).

13.3.4 Production of Mushrooms

Mushrooms, the fruiting bodies of various Basidiomycetes are used as food and as flavoring agents in soups and sauces. *Pleurotus sajor-caju* is a commercially important mushroom grown on paddy straw. Supplementation of oil-cakes from mustard, niger, sunflower, cottonseed and sobean as nitrogen source to the rice straw substrate increased the mushroom yield by 50–100%. This also increased the digestibility of rice straw by mushrooms resulting in a corresponding increase in the amino acid and sugar content and significant decrease in cellulo-hemicellulosics. Oil cake supplementation enhances the secretion of cellulases, hemicellulases and laccases thereby decreasing the cellulose, hemicellulose and lignin content of rice straw significantly (Bano et al. 1993). About 40% rice straw is left behind as spent substrate with each harvest of the mushroom crop. In view of utilizing the large quantities of spent rice straw, Shashirekha et al. 2002 investigated the evaluation of the same supplemented with oil cakes for the production of *Pleurotus sajor-kaju*. A twelve fold enhancement in mushroom yield was reported with cottonseed cake (0.15% N levels) illustrating the ability of the cake to supplement necessary nutrients exhausted in the spent rice straw substrate. This research work discusses the immense potential of oil cakes to support two mushroom harvests utilizing the same substrate without limiting the production yield.

13.3.5 Biological Detoxification of Oil Cakes

Oil-cakes are mainly used as livestock feed worldwide and there is an increasing trend, towards its utilization as feed for poultry, non-ruminants and aquaculture. The presence of certain anti-nutrients in several oil cakes limit its usage as animal feed. Table 13.7 enumerates the spectrum of anti-nutrients present in various oil cakes and its effect on animals fed. Several methods are devised to inactivate or remove toxic components. Physical treatments such as dehulling, heat treatments (cooking, autoclaving, toasting) are largely used for detoxifying legume seeds. Toxic substances are withdrawn when cooked in liquid medium. Autoclaving aims at de-

Table 13.7 Principal antinutritional factors in oil cakes and its effect on fed-animals (Ravindran and Blair 1992)

Oil cake	Antinutrient	Effect on fed-animals
Soybean	raffinose , stachyose, phytate	Indigestibility to fish
Rapeseed	glucosinates, sinapine, erucic acid, tannins, phytates	Glucosinates are goitrogenic & toxic to monogastrics and erucic acid cause heart lesions in animals, toxic to aquaculture
Cottonseed	Gossypol	Inhibits hydrolytic enzymes and reduces palatability
Groundnut	aflatoxins as contaminants	Potent carcinogens and highly toxic to animals
Sunflower	chlorogenic acid	Inhibit hydrolytic enzymes in poultry
Palm kernel cake	high fiber content	Low palatability in monogastrics
Sesame	oxalates, phytates	Lowers mineral availability in animals
Linseed	linatin, cyanogenic glucosides	Toxic to poultry
Safflower	phenolic glucosides	Bitter flavor thereby lowering palatability
Castor	ricin, ricinine, castor bean allergen	Ricin, highly toxic protein, ricinine, an alkaloid, reducing palatability
Kapok	Tannins	Lowers palatability and digestion

naturing biologically active proteins such as ricin, protease inhibitors, goitrogenic factors etc. Chemical treatments include ammoniation, sodium carbonate treatment (glucosinates), addition of ferrous sulphate (gossypol), supplementation of choline or methionine (chlorogenic acid) etc (Delort-Laval 1993). These treatments are highly discouraged as it drastically reduces the suitability of oil cake as animal feed.

Enzymatic and fermentative treatments to enhance the nutritional value of oil cakes are being explored. The negative effect of phytates on mineral availability and protein digestibility can be reduced by the addition of microbial phytase to animal feeds. Fermentative treatment involves the use of micro-organisms for removing anti-nutrients and adding essential nutrients and amino acids. Removal of unavailable carbohydrates, phytates and tannins is well within the capabilities of this approach (Ravindran and Blair 1992).

Fermentation of sesame oil cake with *Lactobacillus acidophilus* completely removed phytates and reduced the tannin content to 50% increasing the suitability of the oil cake as feed to rohu fingerlings (*Labeo rohita*) (Mukhopadhyay 2001). Phosphate excretion studies was investigated in chicks fed with soybean cake fermented with *Aspergillus usamii*. Results indicated that fermentation improved phosphorus bioavailability indicating the complete degradation of phytate and thereby reducing phosphate excretion (Hirabatashi et al. 1998).

Detoxification of gossypol in fermented cottonseed cake using *Diploidia* (class: *Fungi imperfectii*) was evaluated in growing pigs. Liver free and bound gossypol levels were found to be significantly lower for fermented cottonseed cake fed pigs, with no mortality when compared with control (50% mortality) (Kornegay et al. 1972).

Mustard oil cake contains glucosinolates which gets subjected to endogenous degradation by myrosinase contained in the seeds resulting in the production of toxic breakdown metabolites such as nitriles, thiocynates and isothiocyanates. Solid state fermentation of mustard oil cake by *Aspergillus* sp. NR 4201 resulted in complete degradation of glucosinolates without any toxic breakdown product rendering the oil cake for utilization as animal feed (Rakariyatham and Sakorn 2002).

In view of the value addition of palm kernel cake fermented with *Rhizopus stolonifer* NAU 07 under solid state fermentation for the production of fructosyl transferase, chemical and nutrient composition of spent substrate such as crude protein, crude fiber, ash and lipid contents was studied. The protein content was observed to increase by 33.3% while ash content decreased by 44.5% thereby improving the nutritional qualities of the oil cake (Lateef et al. 2008).

13.3.6 Bioconversion of Oil Cakes

Hem et al. (2008) reported the development of a rural self sufficient technology in Republic of Guinea to utilize palm kernel oil cake, the only available agricultural by-product to generate aquaculture feed for Tilapia (*Oreochromis niloticus*). Palm kernel cake was moistened with water (1:2) and kept for fermentation in rectangular iron tanks. The insect *Hermetia illucens* (black soldier fly), attracted by the odour of fermented matter lays its eggs on the fermented matter generating black soldier (BS) larvae. An inverse correlation was obtained between conversion rate (PKOC kg/ larval biomass kg) on Y-axis and duration of culture time (days) on X-axis, indicating efficient bioconversion from the oil-cake to larval biomass. These larvae grown in the fermented medium are harvested after four weeks to use as feed stuff along with rice bran (3:7) for the Tilapia fish.

13.3.7 Other Applications

Oil cakes have an array of roles ranging from being a food product to soil amendment. In Indonesia, peanut oil cake is used as a popular fermented food called Ontjom (lontjom) (Beuchat 1974) and on the contrary it is also used as a promising soil amendment promoting plant growth in India (Bhattacharya and Goswami 1987). Olive oil cake has proved to be a good source of combustible gases such as methane. Combustion efficiency of oil cake is widely investigated for its use as an energy source (Abu-Qudais and Okasha 1996). Investigations on anaerobic treatment of cotton seed cake revealed significant methane generation potential indicating it to be a good source for biogas production (Isci and Demirer 2007). Catalytic steam-pyrolysis of cottonseed cake generated bio-oils, a high quality liquid fuel comparable with gasoline fraction of petroleum (Pütün et al. 2006).

Combinations of neem oil cake, mustard oil cake and castor oil cake were evaluated for their efficiency against plant parasitic nematodes and soil-inhabiting parasitic fungi infesting crops such as mungbean and chickpea. Application of these oil cakes as soil amendment reduced the population of plant-parasitic nematodes,

Meloidogyne incognita, Rotylenchulus reniformis, Tylenchorhynchus brassicae, Helicotylenchus indicus and the frequency of the pathogenic fungi *Macrophomina phaseolina, Rhizoctonia solani, Phyllosticta phaseolina, Fusarium oxysporum* f. *ciceri* significantly, but increased the frequency of saprophytic fungi. Plant growth parameters were observed to be improved by several fold and residual effects of oil cakes were also noted in the subsequent crop, in the next growing season (Tiyagi and Alam 1995).

13.4 Conclusion

Oil cakes being rich in proteins, carbohydrates and minerals find various applications in the bioprocess industry such as production of enzymes, secondary metabolites, biomass etc. It has been mainly recognized as an important nitrogen source in many commercial fermentation studies. Available information indicated the need to evaluate and refine the methods depicting the digestibility of these substrates and availability of nutrients provided for micro-organisms. Research must be focused to develop simple, low-cost and self-sustainable technologies with efficient utilization of indigenous substrates promoting the social development of rural sectors.

Abbreviations

DM: dry matter
CM: crude matter
AA: amino acid
SBOC: soybean oil cake
RSOC: rapeseed oil cake
CSOC: cotton seed oil cake
GOC: groundnut oil cake
SFOC: sunflower oil cake
PKOC: palm kernel oil cake
SOC: sesame oil cake
LSOC: linseed oil cake
SfOC: Safflower oil cake
COC: coconut oil cake
OOC: olive oil cake
FAO: Food and Agricultural Organisation
SSF: solid state fermentation
SmF: submerged fermentation

References

Abu-Qudais M, Okasha G (1996) Diesel fuel and olive-cake slurry: atomization and combustion performance. Appl Energ 54:315–326

Achaya KT (1990) Oilseeds and Oilmilling in India, A Cultural and Historical Survey. Oxford and IBH Publishing Co Pvt Ltd, New Delhi

Arun D, Dharmalingam K (1999) *Streptomyces peucetius* converts anthracycline intermediates efficiently in culture media containing oil cake as carbon source. World J Microbiol Biotechnol 15: 333–334

Bano Z, Shashirekha MS, Rajarathnam S (1993) Improvement of the bioconversion and biotransformation efficiencies of the oyster mushroom (*Pleurotus sajor-kaju*) by supplementation of its rice straw substrate with oil seed cakes. Enzyme Microb Technol 15:985–989

Barbour GW, Sim JS (1991) True metabolizable energy and true amino availability in canola and flax products for poultry. Poultry Sci 72:98–105

Batal A, Dale N, Café M (2005) Nutrient composition of peanut meal. J Appl Poultry Res 14: 254–257

Bell JM (1984) Nutrients and toxicants in rapeseed meal: A review. J Anim Sci 58:996–1010

Benjamin S, Pandey A (1997) Coconut cake – a potent substrate for the production of lipase by Candida rugosa in solid-state fermentation. Acta Biotechnol 17:241–251

Beuchat LR (1974) Fungal fermentation of peanut press cake. Econ Bot 30:227–234

Bhattacharya D, Goswami BK (1987) Comparative efficacy of neem and groundnut oil-cakes with aldicarb against *Meloidogyne incognita* in tomato. Revue Nématol 10:4467–470

Bhuyan B K (1962) Pactamycin production by *Streptomyces pactum*. Appl Environ Microbiol 10:302–304

Boger B, Szakacs G, Pandey A et al. (2003) Production of phytase by Mucor racemosus in solid-state fermentation. Biotechnol Prog 19:312–319

Briggs HM, Heller VG (1942) The effect of adding large amounts of cottonseed meal to a lamb fattening ration. J Anim Sci 1:277–284

Carvalho LPF, Cabrita ARJ, Dewhurst RJ et al. (2006) Evaluation of palm kernel meal and corn distillers grains in corn silage-based diets for lactating dairy cows. J Dairy Sci 89:2705–2715

Castro BSI, Philip LE, Lapierre H et al. (2007) Ruminal degradability and intestinal digestibility of protein and amino acids in treated soybean meal products. J Dairy Sci 90:810–822

Cavalcanti EAC, Gutarra MLE, Freire DMG et al. (2005) Lipase production by solid-state fermentation in fixed-bed bioreactors. Braz Arch Biol Technol 48:79–84

Cordova J, Nemmaoui M, Ismaïli-Alaoui M et al. (1998) Lipase production by solid state fermentation of olive cake and sugar cane bagasse. J Mol Catal B-Enzym 5:75–78

Creswell DC, Brooks CC (1971) Composition, apparent digestibility and energy evaluation of coconut oil and coconut meal. J Anim Sci 33:366–369

Delort-Laval J (1993) Chemical treatments of oilmeals. In: Applewhite TH (ed) Proceedings of the World Conference on Oilseed Technology and Utilization, AOCS Press, Illinois

Di Luccio M, Capra F, Ribeiro NP et al. (2004) Effect of temperature, moisture, and carbon supplementation on lipase production by solid-state fermentation of soy cake by *Penicillium simplicissimum*. Appl Biochem Biotechnol 113:173–180

Ebune A, Al-Asheh S, Duvnjak Z (1995) Production of phytase during solid state fermentation using *Aspergillus ficuum* NRRL 3135 in canola meal. Bioresour Technol 53:7–12

Ertan F, Balkan B, Balkan S et al. (2006) Solid state fermentation for the production of α-amylase from Penicillium chrysogenum using mixed agricultural by-products as substrate. Biologia 61:657–661

Francis F, Sabu A, Nampoothiri KM et al. (2003) Use of response surface methodology for optimizing process parameters for the production of α-amylase by *Aspergillus oryzae*. Biochem Eng J 15:107–115

Gangadharan D, Sivaramakrishnan S, Nampoothiri KM et al. (2005) Solid culturing of *Bacillus amyloliquefaciens* for alpha amylase production. Food Technol Biotechnol 44:269–274

Germano S, Pandey A, Osaku CA et al. (2003) Characterisation of proteases from *Penicillium* sp. produced by solid state fermentation. Enzyme Microb Technol 32:246–251

Gombert AK, Pinto AL, Castilho LR et al. (1999) Lipase production by *Penicillium restrictum* in solid-state fermentation using babassu oil cake as substrate. Process Biochem 35: 85–90

Gupta RC, Vaish SS, Singh RK et al. (2005) Oil cakes as media for growing *Catenaria anguillulae* sorokin, a facultative endoparasite of nematodes. World J Microbiol Biotechnol 21: 1181–1185

Haq I, Idrees S, Rajoka MI (2002) Production of lipases by *Rhizopus oligosporus* by solid-state fermentation. Process Biochem 37: 637–641

Hem S, Toure S, Sagbla C et al. (2008) Bioconversion of palm kernel meal for aquaculture: experiences from the forest region (Republic of Guinea). Afr J Biotechnol 7:1192–1198

Hirabatashi M, Matsui T, Yano H et al. (1998) Fermentation of soybean meal with *Aspergillus usamii* reduces phosphorus excretion in chicks. Poultry Sci 77:552–556

Ichishima E, Yamane A, Nitta T et al. (1973) Production of a new type of acid carboxypeptidase of molds of the *Aspergillus niger* group. Appl Biochem 26:327–331

Isci A, Demirer GN (2007) Biogas production from cotton wastes. Renew Energ 32:750–757

Johnson LE, Dietz A (1969) Lomofungin, a new antibiotic produced by *Streptomyces lomodensis* sp. Appl Microbiol 17:755–759

Jönsson AG (1968) Protease production by species of entomophthora. Appl Microbiol 16:450–457

Joo HS, Chang CS (2003) Production of protease from a new alkalophilic *Bacillus* sp. I-312 grown on soybean meal: optimization and some properties. Process Biochem 40:1263–1270

Kamini NR, Mala JGS, Puvanakrishnan R (1998) Lipase production from *Aspergillus niger* by solid-state fermentation using gingelly oil cake. Process Biochem 33:505–511

Kashyap P, Sabu A, Pandey A et al. (2002) Extra-cellular L-glutaminase production by *Zygosaccharomyces rouxii* under solid-state fermentation. Process Biochem 38:307–312

Kornegay ET, Kelly RF, Campbell TC et al. (1972) Fungal-treated cottonseed meal for swine. J Nutr 102: 1471–1476

Krishna PSM, Venkateswarlu G, Rao LV (1998) Studies on fermentative production of rifamycin using *Amycolatopsis mediterranei*. World J Microbiol Biotechnol 14:689–691

Krishnan T, Chandra AK (1982) Effect of oilseed cakes on α-amylase production by *Bacillus licheniformis* CUMC305. Appl Environ Microbiol 44:270–274

Lateef A, Oloke JK, Kana EBG et al. (2008) Improving the quality of agro-wastes by solid-state fermentation: enhanced antioxidant activities and nutritional qualities. World J Microbiol Biotechnol. doi: 10. 1007/s1127400897498

Loosli JK, Elliot JM, Warner RG (1960) Comparative value of corn distillers dried grains with solubles, soybean oil meal and linseed oil meal for milk production. J Dairy Sci 43:816–820

López-Nieto MJ, Costa J, Peiro E et al. (2004) Biotechnological lycopene production by mated fermentation of *Blakeslea trispora*. Appl Microbiol Biotechnol 66:153–159

Lyon CK, Gumbmann MR, Betschart AA et al. (1979) Removal of deleterious glucosides from safflower meal. J Am Oil Chem Soc 56:560–564

Mala JGS, Edwinoliver NG, Kamini NR et al. (2007) Mixed substrate fermentation for production and extraction of lipase from *Aspergillus niger* MTCC 2594. J Gen Appl Microbiol 53:247–253

Martín García AI, Moumen A, Yáñez Ruiz DR et al. (2003) Chemical composition and nutrients availability for goats and sheep of two-stage olive cake and olive leaves. Anim Feed Sci Technol 107:61–74

Matelová V, Břečka A (1967) Oxygen-transfer demand of *Bacillus licheniformis* under conditions of submerged cultivation. Appl Environ Microbiol 15:1079–1082

Mitra CR, Misra PS (1967) Amino acids of processed seed meal proteins. J Agri Food Chem 15:697–700

Mukhopadhyay N (2001) Effect of fermentation on apparent total and nutrient digestibility of sesame (*Sesamum indicum*) seed meal in rohu, *Labeo rohita* (Hamilton) fingerlings. Acta Ichthyol Piscat 32:19–28

Nara T, Johnson MJ (1959) Production, purification and characterization of synnematin. J Bacteriol 77:217–226

O'Mara FP, Murphy JJ, Rath M (1997) The amino acid composition of protein feedstuffs before and after ruminal incubation and after subsequent passage through the intestines of dairy cows. J Anim Sci 75:1941–1949

Pandey A, Ashakumary L, Selvakumar P (1995) Copra waste- a novel substrate for soild-state fermentation. Bioresour Technol 51:217–220

Phillipchuk GE, Jackson H (1979) Rapeseed oil meal as a nitrogenous substrate for microbial fermentation. J Gen Appl Microbiol 25:117–126

Prasertsan P, H-Kittikul A, Kunghae A et al. (1997) Optimization of xylanase and cellulase production from *Aspergillus niger* ATCC 6275 in palm oil mill wastes and its application. World J Microbiol Biotechnol 13: 555–559

Pütün E, Uzun BB, Pütün AE (2006) Production of bio-fuels from cottonseed cake by catalytic pyrolysis under steam atmosphere. Biomass Bioenerg 30:592–598

Rakariyatham N, Sakorn P (2002) Biodegradation of glucosinates in brown mustard seed meal (*Brassica juncea*) by *Aspergillus* sp. NR-4201 in liquib and solid-state cultures. Biodegradation 13:395–399

Ramachandran S, Patel AK, Nampoothiri KM et al. (2004) Coconut oil cake- a potential raw material for the production of α-amylase. Bioresour Technol 93:169–174

Ramachandran S, Roopesh K, Nampoothiri KM (2005) Mixed substrate fermentation for the production of phytase by *Rhizopus* sp. using oilcakes as substrates. Process Biochem 40: 1749–1754.

Ravindran V, Blair R (1992) Feed resources for poultry production in Asia and the Pacific. II. Plant protein sources. World Poultry Sci J 48:205–231

Roopesh K, Ramachandran S, Nampoothiri KM et al. (2006) Comparison of phytase production on wheat bran and oilcakes in solid-state fermentation by *Mucor racemosus*. Bioresour Technol 97:506–511

Sabu A, Pandey A, Daud MJ et al. (2005) Tamarind seed powder and palm kernel cake: two novel agro residues for the production of tannase under solid state fermentation by *Aspergillus niger* ATCC 16620. Bioresour Technol 96:1223–1228

Sarada I, Sridhar P (1998) Nutritional improvement for cephamycin C fermentation using a superior strain of *Streptomyces clavuligerus*. Process Biochem 33:317–322

Schumacher KD (1993) Worldwide sources of oilseed meals for feed manufacturing. In: Applewhite TH (ed) Proceedings of the World Conference on Oilseed Technology and Utilization, AOCS Press, Illinois

Shashirekha MN, Rajarathnam S, Bano Z (2002) Enhancement of bioconversion efficiency and chemistry of the mushroom Pleurotus sajor-kaju (Berk and Br.) Sacc. produced on spent rice straw substrate supplemented with oil seed cakes. Food Chem 76:27–31

Sivaramakrishnan S, Gangadharan D, Nampoothiri KM et al. (2007) Alpha amylase production by *Aspergillus oryzae* employing solid state fermentation. J Sci Ind Res India 66: 621–626

Smith K (1986) Advances in feeding soybean products. In: Baldwin AR (ed) World Conference on Emerging Technologies in the Fats and Oils Industry. AOCS Press, Illinois

Sundu B, Kumar A, Dingle J (2006) Palm kernel meal in broiler diets: effect on chicken performance and health. World Poultry Sci J 62:316–325

Tan T, Zhang M, Xu J et al. (2004) Optimization of culture conditions and properties of lipase from *Penicillium camembertii* Thom PG-3. Process Biochem 39:1495–1502

Thampan PK (1975) The coconut palm and its products. Green villa publishing house, Cochin

Thavasi R, Jayalakshmi S, Balasubramanian T et al. (2008) Production and characterization of a glycolipid biosurfactant from Bacillus megaterium using economically cheaper sources. World J Microbiol Biotechnol 24:917–925

Tiyagi SA, Alam MM (1995) Efficacy of oil-seed cakes against plant-parasitic nematodes and soil-inhabiting fungi on mungbean and chickpea. Bioresour Technol 51:233–239

Tuli A, Sethi RP, Khanna PK et al. (1985) Lactic acid production from whey permeate by immobilized *Lactobacillus casei*. Enzyme Microb Technol 7:164–168

Villamide MJ, San Juan LD (1998) Effect of chemical composition of sunflower seed meal on its true metabolizable energy and amino acid digestibility. Poultry Sci 77:1884–1892

Vlyssides AG, Loizides M, Karlis PK (2004) Integrated strategic approach for reusing olive oil extraction by-products. J Clean Prod 12:603–611

Yamauci K, Samanya M, Seki K et al. (2006) Influence of dietary sesame meal level on histological alterations of the intestinal mucosa and growth performance of chickens. J Appl Poultry Res 15:266–273

Zhuang YP, Chen B (2006) Medium optimization for meilingmycin production by *Streptomyces nanchangensis* using response surface methodology. Process Biochem 41:405–409

Chapter 14
Biotechnological Potential of Fruit Processing Industry Residues

Diomi Mamma, Evangelos Topakas, Christina Vafiadi and Paul Christakopoulos

Contents

Abstract Fruit juices and derived products such as nectars and drinks have experienced growing popularity within the last years. Orange waste, apple pomace and grape pomace are the solid by-products derived from processing of oranges, apples and grapes, respectively. Due to increasing production, their disposal represents a growing problem since the plant material is usually prone to microbial spoilage, thus limiting further exploitation. On the other hand, costs of drying, storage and shipment of by-products are economically limiting factors. Therefore, agro-industrial by-products are often utilized as feed or as fertilizer. The application of agro-industrial by-products in bioprocesses offers a wide range of alternative substrates, thus helping to solve pollution problems related to their disposal. Attempts have been made to use orange waste, apple pomace and grape pomace to generate several value-added products through microbial transformations or enzymatic modifications, such as enzymes, bioethanol, organic acids, heteropolysaccharides, aroma compounds, protein enriched feeds, prebiotic oligosaccharides and biologically active molecules.

P. Christakopoulos (✉)
Biotechnology Laboratory, School of Chemical Engineering, National Technical University of Athens, 9 Iroon Polytechniou Str., Zografou Campus, 157 80, Zografou, Greece
e-mail: hristako@orfeas.chemeng.ntua.gr

P. Singh nee' Nigam, A. Pandey (eds.), *Biotechnology for Agro-Industrial Residues Utilisation*, DOI 10.1007/978-1-4020-9942-7_14,

Keywords Fruit processing industry residues · Biotechnological applications · Microbial transformations · Enzymatic modifications

14.1 World Availability of Citrus, Apples and Grapes

According to Food and Agriculture Organization (FAO) estimates the world citrus production for the year 2005 was 94.8 million MT (FAOSTAT-FAO Statistical Database, 2005). The genus *Citrus* includes several important fruits (Bampidis and Robinson 2006), with the most important on a worldwide basis being sweet orange (*C. sinensis*: 61.1% of world citrus production), tangerine (*C. reticulata*: 19.9%), lemon and lime (*C. limon* and *C. aurantifolia*: 12.1%) and grapefruit (*C. paradisi*: 5.0%). Minor citrus genuses that comprise the bulk of the remaining 2.0% include sour orange (*C. quarantium*), shaddock (*C. grandis*) and citron (*C. medica*). Table 14.1 presents the major producing countries. About 20.6% of world production of citrus is in the Mediterranean countries of Spain, Italy, Greece, Egypt, Turkey and Morocco, with Brazil (20%), China (16%) and the USA (11%) being major individual citrus producing countries (Table 14.1). Approximately, 27 million MT of the total citrus production for the year 2005 have been processed to yield juice, essential oils and other by-products (FAOSTAT-FAO Statistical Database, 2005). Citrus by-products are the principal solid by-product of the citrus processing industry and constitute about 50% of fresh fruit weight (Garzón and Hours 1992). According to FAO the world citrus production for the year 2006 increased approximately 17% (FAOSTAT-FAO Statistical Database, 2005).

World apples (*Malus* sp., Rosaceae) production accounts for approximately 54.2 million MT (FAOSTAT-FAO Statistical Database, 2005). Table 14.2 presents the major producing countries. About 46% of world production of apples is in China, followed by USA (8%) and Turkey (5%) (Table 14.2). Fruits that are not suitable for consumption *in natura* are processed, generating large amounts of residues.

Approximately 14 million MT of apples will be processed mainly for the production of juice, jelly and pulp. Apple pomace, the solid residue from juice production, represents around 30% of the original fruit and is generated during fruit pressing (Vendruscolo et al. 2008). US Department of Agriculture estimated an increase in apples production for the year 2006 of about 5%. Grape, one of the world's largest fruit crop, with a reported annual production higher than 58 million MT (FAOSTAT-FAO Statistical Database, 2005), is cultivated mainly as *Vitis vinifera*. The economical importance of grapes and products obtained therefrom, such as wine, grape juice, jams and raisins, is therefore obvious. The most important grape producers are Italy (16%), France (12%), Spain (10%), and the USA (11%) (Table 14.2). About 80% of the produced grape quantity is used in wine-making. Wine-making affords grape pomace as a by-product in an estimated amount of 13% by weight of the grapes (Pinelo et al. 2006).

Table 14.1 World citrus production (in thousand tons) (FAOSTAT-FAO Statistical Database, 2005)

	Total	Oranges	Tangerines	Lemon and limes	Grapefruits
Northern Hemisphere					
Algeria	542.7	390.0	111.0	40.0	
China	15227.9	4462.0	8695.0		1903.0
Cuba	216.0	200.0			7.0
Cyprus	178.7	69.5	50.4	20.9	37.9
Egypt	2706.3	1759.3	612.6	331.4	3.0
Greece	861.0	763.2	59.7	31.5	6.6
India	4662.0	3100.0		1420.0	142.0
Indonesia	1311.7	1311.7			
Iran	3037.0	1900.0		1100.0	
Israel	639.9	184.2	122.8	68.0	264.9
Italy	3320.9	2105.1	611.6	597.4	6.8
Japan	1341.0	88.0	1249.0		
Korea Rep	594.0		594.0		
Lebanon	339.0	200.0	42.0	83.0	
Mexico	6910.0	4300.0	360.0	1890.0	360.0
Morocco	1320.9	827.0	463.9	25.0	5.0
Pakistan	504.5				
Spain	6181.3	2835.4	2500.4	809.5	36.0
Tunisia	307.9	174.7	33.2	28.0	72.0
Turkey	2316.8	1040.0	500.0	670.0	106.8
USA	10498.5	8419.1	367.3	789.4	922.7
Southern Hemisphere					
Argentina	2670.0	770.0	430.0	1300.0	170.0
Australia	716.1	571.0		35.0	15.1
Brazil	18902.5	16565.0	1270.0	1000.0	67.5
Chile	312.0			170.0	
Colombia	330.0				
Paraguay	289.6	205.7		16.4	43.3
Peru	754.0	315.0	172.0	220.0	47.0
South Africa	1543.0	1113.0		180.0	250.0

14.2 Composition of Fruit by-Products

14.2.1 Citrus Residue

Citrus fruits are principally consumed by humans as fresh fruit or processed juice, either fresh chilled or concentrated. After juice is extracted from the fruit, a residue comprised of peel (flavedo and albedo), pulp (juice sac residue), rag (membranes and cores) and seeds remained, which represents 50% of fresh citrus fruit weight (Garzón and Hours 1992). The processing of citrus results in approximately 13.5 million MT solid by-products (FAOSTAT-FAO Statistical Database, 2005).

The composition of citrus fruit is affected by factors such as growing conditions, maturity, rootstock, variety and climate. Citrus fruits contain nitrogen, lipids, sugars, acids, insoluble carbohydrates, enzymes, flavonoids, bitter principles, peel oil, volatile constituents, pigments, vitamins and minerals. The content of citrus

Table 14.2 World apples and grapes production (in thousand MT) (FAOSTAT-FAO Statistical Database, 2005)

Country	Apples	Country	Grapes
Northern Hemisphere		**Northern Hemisphere**	
China	25006.5	China	5698.0
Egypt	550.0	Egypt	1300.0
France	2123.0	France	6787.0
Germany	1600.0	Germany	1122.0
Hungary	720.0	Greece	1200.0
India	1470.0	Hungary	815.0
Iran, Islamic Rep of	2400.0	India	1200.0
Italy	2194.9	Iran, Islamic Rep of	2800.0
Japan	870.0	Italy	9256.8
Korea	669.0	Moldova, Republic of	600.0
Poland	2050.0	Portugal	1000.0
Russian Federation	2050.0	Romania	1027.6
Spain	797.7	Spain	5879.8
Turkey	2550.0	Turkey	3650.0
Ukraine	700.0	USA	6414.6
USA	4254.3	**Southern Hemisphere**	
Southern Hemisphere		Argentina	2365.0
Argentina	1262.4	Australia	1834.0
Brazil	843.9	Brazil	1208.7
Chile	1350.0	Chile	2250.0
South Africa	778.6	South Africa	1700.0

by-products is influenced by factors that include the source of the fruit and type of processing (Bampidis and Robinson 2006). Residues of citrus juice production are a source of dried pulp and molasses, fiber-pectin, cold-pressed oils, essences, D-limonene, juice pulps and pulp wash, ethanol, seed oil, pectin, limonoids and flavonoids. Fiber-pectins may easily be recovered from lime peels and are characterized by high fiber contents. The main flavonoids found in citrus species are hesperidin, narirutin, naringin and eriocitrin. Citrus seeds and peels were found to possess high antioxidant activity (Schieber et al. 2001).

14.2.2 Apple Pomace

Apples that are not suitable for consumption *in natura* are processed, generating large amounts of residues named apple pomace. The apple pomace is a heterogeneous mixture consisting of peel, core, seed, calyx, stem and soft tissue. It has high water content and is mainly composed of insoluble carbohydrates such as cellulose, hemicellulose and lignin. Simple sugars, such as glucose, fructose and sucrose, as well as minerals, proteins, vitamins and polyphenols are part of apple pomace composition. The composition varies according to the apple variety used and the type of processing applied for juice extraction, especially regarding how many times the fruits are pressed (Vendruscolo et al. 2008).

Production of pectin is considered the most reasonable way of utilizing apple pomace both from an economical and from an ecological point of view. In comparison to citrus pectins, apple pectins are characterized by superior gelling properties. Apple pomace has been shown to be a good source of polyphenols which are predominantly localized in the peels and are extracted into the juice to a minor extent. Major compounds isolated and identified include catechins, hydroxycinnamates, phloretin glycosides, quercetin glycosides and procyanidins. Since some phenolic constituents have been demonstrated to exhibit strong antioxidant activity *in vitro*, commercial exploitation of apple pomace for the recovery of these compounds seems promising (Schieber et al. 2001).

14.2.3 Grape Pomace

The major part of grapes production is used for wine-making and the major solid by-product generated is grape pomace. Grape pomace consists of three different components, seeds, stalks and skins. Taking into account that about 80% is used in winemaking, approximately 10 million tons of grape pomace arise within a few weeks during harvest. The chemical composition of grape pomace is rather complex: alcohols, acids, aldehydes, esters, pectins, polyphenols, mineral substances, sugars etc. are the most represented classes of compounds (Ruberto et al. 2008). One of the main environmental problems related to the management of the winery and distillery residues is the generation of large amounts during a short period of the year (August–October) (FAOSTAT-FAO Statistical Database, 2005), as well as some polluting characteristics of these residues, such as low pH and a high content of phytotoxic and antibacterial phenolic substances, which resist biological degradation. A great range of products such as ethanol, tartrates, citric acid, grape seed oil, hydrocolloids, and dietary fiber are recovered from grape pomace (Schieber et al. 2001). Anthocyanins, catechins, flavonol glycosides, phenolic acids and alcohols and stilbenes are the principal phenolic constituents of grape pomace (Pinelo et al. 2006).

14.3 Biotechnological Applications of Fruit Wastes

14.3.1 Microbial Transformations

Bacteria, yeast, and fungi have been cultivated under both submerged (SmF) and solid state fermentation (SSF) on orange waste, apple pomace and grape pomace for different purposes. A synopsis of the major strategies for fruit waste utilization for the microbial production of value-added products is following.

14.3.1.1 Enzymes

The most important area of citrus wastes and apple pomace utilization is the production of enzymes, especially pectinolytic ones. Pectinolytic enzymes or pectinases are a heterogeneous group of related enzymes that hydrolyze the pectic substances. Pectinolytic enzymes are of significant importance in the current biotechnological era with their all embracing applications in fruit juice extraction and its clarification, scouring of cotton, degumming of plant fibers, waste water treatment, vegetable oil extraction, tea and coffee fermentations, bleaching of paper, in poultry feed additives and in the alcoholic beverages and food industries (Jayani et al. 2005).

Both SmF and SSF conditions were evaluated by several researchers for the production of the above mentioned enzymes using citrus wastes or apple pomace as carbon sources. Bacteria, yeasts, and fungi under both SmF and SSF conditions were able to produce pectinolytic enzymes using citrus wastes (Garzón and Hours 1992; Fonseca and Said 1994; Ismail 1996; Kapoor et al. 2000; Martins et al. 2002; De Gregorio et al. 2002; Dhillon et al. 2004; Mamma et al. 2008) or apple pomace (Hang and Woodams 1994a; Berovic and Ostroversnik 1997; Pericin et al. 1999; Zheng and Shetty 2000; Joshi et al. 2006) as carbon sources.

Apart from pectinolytic enzymes, several other enzymatic activities have been produced on apple pomace namely β-fructofuranosidase (Hang and Woodams 1995), xylanase (Villas-Bôas et al. 2002; Seyis and Aksoz 2005), β-glucosidase (Hang and Woodams 1994b), manganese-dependent peroxidase and cellulase (Villas-Bôas et al. 2002). Citrus waste on the other hand has been used for the production of α-amylase, neutral and alkaline proteases (Mahmood et al. 1998), xylanase (Ismail 1996; Seyis and Aksoz 2005; Mamma et al. 2008) and cellulase (Ismail 1996; Mamma et al. 2008).

In literature there are a few reports on grape pomace utilization for the production of enzymes, namely pectinase, cellulase and xylanase (Botella et al. 2005, 2007; Díaz et al. 2007) by different *Aspergillus* species.

14.3.1.2 Bioethanol

With the inevitable depletion of the world's petroleum supply and due to increased prices for oil, there has been an increasing worldwide interest in alternative, non-petroleum-based sources of energy. Ethanol is one of the most important renewable fuels contributing to the reduction of negative environmental impacts generated by the worldwide utilization of fossil fuels (Cardona and Sánchez 2007). Application of agro-industrial residues in bioethanol not only provides alternative substrates but also helps solve their disposal problem.

Several researchers have successfully hydrolyzed citrus waste using commercial cellulase and pectinase enzymes to glucose, galactose, fructose, arabinose, xylose, rhamnose, and galacturonic acid (Grohmann and Baldwin 1992; Grohmann et al. 1994a, 1995a; Wilkins et al. 2007a). According to Grohmann et al. (1994a), glucose, fructose and galactose from hydrolyzed citrus peel waste can be fermented to ethanol by *Saccharomyces cerevisiae* yeast. Galacturonic acid from pectin hydrolysis, arabinose, and xylose as well as the sugars mentioned above can be fermented by *Escherichia coli* KO11 to produce ethanol and acetic acid (Grohmann et al. 1994b, 1995b). *E. coli* KO11 is a recombinant bacterial strain developed to ferment arabinose and xylose as well as hexoses to ethanol. However, in order to ferment these sugars, orange peel oil concentration in the hydrolysate must be reduced prior to fermentation (Grohmann et al. 1994a). The inhibitory effect on yeast growth due to orange peel oil and/or D-limonene, a monoterpene that makes up more than 90% of citrus peel oils, has been observed by several researchers (Wilkins et al. 2007b, c). Yields obtained by the previous researchers were quite promising in using citrus wastes for bioethanol production.

It should be noted that due to the high amounts of citrus wastes available in the US, researchers of the US Department of Agriculture worked with commercial enzymes to economically hydrolyze pectin, cellulose and hemicellulose from citrus peel wastes. The goal was to optimize the process and develop a model refinery that would also extract marketable by products (Predd 2006). Florida Power & Light Energy LLC (FPL Energy LCC) planned to develop a commercial scale cellulosic ethanol plant that can produce ethanol using waste citrus peel as feedstocks (O'Sullivan and Stewart 2007), while the southeast Biofuels LLC subsidiary has filed an application with the Florida Department of agriculture and Consumer Services for a \$500.000 grant in concerting citrus peel waste to ethanol (Ames 2008).

A SSF process for the production of ethanol from apple pomace by different strains of *S. cerevisiae* was described by several researchers (Ngadi and Correia 1992; Joshi and Sandhu 1996; Sandhu and Joshi 1997; Khosravi and Shojaosadati 2003). The results indicated that alcohol fermentation from apple pomace is an efficient method to reduce waste disposal, with the concomitant production of ethanol. Nogueira et al. (2005) evaluated the alcoholic fermentation of the aqueous extract of apple pomace with satisfactory yields showing that it is a suitable substrate for alcohol production.

Finally, grape pomace has been evaluated by Hang et al. (1986) as potentially feedstock for bioethanol production under SSF by the naturally occurring yeast flora. The yields obtained were 80% of the theoretical based on the fermentable sugar consumed.

14.3.1.3 Organic Acids

Among the various products obtained through microbial cultivation on agro-industrial residues, organic acids are particularly important. The ratio of carboxylic acids microbiologically manufactured in the bulk of biotechnological products is very high. These compounds are valuable building blocks for chemical synthesis, which can be used in several applications. Among organic acids, citric acid production has been well studied and reported. The amount of citric acid manufactured annually exceeds 800000 MT, and its production is increasing at 5% a year. Citric acid is widely used in several industrial processes, such as in the food and pharmaceutical industries. It is produced mainly by submerged fermentation using *A. niger* or *Candida* sp. from different sources of carbohydrates, such as molasses and starch based media (Vendruscolo et al. 2008).

Several researchers have evaluated different types of citrus waste as carbon source for citric acid producton under both SmF (Aravantinos-Zafiris et al. 1994; Rivas et al. 2008) and SSF (Zhang 1988; Kang et al. 1989) conditions using different *A. niger* strains. Hang and Woodams (1985) used grape pomace as substrate for citric acid production under SSF.

The use of apple pomace as substrate for the production of citric acid by *A. niger* in SSF (column reactors) was reported by Shojaosadati and Babaeipour (2002). Apple pomace has also been used for fatty acid production. Stredansky et al. (2000) evaluated the γ-linolenic acid (GLA) production in *Thamnidium elegans* by SSF.

14.3.1.4 Heteropolysaccharides

Long-chain, high-molecular-mass polymers that dissolve or disperse in water to give thickening or gelling properties are indispensable tools in food product formulation. Such food polymers are also used for secondary effects, which include emulsification, stabilization, suspension of particulates, control of crystallization, inhibition of syneresis, encapsulation, and film formation (Vuyst and Degeest 1999). Xanthan gum is the most important microbial polysaccharide from the commercial point of view, produced by bacteria of the genus *Xanthomonas,* with a worldwide production of about 30000 tons per year. It has widespread commercial applications as a viscosity enhancer and stabilizer in the food, pharmaceutical and petrochemical industries (Galindo 1994). The utilization of agro-industrial by-products for the production of polysaccharides by microorganisms has many advantages, such as reducing production costs and recycling natural resources.

In studies of Bilanovic et al. (1994) and Green et al. (1994) four different fractions of citrus waste were compared as substrates for xanthan fermentation: whole citrus waste, pectic, hemicellulosic and cellulosic extracts, in SmF. Stredansky and Conti (1999) studied the SSF as an alternative strategy for the production of xanthan by *X. campestris*. The choice was based on the observation that solid substrates reproduce the natural habitat of this phytopathogenic bacterium. This technique allows problems connected with broth viscosity to be overcome and utilizes cheap substrates. The exopolysaccharide was produced on a number of agro-industrial residues or by-products such as spent malt grains, apple pomace, grape pomace, and citrus peels. With most of the substrates, the gum production was comparable to those obtained with SmF (Stredansky and Conti 1999).

14.3.1.5 Aroma compounds

Most of the flavouring compounds are presently produced via chemical synthesis or extraction from natural materials. However, recent market surveys have demonstrated that consumers prefer foodstuff that can be labelled as natural. Plants have been major sources of essential oils and flavours but their use depends on natural factors difficult to control, such as weather conditions and plant diseases. An alternative route for flavour synthesis is based on microbial biosynthesis or bioconversion (Janssens et al. 1992). Several micro-organisms including bacteria and fungi, are currently known for their ability to synthesise different aroma compounds. Fungi from the genus *Ceratocystis* produce a large range of fruit-like or flower-like aromas (peach, pineapple, banana, citrus and rose) depending on the strain and the culture conditions (Christen et al. 1997; Meza et al. 1998). Among the genus, *C. fimbriata* has a great potential for ester synthesis. It grows rapidly, has a good ability to sporulate and produces a wide variety of aromas.

Bromarski et al. (1998) evaluated the potential of several agro-industrial residues such as cassava bagasse, apple pomace, amaranth and soybean using a strain of *C. fimbriata*. All media supported fungal growth. While amaranth medium produced pineapple aroma, the medium containing apple pomace produced a strong fruity aroma. This same medium was used by Christen et al. (2000) for the production of volatile compounds by *Rhizopus* strains. Authors found that the production of volatile compounds was related mainly to the medium used, and no difference was observed among the strains studied. The odors detected have a slight alcoholic note, and the apple pomace produced intermediate results, compared with the amaranth grain supplied with mineral salt solution.

Medeiros et al. (1999) cultivated a strain of *Kluyveromyces marxianus* in SSF using different solid substrates such as cassava bagasse, giant palm bran, apple pomace, sugarcane bagasse and sunflower seeds. The feasibility of using cassava bagasse and giant palm bran as substrates to produce fruity aroma was confirmed.

14.3.1.6 Protein Enriched Feeds

Cells of algae, fungi, yeasts, and bacteria are composed of up to 60% high-quality protein. These organisms multiply quickly under different conditions, being able to consume diverse types of industrial residues. Considering that traditional animal protein sources, such as meat and milk, have a higher cost and, as such, are not accessible to a large part of the global population, the production of alternative protein sources, such as those originated by microorganisms, appears to be an attractive solution for raising protein intake.

Furthermore, the use of agro-industrial residues for growing microbial cells as a suitable protein source for human consumption is an interesting approach for adding value to industrial by-products. The use of biotechnological manipulated ingredients for the production of animal feed has been growing each year. During microbial processing, along with the conversion of lignocellulosic waste into foods, an increase in protein content and an improvement in the digestibility of the substrate are observed (Vendruscolo et al. 2008).

Citrus fruit peel and apple pomace have been used for protein enrichment and single cell protein (SCP) by the different fungal species of *Penicillium* (Scerra et al. 1999; Vendruscolo et al. 2008), *Neurospora, Chaetomium, Sporotrichum* (Shojaosadati et al. 1999), *Aspergillus* (Vendruscolo et al. 2008), *Rhizopus* (Soccol and Vandenberghe 2003) and *Trichoderma* (De Gregorio et al. 2002; Vendruscolo et al. 2008) and yeasts, such as *S. cerevisiae, Torula utilis Candida utilis* (Vendruscolo et al. 2008). The recycling of viticulture residues through SSF by *Pleurotus* has great potential to produce human food and yields an available high-fiber feed for limited use in ruminants (Nchez et al. 2002).

14.3.2 Enzymatic Modifications

Numerous studies have been made in the exploitation of fruit waste ingredients to generate several value-added products through enzymatic modification such as prebiotic oligosaccharides and modified biologically active molecules.

14.3.2.1 Prebiotic Oligosaccharides

In recent years a number of oligomers termed prebiotics have been described. These resist digestion in the upper gastrointestinal (GI) tract and are able to modulate the gut microbiota by stimulating indigenous beneficial flora components while suppressing, or not affecting, less desirable bacteria, such as proteolytic bacteroides and clostridia (Gibson and Roberfroid 1995). Prebiotics have also been reported to indirectly lead to a reduction in serum triglyceride levels (Williams and Jackson 2002). In addition, there is evidence showing that prebiotics may indirectly affect mineral absorption in the large bowel and show beneficial effects against inflammatory bowel diseases by stimulating butyrate production and thus accelerating the mucosal cell proliferation and healing processes (Bamba et al. 2002).

Although any dietary material that enters the large intestine can be considered as potentially prebiotic, currently, the most well known prebiotics are non-digestible oligosaccharides (Gibson and Roberfroid 1995). Different oligosaccharides with prebiotic properties are commercially available, such as inulin, fructooligosaccharides (FOS), galacto-oligosaccharides and lactulose, but currently there is increasing interest in the identification and development of new prebiotic compounds, perhaps with added functionality (Menne et al. 2000; Rao 2001; Tuohy et al. 2002).

Pectic substances are hydrolysed by the action of pectinases or pectolytic enzymes that are widely distributed in higher plants and microorganisms (Jayani et al. 2005). Pectic oligosaccharides (POS) were manufactured from commercial pectin in an enzyme membrane reactor (Olano-Martin et al. 2001) and then evaluated for their prebiotic properties (Olano-Martin et al. 2002). These pectic oligosaccharides had a low prebiotic potential compared to FOS, although they were more selectively fermented than were the parent pectins (Olano-Martin et al. 2002). Pectic oligosaccharides also protected colonocytes against *Escherichia coli* verocytotoxins (Olano-Martin et al. 2003a) and stimulated apoptosis in human colonic adenocarcinoma cells (Olano-Martin et al. 2003b).

Recently, it has been demonstrated that POS from orange peel showed prebiotic properties increasing the bifidobacterial and *E. rectale* numbers (Manderson et al. 2005). Orange peel albedo (white part) was also a good source of pectic oligosaccharides with prebiotic properties produced by a microwave and autoclave extraction (Hotchkiss et al. 2003). Incubating bergamot peel for 2 h with a commercial enzyme preparation from *Aspergillus* sp. (pectinase 62 L) produced a material rich in oligosaccharides. The prebiotic effect of a POS rich extract enzymatically derived from bergamot peel was studied using pure and mixed cultures of human faecal bacteria. Addition of the bergamot oligosaccharides (BOS) resulted in a high increase in the number of bifidobacteria and lactobacilli, whereas the clostridial population decreased. A prebiotic index (PI) was calculated for both FOS and BOS after 10 and 24 h incubation. Generally, higher PI scores were obtained after 10 h incubation, with BOS showing a greater value (6.90) than FOS (6.12) (Mandalari et al. 2007).

Furthermore the potential of apple pomace for lactic acid production and oligomeric carbohydrates by simultaneous saccharification and fermentation (SSF) was evaluated (Gulloä et al. 2007). It was found that operating at low cellulase (Celluclast 1.5 L, cellulases from *T. reesei*) and cellobiase (NS50010, *β*-glucosidase from *A. niger*) charges (1 FPU/g and 0.25 IU/FPU, respectively) and short reaction times (10 h), 18.3 kg of oligosaccharides (which can be used as prebiotics) can be produced from 100 kg of dry apple pomace. The distribution of total oligosaccharide components was as follows: 5.8 kg of glucooligosaccharides, 7.4 kg of xylooligosaccharides, and 5.1 kg of arabinooligosaccharides.

14.3.2.2 Esterification of Flavonoids, Phenolic Acids and Terpenoids

Flavonoids (aglycon, glycosylated) are widely used in pharmaceutic, cosmetic and food preparation. They have several physico-chemical properties and biological activities but they are characterized by a low solubility and stability. In order to take advantage of these properties, their enzymatic acylation with fatty and aromatic acids under different operating conditions has been suggested as a promising route by several authors.

Various types of enzymes have been tested for acylation of flavonoids, such as proteases, acyl transferases and lipases, subtilisin (the first enzyme used for flavonoid ester synthesis) and mostly lipase B of *Candida antarctica* (CAL-B) (Chebil et al. 2006). Some reactions were also catalyzed by acyl transferases, but they required the use of either an activated acyl donor (ester of coenzyme A) or the presence *in situ* of a system allowing the generation of these derivatives (Chebil et al. 2006). Different acyl donors (aliphatic acids, aromatic acids and vinyl esters) were used for the enzymatic esterification of isoquercitrin, quercitrin, luteolin-7 glucoside, naringin, rutin, catechin-7-*O*-α-D-glucoside, phloridzin, hesperidin, epigallocatechin, 3-glucoside anthocyanin (delphinidin, cyaniding, pelargonidin), quercetin, catechin in solvent free or added-solvent systems with the reaction time and temperature ranging from a few hours to several days and from 30 to 60°C, respectively (Chebil et al. 2006). These studies include the investigation of several factors that affect the acylation reactions such as the type, origin and concentration of the enzyme, the nature of the reaction, the operating conditions, the composition of the reaction media, and the nature of substrates on regioselectivity (Chebil et al. 2006).

Phenolic acids possess interesting biological properties (antioxidant, chelator, free radical scavenger, UV filter, antimicrobial). Generally, such natural antioxidants are partially soluble in aqueous media. This is limiting their usefulness in oil-based food processing and has been reported as a serious disadvantage if an aqueous phase is also present. Therefore, the modification of these compounds via esterification with fatty alcohols results in the formation of more lipophilic derivatives. There are several reports on the enzymatic lipophilisation of phenolic acids in non-conventional reaction media, such as organic solvent mixtures, ionic liquids and solvent-free systems. Although lipases are the most commonly used biocatalysts for this kind of reactions, feruloyl esterases, tannases, cutinases are also reported (Figueroza-Espinoza and Villeneuve 2005). Several studies report the enzymatic lipophilisation of phenolic acids present in citrus wastes, apple and grape pomace, such as caffeic, cinnamic, ferulic, *p*-hydroxybenzoic, gentistic, gallic, vanillic, syringic, *o*- and *m*-coumaric, *p*-coumaric, chlorogenic and sinapic acid (Figueroza-Espinoza and Villeneuve 2005; Karboune et al. 2005; Safari et al. 2006; Stevenson et al. 2007; LopezGiraldo et al. 2007).

Active components of the essential oils are mostly constituted by alcohols of terpenic nature (e.g. citronellol, carvone, etc.). These compounds, with similar chemical characteristics such as terpenic esters, are being used mainly as fragrances and flavor agents of great industrial importance. These compounds, when chemically synthesized, are not considered as natural products, so enzymes, and especially lipases, can be used as biocatalysts to carry out their esterification and transesterification reactions (Perea et al. 2007). Recent reports on the enzymatic ersterification of terpenoids refer to D-limonene, menthol, geraniol, citronellol, nerol (Kahlow et al. 2001; Dominguez de Maria et al. 2006; Perea et al. 2007) esterification with different acyl donors, using microbial lipases, often leading to optically active products of great importance.

14.3.2.3 Glycosylation of Flavonoids

Functional properties, such as solubility, physicochemical stability, bioactivity, pharmacokinetics and cellular localization of natural products, such as flavonoids, are greatly affected by glycosylation, which is, therefore, an important factor to be considered in industrial applications of these products. Recently, for example, it has been reported that the efficiency of absorption of a flavonoid glucoside, quercetin glucoside, was greater than that of quercetin itself (Morand et al. 2000). Glycosylation occurs widely in plant cells and is considered to be an important method for the conversion of water-insoluble and unstable organic compounds into the corresponding water-soluble and stable compounds (Shimoda et al. 2006, 2007a, b). This reaction can be carried out through enzymatic or chemical methods. Enzymatic synthesis of glycoconjugates has many advantages over conventional chemical methods. The regio- and enantio-selectivity of GTs provide a simple means to synthesise stereospecific glycosides without the requirement of protection and deprotection of other functional groups. These processes are often difficult or impossible for chemical synthesis.

Enzymatic glycosylation usually employs glycosidases or "glycosynthases" and glycosyltransferases (GTs). The use of glycosidases provides an alternative method for biocatalytic glycosylation (Ly and Withers 1999). One of the most powerful approaches to the enzymatic synthesis of glycosides is Withers's "glycosynthase" technology (Jakeman and Withers 2002). Glycosynthases are genetically engineered nucleophile-less mutant glycosidases that can catalyze the formation of glycosidic linkages, primarily but not exclusively by using glycosyl fluoride donors, yet are incapable of hydrolyzing the product. Recently, a non-natural glycosidase mutant (Cel7B-E197S glycosynthase) has been identified that has novel regiospecificity (O4', O6) and activity (disaccharide transfer to flavonoids) with catalytic efficiencies comparable with those of natural GT counterparts (Yang et al. 2007).

Unlike glycosidases and glycosynthases, GTs are enzymes that have evolved naturally for glycosylation reactions. Many mammalian and microbial GTs have been employed for the synthesis of oligosaccharides and antibiotic glycosides (Lim 2005). In contrast, due to the small number of plant GT sequences that were available, their use in biocatalysis has been limited. A GT (UGT73A10) isolated from cDNA of *Lycium barbarum* L. fruits effectively catalyzed the regiospecific glucosylation of (+)-catechin towards to 4-*O*-D-glucopyranoside of (+)-catechin (Noguchi et al. 2008). The reaction product was more stable than aglycon under acidic conditions and at elevated temperatures. Moreover, two *Arabidopsis* cluster F GTs are capable of glycosylating the 3-OH of quercetin in vitro. While these two closely related GTs have been found to recognize the same acceptor molecule displaying the same regioselectivity, they catalyse the glycosidic linkage by means of different sugar donors. GT 78D2 uses UDP-Glc while GT 78D1 prefers UDP-Rha as the sugar donor (Jones et al. 2003; Limet et al. 2004). Another GT in cluster F, ACGaT from *Aralia*, forms quercetin-*3-O*-galactoside with UDP-Gal as the sugar donor (Kubo et al. 2004), indicating that all three GTs have a common ancestor, they have evolved to recognize different sugar donors.

Whilst many plant GTs glycosylate the aglycone moiety of glycosides to form products such as quercetin-3,7-di-*O*-glucoside (Limet et al. 2004), some other GTs (Family 1, CAzY) are found to glycosylate the sugar moiety of glycosides. *Citrus* Cm1,2RhaT is able to transfer a rhamnose molecule onto the 2-OH group of the glucose moiety attached to the 7-OH group of naringenin, resulting in the formation of a bitter tasting glycoside in some *Citrus* species (Frydman et al. 2004). There is also a GT in *Citrus* species such as mandarin, that transfers rhamnose onto the 6-OH group of the glucose moiety of naringenin-7-O-glucoside forming a tasteless glycoside (Frydman et al. 2004).

14.3.2.4 Halogenation of Flavonoids

The effects of flavonoids, on the central nervous system have been considered. They process anxiolytic activity and low sedative or myorelaxant effects (Medina et al. 1997). Among the most active compounds, a number of halogenated flavones have been reported; in particular, 6-bromoflavone and 6-bromo- 3-nitroflavone showed activities close to or higher than that of diazepam, a benzodiazepine derivative which is a classical anxiolytic, anticonvulsant, sedative and skeletal muscle relaxant drug. In order to show these activities, the presence of electro-donating or withdrawing substituents on the aromatic ring of the flavonoids seems to be essential (Sternbach 1978). In the literature, several methods for halogenating aromatic compounds are reported. Direct bromination, e.g. with elemental bromine, is a highly polluting method which, in addition, involves serious difficulties connected with the handling of a highly corrosive agent. Other methods, including NBS-amberlyst (Goldberg and Alper 1994), metal-oxo-catalysed $KBr–H_2O_2$ (Clague and Butler 1995) and $KBr–NaBO_3$ (Roche et al. 2000) suffer from harsh conditions or require complex or laborious work-up.

One of the enzymatic reactions that have been widely studied is a chloroperoxidase-catalyzed halogenation (Franssen et al. 1987). Chloroperoxidase from *Caldariomyces fumago* (CPO; EC 1.11.1.10) is a well-known enzyme, capable of halogenating a great variety of organic compounds such as b-ketoacids, cyclic b-diketones, steroids, alkenes, activated aromatic compounds and heterocylcic compounds (Yaipakdeea and Robertsonb 2001). The reaction mechanism of CPO involves the formation of a halogenium ion (X^+) or hypohalous acid (HOX) as an intermediate which can effect electrophilic substitution with electron-rich substrates (Libby et al. 1992).

The whole cells and the chloroperoxidase enzyme of *Caldariomyces fumago* were capable of halogenating the flavanones, naringenin and hesperetin, at C-6 and C-8 in the presence of either Cl^- or Br^- (Yaipakdeea and Robertsonb 2001). The biohalogenated products of naringenin and hesperetin were isolated and found to be identical to those obtained from chemical reactions using molecular halogen and hypohalous acid.

References

Ames D (2008) Xethanol looks at ethanol from citrus peel waste. Ind Bioproces 30:5

Aravantinos-Zafiris G, Tzia C, Oreopoulou V et al. (1994) Fermentation of orange processing wastes for citric acid production. J Sci Food Agr 65:117–120

Bamba T, Kanauchi O, Andoh A et al. (2002) A new prebiotic from germinated barley for nutraceutical treatment of ulcerative colitis. J Gastroen Hepatol 17:818–824

Bampidis VA, Robinson PH (2006) Citrus by-products as ruminant feeds: A review. Anim Feed Sci Tech 128:175–217

Berovic M, Ostroversnik H (1997) Production of *Aspergillus niger* pectolytic enzymes by solid state bioprocessing of apple pomace. J Biotechnol 53:47–53

Bilanovic D, Shelet G, Green M (1994) Xanthan fermentation of citrus waste. Bioresour Technol 48:169–172

Botella C, de Ory I, Webb C et al. (2005) Hydrolytic enzyme production by *Aspergillus awamori* on grape pomace. Biochem Eng J 26:100–106

Botella C, Díaz A, de Ory I et al. (2007) Xylanase and pectinase production by *Aspergillus awamori* on grape pomace in solid state fermentation. Process Biochem 42:98–101

Bromarski A, Soccol CR, Christen P et al. (1998) Fruity aroma production by *Ceratocystis fimbriata* in solid cultures from agro-industrial wastes. Rev Microbiol 29:208–212

Cardona CA, Sánchez OJ (2007) Fuel ethanol production: Process design trends and integration opportunities. Bioresour Technol 98:2415–2457

Chebil L, Humeau C, Falcimaigne A et al. (2006) Enzymatic acylation of flavonoids. Process Biochem 41:2237–2251

Christen P, Bramorski A, Revah S et al. (2000) Characterization of volatile compounds produced by *Rhizopus* strains grown on agro-industrial solid wastes. Bioresour Technol 71:211–215

Christen P, Meza JC, Revah S (1997) Fruity aroma production of solid state fermentation by *Ceratocystis fimbriata*: influence of the substrate type and the presence of precursors. Mycol Res 101:911–919

Clague MH, Butler A (1995) On the mechanism of cis-Dioxovanadium(V)-catalyzed oxidation of bromide by hydrogen peroxide: Evidence for a reactive, binuclear vanadium(V) peroxo complex. J Am Chem Soc 117:3475–3484

De Gregorio A, Mandalari G, Arena N et al. (2002) SCP and crude pectinase production by slurry-state fermentation of lemon pulps. Bioresour Technol 83:89–94

Dhillon SS, Gill RK, Gill SS et al. (2004) Studies on the utilization of citrus peel for pectinase production using fungus *Aspergillus niger*. Int J Environ Stud 61:199–210

Díaz A, Caro I, de Ory I et al. (2007) Evaluation of the conditions for the extraction of hydrolitic enzymes obtained by solid state fermentation from grape pomace. Enzyme Microb Technol 41:302–306

Dominguez de Maria P, Alcantara AR, Carballeira JD et al. (2006) *Candida rugosa* lipase: A traditional and complex biocatalyst. Cur Org Chem 10:1053–1066

FAOSTAT-FAO Statistical Database, 2005

Figueroza-Espinoza M-C, Villeneuve P (2005) Phenolic acids enzymatic lipophilisation. J Agric Food Chem 53:2779–2787.

Fonseca MJV, Said S (1994) The pectinase produced by *Tubercularia vulgaris* in submerged culture using pectin or orange-pulp pellets as inducer. Appl Microbiol Biotechnol 42:32–35

Franssen MCR, van Boven HG, van der Plas HC (1987) Enzymatic halogenation of pyrazoles and pyridine derivatives. J Heterocycl Chem 24:1313–1316

Frydman A, Weisshaus O, Bar-Peled M et al. (2004) Citrus fruit bitter flavors: isolation and fuctional characterization of the gene Cm1,2RhaT encoding a 1,2 rhamnosyltransferase, a key enzyme in the biosynthesis of the bitter flavonoids of citrus. Plant J 40:88–100

Galindo E. (1994) Aspects of the process for xanthan production. Trans Inst Chem Eng 72:227–237

Garzón CG, Hours RA (1992) Citrus waste: An alternative substrate for pectinase production in solid-state culture. Bioresour Technol 39:93–95

Gibson GR, Roberfroid MB (1995) Dietary modulation of the human colonic microbiota: Introducing the concepts of prebiotics. J Nutr 125:1401–1412

Goldberg Y, Alper H (1994) Electrophilic halogenation of aromatics and heteroaromatics with N-halosuccinimides in a solid/liquid system using an H^+ ion exchanger or ultrasonic irradiation. J Mol Cat 88:377–383

Green M, Shelef G, Bilanovic D (1994) The effect of various citrus waste fractions on xanthan fermentation. Chem Eng J 56:B37–B41

Grohmann K, Baldwin EA (1992) Hydrolysis of orange peel with pectinase and cellulase enzymes. Biotechnol Lett 14:1169–1174

Grohmann K, Baldwin EA, Buslig BS (1994a) Production of ethanol from enzymatically hydrolyzed orange peel by the yeast *Saccharomyces cerevisiae*. Appl Biochem Biotechnol 45/46:315–327

Grohmann K, Baldwin EA, Buslig BS et al. (1994b) Fermentation of galacturonic acid and other sugars in orange peel hydrolysates by the ethanolgenic strain of *Escherichia coli*. Biotechnol Lett 16:281–286

Grohmann K, Cameron RG, Buslig BS (1995a) Fractionation and pre-treatment of orange peel by dilute acid hydrolysis. Bioresour Technol 54:129–141

Grohmann K, Cameron RG, Buslig BS (1995b) Fermentation of sugars in orange peel hydrolysates to ethanol by recombinant *Escherichia coli* K011. Appl Biochem Biotechnol 51/52:423–435

Gulloä B, Garrote G, Alonso JL et al. (2007) Production of L-lactic acid and Oligomeric compounds from apple pomace by simultaneous saccharification and fermentation: A response surface methodology assessment. J Agric Food Chem 55:5580–5587

Hang YD, Lee CY, Woodams EE (1986) Solid-state fermentation of grape pomace for ethanol production. Biotechnol Lett 8:53–56

Hang YD, Woodams EE (1985) Grape pomace: A novel substrate for microbial production of citric acid. Biotechnol Lett 7:253–254

Hang YD, Woodams EE (1994b) Apple pomace: a potential substrate for production of β-glucosidase by *Aspergillus foetidus*. LWT-Food Sci Technol 27:587–589

Hang YD, Woodams EE (1994a) Production of fungal polygalacturonase from apple pomace. LWT-Food Sci Technol 27:194–196

Hang YD, Woodams EE (1995) β-fructofuranosidase production by *Aspergillus* species from apple pomace. LWT-Food Sci Technol 28:340–342

Hotchkiss AT, Olano-Martin E, Grace WE et al. (2003) Pectic oligosaccharides as prebiotics. In: Eggleston G, Côté GL (Eds) Oligosaccharides in Food and Agriculture, ACS Symposium series 849, Oxford University Press, USA

Ismail AS (1996) Utilization of orange peels for the production of multi-enzyme complexes by some fungal strains. Process Biochem 1:645–650

Jakeman DL, Withers SG (2002) Glycosynthases: New Tools for Oligosaccharide Synthesis. Trends Glycosci Glycotechnol 14:13–25

Janssens L, de Pooter HL, Vandamme EJ et al. (1992) Production of flavours by microorganisms. Process Biochem 27:195–215

Jayani RS, Saxena S, Gupta R (2005) Microbial pectinolytic enzymes: a review. Process Biochem 40:2931–2944

Jones P, Messner B, Nakajima J et al. (2003) UGT73C6 and UGT78D1, Glycosyltransferases involved in flavonol glycoside biosynthesis in *Arabidopsis thaliana*. J Biol Chem 278: 43910–43918

Joshi VK, Parmar M, Rana NS (2006) Pectin esterase production from apple pomace in solid-state and submerged fermentations. Food Technol Biotechnol 44:253–256

Joshi VK, Sandhu DK (1996) Composition of distillates from the solid state fermentation of apple pomace by different yeasts. Natl Acad Sci Lett 19:219–224

Kahlow UHM, Schmid RD, Plewiss J (2001) A model of the pressure dependence of the enantioselectivity of *Candida rugosa* lipase towards (±)-menthol. Prot Sci 10:1942–1952

Kang SK, Park HH, Lee JH et al. (1989) Citric acid fermentation from mandarin orange peel by *Aspergillus niger*. Sanop Misaengmul Hakhoechi 17:510–518

Kapoor M, Beg QK, Bhushan B et al. (2000) Production and partial purification and characterization of a thermo-alkali stable polygalacturonase from *Bacillus* sp. MG-cp-2. Process Biochem 36:467–473

Karboune S, Safari M, Lue B-M et al. (2005) Lipase-catalyzed biosynthesis of cinnamoylated lipids in a selected organic solvent medium. J Biotechnol 119:281–290

Khosravi K, Shojaosadati SA (2003) A solid state of fermentation system for production of ethanol from apple pomace. Fanni va Muhandisi-i Mudarris 10:55–60

Kubo A, Arai Y, Nagashima S et al. (2004) Alteration of sugar donor specificities of plant glycsyltransferases by a single point mutation. Arch Biochem Biophys 429:198–203

Libby RD, Shedd AL, Phipps AK et al. (1992) Defining the involvement of HOCl or Cl_2 as enzyme generated intermediates in chloroperoxidase-catalyzed reactions. J Biol Chem 267:1769–1775

Lim E-K (2005) Plant glycosyltransferases: Their potential as novel biocatalysts. Chem Eur J 11:5486–5494

Lim E-K, Ashford DA, Hou B et al. (2004) Arabidopsis glycosyltransferases as biocatalysts in fermentation for regioselective synthesis of diverse quercetin glucosides. Biotechnol Bioeng 87:623–631

Lopez Giraldo L Laguerre M, Lecomte J et al. (2007) Lipase-catalyzed synthesis of chlorogenate fatty esters in solvent-free medium. Enzyme Microb Technol 41:721–726

Ly HD, Withers SG (1999) Mutagenesis of glycosidases. Annu Rev Biochem 68:487–522

Mahmood AU, Greenman J, Scragg AH (1998) Orange and potato peel extracts: analysis and use as *Bacillus* substrates for the production of extracellular enzymes in continuous culture. Enzyme Microb Technol 22:130–137

Mamma D, Kourtoglou E, Christakopoulos P (2008) Fungal multienzyme production on industrial by-products of the citrus-processing industry. Bioresour Technol 99:2373–2383

Mandalari G, Nueno Palop C, Tuohy K et al. (2007) In vitro evaluation of the prebiotic activity of a pectic oligosaccharide-rich extract enzymatically derived from bergamot peel. Appl Microbiol Biotechnol 73:1173–1179

Manderson K, Pinart M, Tuohy KM et al. (2005) In vitro determination of prebiotic properties of oligosaccharides derived from an orange juice manufacturing by-product stream. Appl Environ Microbiol 71:8383–8389

Martins ES, Silva D, Da Silva R et al. (2002) Solid state production of thermostable pectinases from thermophilic *Thermoascus aurantiacus*. Process Biochem 37:949–954

Medeiros ABP, Soccol CR, Pandey A et al. (1999) Aroma production by *Kluyveromyces marxianus* in solid state fermentation using cassava bagasse as substrate. IX European Congress on Biotechnology, July 11–15, Brussels, Belgium

Medina JH, Vioila H, Wolfman C et al. (1997) Overview-Flavonoids: A new family of benzodiazepine receptor ligands. Neurochem Res 22:419–425

Menne E, Guggenbuhl N, Roberfroid M (2000) Fn-type chicory inulin hydrolysate has a prebiotic effect in humans. J Nutr 130:1197–1199

Meza JC, Christen P, Revah S (1998) Effect of added amino acids on the production of a fruity aroma by Ceratocystis fimbriata. Sci Aliments 18:627–636

Morand C, Manach C, Crespy V et al. (2000) Quercetin 3-O beta-glucoside is better absorbed than other quercetin forms and is not present in rat plasma. Free Radical Res 33: 667–676

Nchez AS, Ysunza F, Beltraän M et al. (2002) Biodegradation of viticulture wastes by *Pleurotus*: a source of microbial and human food and its potential use in animal feeding. J Agric Food Chem 50:2537–2542

Ngadi MO, Correia LR (1992) Kinetics of solid state ethanol fermentation from apple pomace. J Food Eng 17:97–116

Noguchi A, Sasaki N, Nakao M et al. (2008) cDNA cloning of glycosyltransferases from Chinese wolfberry (*Lycium barbarum* L.) fruits and enzymatic synthesis of a catechin glucoside using a recombinant enzyme (UGT73A10). J Mol Catal, B Enzym 55:3921–3933

Nogueira A, Santos LD, Paganini C et al. (2005) Evaluation of alcoholic fermentation of aqueous extract of the apple pomace. Semina: Ciências Agrárias, Londrina 26:179–193

O'Sullivan M, Stewart D (2007) FPL will make ethanol from waste citrus peel. Ind Bioproces 29:1–2

Olano-Martin E, Gibson GR, Rastall RA (2002) Comparison of the in vitro bifidogenic properties of pectins and pectic-oligosaccharides. J Appl Microbiol 93:505–511

Olano-Martin E, Mountzouris KC, Gibson GR et al. (2001) Continuous production of oligosaccharides from pectin in an enzyme membrane reactor. J Food Sci 66:966–971

Olano-Martin E, Rimbach GH, Gibson GR et al. (2003b) Pectin and pectic-oligosaccharides induce apoptosis in in vitro human colonic adenocarcinoma cells. Anticancer Res 23: 341–346

Olano-Martin E, Williams MR, Gibson GR et al. (2003a) Pectins and pectic-oligosaccharides inhibit *Escherichia coli* O157:H7 Shiga toxin as directed towards the human colonic cell line HT29. FEMS Microbiol Lett 218:101–105

Perea A, Torres R, Rodrıguez E (2007) Enzymatic synthesis of citronellol esters: A kinetics study. J Biotechnol Abst 131S:S98–S121

Pericin DM, Antov MG, Popov SD (1999) Simultaneous production of biomass and pectinases by *Polyporus squamosus*. Acta Periodica Technol 29:183–189

Pinelo M, Arnous A, Meyer AS (2006) Upgrading of grape skins: Significance of plant cell-wall structural components and extraction techniques for phenol release. Trends Food Sci Tech 17:579–590

Predd P (2006) Fueling the future with citrus waste. Environ Sci Technol 40:5170–5171

Rao VA (2001) The prebiotic properties of oligofructose at low intake levels. Nutr Res 6:843–848

Rivas B, Torrado A, Torre P et al. (2008) Submerged citric acid fermentation on orange peel autohydrolysate. J Agric Food Chem 56:2380–2387

Roche D, Prasad K, Repic O et al. (2000) Mild and regioselective oxidative bromination of anilines using potassium bromide and sodium perborate. Tetrahedron Lett 41:2083–2086

Ruberto G, Renda A, Amico V et al. (2008) Volatile components of grape pomaces from different cultivars of Sicilian *Vitis vinifera* L. Bioresour Technol 99:260–268

Safari M, Karboune S, St-Louis R et al. (2006) Enzymatic synthesis of structured phenolic lipids by incorporation of selected phenolic acids into triolein. Biocatal Biotransfor 24: 272–279

Sandhu DK, Joshi VK (1997) Solid state fermentation of apple pomace for concomitant production of ethanol and animal feed. J Sci Ind Res 56:86–90

Scerra V, Caridi A, Foti F et al. (1999) Influence of dairy *Penicillium* sp. on nutrient content of citrus fruit peel. Anim Feed Sci Technol 78:169–176

Schieber A, Stintzing FC, Carle R (2001) By-products of plant food processing as a source of functional compounds – recent developments. Trends Food Sci Tech 12:401–413

Seyis I, Aksoz N (2005) Xylanase production from *Trichoderma harzianum* 1073 D3 with alternative carbon source and nitrogen sources. Food Technol Biotechnol 43:37–40

Shimoda K, Harada T, Hamada H et al. (2007a) Biotransformation of raspberry ketone and zingerone by cultured cells of *Phytolacca americana*. Phytochemistry 68:487–492

Shimoda K, Kondo Y, Nishida T et al. (2006) Biotransformation of thymol, carvacrol, and eugenol by cultured cells of *Eucalyptus perriniana*. Phytochemistry 67:2256–2261

Shimoda K, Kwon S, Utsuki A et al. (2007b) Glycosylation of capsaicin and 8-nordihydrocapsaicin by cultured cells of *Catharanthus roseus*. Phytochemistry 68:1391–1396

Shojaosadati SA, Babaeipour V (2002) Citric acid production from apple pomace in multi-layer packed bed solid-state bioreactor. Process Biochem 37:909–914

Shojaosadati SA, Faraidouni R, Madadi-Nouei A et al. (1999) Protein enrichment of lignocellulosic substrates by solid state fermentation using *Neurospora sitophila*. Resour Conserv Recycl 27:73–87

Soccol CR, Vandenberghe LPS (2003) Overview of applied solid-state fermentation in Brazil. Biochem Eng J 13:205–218

Sternbach LH (1978) The benzodiazepine story. Prog Drug Res 22:229–266

Stevenson DE, Parkar SG, Zhang J et al. (2007) Combinatorial enzymic synthesis for functional testing of phenolic acid esters catalysed by *Candida antarctica* lipase B (Novozym 435®). Enzyme Microb Technol 40:1078–1086

Stredansky M, Conti E (1999) Xanthan production by solid state fermentation. Process Biochem 34:581–587

Stredansky M, Conti E, Stredanska S et al. (2000) γ-Linolenic acid production with *Thamnidium elegans* by solid state fermentation on apple pomace. Bioresour Technol 73:41–45

Tuohy KM, Ziemer CJ, Klinder A, et al. (2002) A human volunteer study to determine the prebiotic effects of lactulose powder on human colonic microbiota. Microbial Ecol Health Dis 14: 165–173

Vendruscolo F, Albuquerque AM, Streit F et al. (2008) Apple pomace: A versatile substrate for biotechnological applications. Crit Rev Biotechnol 28:1–12

Villas-Bôas SG, Esposito E, Mendonca MM (2002) Novel lignocellulolytic ability of *Candida utilis* during solid state cultivation on apple pomace. World J Microbiol Biotechnol 18: 541–545

Vuyst LD, Degeest B (1999) Heteropolysaccharides from lactic acid bacteria. FEMS Microbiol Rev 23:153–177

Wilkins MR, Suryawati L, Maness NO et al. (2007b) Ethanol production by *Saccharomyces cerevisiae* and *Kluyveromyces marxianus* in the presence of orange-peel oil. World J Microbiol Biotechnol 23:1161–1168

Wilkins MR, Widmer WW, Grohmann K (2007c) Simultaneous saccharification and fermentation of citrus peel waste by *Saccharomyces cerevisiae* to produce ethanol. Process Biochem 42:1614–1619

Wilkins MR, Widmer WW, Grohmann K et al. (2007a) Hydrolysis of grapefruit peel waste with cellulase and pectinase enzymes. Bioresour Technol 98:1596–1601

Williams CM, Jackson KG (2002) Inulin and oligofructose: Effects on lipid metabolism from human studies. Br J Nutr 87:S261–S264

Yaipakdeea P, Robertsonb LW (2001) Enzymatic halogenation of flavanones and flavones. Phytochemistry 57:341–347

Yang M, Davies GJ, Davis BG (2007) A glycosynthase catalyst for the synthesis of flavonoid glycosides. Angew Chem Int Ed 46:3885–3888

Zhang Q (1988) Utilization of citrus wastes in production of citric acid. Shipin Kexue (Beijing, China) 104:21–24

Zheng Z, Shetty K (2000). Solid state production of polygalacturonase by *Lentinus edodes* using fruit processing wastes. Process Biochem 35:825–830

Chapter 15
Wine Industry Residues

Bo Jin and Joan M. Kelly

Contents

Abstract The wine industry produces a substantial quantity of organic residues that are both highly polluting and costly to treat. These residues are mainly carbohydrate-rich organics such as sugars and cellulose, which are biodegradable and naturally rich in nutrients, making them suitable substrates for biotechnological production. This chapter briefly introduces potential utilization of winery residues for produc-

B. Jin (✉)
School of Earth and Environmental Sciences, School of Chemical Engineering, The University of Adelaide, Adelaide, SA 5005, Australia
e-mail: bo.jin@adelaide.edu.au

P. Singh nee' Nigam, A. Pandey (eds.), *Biotechnology for Agro-Industrial Residues Utilisation*, DOI 10.1007/978-1-4020-9942-7_15,

tion of bioenergy and biomaterials through bioconversion processes. We highlight advanced molecular biotechnology for strain development and integrated biotechnological processes, which can lead the bioconversion of winery residues into renewable bioenergy and biomaterials as a sustainable solution.

Keywords Winery residues and effluent · Simultaneous saccharification and fermentation · Bioenergy · Organic acid · Molecular biotechnology · Microbial biomass

15.1 Introduction

Wine production is one of the most important industries in many countries worldwide. Distilled grape marc, a complex lignocellulosic material, is one of the most abundant organic residues. The winery residue is produced after pressing the crushing grapes in white wine processing or after fermentation and maceration in red wine production. Grape marc is usually distilled in wineries to recover ethanol which is further used to produce spirituous liquors, leaving huge amounts of distilled grape marc unused after the winemaking process. The treatment of these waste residues requires many successive and costly steps, which pose increasing disposal and pollution problems. Landfill and incineration are popular methods to deal with winery residues, but both contribute to production of greenhouse gas. Most of the existing winery waste treatment processes cause large losses of nutrient resources. Grape marc contains cellulosic and hemicellulosic material that can be hydrolyzed to produce liquors containing monomers of glucose from cellulose and xylose and other sugars from hemicellulose, which are biodegradable and naturally rich in nutrients, making them suitable substrates for biotechnological production by microorganisms.

Bioconversion of the carbohydrate wastes to value-added products is of great importance for the production of renewable resources in a sustainable society. With industrial development growing rapidly, there is general agreement that sustainable environmental protection can only be achieved by integrating a general environmental awareness into a company's business functions. Recently, bioconversion of carbohydrate wastes is receiving increased attention in view of the fact that these wastes can act as a substrate for the production of useful biomaterials and chemical intermediates.

In this Chapter, the winery residues are regarded as a valuable biomass or potential substrates for biotechnological production. Therefore, chemical constituents of the winery residues and effluent will be described. Recently developed biotechnological processes which integrate saccharification with fermentation steps will be introduced as potential bioprocesses for bioconversion of winery residues into bioenergy and biomaterials. Advanced molecular biotechnology for strain development to meet specific bioprocess and bioproducts will be discussed. As potential bioproducts from winery residues, representative bioenergy (ethanol and hydrogen), organic acid (lactic acid and citric acid), enzyme, polymers and microbial biomass will be briefly discussed. It is expected that this work gives guidance for development of sustainable technology to make winery residues suitable for development of renewable energy and bioproducts, thereby offering biotechnology an opportunity to assist in maintaining environmental quality.

15.2 Description of Chemical Characteristics of Winery Residues

15.2.1 Production of Winery Residues

The main solid by-products and residues from the winery production are grape stalk, grape marc, wine lees and winery sludge (biosolids). Figure 15.1 shows the process flow sheet for generation of winery residues and wastewater from the winery and distillery industries. The main wastes from the viticulture activities are the vine stalks generated during the pruning of the grapevine. The principle by-product is the grape marc, which comprises grape stalks, seeds and skins left after the crushing, draining and pressing stages of wine production. Grape marc is commonly processed to produce alcohol and tartaric acid, which results in a new lignocellulosic by-product, spent grape marc. The wine lees are accumulated in the bottom of grape-juice or wine fermentation tanks. The distillation of the alcohol from low-quality wine, wine lees and grape marc produces a large quantity of a viscose and acidic wastewater known as vinasse. In many winery industries, an aerobic depuration process is operated after the distillation to treat the winery effluents, vinasse and winery wastewater, therefore, generating waste biosolids. A large proportion of the wastewater comes from cleaning and wine production and cooling processes.

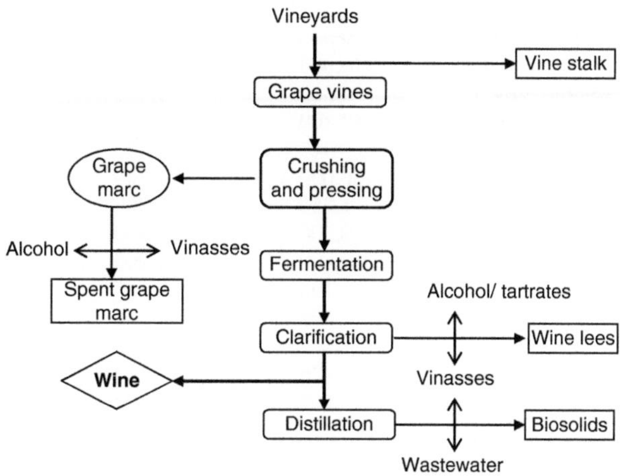

Fig. 15.1 Wine and distillation waste products. Sketch flow sheet showing the generation of by-products, residues and wastewater from the wine and distillation industries

15.2.2 Characteristics of Residues from Winery Industry

Grape marc (pomace) typically contains approximately 50% skins, 25% seeds and 25% stalks. Most grape marc is processed to extract commercial products such as alcohol by fermentation and distillation, grape tannins, and anti-oxidants. Grape marc is also used as feedstock for cattle and livestock but is limited to 30% of the total food for ruminants due to the very low nutritional value, or composted and returned to mulch vines. Winery residues are lingocellulosic biomass, which typically consists of three basic polymers: 50% cellulose $(C_6H_{10}O_5)_x$-glucose polymer, 35% hemicellulose $(C_5H_8O_4)_m$-xylose polymer and 15% lignin $[C_9H_{10}O_3.(OCH_3)_{0.9-1.7}]_n$. Table 15.1 summarizes the typical chemical compositions of grape stalk, grape marc, wine lees and winery biosolids. These data represent the analytical results from 87 samples of winery and distillery residues from different Spanish wineries and distilleries (adapted from Bustamante et al. 2008).

15.2.3 Chemical Characteristics of Winery Wastewater

The chemical composition of winery wastewater collected in vintage and non-vintage periods in Adelaide, Australia are shown in Table 15.2 (adapted from Chapman 1995; Chapman et al. 2001). The winery wastewater is rich in organic compounds, mainly carbohydrate-rich organics including sugars and organic acids, which are ideal substrates for biotechnological production.

15.3 Advances in Molecular Microbiology

As outlined above, much of the waste produced in the wine industry, like other food and beverage industries, is high in nutrients, and this represents a problem if the

Table 15.1 Typical chemical composition of winery residues

Parameter	Vine stalk	Grape marc	Wine lees	Winery sludge
Organic matter	920,000	915,000	759,000	669,000
Organic carbon*	316,000	280,000	300,000	257,000
Soluble carbon	74,500	37,400	87,800	26,900
Total nitrogen	12,400	20,300	35,200	44,300
P	940	1,150	4,940	10,900
K	30,000	24,200	72,800	20,700
Na	1,250	1,200	1,385	3,390
Ca	9,500	9,400	9,600	81,600
Mg	2,100	1,200	1,600	2,700
Fe	128	136	357	3,128
Mn	25	12	12	91
Cu	22	28	189	262
Zn	26	24	46	227
Polyphenols	23,900	10,600	8,900	1,100

Amounts are expressed on a dry weight basis, in mg per Kg (Adapted from Bustamante et al. 2008. The * indicates oxidisable organic carbon.

Table 15.2 Nutrient composition of wastewater produced by the wine industry

	Fresh effluent		After separation of solids in storage lagoons	
Parameter (mg l^{-1})	Vintage	Non-vintage	Vintage	Non-vintage
Total organic carbon	1,400	900	800	300
Total K-nitrogen	47	31	31	19
Total P	0.020	0.010	0.002	0.002
C : N : P	30:2.8:0.3	30:6.1:1.3	30:1.4:0.1	30:5.2:0.5
Sulfur	349	209	6.7	30.6
Fe	2.8	2.3	0.95	1.2
Na	250	328	148	219
K	131	204	98	55
Ca	26	30	25	22
Mg	29	15	10	10
Tartaric acid	530	350	200	20
Lactic acid	350	120	150	10
Acetic acid	100	50	220	370
Ethanol	3,130		1,490	
Glycerol	190	120	60	20
Glucose	300	280	trace	Trace
Fructose	530	270	80	trace
pH	4–8	6–10	4–6	4–8
Total solids*	1,570	1,640	2,430	2,800
Suspended solids*	190	410	230	410

The apparent increase in solids after separation (*) is due to algal biomass growth during storage (adapted from Chapman 1995).

waste is released into the environment, but opens up an opportunity for the waste to be turned into valuable secondary products. Recent reviews of methods for treatment of wine industry waste (Arvanitoyannis et al. 2006b), as well as potential uses and applications for the waste materials (Arvanitoyannis. 2006a), have concentrated on engineering and on classical microbiological approaches. Microorganisms have been used in a number of different treatments, either to decontaminate the waste by removing either toxic compounds or nutrients, or to produce valuable products such as enzymes and organic acids including citric and lactic acid, ethanol, or biomass for use in animal production. Recent advances in genomics have opened up new and exciting possibilities for genetic and metabolic engineering of microorganisms to achieve desirable outcomes (Abe et al. 2006; Martinez et al. 2008).

15.3.1 Bacteria

Many bacteria can ferment sugar to ethanol, and thus can potentially convert cellulosic biomass to ethanol. Strains of *Zymomonas mobilis* ferment only sucrose, glucose and fructose to ethanol, and this restricts their use with more complex substrates. However, strains of *Escherichia coli* have been engineered such that they contain genes to express *Z. mobilis* pyruvate decarboxylase and alcohol dehydrogenase, and these show enhanced ethanol production. *Klebsiella oxytoco* has also been manipulated to express the *Z. moblis* ethanol fermentation enzymes, and further changed to express a celliobiose uptake system and cellulase genes from *Erwinia chrysanthemi* (Zhou et al. 2001).

However, considerable mechanical and chemical pre-treatments of lignocellulosic compounds are required, as the bacteria do not directly degrade lignin (Alterthum and Ingram 1989). In none of these cases has the ethanol production reached levels that can be achieved using yeast, however, further engineering may be possible now that genome sequences are available for a wide range of prokaryotes, and a greater range of pathways, as well as the genes encoding enzymes for steps in the pathways, become known.

Many anaerobic bacteria, including thermophilic *Clostridium thermocellum,* can degrade cellulose and hemicellulose via a multi-enzyme complex, the cellulosome, which performs a variety of cellulolytic and hemicellulolytic activities (Masai et al. 2007). However *C. thermocellum* cannot use glucose (Lynd et al. 2002), and thus this species has potential for mixed culture fermentations.

15.3.2 Filamentous Fungi

Although the yeast, *Saccharomyces cerevisiae,* has been extensively used in the wine, brewing and baking industries, it is not effective for use with lignocellulosic substrates, as it does not produce a large range of enzymes necessary to break down lignin and cellulose, it does not have a highly expressing secretory pathway, and pre-treatments of lignocellulosic material often produces inhibitors to growth. However, progress has been made in engineering yeast to ferment pentose sugars (Hahn-Hagerdal et al. 2007). Filamentous fungi, on the other hand, can produce a large range of enzymes necessary to break down lignin and cellulose, and have highly efficient secretory pathways, and thus have more potential for winery waste conversion.

There are numerous examples where filamentous fungi have been used to produce useful end-products in industries that produce cellulosic and other potentially nutrient rich waste materials, and these have been the subject of a number of recent reviews (Kumar et al. 2008). The naturally occurring filamentous fungi and yeasts have been enhanced by generations of mutagenesis and selection, or by using recombinant DNA techniques to alter one or a few enzymes, such as various cellulolytic enzymes, or by taking advantage of synergistic reactions by cultivating a number of different microorganisms with complementary properties together.

However, several significant problems remain and must be overcome to make the process ready for large-scale use for the production of products for the food and pharmaceutical industries, and for the production of biomass and bio-fuels (Somerville 2006). At present these problems include the accessibility of the cellulose in lignin requiring pre-treatment that is both expensive and potentially polluting, the amounts of enzymes required to convert cellulose to sugars, and in the case of bio-fuel production, the fact that highly cellulolytic fungi do not convert sugars to ethanol at high levels in aerobic conditions, and fungi that do convert sugars to ethanol are not naturally highly cellulolytic. Because filamentous fungi are better able to access and degrade lignin and cellulosic material, these represent the best prospect for the treatment of solid winery wastes.

There are some recent examples, specifically in the wine industry, where filamentous fungi have been used to convert wastes to valuable products. Investigations using *Aspergillus awamori* with grape marc as the substrate showed that it provided all the nutrients necessary to produce hydrolytic enzymes including cellulases, pectinases and xylanases, in a solid state fermentation process. Further, the solid grape waste was competitive with other agricultural wastes when used as the substrate in such processes (Botella et al. 2005). *Phanerochaete chrysosporium*, a white rot fungus that degrades lignin but leaves the cellulose of the wood un-degraded and thus useful for further processing, has been used with grape seed and stem waste for laccase production (Couto et al. 2006; Moldes et al. 2007). *Trichoderma viride* is used as a bio-control agent against phytopathogens, and solid state fermentation using grape marc and wine lees as a substrate resulted in high yields of *T. viride* conidia (Bai et al. 2008). *Rhizopus oligosporus* has been used in solid state phenolic antioxidant production using cranberry pomace as a substrate (Vattem and Shetty 2002). Strains of *T. viride*, *A. niger* and *A. oryzae* were assessed for efficiency in biomass mass production and nutrient clearing of winery waste water (Zhang et al. 2008). Efficiencies of 5 gm per litre were found for *T. viride* without nutrient supplementation, accompanied by 84–90% chemical oxygen demand reduction, indicating that a biotechnological treatment process integrated with fungal biomass production from the winery waste streams could be effectively developed.

Pre-treatments can increase the efficiency of bioreactors, and *Trametes pubescens* has been used effectively with wine distillery waste water to provide additional nutrients for altered flux and the removal of components of wastewater leading to a lower overall contamination measured by chemical oxygen demand (Melamane et al. 2007). Notwithstanding these early successful applications, full adoption of fungal conversion of winery waste to useful end-products is technically problematic and still some way off. The information emerging from fungal genomic, metabolomic and proteomic projects will provide the necessary platform to enable the next step towards successful biological waste conversion in the wine and other industries.

15.3.3 Prospects Arising from Fungal Genome Projects

With the advent of publicly available whole genome sequences, new approaches to strain development and assessment are being developed.

15.3.3.1 Aspergillus oryzae

Aspergillus oryzae has a long history of use in the food industry, and more recently it has been modified by using recombinant DNA techniques to overproduce both native and heterologous proteins of importance in the food, chemical and pharmaceutical industries (Machida 2002). This fungus has a very high secretory ability, and there are a large number of examples where the safety of such products for human use has been verified, and GRAS (Generally Regarded As Safe) status has been granted (Abe et al. 2006). Much of the production of high value added products has been from growth on valuable compounds including glucose and starch (Ward et al. 1997), and thus modification of these processes to utilize waste materials will be beneficial. Analysis of highly expressed genes identified highly expressed pathways (Maeda et al. 2004), but a full description of the metabolic capability of the organism was uncovered by whole genome sequencing. Analysis of the genome identified that *A. oryzae* contains a greater number of metabolic and transporter genes than in other fungi, including the closely related *Aspergillus nidulans* and *Aspergillus fumigatus*, and*Neurospora crassa* (Kobayashi et al. 2007; Machida et al. 2005), particularly the hydrolytic and proteolytic enzymes that degrade high molecular weight bimolecular such as carbohydrates, polypeptides and nucleic acids, consistent with its use in applied processes (Abe et al. 2006).

The full sequence of the genome has accelerated the development of oligonucleotide DNA microarray technology, and arrays are commercially available (FarmLab Inc, Tokoyo Japan). This microarray technology opened up a number of new and promising areas for development. At the production level, microarrays have been used to monitor gene expression levels on a genome wide scale, both during liquid fermentations and on solid media including refuse material (Maeda et al. 2004), allowing a full understanding of the metabolic flux during the fermentation, and the tweaking of growth conditions to achieve maximum production. This full understanding of the metabolic flux at all stages of fermentations will provide the impetus for further genetic engineering, with the possibility to more cleanly analyze all the effects of changes made. At the research level, knowledge of the *A. oryzae* genome sequence has led to the discovery of new proteins and biochemical pathways for further analysis. One example is in the degradation of biodegradable plastics, where a theoretical analysis of the *A. oryzae* genome combined with microarray analysis uncovered genes that were responsible for degrading one such compound (Maeda et al. 2005).

15.3.3.2 Aspergillus niger

Aspergillus niger has been used as a biological factory in the production of chemicals, commercial enzymes and pharmaceuticals. *A. niger* has been used for the production of organic acids, and citric acid fermentation has been honed over the years to such an extent that 90% of sugar is converted. *A. niger* is also highly efficient for secreted protein production, with yields up to 10 gm per litre. Metabolic engineering has proved a very powerful approach to optimising industrial fermentation processes by making specific genetic changes using recombinant DNA technology for the construction of improved production strains (Patil et al. 2004). In a 2001 review, Nielsen has set out examples of such approaches, including where heterologous proteins were produced, where substrate range was extended, where new products were formed, where xenobiotics were degraded, where overall cellular physiology was improved, where by-products were reduced or eliminated, and where yield or productivity was improved (Nielsen 2001).

Mathematical models have been used to design optimal metabolic network structures, and the post-genomic era has led to genome-scale stoichiometric models of microorganisms (Patil et al. 2004). A comprehensive analysis of *A. niger* central carbon metabolism has been developed, and used to quantify the flux through the branches of the metabolic network (David et al. 2003). This framework allows the *in silico* analysis of environmental and genetic changes, an important predictive tool in the investigation of metabolite production in newly designed production processes and strains. The availability of the full genome sequence in the model filamentous fungus, *A. nidulans*, has allowed the full potential of this approach to become apparent (David et al. 2006, 2005).

A genome wide transcriptional analysis of gene expression in glucose, glycerol and ethanol growth conditions has been established (David et al. 2006), and a complete metabolic network has been defined. In addition, this approach has allowed the genome-wide examination of the effects of a mutation in *creA*, which encodes the major repressor protein in carbon catabolite repression (Dowzer and Kelly 1991), by the quantification of the fluxes in the central carbon metabolism for different conditions of glucose repression (David et al. 2005). These conditions included growth on xylose in the presence of glucose, and as these are components of lignocellulose, an abundant and renewable carbon source, they are relevant to industrial production processes.

15.3.3.3 Trichoderma reesei

Trichoderma reesei is adapted to a nutrient poor environment, using extracellular hydrolyases to extract glucose from polysaccharides, and enzymes produced from *T. reesei* are used in the paper, textile and food industries. Intensive analysis of the regulation of transcription of cellulase and hemicellulase genes has been undertaken (Stricker et al. 2008), and these have been the target of genetic manipulation to increase production.

15.3.3.4 Other Fungi

The genome sequencing efforts have increased exponentially as methods of sequencing and data analysis have been refined, and it can be expected that the genomes of more esoteric non-model fungi with a widely diverse metabolic repertoires will be determined. The future of value adding industrial waste materials, including those of the wine industry, lies with applying these techniques of genetic and metabolic engineering to a broad range of microorganisms, as their DNA sequences rapidly become available. For example, *P. chrysosporium* has superior lignin degrading properties, and its genome sequence is about to be released (Martinez et al. 2005). Value-added products including fermentable sugars, organic acids, solvents and bio-fuels may efficiently be produced from lignocellulosic biomass using the understandings that arise from genome sequences of presently intractable microorganisms and advances in engineering technology.

15.4 Biotechnological Processes for Bioconversion of Winery Residues to Bioenergy and Biomaterials

15.4.1 Process Synthesis

The bioconversion of cellulose and hemicellulose materials such as winery residues to bioproducts generally takes place in four operation phases: pre-treatment, hydrolysis, fermentation and product separation/distillation. According to Cardona and Sanchez's report (Cardona et al. 2007), a schematic flow sheet for the bioconversion process for winery residues is proposed in Fig. 15.2. The first step is size reduction and pre-treatment by delignification of the lignocellulosic feedstock to liberate cellulose and hemicellulose from lignin. The pre-treatment is to alter or remove structural and compositional impediments to hydrolysis in order to improve the rate of enzyme hydrolysis and increase yields of fermentable sugars from cellulose and hemicellulose. The second phase is depolymerization of the carbohydrate polymers (cellulose and hemicellulose) using acid and enzymatic hydrolysis to produce free sugars. In the fermentation operation, the sugars from the pre-treatment and hydrolysis steps are fermented by bacteria, yeast or filamentous fungi. Depending on the operation system and products involved in the fermentation, the target products will be separated and recovered in a downstream process. To develop a viable biotechnological process for producing market products from winery residues, the following features need to take into a consideration (Hahn-Hagerdal et al. 2006):

- Efficient de-polymerization of cellulose and hemicellulose to soluble sugars.
- Efficient fermentation of a mixed-sugar hydrolysate containing six-carbon (hexoses) and five-carbon (pentoses) sugars as well as fermentation inhibitory compounds.
- Advanced process integration to minimize process energy demand.
- Cost-efficient removal of lignin.

The sequential bioprocess configuration employed to convert winery residues into bioproducts implies that the solid fraction of pre-treated lingocellulosic materials undergo acid or enzymatic hydrolysis (saccharification). Once the hydrolysis is completed, the resulting cellulose hydrolysate is fermented and converted into the target bioproducts. This process is called separate hydrolysis and fermentation (SHF). In recent decades, advanced biotechnological processes have been developed to minimize the process steps and enhance the production efficiency. The most promising of these is to combine the hydrolysis and fermentation into a single stage. The solid fraction from the pre-treatment that contains cellulose and lignin undergoes a simultaneous saccharification and fermentation (SSF) process or a simultaneous saccharification and co-fermentation (SSCF), as described in Fig. 15.2.

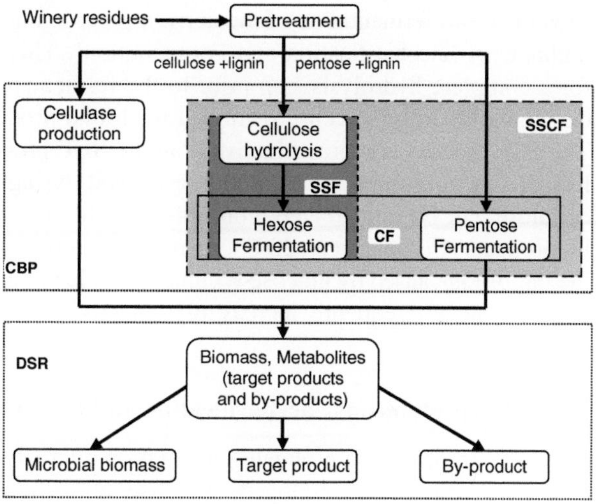

Fig. 15.2 Process diagram for bioconversion of winery residues to bio products. SSF: simultaneous saccharification and fermentation; SSCF: simultaneous saccharification and co-fermentation; CF: co-fermentation; CBP: consolidated bio processing; DSR: downstream separation and recovery

15.4.2 Process Configuration and Integration

15.4.2.1 Pre-treatment

The main processing challenge in the bioproduction from winery residues is the pre-treatment step. The lignocellulosic biomass is made up of a matrix of cellulose and lignin bound by hemicellulose chains. During the pre-treatment, this matrix should be broken in order to reduce the degree of crystallinity of the cellulose and increase the fraction of amorphous cellulose, which is the most suitable form for enzymatic attack. A successful pre-treatment must meet the following requirements (Patel et al. 2007): (1) to improve formation of sugars or the ability to subsequently form sugars by hydrolysis, (2) to avoid loss of carbohydrate, (3) to avoid formation of by-products inhibitory to subsequent hydrolysis and fermentation processes, and (4) to be cost effective.

The pre-treatment stage provides the physical disruption of the lignocellulose matrix in order to facilitate acid-or enzymatic-catalyzed hydrolysis. The pre-treatment can have significant implications on the configuration and efficiency of the process and, ultimately the economics. The pre-treatment can be carried out by physical (mechanical attack, steam explosion), chemical (ammonia fibber explosion, supercritical CO_2 treatment) and biological methods. Depending on the pre-treatment and hydrolysis processes involved, the lignocellulosic biomass of the winery residues can be saccharified into mainly glucose followed by pentoses and other hexoses, and lignin (van Wyk 2001). The glucose and pentose and other hexose sugars can be used for bioconversion processes.

15.4.2.2 Simultaneous Saccharification and Fermentation Process

Alternatively, microbial conversion of lignocellulosic materials to bioproducts can be made much more effective by coupling the enzymatic hydrolysis of lignocellulosic substrates and fermentation of the derived glucose into a single step, known as 'Simultaneous Saccharification and Fermentation'. This method eliminates the need for a complete hydrolysis step prior to the fermentation step. In the SSF process, enzymatic hydrolysis, cell growth and microbial production occur simultaneously. The SSF process is more attractive than the SHF process due to higher yields and less energy consumption. The glucose formed during the enzymatic hydrolysis of cellulose can be immediately consumed by the microbial cells converting it into bioproducts (Cardona et al. 2007). Therefore, a direct benefit of SSF is a decrease in the inhibition caused by glucose accumulation, leading to an increase in the saccharification rate, consequently increasing productivity and reducing reactor volume and capital costs.

15.4.2.3 Simultaneous Saccharification and Co-fermentation

The SSCF process is a promising integration alternative, in which co-fermentation of mixed cultures is employed. The use of mixed culture has advantages for allowing the microorganisms access multi-carbon sources, enhances enzymatic saccharification and fermentation efficiency, and consequently promotes product yield and productivity. In a SSCF configuration, it is necessary that both fermenting microorganisms be compatible in terms of operating condition and substrate consumption.

15.4.2.4 Consolidated Bio-Processing

The consolidated bio-processing (CBP), known as direct microbial conversion, is a reaction-reaction integration for the transformation of biomass into bioproducts. The key advance of CBP over other process strategies for biomass conversion is that only one microbial community is employed in the processing system for the production, cellulose hydrolysis, and fermentation, which are carried out in a single step. The important consequence of the CBP operation is that no-capital or operation expenditures are required for enzyme production within the process. Lynd et al. (2005) reported the comparative simulation of SSCF and CBP processes assuming aggressive performance parameters intended to be representative of a mature technology. The feasibility of a CBP process can be established when a microorganism or microbial consortium can be developed according to enzymatic and metabolic activities and process strategies (Sun and Cheng 2002).

The process design has been boosted due to process reaction-reaction integration. The development of a simultaneous process is a promising approach to make the biotransformation of winery residues technically and economically viable as an industrial process. The simultaneous processes such as SSCF and CBP allow synergistic interactions as well as the development of more compact technological schemes. In general, the reaction-reaction integration can be proposed for the integration of different microbial transformation taking place during the metabolic production. This integration is oriented to the complete assimilation by the microorganisms of all the sugars released during the pre-treatment and hydrolysis of lingocellulosic biomass. Besides the understanding of the enzymatic and metabolic aspects related to the molecular biology of these microorganisms and process integration, there is a need to study the relationships of this co-fermentation process through process modelling and simulation in order to define the optimal cultivation conditions for increasing yield and productivity of the target product. One of the major challenges is to optimize the integration of process engineering, fermentation technology, enzyme engineering and metabolic engineering (Hahn-Hagerdal et al. 2006).

15.5 Renewable Energy and Biomaterials from Winery Residues

15.5.1 Renewable Energy

Energy demand is increasing continuously due to speedy growth in population and industrial development. Two significant challenges associated with using conventional fuels, such as coal, oil and natural gas, are depletion of fossil fuels and deterioration of the environment. As concerns increase over the supply of oil and changes in global climate, there is an urgent need to develop clean and renewable alternative energy resources. Ethanol and hydrogen are the two most important fuels contributing to the reduction of negative environmental impacts. Ethanol has already been introduced on a large scale, and it is expected to be one of the dominating renewable bio-fuels in the transport sector within the coming years. Due to its environmentally friendly nature and high-energy yield, hydrogen (H_2) offers a tremendous candidate as an ideal clean and sustainable energy in the future. Hydrogen has the highest gravimetric energy density of any known fuel without producing polluting emissions. In addition, H_2 has great potential for chemical synthesis or for electrical storage and generation with fuel cells. H_2 fuel cells and related technologies provide the essential link between renewable energy sources and sustainable energy services.

15.5.1.1 Ethanol

Ethanol production utilizing lingocellulosic materials such as winery residues, generally takes place in three phase: (1) delignification of the lingocellulosic feedstock to liberate cellulose and hemicellulose from lignin, (2) depolymerization of the carbohydrate polymers (cellulose and hemicellulose) to produce free sugars, and (3) fermentation of mixed hexose and pentose sugars to produce ethanol. Alzate and Toro (2006) and Hamelinck et al. (2005) summarized the most commonly used and proposed process configurations, including SHF, SSF, SSCF and CBP. The SHF has been employed in most industrial processes. SSF has been recognized as an effective process configuration, which effectively removes glucose, an inhibitor to cellulase activity, thereby increasing the yield and rate of cellulose hydrolysis (Sun and Cheng 2002).

As the centre of the ethanol production process, fermentation is carried out under anaerobic conditions. According to the biochemical reactions, the theoretical maximum yield is 0.51 kg ethanol and 0.49 kg CO_2 produced from 1 kg of xylose and glucose. However, in practice, the microorganisms use some of the glucose for growth and the actual yield is less than 100% theoretical yield. Xylose-fermenting microorganisms for ethanol production are found among bacteria, yeast and filamentous fungi. Bacteria include both native and genetically engineered organisms, and many have characteristics useful for SSF process. The ethanologenic bacteria that demonstrate the most promise for industrial exploitation are *E. coli, K. oxytoca* and *Z. mobilis* (Dien et al. 2003). *Z. mobilis* is well recognized for its ability to produce ethanol rapidly and efficiently from glucose-based feed stocks. One of the most effective ethanol producing yeasts, *S. cerevisiae*, has several advantages owing to its high yield from hexoses and high tolerance to ethanol and other inhibitory compounds (Hamelinck et al. 2005).

15.5.1.2 Hydrogen

H_2 production through bioprocesses represents an exiting new area of technology development for bioenergy generation. Hydrogen can be produced by anaerobic bacterial growth on carbohydrate rich substrates giving organic fermentation end products, hydrogen and CO_2. The bacteria found to produce hydrogen from carbohydrate containing organic materials include species of *Enterobactor, Bacillus* and *Clostridium.* Most biologically produced H_2 in the biosphere is produced by microbial fermentation processes. Many microorganisms contain enzymes, known as hydrogenases, that either oxidize H_2 to protons and electrons or reduce protons and thus release molecular H_2. The physiological role and biochemical characteristics of these hydrogenases are variable. The metabolic flexibility of these microorganisms allows the production of a range of end products. It has been widely accepted that the highest theoretical yield of 4 mol H_2/mol glucose can be obtained if acetate is the sole by-product (Eq. 15.1), while a maximum 2 mol H_2/mol glucose is associated with butyrate as the end by-product (Eq. 15.2) (Hallenbeck and Benemann 2002). In practice, however, high H_2 yields are associated with a mixture of acetate and butyrate fermentation products, and low H_2 yields are associated with propionate and reduced end products.

$$C_6H_{12}O_6 + 2H_2O = 2CH_3COOH + 2CO_2 + 4H_2 \quad (15.1)$$

$$C_6H_{12}O_6 = CH_3CH_2CH_2COOH + 2CO_2 + 2H_2 \quad (15.2)$$

Clostridium thermocellum is a thermophilic bacterium that utilizes cellulose as a sole carbon source and carries out mixed product fermentation, synthesizing various amounts of 36 lactate, formate acetate, ethanol, H_2, and CO_2, under different growth conditions (Demain et al. 2005). Hydrogen and soluble end-product synthesis patterns by *C. thermocellum* in batch cultures were investigated by Levin et al. (2006) in batch cultures, using either cellobiose or cellulosic substrates, providing an average yield of 1.6 mol H2 mol/glucose equivalent. Fermentative H_2 production is an exciting R&D area that offers a potential means of producing H_2 from a variety of renewable resources or wastewaters via a low energy required continuous bioprocess. However, a typically low H_2 yield associated with the H_2 fermentation process continues to be a challenge for developing a viable industrial process. Research is needed to update advance knowledge of bioprocess engineering, providing a better understanding of the regulation of metabolisms underlying the metabolic flux network of the H_2 fermentative system, consequently improving H_2 productivity and yield.

15.5.2 Organic Acids

Microbial production of organic acids is a promising approach for obtaining building-block chemicals from renewable carbon sources. Most of the organic acids are natural products of microorganisms, or at least natural intermediates in major metabolic pathways. Because of their functional groups, organic acids are extremely useful as starting materials for the chemical industry. Here, we introduce the recent developments exemplified by two organic acid products: lactic acid and citric acid.

15.5.2.1 Lactic Acid

Lactic acid (2-hydroxypropionic acid) is the most widely occurring multifunctional organic acid, and has been widely used in food and pharmaceutical industries. One of its most promising applications is its use for biodegradable and biocompatible polylactate polymers, an environmentally friendly alternative to non-biodegradable plastics derived from petrochemicals. Lactic acid can be produced using bacteria and fungi. Lactic acid producing bacteria (LAB) have received wide interest because of their high growth rate and product yield. However, LAB have complex nutrient requirements because of their limited ability to synthesize B-vitamins and amino acids, making supplementation of sufficient nutrients such as yeast extracts to production media necessary. In addition, the difficulty in separating fermentation broth containing lactic acid from bacterial biomass increases the overall cost of production process. Fungal *Rhizopus* species have attracted attention in recent decades, and have been recognized as suitable candidates for lactic acid production. Unlike the LAB, *Rhizopus* strains generate L-lactic acid as a sole isomer of lactic acid (Jin et al. 2003; Yin et al. 1997). *Rhizopus* strains grow better under nitrogen-limited environments than the LAB. Lactic acid production from lignocellulosic agricultural biomass using lactic acid bacteria has been explored (Venkatesh 1997), but a SSF process for lactic acid production using lignocellulosic material and *Rhizopus* species has not been reported. However, Ruengruglikit and Hang (2003) and Miura et al. (2004) developed a process which integrated lignocellulose hydrolysis with lactic acid fermentation. Ruengruglikit and Hang carried out the fermentation in the presence of carboxyl methyl cellulase and xylanase, while Miura et al. used *Acremonium thermophilus* ATCC 24622.

15.5.2.2 Citric Acid

The production of citric acid (also known as the tricarboxylic acid), by the filamentous fungus *A. niger* represents the oldest microbial process for production of a high volume and low cost organic acid. A fermentation process for the production of citric acid can be operated in either a solid state fermentation system or a submerged bioreactor process. *A. niger* can access glucose, starch and cellulose-based carbohydrate materials as carbon sources. The crucial parameters resulting in efficient production of citric acid by *A. niger* have been determined empirically and include high substrate concentration, low and finite content of nitrogen and certain trace metals, through maintenance of high dissolved oxygen and low pH. The exact definition of these parameters enabled the development of a highly efficient biotechnological process, such that citric acid concentration of 140 g/L are now easily reached (Forster et al. 2007). In addition to the well-established filamentous fungal species, yeast *Yarrowia lipolytica* has been recently developed as a microbial cell factory for citric acid production (Papanikolaou et al. 2006). *Y. lipolytica* proved efficient in the production of citric acid from carbon sources such as glucose and sucrose.

15.5.3 Microbial Biomass Protein

Winery wastes are suitable substrates for production of microbial biomass protein (MBP). The MBP products could be used for human or animal consumption, which are marketable products and may offset the operating costs of the treatment process (Jin et al. 2002). Both fungi and yeast can be used to produce MBP. However, fungi seem more attractive because filamentous or pellet morphology of fungi permits low cost separation and recovery of the MBP from the culture media, which makes up a significant fraction of the capital and operating costs for MBP production. Fungi can be grown using almost any organic waste products that contain carbohydrates, such as confectionery and distillery waste, vegetable waste and wood processing effluents.

A biotechnological treatment process integrated with fungal biomass production (FBP) from the winery waste streams was investigated by Zhang et al. (2008). They reported that it could be possible to develop a hybrid biotechnological process, integrating the production of fungal biomass protein with treatment of winery wastewater. Three filamentous fungi, *T. viride* WEBL0702, *A. oryzae* WEBL0401 and *A. niger* WEBL090, demonstrated a high capability for over 80% COD reduction and fungal biomass production. *T. viride* had a lower nitrogen requirement compared to *A. oryzae* and *A. niger*, indicating that *T. viride* could tolerate the fluctuation of nutrient change in winery wastewater.

15.5.4 Polymers

There has been a considerable interest in using low cost carbon substrates for the production of polymers including Poly(3-hydroxybutyrate), and food industry waste water has been shown to be a viable alternative carbon source in high cell density PHB production (Nath et al. 2008). There are no examples to date where winery wastes have been used, but the prospect is that further development of strains and processes will allow winery wastes to be converted to polymers.

15.5.5 Enzymes

There have been few examples to date where winery wastes have been used as the substrate for enzyme production, but it is likely that this waste stream will become increasingly important as economic and environmental necessities drive the development of techniques for processing a broad range of nutrient rich industrial waste products. Winery wastes have been used with *P. chrysosporium*, a white rot fungus, for lignolytic enzyme production (Rodriguez-Navarro et al. 2004), and grape seed and stem waste was the substrate for *T. versicolor* in laccase production (Couto and Toca-Herrera 2007). Grape pomace has been used with*A. awamori* to produce hydrolytic enzymes including cellulases, pectinases and xylanases (Botella et al. 2005).

References

Abe K., Gomi K, Hasegawa F et al. (2006). Impact of *Aspergillus oryzae* genomics on industrial production of metabolites. Mycopathologia 162:143–153

Alterthum F, Ingram LO (1989). Efficient ethanol production from glucose, lactose, and xylose by recombinant *Escherichia coli*. Appl Environ Microbiol 55:1943–1948

Alzate CAC, Toro OJS (2006). Energy consumption analysis of integrated flowsheets for production of fuel ethanol from lignocellulosic biomass. Energy 31:2447–2459

Arvanitoyannis IS, Ladas D, Mavromatis A (2006a). Potential uses and applications of treated wine waste: a review. Int J Food Sci Tech 41:475–487

Arvanitoyannis IS, Ladas D, Mavromatis A (2006b). Wine waste treatment methodology. Int J Food Sci Technol 41:1117–1151

Bai ZH, Jin B, Li Y J et al. (2008). Utilization of winery wastes for *Trichoderma viride* bio-control agent production by solid state fermentation. J Environ Sci 20:353–358

Botella C, de Ory I, Webb C et al. (2005). Hydrolytic enzyme production by *Aspergillus awamori* on grape pomace. Biochem Eng J 26:100–106

Bustamante MA, Moral R, Paredes C et al. (2008). Agrochemical characterisation of the solid by-products and residues from the winery and distillery industry. Waste Manag 28:372–380

Cardona F, Carrasco P, Perez-Ortin JE et al. (2007). A novel approach for the improvement of stress resistance in wine yeasts. Int J Food Microbiol 114:83–91

Chapman J (1995). Land Disposal of Winery and Distillery Wastewaters. PhD thesis, University of Adelaide, Australia.

Chapman J, Baker P, Wills S (2001). Winery Wastewater Handbook, Winetitles, Adelaide.

Couto SR, Lopez E, Sanroman MA. (2006). Utilisation of grape seeds for laccase production in solid-state fermentors. J Food Eng 74:263–267

Couto SR, Toca-Herrera JL (2007). Laccase production at reactor scale by filamentous fungi. Biotechnol Adv 25:558–569

David H, Akesson, M, Nielsen J (2003). Reconstruction of the central carbon metabolism of *Aspergillus niger*. Eur J Biochem 270:4243–4253

David H, Hofmann, G, Oliveira, AP et al. (2006). Metabolic network driven analysis of genome-wide transcription data from *Aspergillus nidulans*. Genome Biol 7:R108

David H, Krogh AM, Roca C et al. (2005). CreA influences the metabolic fluxes of *Aspergillus nidulans* during growth on glucose and xylose. Microbiol. 151:2209–2221

Demain AL, Newcomb M, Wu JHD (2005). Cellulase, clostridia, and ethanol. Microbiol Mol Biol Rev 69:124–154

Dien BS, Cotta MA, Jeffries TW (2003). Bacteria engineered for fuel ethanol production: Current status. Appl Microbiol Biotechnol 63:258–266

Dowzer CEA, Kelly JM (1991). Analysis of the *creA* gene, a regulator of carbon catabolite repression in*Aspergillus nidulans*. Mol Cell Biol 11:5701–5709

Forster A, Aurich A, Mauersberger S et al. (2007). Citric acid production from sucrose using a recombinant strain of the yeast Yarrowia lipolytica. Appl Microbiol Biotechnol 75: 1409–1417

Hahn-Hagerdal B, Karhumaa K, Fonseca C et al. (2007). Towards industrial pentose-fermenting yeast strains. Appl Microbiol Biotechnol 74:937–953

Hahn-Hagerdal B, Galbe M, Gorwa-Grauslund M et al. (2006). Bioethanol-the fuel of tomorrow from the residues of today. Trends Biotechnol 24:551–556

Hallenbeck PC, Benemann JR (2002). Biological hydrogen production; fundamentals and limiting processes. Int J Hydrogen Energ 27:1185–1193

Hamelinck CN, Van Hooijdonk G, Faaij APC (2005). Ethanol from lignocellulosic biomass: techno-economic performance in short-, middle- and long-term. Biomass Bioenerg 28:384–410

Jin B, Huang LP, Lant P (2003). *Rhizopus arrhizus* – a producer for simultaneous saccharification and fermentation of starch waste materials to L(+)-lactic acid. Biotechnol Letts 25:1983–1987

Jin B, Yan XQ, Yu Q et al. (2002). A comprehensive pilot plant system for fungal biomass protein production and wastewater reclamation. Adv Environ Res 6:179–189

Kobayashi T, Abe K, Asai K et al. (2007). Genomics of *Aspergillus oryzae*. Biosci Biotechnol Biochem 71:646–670

Kumar R, Singh S, Singh OV (2008). Bioconversion of lignocellulosic biomass: biochemical and molecular perspectives. J Ind Microbiol Biotechnol 35:377–391

Levin D B, Islam R, Cicek N et al. (2006). Hydrogen production by *Clostridium thermocellum* 27405 from cellulosic biomass substrates. Int J Hydrogen Energ 31:1496–1503

Lynd LR, van Zyl W H, McBride JE et al. (2005). Consolidated bio processing of cellulosic biomass: an update. Curr Opin Biotechnol 16:577–583

Lynd LR, Weimer P J, van Zyl WH et al. (2002). Microbial cellulose utilization: fundamentals and biotechnology. Microbiol Mol Biol Rev 66:506–577

Machida M. (2002). Progress of *Aspergillus oryzae* genomics. Adv Appl Microbiol 51:81–106

Machida M, Asai K, Sano M et al. (2005). Genome sequencing and analysis of *Aspergillus oryzae*. Nature 438:1157–1161

Maeda H, Sano M, Maruyama Y (2004). Transcriptional analysis of genes for energy catabolism and hydrolytic enzymes in the filamentous fungus *Aspergillus oryzae* using cDNA microarrays and expressed sequence tags. Appl Microbiol Biotechnol 65:74–83

Maeda H, Yamagata Y, Abe K et al. (2005). Purification and characterization of a biodegradable plastic-degrading enzyme from *Aspergillus oryzae*. Appl Microbiol Biotechnol 67:778–788

Martinez AT, Speranza M, Ruiz-Duenas FJ et al. (2005). Biodegradation of lignocellulosics: microbial, chemical, and enzymatic aspects of the fungal attack of lignin. Int Microbiol 8: 195–204

Martinez D, Berka RM, Henrissat B et al. (2008). Genome sequencing and analysis of the biomass-degrading fungus *Trichoderma reesei* (syn. *Hypocrea jecorina*). Nature Biotechnol 26:553–560

Masai E, Katayama Y, Fukuda M (2007). Genetic and biochemical investigations on bacterial catabolic pathways for lignin-derived aromatic compounds. Biosci Biotechnol Biochem 71: 1–15

Melamane XL, Tandlich R, Burgess JE (2007). Treatment of wine distillery wastewater by high rate anaerobic digestion. Water Sci Technol 56:9–16

Miura S, Arimura T, Itoda N (2004). Production of L-lactic acid from corncob. J Biosci Bioeng 97:153–157

Moldes AB, Vazquez M, Dominguez JM et al. (2007). Evaluation of mesophilic biodegraded grape marc as soil fertilizer. Appl Biochem Biotechnol 141:27–36

Nath A, Dixit M, Bandiya A et al. (2008). Enhanced PHB production and scale up studies using cheese whey in fed batch culture of *Methylobacterium* sp. ZP24. Bioresour Technol 99: 5749–5755

Nielsen J (2001). Metabolic engineering. Appl Microbiol Biotechnol 55:263–283

Papanikolaou S, Galiotou-Panayotou M, Chevalot I et al. (2006). Influence of glucose and saturated free-fatty acid mixtures on citric acid and lipid production by *Yarrowia lipolytica*.Curr Microbiol 52:134–142

Patel S J, Onkarappa R, Shobha K S (2007). Fungal pre-treatment studies on rice husk and bagasse for ethanol production. Electron J Environ Agr Food Chem 6:1921–1926

Patil KR, Akesson M, Nielsen J (2004). Use of genome-scale microbial models for metabolic engineering. Curr Opin Biotechnol 15:64–69

Rodriguez-Navarro S, Fischer T, Luo MJ et al. (2004). Sus1, a functional component of the SAGA histone acetylase complex and the nuclear pore-associated mRNA export machinery. Cell 116:75–86

Ruengruglikit C, Hang YD (2003). L(+)-lactic acid production from corncobs by *Rhizopus oryzae* NRRL-395. LWT Food Sci Technol 36:573–575

Somerville C (2006). The billion-ton bio-fuels vision. Science 312:1277

Stricker AR, Mach RL, de Graaff LH (2008). Regulation of transcription of cellulases- and hemicellulases-encoding genes in *Aspergillus niger* and *Hypocrea jecorina (Trichoderma reesei)*. Appl Microbiol and Biotechnol 78:211–220

Sun Y, Cheng JY (2002). Hydrolysis of lignocellulosic materials for ethanol production: A review. Bioresour Technol 83:1–11

van Wyk JPH (2001). Biotechnology and the utilization of biowaste as a resource for bioproduct development. Trends Biotechnol 19:172–177

Vattem DA, Shetty K (2002). Solid-state production of phenolic antioxidants from cranberry pomace by *Rhizopus oligosporus*. Food Biotechnol 16:189–210

Venkatesh KV (1997). Simultaneous saccharification and fermentation of cellulose to lactic acid. Bioresour Technol 62:91–98

Ward PP, Chu HP, Zhou XD et al. (1997). Expression and characterization of recombinant murine lactoferrin. Gene 204:171–176

Yin PM, Nishina N, Kosakai Y et al. (1997). Enhanced production of L(+)-lactic acid from corn starch in a culture of *Rhizopus oryzae* using an air-lift bioreactor. J Ferment Bioeng 84:249–253

Zhang ZY, Jin B, Bai ZH et al. (2008). Production of fungal biomass protein using microfungi from winery wastewater treatment. Bioresour Technol 99:3871–3876

Zhou S, Davis FC, Ingram LO (2001). Gene integration and expression and extracellular secretion of *Erwinia chrysanthemi* endoglucanase CelY (*celY*) and CelZ (*celZ*) in ethanologenic *Klebsiella oxytoca* P2. Appl Environ Microbiol 67:6–14

Chapter 16
Biotechnological Potential of Brewing Industry By-Products

Solange I. Mussatto

Contents

Abstract The manufacture of beer inevitably involves generation of various residues and by-products. The most common by-products are spent grains, spent hops and surplus yeast, which are generated from the main raw materials used for beer elaboration, the barley malt, hop and yeast, respectively. These three brewery by-products are available in large quantities throughout the year, but their use has still been limited, being basically sold to local dairy farmers to be used as cattle feed, or simply as a land fill. However, they represent large potential resources for use in biotechnological processes, in consequence of their complex compositions containing carbon, nitrogen and minerals. Several attempts have been made to use them in biotechnological processes, as for example in fermentative processes for the production of value-added compounds (xylitol, arabitol, ethanol, lactic acid, among others); as substrate for microorganisms cultivation, or simply as raw material for extraction of compounds such as sugars, proteins, acids and antioxidants. From an environmental viewpoint, the elimination of industrial by-products represents a solution to pollution problems, and merits thus large attention. In this chapter, the main

S.I. Mussatto (✉)
Institute for Biotechnology and Bioengineering, Centre of Biological Engineering, University of Minho. Campus de Gualtar, 4710-057, Braga, Portugal
e-mail: solange@deb.uminho.pt

P. Singh nee' Nigam, A. Pandey (eds.), *Biotechnology for Agro-Industrial Residues Utilisation*, DOI 10.1007/978-1-4020-9942-7_16,

characteristics and potential applications of the three main brewery by-products are reviewed focusing on their use in biotechnological processes.

Keywords Brewing · Spent grains · Spent hops · Trub · Surplus yeast · Fermentation

16.1 By-Products Generation During the Brewing Process

Brewing process consists initially mixing of the milled barley malt with water in a mash tun, and the temperature is slowly increased (from 37 to 78°C) to promote enzymatic hydrolysis of malt constituents. During this process, malt starch is converted to fermentable sugars (mainly maltose, and maltotriose) and non-fermentable sugars (dextrins), and proteins are partially degraded to polypeptides and amino acids. This enzymatic conversion stage (mashing) produces a sweet liquid called wort. The insoluble, undegraded part of the malted barley grain is allowed to settle to form a bed in the mash tun and the sweet wort is filtered through it (lautering). The residual solid fraction obtained after this stage is constituted by the spent grains (Mussatto et al. 2006a). In some cases, due to economic reasons or aiming to produce beers of distinct qualities, part of the barley malt – usually 15 to 20% – is replaced by unmalted cereal like corn (maize), rice, wheat, oats, rye or sorghum, called adjuncts (Kunze 1996). In these cases, at the end of the mashing process, the insoluble part of these grains is separated with the undegraded part of the malted barley grain (spent grains).

After filtered, the wort is transferred to the brewing kettle (also known as *cooper*) where it is boiled during at least one hour with the addition of hops. During this process, the bitter and aromatic hop components are transferred into the wort. Such substances will confer typical beer qualities, such as bitter taste, flavour, and foam stability. At the boiling end, the medium is cooled and the liquid extract is then separated from the hop residues to be further processed (Keukeleire 2000). The hop residues (spent hops), which are then useless are dumped directly as being of no further value. A fraction of the hop components end up in the trub (a precipitation product of the wort boiling process that may include insoluble hop materials, condensation products of hop polyphenols and wort proteins, and isomerized hop acids adsorbed onto trub solids (Huige 2006)). Hops can be used in the natural form (cone), or in the forms of powder (hammer-milled whole hops), pellets (powered hops packed as pellets), or extract (the main components of hops to the beer quality are extracted by means of a suitable solvent) (Kunze 1996).

After removal of the precipitate produced during boiling, the cooled hopped wort is pitched with yeast in a fermentation vessel, where the yeast cells will convert the fermentable sugars to ethanol and carbon dioxide. During fermentation, cell mass increases three- to six fold. At the end of this stage, most of the cells are collected as surplus yeast, at the top or at the bottom of the fermentor (according to the nature of yeast used) (Keukeleire 2000). Figure 16.1 is a schematic representation of the brewing process and the points where the main by-products are generated.

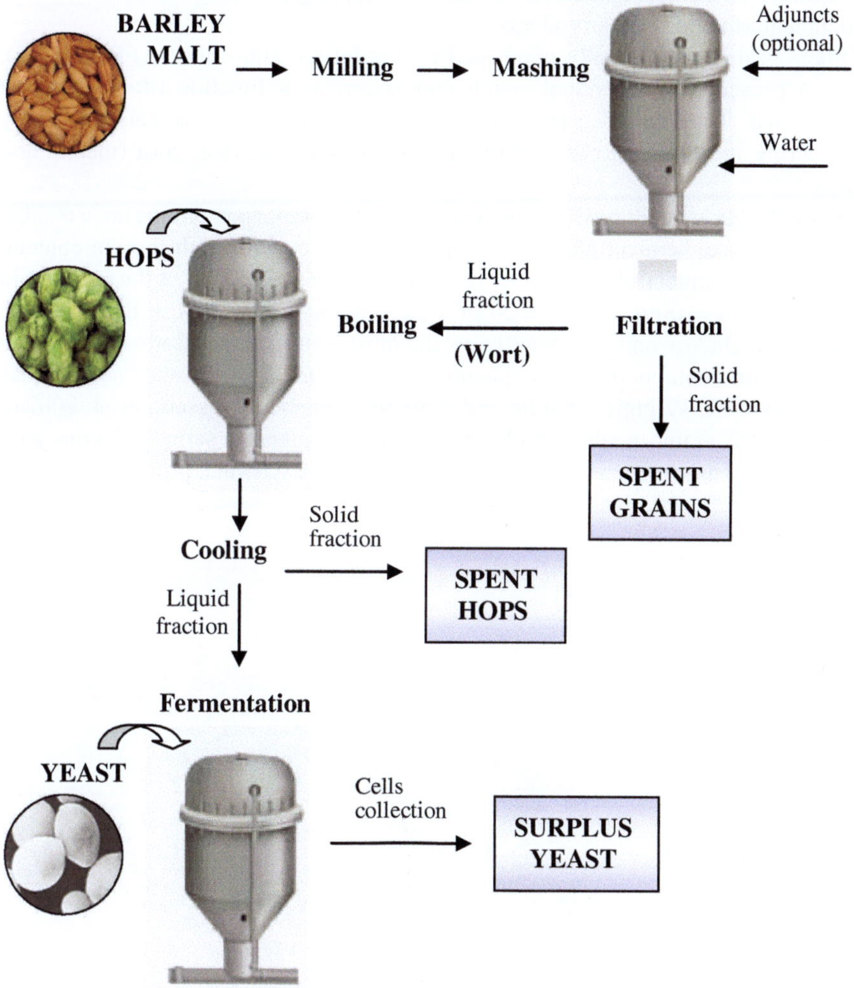

Fig. 16.1 Schematic representation of the brewing process and points where the main by-products are generated

16.2 Spent Grains

16.2.1 Chemical Composition

Spent grains are the most abundant brewing residue, corresponding to around 85% of total generated. It is estimated that about 200 t of wet spent grains with 70 to 80% water content are produced per 10,000 hl of produced beer (Kunze 1996). It is therefore of particular interest to sell these spent grains or reuse them for the obtainment of added-value products.

Spent grains are basically composed by the barley grain husks (Figure 16.2), but can present some residual starch endosperm as a function of the brewing regime used. They can also present some residues from other cereals eventually used during mashing together with barley, such as wheat, rice, corn (maize) and others.

Since the barley malt husk is a lignocellulosic material, spent grains are a residue rich in cellulose, hemicellulose and lignin, and also contain high protein content. Cellulose and hemicellulose together comprise almost 50% (w/w) of the spent grain composition, revealing the presence of a large amount of sugars in this material, with xylose, glucose and arabinose being the most abundant (Mussatto and Roberto 2006). Lignin contains numerous phenolic components, mainly acids such as ferulic, *p*-coumaric, syringic, vanillic and *p*-hydroxybenzoic (Mussatto et al. 2007a). Protein bound amino acids include leucine, valine, alanine, serine, glycine, glutamic acid and aspartic acid in the largest amounts, and tyrosine, proline, threonine,

Fig. 16.2 Appearance of spent grains (**A**) and micrograph by scanning electron microscopy (**B**) with 45-fold magnification (Mussatto et al. 2006b)

arginine, and lysine in smaller amounts. Cystine, histidine, isoleucine, methionine, phenylalanine, and tryptophan can also be present (Huige 2006, Mariani 1953).

Spent grains also contain considerable amounts of minerals present in ashes (specially silicon); and extractives, a fraction consisting of waxes, lipids, gums, starches, resin, tannins, essential oils and various other cytoplasmatic constituents. Lipids include triacylglycerols, diacylglycerols, fatty acids (palmitic, oleic and linoleic acids), sterols, sterol esters and sterol glycosides, plus various hydrocarbons (including alkanes and carotenoids) (Briggs et al. 1986). Vitamins such as: biotin, choline, folic acid, niacin, pantothenic acid, riboflavin, thiamine and pyridoxine, are also present in this material.

Nevertheless, the chemical composition of spent grains vary according to barley variety, harvest time, malting and mashing conditions, and the quality and type of adjuncts (other cereal grains) added in the brewing process (Huige 2006). Some analysis of spent grains chemical composition is shown in Table 16.1

Table 16.1 Chemical composition of spent grains

	Reference [a]			
Components (% dry weight basis)	1	2	3	4
Cellulose (glucan)	16.8	25.4	21.9	25.3
Hemicellulose	28.4	21.8	29.6	41.9
xylan	19.9	nr	20.6	nr
arabinan	8.5	nr	9.0	nr
Lignin	27.8	11.9	21.7	16.9
Acetyl groups	1.3	nr	1.1	nr
Proteins	15.3	24.0	24.6	nr
Ashes	4.6	2.4	1.2	4.6
Extractives	5.8	nr	nr	9.5
Minerals (mg/kg dry weight basis)				
Calcium	3515.0	nr	nr	nr
Sodium	309.3	nr	nr	nr
Potassium	258.1	nr	nr	nr
Magnesium	1958.0	nr	nr	nr
Aluminum	36.0	nr	nr	nr
Iron	193.4	nr	nr	nr
Barium	13.6	nr	nr	nr
Strontium	12.7	nr	nr	nr
Manganese	51.4	nr	nr	nr
Copper	18.0	nr	nr	nr
Zinc	178.0	nr	nr	nr
Phosphorus	5186.0	nr	nr	nr
Sulfur	1980.0	nr	nr	nr
Chromium	5.9	nr	nr	nr
Silicon	10740.0	nr	nr	nr

[a] 1 From Mussatto and Roberto et al. 2006; 2 From Kanauchi et al. 2001; 3 From Carvalheiro et al. 2004; 4 From Silva et al. 2004. nr = non reported

16.2.2 *Potential Uses of Spent Grains in Biotechnological Processes*

The main application of spent grains until nowadays has been as animal feed, due to its high content of protein and fibre. However, because the chemical composition is rich in sugars, proteins and minerals, several attempts have been made to use them in biotechnological processes. When compared to the other two main brewery by-products, the spent grains are the most evaluated for reuse in biotechnological purposes. The great interest in this material is probably due to the large amount generated and rich chemical composition that permit its reuse in different areas.

16.2.2.1 Substrate for Microorganisms Cultivation and Enzyme Production

Wet spent grains from a lauter tun contain high moisture 70–80% (w/w) and fermentable sugar contents, characteristics that make it liable to deteriorate rapidly due to microbial activity. Owing to these characteristics and the high protein content, several studies have been performed aiming to use the spent grains as substrate for cultivation of microorganisms. The fungus species *Pleurotus*, *Agrocybe*, *Lentinus*, *Aspergillus* and *Trichoderma*, and the bacteria *Bacillus* and *Streptomyces* (a soil actinobacteria), are some of the microorganisms successfully cultivated in spent grains (Wang et al. 2001, Szponar et al. 2003, Bartolomé et al. 2003, Bogar et al. 2002, Sim and Oh 1990).

The production of several enzymes was also verified from cultivation of microorganisms in spent grains, including xylanase by *Aspergillus awamori* (Bhumibhamon 1978) and *Streptomyces avermitilis* (Bartolomé et al. 2003), feruloyl esterase by *Streptomyces avermitilis* (Bartolomé et al. 2003), alpha-amylase by *Bacillus subtilis* (Duvnjak et al. 1983), *Bacillus licheniformis* (Okita et al. 1985) and *Aspergillus oryzae* (Bogar et al. 2002, Francis et al. 2003), and cellulase by *Trichoderma reesei* (Sim and Oh 1990).

16.2.2.2 Source of Value-Added Compounds

As before mentioned, spent grain is a lignocellulosic material and thus, it is constituted by several polysaccharides, which can be degraded into their corresponding constituents by hydrolytic procedures (hydrothermal, enzymatic or acidic). On hydrolysis, cellulose yields glucose, while the hemicellulose yields xylose, arabinose, mannose, galactose and the acids acetic and hydroxycinnamic (ferulic and *p*-coumaric) (Mussatto and Roberto 2004, Palmqvist and Hahn-Hägerdal 2000). Additionally, a wide variety of arabino-oligoxylosides with different structural features can be obtained according to the hydrolysis process used (Kabel et al. 2002). As a whole, the compounds obtained by hydrolysis of spent grain are of industrial interest, mainly in the food industry. Such compounds can be purified for use as is, or as substrate in different fermentative processes for obtainment of value-added products. Considering also the low cost, spent grains appears to be an attractive raw material for industrial purposes.

16.2.2.3 Use in Fermentative Processes

Due to the composition rich in sugars and nutritional factors, hydrolysates produced from spent grains can be used in fermentative processes to produce several compounds of industrial interest. Some examples include the use of the sugar rich hydrolysate as fermentation medium for the production of ethanol by *Saccharomyces cerevisiae* (Laws and Waites 1986), xylitol by *Candida guilliermondii* (Mussatto et al. 2008), xylitol, arabitol, ethanol and glycerol by *Debaryomyces hansenii* (Carvalheiro et al. 2005, Duarte et al. 2004), and lactic acid by *Lactobacillus delbrueckii* (Mussatto et al. 2007b), *Lactobacillus pentosus* or *Lactobacillus rhamnosus* (Cruz et al. 2007).

Besides the use as substrate in fermentative processes, spent grain can also be employed as carrier for cell immobilization. It has been suggested its use as cell immobilization carrier during the production of pectinase by *Kluyveromyces marxianus* CCT 3172 (Almeida et al. 2003), straightdough and sourdough bread using baker's yeast, kefir and *Lactobacillus casei* (Plessas et al. 2007), or during beer elaboration by *Saccharomyces yeast* (Brányik et al. 2001, Dragone et al. 2007). In the last case, the material must be initially pre-treated by a sequence of HCl and NaOH solutions to obtain a cellulose-based carrier, which due to the irregular shape and non-homogeneous chemical composition, provide "active sites" that are readily colonized by brewing yeasts.

There are also other alternatives for the spent grains reuse in breweries, such as the use of an extract obtained from spent grain pressing as an antifoaming agent in the fermentor. Besides effective for such purpose, the extract addition improves the hop utilization and not affects the properties of the final beer (Roberts 1976). When neutralized and added to wort, the spent grains extract enhances the yeast performance and also not affects the quality of the produced beer.

16.2.2.4 Other Uses

A range of other uses for spent grains has also been proposed, as an adsorbent for removing VOC emissions or organic material from effluent, as a source of biogas and soil conditioner produced by anaerobic digestion, as a medium for growing earthworms to use in poultry food, and as raw material for the production of charcoal bricks or bleached cellulose pulps that can be used in the manufacture of specific kind of papers (Mussatto et al. 2006b, 2008, Briggs et al. 1986).

Additionally, the water pressed from spent grains has a high biological demand (BOD) and constitutes an undesirable effluent. Some efforts to utilize this liquor include: (1) recovering its extract by incorporating it in a subsequent mash (with or without prior treatments to remove lipids and polyphenols); (2) using it as an antifoaming agent in deep fermentors; (3) using it as a nutrient medium for supporting useful microorganisms; (4) collecting the solids from it by centrifugation and adding to animal feeds (Briggs et al. 1986).

16.3 Spent Hops

16.3.1 Chemical Composition

The hop, *Humulus lupulus*, is an agricultural crop essentially used for brewing. It is rich in bitter constituents (α-acids (humulones), β-acids (lupulones), soft and hard resins) and ethereal oils, which supply bittering and aroma components to beer. However, only 15% of the hops constituents end up in the beer, 85% will became spent hop material (Huige 2006). The lupulones, for example, are insoluble at the normal pH value of the wort and not isomerizes during boiling, being largely removed with the spent hops and the trub. The hop phenolic components (e.g., *p*-hydroxycoumaric, gallic, ferulic, protocatechinic, and caffeic acids; catechins, flavones, and anthocyanidines, among others) are also precipitated with proteins during wort boiling (Esslinger and Narziss 2005).

Spent hop is a material with high amounts of nitrogen free extract, fibres and proteins (Table 16.2). Crude fibre is constituted by several sugars (rhamnose, arabinose, xylose, mannose, galactose and glucose), among of which, glucose and xylose are the most abundant. Pectic sugars, uronic acid, rhamnose, arabinose and galactose account for 46% of the polysaccharides in spent hops (Oosterveld et al. 2002). Mono- and multifunctional aliphatic carboxylic acids in spent hops include oxalic, glucaric, gluconic, threonic, glyceric, glycolic, lactic and acetic acid (Fischer and Bipp 2005).

As previously mentioned, a fraction of the hop components end up in the trub, mainly when hop powdered, pellets or extracts are used in the brewing process. Hot trub consists of protein (40–70%), bitter substances (7–15%), other organic compounds, such as polyphenols, and mineral substances (20–30%); while the cold trub consists of protein (50%), polyphenols (15–25%) and carbohydrates of high molecular mass (20–30%) (Esslinger and Narziss 2005).

Table 16.2 Chemical composition of spent hops

	Reference [a]	
Component (% dry matter basis)	1	2
Protein	23.0	22.4
Lipid	4.5	nr
Ash	6.5	6.0
Crude fiber [b]	26.0	23.6
Nitrogen free extract	40.0	nr

[a] 1 From Huige 2006; 2 From Briggs et al. 1982
[b] cellulose, hemicellulose and lignin nr = non reported

16.3.2 Potential Uses of Spent Hops in Biotechnological Processes

In spite of the chemical composition rich in nitrogen, carbon and protein, spent hops have been few explored as substrate in biotechnological processes. Basically, the addition to spent grains is the most prevalent method for disposal of spent hops at the brewery site, but it compromises quality of the mixture, mainly if it is desired the spent grains use as animal feed. Unlike spent grains, the direct use of spent hops as feed supplement is not desirable due to the presence of bitter substances in this residue. Animals unwillingly eat bitter fodder and they are discouraged by sedative–hypnotic properties of 2-methyl-3-buten-2-ol, which is the product of bitter acid degradation. Therefore, to be used as feed supplement, it is firstly necessary to remove or degrade the spent hops bitter acids that can be made by selected fungi or yeasts like *Candida parapsilosis* (Huszcza and Bartmanska 2008, Huszcza et al. 2008).

When obtained separately from spent grains, an alternative frequently used for spent hops disposal is as mulch or as soil conditioner and fertilizer, due to the high nitrogen content (Table 16.2) (Huige 2006). Additionally, some alternative ways of utilizing spent hops in biotechnological processes have been proposed, which are listed below.

16.3.2.1 Source of Value-Added Compounds

Several compounds of industrial interest can be recovered from spent hops, such as flavours, saccharides and organic acids, which can be obtained after oxidation or hydrolysis of this material (Oosterveld et al. 2002, Fischer and Bipp 2005, Laufenberg et al. 2003, Vanderhaegen et al. 2003, Krishna et al. 1986). Among these compounds, the hop acids, particularly, have potential as natural antibacterial in distillery mashes for alcoholic fermentation, being a safe alternative to control bacteria in ethanol fermentations, able to efficiently replace antibiotics in ethanol production (Ruckle and Senn 2006).

Pectins, compounds widely used as ingredient for the food industry as gelling agent and thickening agent, represent a large part of the polysaccharides in spent hops and can be recovered from this material by acid extraction conditions. Spent hops pectins include homogalacturonans, and arabinogalactan-proteins with a protein part rich in cystein, threonin, serinin, alanin, and hydroxyprolin (Oosterveld et al. 2002). Residual hops resins can also be recovered by extraction with acetone, obtaining an unsaturated drying oil for paints (Huige 2006).

16.3.2.2 Use in Fermentative Processes

Trub addition to the pitching wort was found to increase yeast vitality and yield as well as fermentation performance of *Saccharomyces cerevisiae*, and the more trub was added, the greater the effect. The influence of hot trub on yeast vitality and fermentation performance is associated with several components of hot trub, e.g. lipids, zinc and particulate properties. Although some authors underline the nutritive importance of lipids and particularly of unsaturated long-chain fatty acids, others believe zinc might be the most effective component of trub (Kühbeck et al. 2007).

16.4 Surplus Yeast

16.4.1 Chemical Composition

The surplus yeast is another brewing by-product that merits considerable attention, due to the large quantity produced (is the second largest by-product from breweries) and rich chemical composition.

The most abundant element in yeast cells is carbon, which accounts for just under 50% of the dry weight. Other major elemental components are oxygen (30–35%), nitrogen (5%), hydrogen (5%) and phosphorus (1%). The most abundant classes of macromolecules are proteins and carbohydrates (Table 16.3). However, the precise composition of each class of macromolecules within a given cell varies as a function of physiological condition and phase in growth cycle (Briggs et al. 2004).

The protein content present several bound amino acids, including arginine, cystine, glycine, histidine, isoleucine, leucine, lysine, methionine, phenylalanine, threonine, tryptophan, tyrosine, and valine; leucine, lysine and tyrosine being the most

Table 16.3 Chemical composition of surplus yeast

	Reference [a]		
Components (% dry weight basis)	1	2	3
Protein	48	nr	50
Lipid	nr	1	nr
Ash	7	8	7
Crude fiber [b]	3	nr	nr
Carbohydrates	nr	36	42
Minerals in ash (%)			
Calcium	0.12	nr	nr
Chlorine	0.12	nr	nr
Iron	0.01	nr	nr
Magnesium	0.24	nr	nr
Phosphorus	1.43	nr	nr
Potassium	1.71	nr	nr
Sodium	0.09	nr	nr
Sulfur	0.38	nr	nr
Vitamins (mg/100 g)			
Niacin	nr	nr	50
Thiamin	nr	nr	15
Panthotenate	nr	nr	10
Riboflavin	nr	nr	7
Folic acid	nr	nr	4
Pyridoxine	nr	nr	3
Biotin	nr	nr	0.2

[a] 1 From Huige 2006; 2 From Lamoolphak et al. 2006; 3 From Lewis and Young 1995

[b] glucans, mannans, and polymeric hexosamines. nr = non reported

abundant. It is also rich in vitamins (biotin, choline, folic acid, niacin, pantothenic acid, riboflavin, thiamin, and vitamin B-6), mainly niacin (Lewis and Young 1995; Huige 2006).

The total mineral content of yeast is approximately 5–10% of the cell dry weight. This fraction comprises a multitude of elements, specially potassium and phosphorus. The composition of some of them is shown in Table 16.3. Besides the mentioned minerals, cobalt, copper, manganese, and selenium can also be found in lower extends (ppm) (Huige 2006).

16.4.2 Potential Uses of Surplus Yeast in Biotechnological Processes

When compared to spent grains, surplus yeast has a much higher content of protein, vitamins and amino acids; reason by which it has been currently utilized for animal feed and nutritional supplement after drying (Chae et al. 2001). However, the inclusion of yeast in food products is limited by the amount of nucleic acid, primarily ribonucleic acid (RNA); since in humans, RNA is metabolized to uric acid, which can lead to gout (Huige 2006). Due to the composition rich in protein, amino acids, minerals, and other compounds of interest, several attempts have been done aiming to reuse the surplus yeast in biotechnological processes.

16.4.2.1 Source of Value-Added Compounds

Several compounds of industrial interest can be isolated from brewer's yeast, such as enzymes, proteins, vitamins, amino acids, cytochromes, the purine components of DNA and RNA, among others (Huige 2006). Protein and amino acids, for example, can be recovered by employing various processes such as autolysis, plasmolysis in organic salt solution or non-polar organic solvent, acid or alkali catalyzed hydrolysis, enzymatic hydrolysis, or hydrothermal decomposition (Lamoolphak et al. 2006).

β-glucan, a hydrocolloid of large interest by the pharmaceutical and functional food industries, can also be extracted from brewer's yeast. This compound is of great interest because it has potential in improving the functional properties of food products, being used as a thickening, water-holding, or oil-binding agent, and emulsifying or foaming stabilizer (Romero and Gomez-Basauri 2003, Thammakiti et al. 2004). It has also, a potentially valuable addition to starch-based foods to restrict retrogradation of the starch (Satrapai and Suphantharika 2007). It is expected that β-glucan from brewer's yeast, with low cost of production, simple extraction technology, and potential infinite supply will dominate the market for the foreseeable future (Zekovic et al. 2005).

Yeast extract (a mixture of amino acids, peptides, nucleotides and other soluble components of yeast cells) is produced by the breaking down of yeast cells using endogenous or exogenous enzymes. Such compound is of particular interest for use in the food industry as a flavouring agent in soup, sauces, gravies, stews, snack food and canned food. Other application is as vitamin supplements in health food (Chae et al. 2001).

16.4.2.2 Use in Fermentative Processes

The use of brewer's yeast autolysate during fermentation of vegetable juices by *Lactobacillus acidophilus*, favourably affects the increase of the number of lactic-acid bacteria, reduction of fermentation time and enrichment of vegetable juices with amino acids, vitamins, minerals and antioxidants (Rakin et al. 2007). The use of yeast extract in microbiological media is also well known and largely used as source of nutrients (Chae et al. 2001).

References

Almeida C, Brányik T, Moradas-Ferreira P et al. (2003) Continuous production of pectinase by immobilized yeast cells on spent grains. J Biosci Bioeng 96:513–518

Bartolomé B, Gómez-Cordovés C, Sancho AI et al. (2003) Growth and release of hydroxycinnamic acids from brewer's spent grain by *Streptomyces avermitilis* CECT 3339. Enzyme Microb Technol 32:140–144

Bhumibhamon O (1978) Production of acid protease and carbohydrate degrading enzyme by *Aspergillus awamori*. Thai J Agr Sci 11:209–222

Bogar B, Szakacs G, Tengerdy RP et al. (2002) Production of α-amylase with *Aspergillus oryzae* on spent brewing grain by solid substrate fermentation. Appl Biochem Biotechnol 102–103: 453–461

Brányik T, Vicente A, Cruz JM et al. (2001) Spent grains – A new support for brewing yeast immobilisation. Biotechnol Lett 23:1073–1078

Briggs DE, Hough JS, Stevens R, Young TW (1982) Malting and Brewing Science, vol 2, Hopped Wort and Beer. Chapman and Hall, New York

Briggs DE, Hough JS, Stevens R, Young TW (1986) Malting and Brewing Science, vol 1, Malt and Sweet Wort, 2nd edn. Chapman and Hall, London

Briggs DE, Boulton CA, Brookes PA, Stevens R (2004) Brewing: Science and Practice, Woodhead Publishing, Cambridge

Carvalheiro F, Esteves MP, Parajó JC et al. (2004) Production of oligosaccharides by autohydrolysis of brewery's spent grain. Bioresource Technol 91:93–100

Carvalheiro F, Duarte LC, Lopes S et al. (2005) Evaluation of the detoxification of brewery's spent grain hydrolysate for xylitol production by *Debaryomyces hansenii* CCMI 941. Process Biochem 40:1215–1223

Chae HJ, Joo H, In M-J (2001) Utilization of brewer's yeast cells for the production of food-grade yeast extract. Part 1: Effects of different enzymatic treatments on solid and protein recovery and flavor characteristics. Bioresour Technol 76:253–258

Cruz JM, Moldes AB, Bustos G et al. (2007) Integral utilisation of barley husk for the production of food additives. J Sci Food Agr 87:1000–1008

Dragone G, Mussatto SI, Almeida e Silva JB (2007) High gravity brewing by continuous process using immobilised yeast: Effect of wort original gravity on fermentation performance. J I Brewing 113:391–398

Duarte LC, Carvalheiro F, Lopes S et al. (2004) Comparison of two posthydrolysis processes of brewery's spent grain autohydrolysis liquor to produce a pentose-containing culture medium. Appl Biochem Biotechnol 113–116:1041–1058

Duvnjak Z, Budimir A, Suskovic J (1983) Effect of spent grains from beer production on production of α-amylase by *Bacillus subtilis* 21+. Prehram Technol Rev 21:97–101

Esslinger HM, Narziss L (2005) Beer. Wiley-VCH Verlag GmbH & Co. KGaA, Weinheim

Fischer K, Bipp H-P (2005) Generation of organic acids and monosaccharides by hydrolytic and oxidative transformation of food processing residues. Bioresour Technol 96: 831–842

Francis F, Sabu A, Nampoothiri KM et al. (2003) Use of response surface methodology for optimizing process parameters for the production of α-amylase by *Aspergillus oryzae*. Biochem Eng J 15:107–115

Huige NJ (2006) Brewery by-products and effluents. In: Priest FG, Stewart GG (eds) Handbook of Brewing, 2nd edn. CRC Press, Boca Raton

Huszcza E, Bartmanska A (2008) The implication of yeast in debittering of spent hops. Enzyme Microb Technol 42:421–425

Huszcza E, Bartmanska A, Anioł M et al. (2008) Degradation of hop bitter acids by fungi. Waste Manage 28:1406–1410

Kabel MA, Carvalheiro F, Garrote G et al. (2002) Hydrothermally treated xylan rich by-products yield different classes of xylo-oligosaccharides. Carbohyd Polym 50:47–56

Kanauchi O, Mitsuyama K, Araki Y (2001). Development of a functional germinated barley foodstuff from brewers' spent grain for the treatment of ulcerative colitis. J Am Soc Brew Chem 59:59–62

Keukeleire DD (2000) Fundamentals of beer and hop chemistry. Quim Nova 23:108–112

Krishna G, Czerkawski JW, Breckenridge G (1986) Fermentation of various preparations of spent hops (*Humulus lupulus L.*) using the rumen simulation technique (Rusitec). Agr Wastes 17: 99–117

Kühbeck F, Müller M, Back W et al. (2007) Effect of hot trub and particle addition on fermentation performance of *Saccharomyces cerevisiae.* Enzyme Microb Technol 41:711–720

Kunze W (1996) Technology Brewing and Malting. VLB, Berlin

Lamoolphak W, Goto M, Sasaki M et al. (2006) Hydrothermal decomposition of yeast cells for production of proteins and amino acids. J Hazard Mater B 137:1643–1648

Laufenberg G, Kunz B, Nystroem M (2003) Transformation of vegetable waste into value added products: (A) the upgrading concept; (B) practical implementation. Bioresour Technol 87:167–198

Laws DRJ, Waites MJ (1986) Utilization of Spent Grains. Patent number 85-305109 169068, Brewing Research Foundation, UK. 24pp.

Lewis MJ, Young TW (1995) Brewing. Chapman and Hall, London

Mariani E (1953). Chromatographic examination of the amino acids of beer and spent grains. Brass Malt Belg 3:50–53

Mussatto SI, Roberto IC (2004) Alternatives for detoxification of dilute-acid lignocellulosic hydrolyzates for use in fermentative processes: A review. Bioresour Technol 93:1–10

Mussatto SI, Roberto IC (2006) Chemical characterization and liberation of pentose sugars from brewer's spent grain. J Chem Technol Biotechnol 81:268–274

Mussatto SI, Roberto IC (2008) Establishment of the optimum initial xylose concentration and nutritional supplementation of brewer's spent grain hydrolysate for xylitol production by *Candida guilliermondii*. Process Biochem 43:540–546

Mussatto SI, Dragone G, Roberto IC (2006a) Brewer's spent grain: Generation, characteristics and potential applications. J Cereal Sci 43:1–14

Mussatto SI, Dragone G, Rocha GJM et al. (2006b) Optimum operating conditions for brewer's spent grain soda pulping. Carbohyd Polym 64:22–28

Mussatto SI, Dragone G, Roberto IC (2007a) Ferulic and *p*-coumaric acids extraction by alkaline hydrolysis of brewer's spent grain. Ind Crop Prod 25:231–237

Mussatto SI, Fernandes M, Dragone G et al. (2007b) Brewer's spent grain as raw material for lactic acid production by *Lactobacillus delbrueckii*. Biotechnol Lett 29:1973–1976

Mussatto SI, Rocha GJM, Roberto IC (2008) Hydrogen peroxide bleaching of cellulose pulps obtained from brewer's spent grain. Cellulose 15:641–649

Okita H, Yamashita H, Yabuuchi S (1985) Production of microbial enzymes using brewers' spent grain. Hakko Kogaku Kaishi – J Ferment Technol 63:55–60
Oosterveld A, Voragen AGJ, Schols HA (2002) Characterization of hop pectins shows the presence of an arabinogalactan-protein. Carbohyd Polym 49:407–413
Palmqvist E, Hahn-Hägerdal B (2000) Fermentation of lignocellulosic hydrolysates. II. inhibitors and mechanisms of inhibition. Bioresour Technol 74:25–33
Plessas S, Trantallidi M, Bekatorou A et al. (2007) Immobilization of kefir and *Lactobacillus casei* on brewery spent grains for use in sourdough wheat bread making. Food Chem 105:187–194
Rakin M, Vukasinovic M, Siler-Marinkovic S et al. (2007) Contribution of lactic acid fermentation to improved nutritive quality vegetable juices enriched with brewer's yeast autolysate. Food Chem 100:599–602
Roberts RT (1976) Use of an extract of spent grains as an antifoaming agent in fermentors. J I Brewing 82:96–96
Romero R, Gomez-Basauri J (2003) Yeast and yeast products, past, present and future: From flavors to nutrition and health. In: Lyons TP, Jacques KA (eds) Nutritional Biotechnology in the Food and Feed Industries. Proceedings of Alltech's 19th International Symposium. Nottingham University Press, Loughborough, Leics, UK. (pp. 365–378)
Ruckle L, Senn T (2006) Hop acids can efficiently replace antibiotics in ethanol production. Int Sugar J 108:139–147
Satrapai S, Suphantharika M (2007) Influence of spent brewer's yeast β-glucan on gelatinization and retrogradation of rice starch. Carbohyd Polym 67:500–510
Silva JP, Sousa S, Rodrigues J et al. (2004) Adsorption of acid orange 7 dye in aqueous solutions by spent brewery grains. Sep Purif Technol 40:309–315
Sim TS, Oh JCS (1990) Spent brewery grains as substrate for the production of cellulases by *Trichoderma reesei* QM9414. J Ind Microbiol 5:153–158
Szponar B, Pawlik KJ, Gamian A et al. (2003) Protein fraction of barley spent grain as a new simple medium for growth and sporulation of soil Actinobacteria. Biotechnol Lett 25:1717–1721
Thammakiti S, Suphantharika M, Phaesuwan T et al. (2004) Preparation of spent brewer's yeast β-glucans for potential applications in the food industry. Int J Food Sci Technol 39:21–29
Vanderhaegen B, Neven H, Coghe S et al. (2003) Bioflavoring and beer refermentation. Appl Microbiol Biotechnol 62:140–150
Wang D, Sakoda A, Suzuki M (2001) Biological efficiency and nutritional value of *Pleurotus ostreatus* cultivated on spent beer grain. Bioresour Technol 78:293–300
Zekovic DB, Kwiatkowski S, Vrvic MM et al. (2005). Natural and modified (1 → 3)-beta-D-glucans in health promotion and disease alleviation. Crit Rev Biotechnol 25:205–230

Chapter 17
Biotechnological Potential of Cereal (Wheat and Rice) Straw and Bran Residues

Hongzhang Chen, Ye Yang and Jianxing Zhang

Contents

Abstract Cereal straw, one of the most abundant renewable lignocellulose resources which possess valuable components, has gradually become the research hot spot as a promising substitute for both the fossil fuel resource and petroleum-based industry with the increasing calling for bio-fuel and green chemistry. However, existing technologies of straw utilization unilaterally emphasize the primary utilization of the whole plant or some certain components, which not only result in low technical content of corresponding products but also fail to make full use of the lignocellulose resources. Based on the decades of research work, we find out that the bio-structural inhomogeneities of straw, both in the chemical composition and molecular structure between each part of straw, are the ultimate reasons why straw can not be utilized in a whole.Thus, the concept of fractionated conversion of straw emerges as the time requires. In this chapter, this innovative concept is explained

H. Chen (✉)
State Key Laboratory of Biochemical Engineering, Institute of Process Engineering, Chinese Academy of Sciences, Beijing 100190, PR China
e-mail: hzchen@home.ipe.ac.cn

P. Singh nee' Nigam, A. Pandey (eds.), *Biotechnology for Agro-Industrial Residues Utilisation*, DOI 10.1007/978-1-4020-9942-7_17,

in detail by taking the fractionated conversion of the corn straw, rice straw and rice husk as examples. Only through utilizing different parts of straw in the guidance of its structures and characteristics we can make full use of the straw resources.

Keywords Fractionated conversion · Biomass total utilization · Cereal straw · Steam explosion · Inhomogeneity · Biorefinery

17.1 Introduction

Along with the fast growing of world's population and great development of social economy, though the petroleum-based agriculture and industry have greatly boosted the development of both the society and people's living standard, in less than 400 years, we human beings have almost used up the fossil fuels such as coal, oil, and natural gas formed and accumulated through 2.5 billion years and caused a series of problems such as energy crisis, resource exhaustion and environment deterioration which not only threaten the living of ourselves but also confront the chemical industry with new challenge of human beings' sustainable development. Meanwhile, the traditional starch–based fermentation industry, however, cost as many as several billion tons of foodstuff per year, hence the limitation of both arable land and foodstuff set huge barrier to its development (Wyman 2007).

Energy shortage, food shortage and the call for developing biomass resource as chemical raw material in Green Chemistry starting from 1990s stimulate the countries worldwide to notice that the utilization of natural cellulose material like straw holds the strategic significance to their development. European Union has brought forward a short-term goal of alleviating the fossil energy dependence on each member country by cutting down 20–30% cost of bio-fuel production and actualizing 27–48% motor vehicle using bio-fuel (Council 2006).

The United States has invested several billion US dollars in the research work of substitutable energy and clean energy since the announcement of "Advanced Energy Initiative" (Milliken et al. 2007) in 2006, including non-grain crop based ethanol production, and reducing the technical costs of renewable energy such as wind energy, solar energy, geothermal energy and biomass energy. They have already invested 354 million US dollars, to reach the final goal of replacing 75% petroleum imported from Middle East by 2025 (President Bush 2006; Schell et al. 2008). Chinese government has pointed out clearly to strengthen the utilization of biomass resource and exploit the technology of biomass-based clean liquid fuel production in the "China's Agenda 21 —White Paper on China's Population, Environment, and Development in the 21st Century".

Cereal straw is one of the most abundant, annually renewable resources in the world. According to a valid data, there are as many as 2.9 billion tons of cereal straw produced per year all over the world, and only in China there is 0.7 billion tons cereal straw produced per year. However, such abundant resource has not attracted enough attentions and thus has not been utilized reasonably. In fact, cereal straw is the production of plants' photosynthesis, which is constituted by high percentage of macromolecule compounds such as cellulose, hemi-cellulose and lignin. Both cellulose and hemi-cellulose are polymers made up of fermentable sugar which can be fermented into chemical materials and liquid fuel such as ethanol, acetone, acetic acid, as well as be used as the fermenting materials of antibiotics, organic acid and enzyme after hydrolysis. Lignin, comprised of phenylpropane derivatives, can be further transformed to other chemicals used as the raw material in organic chemistry industry (Chen 2005).

17.2 The Utilization of Cereal Straw and Bran Residues

Cereal straw is one of the most abundant renewable resources in the world, long before human have been utilizing it in various different forms.

17.2.1 Firewood Fuel

As a traditional energy transforming mode, direct combustion is economic, low cost and easy to promote. However, according to the research, using natural straw as the fuel to combust directly has pretty low combustive efficiency because it is very difficult to be combusted completely. As a result, the heat loss usually varies from 30% to 90%, which not only wastes the resource but also causes serious pollution problem to the environment. Presently, relative research of straw as one kind of fuel is concentrated on the improvement of the low caloric value, and researches such as central gas supply of straw gasification, technology of methane or ethanol production from straw fermentation are on the way.

17.2.2 Paper Making

The utilization history of lignocellulose and fiber material has much to do with paper making industry which can date back to 3rd Century BC (Kamm et al. 2005; Kamm et al. 2007). Plant fiber is the raw material in the pulp and paper industry. Nowadays, wood fiber accounts for more than 90% of the world paper, nevertheless, to those countries which lack in wood fiber, fiber material such as straw is a good substitute. China is the largest straw pulp -producing country in the world, providing more than 75% of the world's non-wood pulp (Chen 2008). Pulp and paper industry all focus on the utilization of the cellulose component in the fiber material and removal of both the hemi-cellulose and lignin components which accounts for the formation of black liquor. This process not only wastes the hemi-cellulose and lignin components, but the removal step dramatically generates the increment of the cost, and the black liquor pollutes the nearby environment especially the water resources. Obviously, it is urgent to develop new technology for straw utilization in solving the problems mentioned above. Fortunately, there are researchers who bring out biotechnologies such as bio-pulping (Chen et al. 2002), bio-bleaching and enzymatic deinking (Chen 2005) to tackle the pollution problem.

17.2.3 Animal Feed

The development of animal husbandry depends on the sufficient supply of feedstuff. Countries like New Zealand, Australia have abundant meadow which can afford forage-based animal husbandry while other developed countries like America can afford grain-based animal husbandry. China is a country which has a large population and relatively less cultivatable land, in other words, neither forage nor grain is sustainable for the country's fast growing animal husbandry. The abundant straw resource is undoubtedly the most suitable choice considering the national conditions. In fact, the crop straw occupies about 30% of the livestock feedstuff. However, limited by the straw structure itself, natural straw has high percentage of lignin and ash components while only low percentage of raw protein, and poor palatability, which together lead to the low digestion and insufficient amount of nourishment. Ruminant such as cattle and sheep can digest about 40–50% of the straw on average while pig can only digest 3–25% and chicken are the worst, almost can't digest it at all (Liu 2006). Therefore, it is necessary to develop straw processing technology in order to decrease or even eliminate its limitation of digestion and nourishment. Now, the feed industry has made great progress due to the straw processing technologies like silage and straw ammoniation.

17.2.4 Returning Straw to Soil

It is a tradition in the agriculture history to use organic fertilizer, and the easiest and most traditional method is returning straw which mainly contains both straw mulching and straw incorporation to the soil. According to straw returning application, the soil can gain more organism and nutrient which can bring soil fertility betterment as well as adjustment of the physical properties, and finally optimize the environment of farmland. However, the problem is that returning natural straw to soil directly needs multiple kinds of microorganisms in the soil to function together to decompose which may take as long as several years. Obviously this kind of returning is so slow to the extent that the undecomposed straw can not function as fertilizer but also set some barriers to the new shoots. As a result, nowadays returning the manure of ruminant to soil and other compost are more commonly used.

17.2.5 Lignocellulose Chemical Industry

Up to now, it has been more than 200 years since the lignocellulose and cellulose fiber chemical industry was founded (Kamm et al. 2005). Utilizing the complex biorefinery technologies, through corresponding chemical processing, the lignocellulose can be decomposed into different fractions such as cellulose, hemi-cellulose, lignin, extraction and ash, from which the product line is based (Kamm et al. 2007). For example, to produce vanillin from lignin, gain carbohydrates from cellulose, and prepare furfural from hemi-cellulose. Wood has gained wide research and industry utilization because of its high content of homogeneous component of cellulose. As the substitute material of wood, straw catches more and more attention from worldwide researchers and becomes the research hotspot.

In general, state of the art of straw utilizing mainly focuses on the primary utilization of straw which are weak in foundation, integrity and system. The reports unilaterally emphasize on the utilization of the whole plant or some certain component and fail to reach the goal of making full use of the three main components (cellulose, hemi-cellulose and lignin) of the lignocellulose resources which accounts for the problems of the low technical content on the using of biomass resource and the poor quality of the corresponding products. In order to solve problems mentioned above fundamentally, the most crucial point is to realize that the bio-structural inhomogeneity of straw, that is to say, the differences, both in the chemical composition and structure between each part of straw, are the ultimate reason that straw can not be used in a whole.

The three main components—cellulose, hemi-cellulose and lignin crosslink tightly in the unpretreated straw, and due to their totally different chemical structures and properties, none of them can be utilized efficiently. Therefore, it is necessary to separate each component apart while maintaining the macromolecule's integrality as much as possible. Only through this processing, the different fractions can be utilized in an optimized way and fulfill their greatest value. Meanwhile, we should cultivate a clear sense that not only the cellulose component is a valuable resource, the other components are also potential resources rather than wastes which await the future industrial utilization. Secondly, the breakthrough of straw transformation technology calls for new development of corresponding process engineering theory on the solid phase complex materials.

Finally, it is more important to investigate the characteristics of straw utilization and technical bottle-neck, and build up systematic theory on the straw transformation. As for the utilization of biomass resource, applying mechanically existing knowledge is far from enough to solve the technical problems and it is necessary to build up more updating, comprehensive and systematic theory to guide the development of cellulose science.

In this chapter we will discuss about corn straw, rice straw and rice husk as examples to introduce the progress of research work done for the fractionated conversion of cereal straw in our laboratory.

17.3 Fractionated Conversion of Cereal Straw

17.3.1 Bringing Out the Concept of Fractionated Conversion Process

Since the 1970s, the transformation and utilization technology of biomass have made great progress, however, from the aspect of current utilization and development of biomass resource, there are still large amount of barriers and problems, one of which is the unilateral emphasis on the biological or thermal chemical technology without the sense of total utilization of natural solid phase organic material through fractionating and recycling. The single utilization technology not only wastes much of the biomass resource but also causes pollutions to the environment, therefore, looking for new effective technology of biomass utilization becomes the inevitable trend. Under such circumstances, the concept of fractionated conversion emerges as the time requires.

Based on the summarization of our many years research experience, we bring out the concept of fractionated conversion of cereal straw. Now the question arises what is this so-called fractionated conversion of cereal straw? In fact it is a process in which preparation of biomass products depends on the compositions and characteristics of the raw materials. The main route is biotransformation and thermal chemical transformation, biotechnology and other physical or chemical methods will be used as well if necessary in this conversion.

17.3.2 Flow Diagram of Ecological Industry Chain

Technical process scheme (shown in Fig. 17.1) is as follows:

(1) Cutting up straw into 5 cm or so pieces, and adding water to adjust the moisture to about 35% (Chen and Liu 2007).

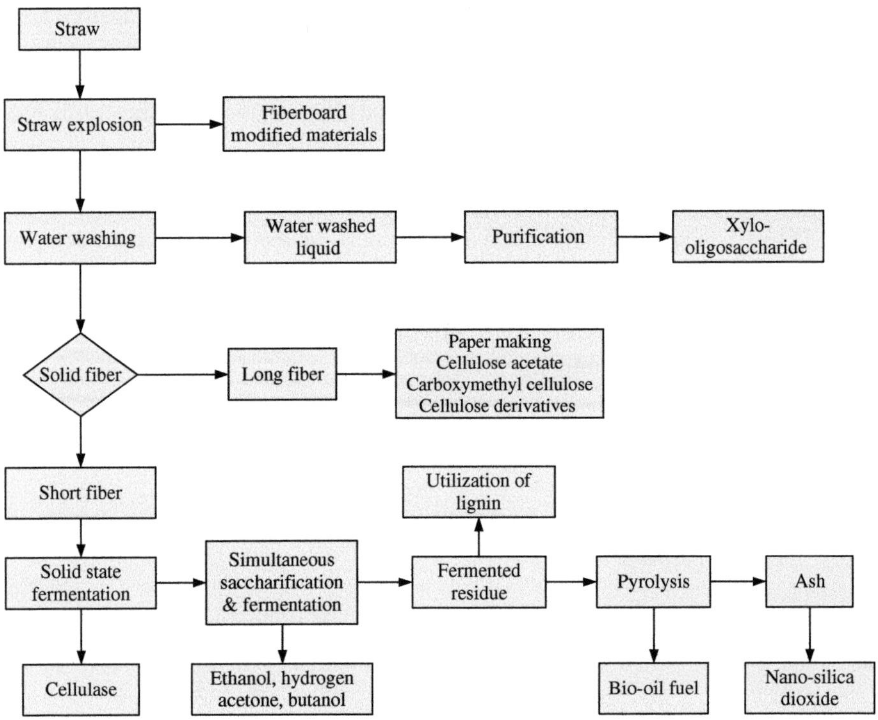

Fig. 17.1 Brief introduction of flow diagram of ecological industry chain

(2) Using steam explosion pre-treatment to activate the raw straw, and the condition of steam explosion is 1.5 MPa, maintaining for 3.5 min (Chen and Liu 2007).
(3) Fibers separated from the steam exploded (SE) production can produce medium and high density strawboard directly. SE production can also be washed with water. The water washed liquor can be used as stuff to produce xylo-oligosaccharide, and the solid fiber can be separated into two main parts: the long fiber part and the short fiber part through the carding equipment.
(4) Long fiber part can be used in ethanol autocatalytic pulping, preparation of cellulose acetate (Zhang and Chen 2007) and carboxymethyl cellulose.
(5) The separated short fiber part can be the appropriate substrate to produce cellulase in solid state fermentation (Xu et al. 2002) which can be used in the later fermentation of ethanol (Chen et al. 2007), hydrogen (Li and Chen 2007) and acetone butanol (Qureshi et al. 2007).
(6) Simultaneous saccharification and fermentation (SSF) (Han and Chen 2008) for fuel ethanol: use the short fiber as the substrate, adding cellulase obtained from the solid state fermentation and activated yeast (Chen and Jin 2006; Rudolf et al. 2005) to conduct SSF.
(7) Utilization of fermentation residues: they can be used in generating electricity or preparing bio-fuel (Luo et al. 2004).
(8) Utilization of straw ash: reclaim the valuable nano-silicon dioxide from ash.

17.3.3 Fractionated Conversion for Various Products

17.3.3.1 Fractionated Utilization of SE Corn Straw

Cut up the corn straw into 5 cm or so pieces and then add 30% (wt. %) water to the material and mix them up. Put them into a 0.5 m^3 steam explosion reactor, keep the steam pressure at 1.5 MP in the reactor and maintain for 3, 4, 5 min respectively, and then release the pressure in a quick shot. Then the steam exploded straw is extracted with 5 times water (wt. %) and both the water washed liquor and the solid material are collected. The sugar content of water washed liquor can be determined by HPLC. The solid material then is separated into long fiber part and short fiber part through the carding equipment. The utilization of each fraction are as follows: xylo-oligosaccharide can be distilled from the water washed liquor and then prepare levulinic acid; long fiber part can be used as paper pulping and the short fiber part can ferment ethanol and acetone butanol.

Content of the long fiber part of the SE products decreases along with the intensity of steam explosion. In the same steam pressure of 1.5 MPa, maintaining as long as 3 min, 4 min and 5 min, the contents of long fiber are 69.71% , 46.53% , 45.39%, respectively. And the weight proportion of long fiber and short fiber are 1:0.45, 1:1.15, 1:1.17, consequently.

Compared with the natural corn straw, the cellulose content of SE product can reach 40% or so which is higher than that of natural corn straw; the content of hemi-cellulose decreases from 30% to 20%. Under the same SE condition, the

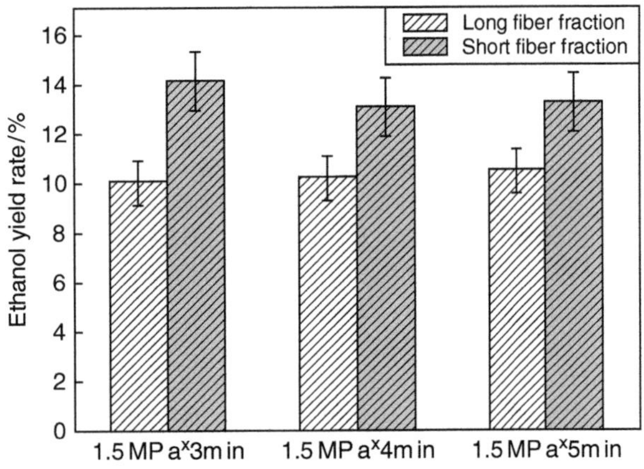

Fig. 17.2 The ethanol yield of fractionated corn straw

content of neutral washing component in long fiber part and short fiber part is different and the former one has a lower percentage; the contents of ash and lignin have less discrepancy.

The ethanol fermentation test using fractionated SE corn straw as substrate was conducted. In general, all ethanol yields (shown in Fig. 17.2) were above 10% and the short fiber part had the highest yield of above 14%.

17.3.3.2 Fractionated Conversion at Morphological Organic Level of Rice Straw

a. Inhomogeneity of rice straw structure
   In the angle of configuration, rice straw can be divided into four main different organs: root (seed root, adventitious root, branch root), stalk (node and internode), leaf (leaf blade and leaf sheath) and panicle (threshing panicle), and the straw roots (Jin and Chen 2006) are usually kept in the soil after harvest. In Fig. 17.3, it is clearly shown that the differences in chemical compositions result in the different transforming capacity of straw's morphological fractions; therefore the utilization of rice straw in a fractionated way is not only important but also necessary.
b. Inhomogeneity of capacity of ethanol fermentation of rice straw's morphological fractions

Using abundant agricultural cereal straw to prepare fuel ethanol in order to partly replace fossil fuel is a significant developmental trend. However, the structural inhomogeneity and large amount of non-fiber cells of straw directly lead to poor ethanol yield and high cost. Under such circumstances, investigation on the capacity of ethanol fermentation of rice straw's morphological fractions will turn out to be the pertinent basic guidance of appropriate transformation of fractionated rice straw. Under the same SSF condition, the different ethanol yields of leaf sheath, leaf blade, node, internode and panicle respectively are shown in the Fig. 17.4 (Jin 2007).

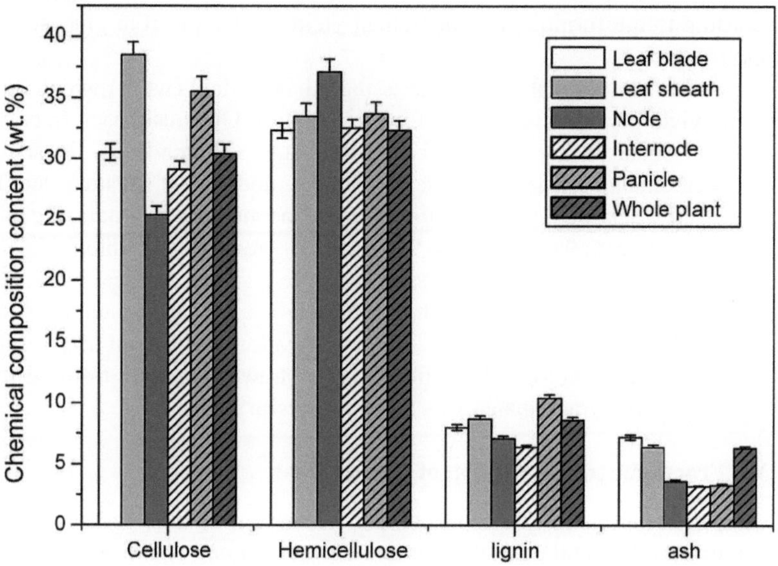

Fig. 17.3 Chemical composition of each morphological fraction of rice straw lower than 40 mesh

Cellulose component of the straw will firstly be hydrolyzed into glucose which then transformed to ethanol via microorganism fermentation, the biochemical reaction is as formula (17.1):

$$(C_6H_{10}O_5)_n \rightarrow n\,C_6H_{12}O_6 \rightarrow 2n\,C_2H_5OH + 2n\,CO_2 \tag{17.1}$$

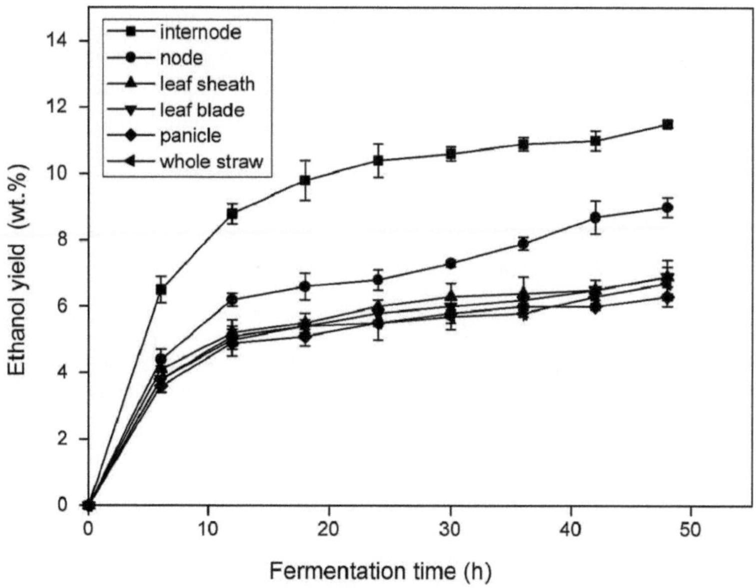

Fig. 17.4 Ethanol yield of each morphological fraction of rice straw (Jin 2007)

According to the formula, the theoretical yield of ethanol from glucose is 0.51 g-ethanol/g-glucose.

In Fig. 17.4, the vertical coordinate is the ethanol yield (wt.%) which denotes the ethanol yield per gram of substrate in the SSF test. Obviously each fraction has distinguished transforming capacity. Under the same SSF condition, the sequence of capacity of ethanol fermentation is: internode > node >leaf sheath > leaf blade > panicle, and when the fermenting time reaches as much as 48 hours, the ethanol yield can reach 11.5%, 9.0%, 6.9%, 6.9% and 6.3%, respectively. Internode has the highest ethanol yield which towers above that of leaf sheath and leaf blade 66.7%. This law has positive correlativity with the enzymatic capacity of each morphological fraction of rice straw basically. Via the investigation of different transformation capacity of each morphological fraction of rice straw, it is more testified that to utilize rice straw in a fractionated way is very necessary.

17.3.3.3 Fractionated Conversion of Cereal Husk

Due to the abundant published papers on the research using bran and chaff, we will mainly introduce the total utilization of rice husk based on our research work.

Rice husk is characteristic of small cumulus density, hard husk and poor degradation capacity which directly lead to the ignorance of its utilization. The main components of rice husk are cellulose, lignin and silicon derivatives which may vary with the different breeds and producing areas. The average contents are: raw fiber 35.5% ~ 45% (polycondensed pentose 16% ~ 22%), lignin 21% ~ 26%, ash 11.4% ~ 22%, and silicon dioxide 10% ~ 21%.

Based on the characteristics of each component of rice husk and our research work, we bring forward the new technical process of multilevel utilization of rice husk which includes classifying rice husk into rich in silicon part, hemi-cellulose part and short fiber part and then adopting corresponding feasible means to utilize each part. The detailed processing is as follows: the rich in silicon part can prepare nano-silicon dioxide directly, and the hemi-cellulose part can prepare furfural while the short fiber part can be hydrolyzed to ferment fuel ethanol, together these can realize the effective total utilization of rice husk.

1. Project scheme

Technical process scheme (shown in Fig. 17.5) is as follows:

(1) SE pre-treatment of rice husk.
(2) Mechanical carding of rice husk: using mechanical carding equipment to separate SE rice husk into 3 parts, hemicellulose, component rich in silicon and short fiber.
(3) Preparation of fuel ethanol in SSF: using cellulose enzymatic hydrolysis-fermentation coupling method to prepare fuel ethanol.
(4) Preparation of furfural: to prepare furfural either by catalyzed with solid superacid or in acid free autocatalysis.
(5) Preparation of nano-silicon dioxide with component rich in silicon.

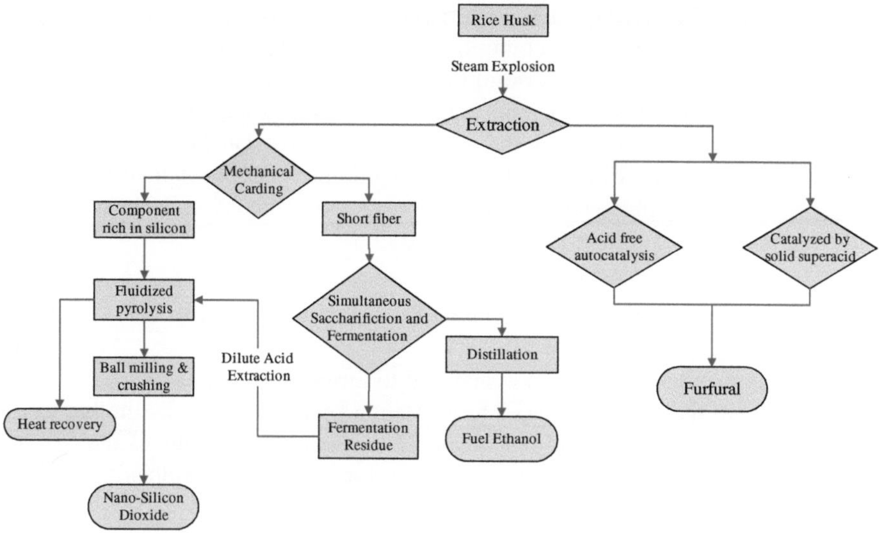

Fig. 17.5 Flow diagram of rice husk's utilization

2. Preparation of nano-silicon dioxide from SE rice husk's rich in silicon part

Compared with other biomass resources like wood, straw has much higher ash content, above 60% of which is silicon dioxide. Among all the straws, rice straw has the highest content of both the ash content and the percentage of silicon dioxide in the ash. In the plant, silicon dioxide usually appears as amorphous global nano-conglomeration (10 nm or so) which is comprised of $SiO_n(OH)_{4-2n}$. For a long time, the effect of ash on the cellulose enzymatic hydrolysis has always been ignored, however, according to the research into the correlation between the physicochemical characteristic of silicon and cellulose hydrolysis in the cell wall of rice straw, a new discovery emerges that the content of lignin and insoluble silicon has distinct synergistic relation which uncovers their synergistic effect on blocking the cellulose from hydrolysis.

Using steam explosion to pretreat straw, then to separate fiber tissue from parenchyma cell in carding to obtain most of the ash part, in this way the decrease of ash amount can reduce its inhibiting effect on hydrolysis as well as collect the silicon for better utilization. Research result shows that silicon dioxide still exists in the residues after enzymatic hydrolysis and fermentation, and SE pre-treatment causes no effect on the configuration of silicon dioxide which is still at amorphous state. Besides, the extremely low impure heavy metal ion content in the straw facilitates the preparation of high purity amorphous nano-silicon dioxide which not only cuts down the production cost but also enhances the added value of straw. Such obtained nano-silicon dioxide is white powder which has an average granular diameter of 50 nm, and the total impure ion content is below 5.5 mg/kg (< 10 ppm) which reaches the high purity level.

X-ray Diffraction (XRD) study on the structure of silicon dioxide in the rice husk under different SE pretreating conditions, comparing with crystalloid SiO_2 map, indicates some important discoveries. From the comparison map, except the position and intensity coincidence of several apices, the positions and intensities of other apices are distinct, in other words, the produced SiO_2 still existed in amorphous state which testifies that the SE pre-treatment and carding process have no effect on the SiO_2 structure. SiO_2 obtained after the acid bath and combustion is not nano-level SiO_2 yet, and it needs further disposal with Ball Mill.

3. Fermentation of SE rice husk's short fiber part to ethanol in SSF

Using plant cellulose as the raw material to ferment ethanol is an effective approach to solve the current ethanol fermentation problems, such as the high cost and limitations of raw material resources. The fermentation results of the SE rice husk are shown in the Table 17.1 clearly in which the steam explosion and later carding cause similar effects on the fermentation yield as enzymatic hydrolysis, and the short fiber parts obtained from the mechanical carding are more suitable to ferment ethanol and also gain much higher yield because of the increase of material accessibility to enzyme.

Table 17.1 Effects of steam explosion and carding on the ethanol fermentation yield

Carded SE material		Ethanol yield(%)
1.5 MPa 6 min*	Rich in silicon part	7.08
	Short fiber part	11.46
1.6 MPa 5 min*	Rich in silicon part	7.85
	Short fiber part	12.48

* the SE condition of raw material.

4. Preparation of fufural from SE rice husk hemi-cellulose part

Presently, the main industrial raw materials for industrial furfural production are corncob and bagasse. And the industrial process is to add sulfuric acid as catalyst, and then after two reactions the hemi-cellulose can be hydrolyzed to pentose which can be finally hydrolyzed into furfural. Steam explosion pre-treatment can decompose most of the hemi-cellulose into xylose and part of them can be further hydrolyzed into furfural, meanwhile there is some acid produced in the steam explosion process which can act as catalyst in the later furfural preparation.

A series of tests were conducted on the preparation of furfural from SE rice husk's hemi-cellulose especially on the preparation process from SE extraction liquor and optimize the reaction conditions. The results show great prospect in developing innovative process for the industrial furfural production and perfect the total utilization of rice husk.

17.4 Conclusion

Agricultural straws may vary in component contents with the breeds and environmental conditions, and even the same breed of straws can have distinct differences on the chemical compositions and structures on each morphological fraction. Therefore, unless utilize the straw resource by combining multiple subjects of knowledge, technology and industrial information, none of one single technology can realize the development and utilization of straw resource.

To optimize cellulose processing by refining biomass pre-treatment and utilizing the residues is an advisable method to reduce cost and enhance comprehensive utilization of straw. According to genetic engineering technology, we can design or cultivate specific microorganism which can ferment both C-5 and C-6 sugars and bear some inhibitions to promote current fermentation technology. And with the increasing knowledge of the relation between straw's cell wall structure and inner mechanism of enzymatic hydrolysis, the gene modified straw can be designed to meet the wanted needs(Energy 2006; Sedlak et al. 2003). The natural macromolecule and high polymer, and those high polymers obtained from microorganism fermentation, can also be the ideal raw material for the production of bio-based material and chemicals.

In a word, to establish the biomass transforming and utilizing technology network with the fractionated conversion of lignocellulose in the core can not only set new ecological balanced system but also can form new biorefining product chain, thereby initiate an innovative way to realize the farthest utilization of lignocellulose and ultimately substitute the petroleum product chain.

Acknowledgments This work was financially supported by National Basic Research Program of China (973 Program, 2004CB719700), National key technology R&D program (2007BAD39B01) and Knowledge Innovation Program of Chinese Academy of Sciences (KSCX1-YW-11A1).

Abbreviations

SE: Steam Exploded
SSF : Simultaneous saccharification and fermentation
HPLC: High Pressure Liquid Chromatography

References

Chen H (2005) Cellulose Biotechnology. Chemical Industry Press, Beijing
Chen H (2008) Biomass Science and Engineering. Chemical Industry Press, Beijing
Chen H, Jin S (2006) Effect of ethanol and yeast on cellulase activity and hydrolysis of crystalline cellulose. Enzyme Microb Technol 39:1430–1432
Chen H, Liu L (2007) Unpolluted fractionation of wheat straw by steam explosion and ethanol extraction. Bioresour Technol 98:666–676
Chen H, Xu F, Li Z (2002) Solid-state production of biopulp by Phanerochaete chrysosporium using steam-exploded wheat straw as substrate. Bioresour Technol 81:261–263

Chen H, Xu j, Li Z (2007) Temperature cycling to improve the ethanol production with solid state simultaneous saccharification and fermentation. Appl Biochem Microbiol 43:57–60

Council BRA (2006) Biofuels in the European Union-A Vision for 2030 and Beyond. http://www.managenergy.net/products/R1275.htm. Accessed 14 March 2006

Energy UDo (2006) Breaking the Biological Barriers to Cellulosic Ethanol: A Joint Research Agenda. www.doegenomestolife.org/biofuels. Accessed 7 December 2006

Han Y, Chen H (2008) Characterization of β-glucosidase from corn stover and its application in simultaneous saccharification and fermentation. Bioresour Technol 99:6081–6087

Jin SY (2007) Fractionation of Straw by Steam Explosion and Superfine Grinding and Preparation of Levulinic Acid by Solid Acid Catalyst [doctoral dissertation]. In. Institute of Process Engineering, Chinese Academy of Sciences, Beijing

Jin S, Chen H (2006) Structural properties and enzymatic hydrolysis of rice straw. Process Biochem 41:1261–1264

Kamm B, Gruber PR, Kamm M (2005) Biorefineries – Industrial Processes and Products. Wiley-VCH Verlag, Weinheim (Germany)

Kamm B, Gruber PR, Kamm M (2007) Biorefineries – Industrial Processes and Products (Translated to Chinese by Pingkai Ouyang). Chemical Industry Press, Beijing

Li D, Chen H (2007) Biological hydrogen production from steam-exploded straw by simultaneous saccharification and fermentation. Int J Hydrogen Energ 32:1742–1748

Liu L (2006) The Fractionation of Straw and Its High Value Conversion [doctoral dissertation]. In. Institute of Process Engineering, Chinese Academy of Sciences, Beijing

Luo Z, Wang S, Liao Y, Zhou J, Gu Y, Cen K (2004) Research on biomass fast pyrolysis for liquid fuel. Biomass Bioenerg 26:455–462

Milliken JA, Joseck F, Wang M, Yuzugullu E (2007) The advanced energy initiative. J Power Sour 172:121–131

President Bush GW (2006) State of the Union Address. Washington, DC. http://www.state.gov/r/pa/ei/pix/prsdnt/60171.htm Accessed 31 January 2006

Qureshi N, Saha BC, Cotta MA (2007) Butanol production from wheat straw hydrolysate using Clostridium beijerinckii. Bioprocess Biosyst Eng 30:419–427

Rudolf A, Alkasrawi M, Zacchi G, Lidén G (2005) A comparison between batch and fed-batch simultaneous saccharification and fermentation of steam pretreated spruce. Enzyme Microb Technol 37:195–204

Schell C, Riley C, Petersen GR (2008) Pathways for development of a biorenewables industry. Bioresour Technol 99:5160–5164

Sedlak M, Edenberg HJ, Ho NWY (2003) DNA microarray analysis of the expression of the genes encoding the major enzymes in ethanol production during glucose and xylose co-fermentation by metabolically engineered Saccharomyces yeast. Enzyme Microb Technol 33:19–28

Wyman CE (2007) What is (and is not) vital to advancing cellulosic ethanol. Trends Biotechnol 25:153–157

Xu F, Chen H, Li Z (2002) Effect of periodically dynamic changes of air on cellulase production in solid-state fermentation. Enzyme Microb Technol 30:45–48

Zhang J, Chen H (2007) Preparation of cellulose acetate from crop straw. J Chem Ind Eng (China) 58:2548–2553

Chapter 18
Palm Oil Industry Residues

Mynepalli K.C. Sridhar and Olugbenga O. AdeOluwa

Contents

Abstract Oil palm industry generates a large quantity of residues and wastes in the form of empty fruit bunch, palm kernel shells, trunk of the plant, fibre, leaves and others. When palm oil is extracted and processed, it also produces effluents with high organic matter, suspended matter and oil and grease. These wastes cause ecosystem degradation and affects health of the communities. This chapter describes the methods of managing or treating the residues and wastes as these are untapped resources. Some of the byproducts derivable from these wastes when appropriately managed are energy, mulch, compost or organic fertilizer from empty fruit bunches shells and sludges from effluent treatment. Empty fruit bunches and palm kernel shells were successfully converted into compost by enriching with goat manure or poultry manure and were useful in developing oil palm nurseries and other food crops. Biogas and electricity are generated from effluent management, and several biochemicals such as ethanol, fatty acids, waxes and others which could be obtained through application of biotechnology. Palm oil wastes contribute to Green House Gases (GHG) and conversion to energy is a good means of obtaining carbon credit facility for sustainable management. The spent materials are also used in cultivating mushrooms. These technologies find wider application in developing African and Asian countries where oil palm plantations are major economic resource.

M.K.C. Sridhar (✉)
Division of Environmental Health Sciences, Faculty of Public Health, University of Ibadan, Ibadan, Nigeria
e-mail: mkcsridhar@yahoo.com

P. Singh nee' Nigam, A. Pandey (eds.), *Biotechnology for Agro-Industrial Residues Utilisation*, DOI 10.1007/978-1-4020-9942-7_18,

Keywords Palm oil wastes · Energy · Mulch · Fertilizer · Palm kernel shells · Elaeis guineensis Jacq

18.1 The Oil Palm

Oil palm cultivation originated in West Africa, where oil palm trees were originally interplanted in traditional agricultural production systems along with other annual and perennial crops. Some 5,000 years ago it was said to have been domesticated in Nigeria. Production was for subsistence or trade within the region. By 1961 trade in palm oil had increased substantially, and Nigeria had 74 per cent of the world's plantations (Poku 2002). By the early 1970s monocrop plantations of oil palm had increased dramatically in Malaysia and Indonesia and by 2000, these two countries accounted for just over half of the world's total plantation area, and Nigeria accounted for just over 30 per cent.

Oil palm (*Elaeis guineensis* Jacq.) is the most productive oil producing plant in the world, with one hectare of oil palm producing between 10 and 35 tonnes of 'fresh fruit bunch (FFB)' per year which is the harvested part. FFB can be harvested generally after three years from planting. Largest amount of FFB is harvested about 10 years after planting. There are three main varieties of the West African oil palm: *Dura, Pisifera,* and *Tenera.* The *Tenera* palm produces the highest oil content of the three, but is actually a hybrid between the *Dura* and *Pisifera.* Over 40 per cent of an individual palm fruit and over 20 per cent of a fruit bunch from a typical *Tenera* palm can be extracted as palm oil. In its lifespan of 200 years, oil palm's economic life is 20–25 years. Of this, the plant spends its first 11–15 months in the nursery, first harvest comes up in 32–38 months from planting, and peak yield is 5–10 years from planting. Normally, Oil palm grows in the lowlands of the humid tropics, 15°N–15°S with evenly distributed rainfall of 1800–5000 mm/year. According to Hartley (1988), palm has a wide adaptability range of soils and low pH, but sensitive to high pH (> 7.5), and to stagnant water. Oil palms are cultivated on large plots of land with planting density of 128–148 plants per hectare. They are largely dependent on the planting materials, soil, temperature and climate.

Oil is obtained from the fleshy mesocarp of the fruit and the yield is at least 45–56% of fresh fruit bunch, while in kernel, it accounts for at least 40–50% (Kittikun et al. 2000). The theoretical potential yield for oil palm from both mesocarp and kernel has been estimated at 17 t oil ha^{-1} yr^{-1} (Corley 1983). At present, average plantation yields in favourable environments are about 6 t oil ha^{-1} yr^{-1} which is considerably greater than yields from any other oil crop (Mutert and Fairhurst 1999). Red palm oil is increasing in popularity, as it contains large quantities of carotenoids. Palm wine is made by tapping the male inflorescence of the oil palm and fermenting the resulting sap. Alternatively, entire trees can be felled and the meristem tapped, which is often done where old plantations are being replanted. Palm wine has been an important part of West African culture, and is still made today in large quantities, fetching good prices.

Usually large quantities of fertilizers are required to cultivate oil palm. Palm oil is used as a cooking oil, is the ingredient for most margarine, is the base for most liquid detergents, soaps, and shampoos, and in its most dense form, serves as the base for lipstick, waxes, and polishes. It is also used to reduce friction during the manufacture of steel. Oil palm plantations are known in most countries for their wide range of negative environmental impacts. These are land and water contamination, loss of land and resources to local inhabitant and loss of biodiversity brought about by cultivation and management.

18.2 Oil Palm Residues and Wastes

Processing of Fresh Fruit Bunches (FFB) for oil extraction involves several steps:

i. Steam sterilization of bunches (inactivates lipase enzymes and kills microorganisms that produce free fatty acids, reducing oil quality)
ii. Stripping fruit from bunches
iii. Crushing, digestion, and heating of the fruit
iv. Oil extraction from macerated fruit (hydraulic pressing)
v. Palm oil clarification
vi. Separating fibre from the endocarp
vii. Drying, grading, and cracking of the endocarp
viii. Separating the endocarp from the kernel
ix. Kernel drying and packing

Oil palm industry produces a wide variety of wastes in large quantities. Liquid wastes arise from oil extraction and processing. The solid wastes are the leaves, trunk, empty fruit bunches, seed shells and fibre from the mesocarp. Many byproducts are made from theese wastes with a view of conservation of resources and to safeguard environment (Figs. 18.1 and 18.2). Composting and biogas have been produced for many years. In recent years, some of the notable biotechnological ventures have been production of organic chemicals, ethanol and other organic solvents, fatty acids, polysaccharides, and polymers. In many developing countries, every part of oil palm is utilized either for housing and fencing, energy source or making local handicrafts (e.g. Raphia palm). Trunks of mature trees that have been cut to allow replanting should be recycled more effectively. They contain up to 1,000 Kg per hectare of potassium. Windrowing the trunks gives a slow breakdown of the material. This is the best way to release the nutrients. This chapter reviews some of the methods and practices of oil palm residue utilization and waste management.

18.2.1 Utilization of Empty Fruit Bunch (EFB)

EFB is a major waste product produced in oil palm plantations which need to be managed. Every 25 metric tonnes of full fruit bunches yield 16 metric tonnes of empty fruit branches. This can be returned to the fields. If applied at 6 metric tons per hectare per year, given average yields, eventually it can return half of the nutrients originally harvested in the bunches once it decomposes. In the past, EFB was often

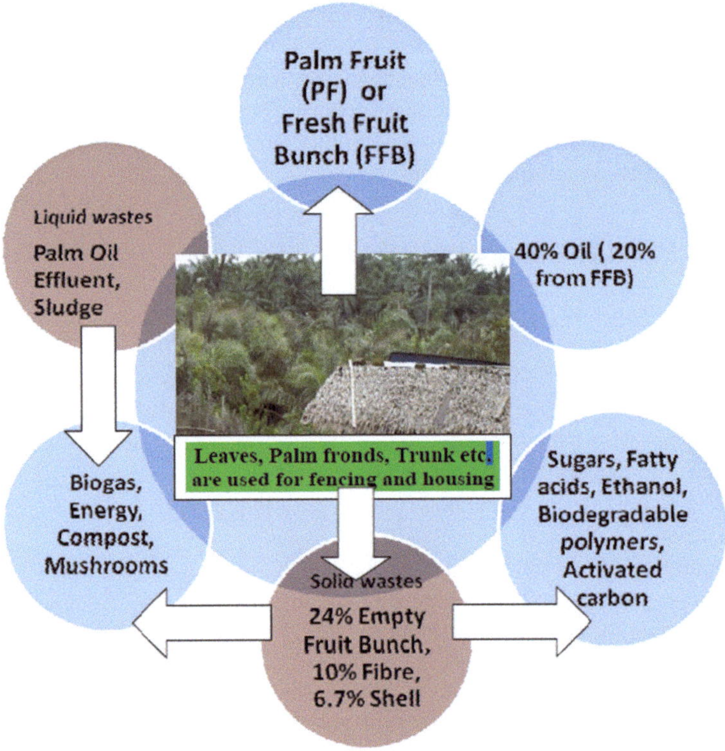

Fig. 18.1 Wastes, residues and byproducts from palm tree

used as fuel to generate steam at the palm oil mills and many farmers still practice it (Fig. 18.3). Currently, burning is prohibited globally to prevent air pollution (Ma et al. 1993) and alternate ways of utilization are sought. The ash, with a potassium content of about 30 per cent (Lim 2000) was used as fertilizer. The EFB is now used mainly as mulch. Empty fruit branches and trunks can be chipped and used as mulch if they are free of diseases (Hamdan et al. 1998). When used in circles around mature trees, the mulch can reduce herbicide requirements, but it may be a less efficient way to recycle nutrients than when spread over a larger area. Placed around young palms, EFB helps to control weeds, prevent erosion and maintain soil moisture. However, due to the current labour shortage, the transportation and distribution of EFB in the field is getting more expensive. Empty fruit bunches have also been used successfully in a joint Finnish/Indonesian project to produce paper. These efforts should be evaluated and, to the extent that they are appropriate, should be encouraged. Studies have shown the production of edible mushrooms is also a financially viable possibility (WWF 2005).

There is a growing interest in composting EFB, in order to add value, and also to reduce the volume to make application easier (Yusri et al. 1995, Thambirajah et al. 1995, Danmanhuri 1998). Aisueni and Omoti (1999) reported that oil palm

Fig. 18.2 (**1**) Palm tree; (**2**) Palm fruits for oil extraction of red palm oil and palm kernel oil; (**3**) Empty Fruit Bunch (EFB) used for energy or composting; (**4**) Pulp from palm fruits after extractopm pf po; ised as emergu spirce' (**5–7**) Raphia palm leaves and fronds used for fencing, roofing or weaving of mats

industry is one of the best sources of agricultural wastes that can be used as organic fertilizers. According to the authors, the palm industry in Nigeria produces 40 million metric tonnes FFB (fresh fruit bunch) annually of which about 16 million tonnes can be gotten for composting into organic manure. Apart from the EFB, palm kernel also has great potential for organic fertilizer production.

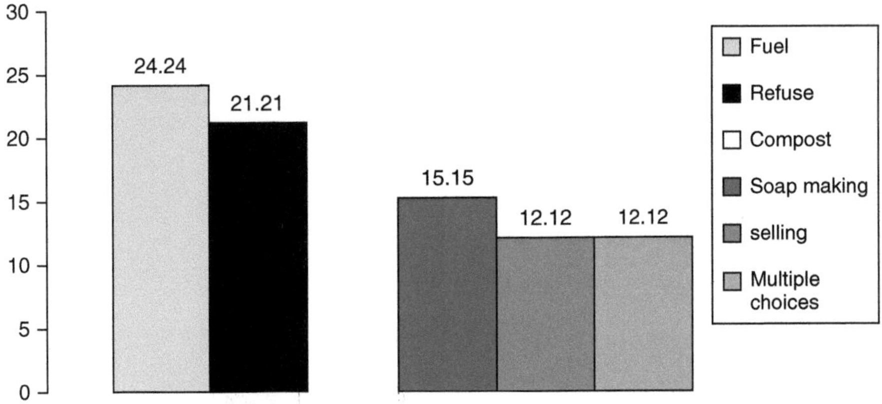

Fig. 18.3 Percentage of oil palm farmers in Nigerial managing the EFB

EFB is composed of 45–50 per cent cellulose and about 25–35 per cent of hemicellulose and lignin (Deraman 1993). It is fibrous, and the fibres stick together to form vascular bundles. Short, uniform, fibres can be obtained by mechanical cutting of the EFB. The fibrous nature of the material promotes aerobic conditions and therefore is considered suitable for the production of good quality compost (Suhaimi and Ong 2001). During the composting process, nitrogen is immobilized.

The natural decomposition of EFB in oil palm plantations after the fruits have been harvested was studied by Hamdan et al. (1998). The EFB was spread in the field as mulch on top of nylon net, at a rate of 30, 60 and 90 mt/ha/year. At each EFB application rate, spots were selected for nitrogen supplementation to meet a C: N ratio of 15, 30 and 60 (Control). Decomposition was estimated by the weight of EFB remaining in the nylon net. The EFB was found to be completely decomposed after 10 months of application.

Composting of EFB is being extended to farmers by the Department of Agriculture of Malaysia (Danmanhuri 1998). The initial method adopted was to mix the EFB with 20% chicken manure, heap it in 3 × 3 × 0.7 m boxes and cover with plastic. It took 11 to 12 months to mature. In 1995, this method was modified by first exposing big piles of EFB in the open for two months. Then mixed with 20% chicken manure and heaped in sheds measuring 12 × 36 × 3 m, the heap was mixed regularly at monthly intervals. It took about 4 months to reach maturity. Maturity was determined when the temperature of heap stabilized at 30°C and the pH reading 4.5 to 6.0. Agharan (1984) composted EFB of varying fractions (bunches cut into 2, 4, 8 portions, and whole bunch as control), mixed with chicken dung at a ratio of 3 parts of EFB to 1 part of chicken dung. The moisture content, pH, C: N ratio, Mg, Ca, and P content of finished compost showed a rise in every heap when the 1st week's value was compared with the 10th week's value. For the K content, heaps showed a difference in % value, no difference in nutrient levels between the different fractions of cuts studied. After the 10th week, the heap made up of bunches cut into 8 became crumbly and loose and the colour turned slightly black, while the others partly disintegrated with dark brown colour.

Danmanhuri (1998) reported that about 15 m^3 of compost could be turned in one hour, using a tractor equipped with a backhoe to mix the compost. Composted material was shredded at the end of composting period and was left for 1 to 2 weeks prior to packing. The final product contained (%) total nitrogen 3.3, phosphate 0.05, K 0.2, Ca 1.0 and Mg 0.2. Different organic N sources like the manure of goat, cattle and chicken, have also been evaluated as N additives for the composting of EFB (Thambirajah et al. 1995) adding 25 kg of manure per 90 kg shredded EFB reduced the C: N ratio. EFB composted with goat, cattle and chicken manure had a C:N ratio of 14:1, 18:1, and 12:1 respectively, after 60 days of composting, while the control without manure had C:N ratio of 1:24.

Zaharah and Lim (2002) investigated the decomposition of nutrients released from EFB component parts (spikelet, stalk and mixture of stalk and spikelet) with and without mineral N fertilizers under field condition. First part of the experiment was observed for a period of 9 months and in the second part, a field experiment was also conducted where 17 year-old palms were treated with different N and K inorganic fertilizer rates, with and without EFB mulching. It was observed that the EFB component parts decomposed at a significantly different rates in the order of stalk > mixture > spikelet Lignin, polyphenol, C and N contents in EFB showed good correlation with soil N dynamics. In mature oil palm, application of inorganic N, EFB and N + EFB significantly increased the yield. The soil chemical characteristics, pH value, organic C and exchangeable K were significantly improved with EFB application, making EFB a suitable ameliorant in improving soil quality for sustainable oil palm production.

The composition of EFB from Nigerian studies by AdeOluwa (2005) is shown in Table 18.1. Here, EFB was converted to compost and used the product to grow oil palm nurseries. EFB, EFB soaked in water (for 5 days) and cow dung were used in the development of a composting process. The EFB and soaked EFB were mixed with varying proportions of cow dung and composted aerobically. The mixing of

Table 18.1 Composition of unsoaked and soaked EFB, and Cow dung used as raw materials to produce compost

	Raw materials used for compost		
Composition	Unsoaked EFB	Soaked EFB	Cow dung
N, %	0.97	1.08	2.26
C, %	45.0	44.7	36.05
C: N	46.48	41.39	15.99
P, %	0.05	0.05	0.3
K, %	0.97	0.43	2.96
Ca, %	0.3	0.27	1.28
Mg, %	0.32	0.26	0.8
S, %	0.14	0.15	0.37
Zn, mg kg^{-1}	49	40	98
Cu, mg kg^{-1}	17	12	25
Mn, mg kg^{-1}	118	107	323
Fe, mg kg^{-1}	706	597	1842

Table 18.2 Composition of the compost obtained after 5 weeks from unsoaked and soaked EFB

Composition	60% EFB + 40% cow dung	60% SEFB + 40% cow dung
N, %	1.19	1.50
C, %	44.45	43.75
C: N ratio	37.89	30.45
P, %	0.11	0.13
K, %	1.19	1.10
Ca, %	0.47	0.44
Mg, %	0.40	0.37
S, %	0.21	0.21
Zn, mg kg^{-1}	99	85
Cu, mg kg^{-1},	19	18
Mn, mg kg^{-1},	144	153
Fe, mg kg^{-1}	650	644

EFB and cow dung in the ratio of 60:40 yielded good quality comost which has matured within 5 weeks (Table 18.2).

The unsoaked EFB treatments had higher nutrient elements than the soaked EFB, except for N, P and S. Lower nutrients in soaked pre-treatments should have been caused by the leaching of these nutrients in water used for soaking the EFB.

Application of EFB composts at the rate of 100 g/10 kg pot indicated significant growth of palm seedlings in th enursery. The results showed: girth (0.71 cm) and index leaf dry weight (0.33 g) of oil palm seedlings under the application of soaked oil palm EFB and cow dung (60:40) were significantly ($p < 0.05$) higher than those from the urea treatment (0.44 cm girth and 0.13 g index leaf dry weight) in the pre-nursery. Unsoaked EFB and cow dung also in the same ratio 60:40 resulted in the best seedling growth in the nursery ($p < 0.05$). Oil palm seedling growth in the nursery was not significantly influenced by the quantity of EFB combined with mineral fertilizer. However, combination of 21 g $plant^{-1}$ NPKMg (1:1:1:1) with unsoaked EFB: cow dung (60:40) significantly ($p < 0.05$) increased the number of leaves of oil palm seedling compared to other levels of mineral fertilizer investigated.

18.2.2 Utilization of Palm Kernel Wastes

Palm kernel oil (white palm oil) is obtained from the seed known as kernel or endosperm. When the oil has been extracted, the residue known as 'palm kernel cake' (PKC) is rich in carbohydrate (48 per cent) and protein (19 per cent) and is used as cattle feed (Onwueme and Sinha 1991). The ash contains large amounts of potassium. When the PKC is further solvent extracted to remove oil, it becomes 'palm kernel de-oil cake' which has little or no nutritional value (carbon 42.73%, nitrogen 0%, volatile matter 67.71% and calorific value 4031 Kcal/Kg) and is mostly used as fuel source in industry. As PKC is deficient in nitrogen, there is need to amend it with additional nitrogen if it has to be converted into compost. In the West African communities livestock wastes are readily available and thus become a rich source

Table 18.3 Composition of wastes used in the composting (Dry weight basis)

Waste	Moisture content, %	Carbon (C%)	Nitrogen (N %)	Phosphorus (P)	Potassium (K)
Palm Kernel Cake	58.92	96.21	2.88	0.60	0.19
Goat dung	29.79	75.94	3.62	0.51	0.18
Poultry droppings	49.11	78.83	2.83	3.29	0.16

for nitrogen amendment. The most common livestock which are reared around the farms and residences are goats and sheep, poultry and to some extent piggery. Cows are common in certain parts.

Kolade et al. (2006) developed a process of converting PKC into compost using poultry manure, and goat manure as supplements. Composting was carried out using combinations of PKC and poultry manure (3:1 ratio) and PKC and goat/sheep manure (3:1 ratio). The composting was carried out in locally made woven baskets which facilitate natural aeration of the composting material. The amount of waste in each basket was 10 Kg (7.5 Kg PKC and 2.5 Kg livestock waste) and kept in a green house at the University. A clean plastic sheet was spread under each basket to collect the leachates and putting back on to the composting material. The composting was carried out for six weeks. The composition of the raw materials and the finished product are given in Tables 18.3 and 18.4.

The quality of the finished composts was assessed by following the nutrient levels, C: N ratio, moisture level and texture. The compost quality was within the acceptable limits (Table 18.2). The compost made from PKC and goat manure however showed higher nitrogen and phosphorus levels which are needed for crop production in Nigeria. Increasing the nitrogen levels in the composts prepared from wastes is a challenge and supplementing with natural sources of nitrogen is more environmentally friendly than opting for mineral sources (Sridhar et al 2001). The results of growth experiments using the test crop *Amaranthus* spp indicate that the performances of composts prepared from PKC + poultry manure and PKC + goat manure were comparable when applied at 4 tons/Ha with those of Organo-mineral fertilizer or chemical fertilizer (NPK 15-15-15).

Referring to the paper by Kolade et al. (2006), Jim Vlahos, a businessman (jimvlahos@comcast.net) commented thus: "Creating charcoal from spent (or waste) palm kernels may have various forms of new benefit and value. Two interesting applications I can see for this charcoal is, first, a soil amendment instead of

Table 18.4 Composition of the composts prepared from palm kernel cake (Composting period of 6 weeks)

Treatment	pH value	Carbon (C %)	Nitrogen (N %)	Phosphorus (P%)	Potassium (K%)
PKC+ poultry manure (3:1 ratio by dry weight)	7.35	81.56	3.52	0.188	0.277
PKC + goat manure (3: 1 ratio by dry weight)	7.46	80.64	4.63	0.195	0.149

composting and, second, activated carbon. For the soil application, a great deal of information can be found at this website: http://www.css.cornell.edu/faculty/lehmann/index.htm. In addition to improving crop yields by putting the charcoal in the soil, the CO_2 that would have been released by burning the palm kernels in a boiler would be sequestered as charcoal in the soil. . . . palm kernel is very dense and can be used to create a quality activated carbon, which is used around the globe in filtering pollutants from water as well as power-plant gaseous emissions. These also represent significant global challenges. As a businessman, I am trying to identify an abundant, low-cost source of palm kernels that have already had the oils extracted from them. In addition to working to solve the global challenges I mentioned, your paper indicated that there is also a more specific problem around palm kernel waste" (Jim Vlahos 2008, Personal communication).

18.2.3 Miscellaneous Organic Chemicals

Kouichi Miura (2001) reported researches on the utilization of oil palm wastes in which a variety of chemicals were derived:

(a) molecular sieving carbon by the phrolysis of oil palm shell impregranted with $ZnCl_2$,
(b) treatment with acetosolv- and ethanosolv-processes to obtain pulp of high quality,
(c) recovery of 30% of small molecule fatty acids and pure cellulose were successfully recovered from oil palm shell through hot water and solvent extraction,.
(d) using a new catalyst, zirconia supporting FeOOH, developed recovery of valuable chemicals from the waste water containing ligno-cellulose, and
(e) developed new methods to produce a bacterial biodegradable plastic from Palm Oil Mill Effluent (POME).

18.2.4 Palm Oil Mill Effluent (POME) Management

Wastewater effluent from palm oil mills is a mixture of water, crushed shells, and fat residue resulting from initial processing of crude palm oil from fresh palm fruits, which must be crushed within 24 hours of harvest. For every one tonne of Crude Palm Oil (CPO) produced, 2.5 m^3 of POME is generated which is a universally accepted figure. POME generation rate per FFB will be 0.5 m^3-POME/t-FFB. In Malaysia some 10 million tonnes of POME is produced as against 9.9 million tones of solid waste from palm oil processing annually.

Igwe and Onyegbado (2007) reviewed the existing technologies that are found applicable in treating high BOD effluents such as POME.

1. **Tank digestion and facultative ponds:** In this system, raw effluent after oil trapping is pumped to a closed tank which has a retention time of about 20 days. The liquid is mixed by means of horizontal stirrers. The methane gas (CH_4)

generated is flared off into the atmosphere. Digested liquid is discharged into a holding pond before it is disposed on to land (Songeha 1974).

2. **Tank digestion and mechanical aeration:** This includes digestion tank and an aeration pond. Raw effluent after oil trapping is pumped to the acidification pond through a cooling tower and retained for one to two days. It is then mixed with an equal volume of liquid from the anaerobic digester before it is fed back to the digester. The hydraulic retention time of the digester is about twenty days. The digested liquid is discharged to an aeration pond with two floating aerators. The liquid is aerated for twenty days before it is discharged.
3. **Decanter and facultative ponds:** In a few mills, decanters are used to separate the fruits juice after pressing into liquid and solid phase, the liquid which is mainly oil is fed to the conventional clarification process. The water resulting from the clarification station is recycled. The solid is either disposed off on land or is dried in a rotary drier to about 10 per cent moisture and then used as fuel. The effluent thus consists of only the sterilizer condensate and waste from the hydro cyclone and is greatly reduced in volume to be treated in a series of ponds (Wood 1984).
4. **Anaerobic and facultative ponds:** This system consists of a series of ponds connected in series for different purposes. The effluent after oil trapping is retained in an acidification buffering pond for about two or three days and the resultant effluent is treated in an anaerobic pond with a hydraulic retention time of thirty to eighty days depending on the mills. This digested liquid is further treated in a series of facultative ponds before it is discharged. In some cases, part of the digested liquid is recycled to the acidification and buffering pond. The total hydraulic retention time of the system ranges from 75 to 120 days.
5. **Antra system:** The treatment consists of a combination of mechanical, chemical process and ponds. The raw effluent after oil trapping is separated into water and solid phases using a three- phase decanter. The oil is returned to the main line while the solid is dried in a rotary drier after the filter press. The water containing dissolved and suspended solids is treated with coagulants and flocculants to remove as much solids as possible before it is fed to an anaerobic digester which has a hydraulic retention time of about ten days. The digested liquid is further treated in an aeration tower and then oxidized (Sinnappa 1978).

Some of the more recent tested technologies to manage the POME are anaerobic digestion, anaerobic digestion with methane tapping, conversion of methane to electricity generation, aerobic lagooning, conventional waste treatment using biological methods e.g. activated sludge process, extended aerobic process and bio reactor system. Some of these practices are:

(A) **Up-flow anaerobic sludge fixed film (UASFF) reactor:** Zinatizadeh et al. (2006) studied the interactive effects of feed flow rate (Q_F) and up-flow velocity (V_{up}) on the performance of an 'Up-flow Anaerobic Sludge Fixed Film (UASFF) reactor treating palm oil mill effluent (POME). Long-term performance of the UASFF reactor was first examined with raw POME at a hydraulic loading rate (HRT) of 3 d and an influent COD concentration of 44,300 mg/l. Twelve dependent parameters were either directly measured or calculated as response. These parameters were total COD removal, soluble COD removal, effluent pH, effluent total volatile fatty acid, effluent bicarbonate alkalinity, effluent total suspended solids, CH_4 percentage in biogas, methane yield (Y_M), specific methanogenic activity (SMA), food-to-sludge ratio (F/M), sludge height in the UAS Bioreactor portion and solids retention time (SRT). The optimum conditions for POME treatment were found to be 2.45 l/d and 0.75 m/h for Q_F and V_{up}, respectively (corresponding to HRT of 1.5 d and recycle ratio of 23.4:1). The present study provides valuable information about interrelations of quality and process parameters at different values of the operating variables.

(B) **Pond system:** Chin et al. (1996) treated POME containing COD in the range of 45,000 to 65,000 mg/l, 5-day BOD 18,000 to 48,000 mg/l and oil and grease greater than 2,000 mg/l. The COD:N:P ratio is around 750:7.3:1. The nutrient content is low for aerobic treatment processes but sufficient for anaerobic treatment processes. They evaluated the treatment efficiency of a pond system consisting of 8 ponds in series treating 600 cu m/day of waste water. The pond system has been in operation since the mid 1980's. Effluent COD was 1,725 mg/l, BOD 610 mg/l, ammonia-N 115 mg/l, nitrate nitrogen 5 mg/l, TKN 200 mg/l and phosphate 60 mg/l. However, the effluent was not able to meet the discharge standard of 50 mg/l BOD.

(C) **Use of a marine yeast:** A study carried out by Oswal et al. (2002) showed that palm oil mill effluent (POME), from a factory site in India contained about $250,000\,\mathrm{mg\,l}^{(-1)}$ chemical oxygen demand (COD), $11,000\,\mathrm{mg}\,l^{(-1)}$ biochemical oxygen demand, $65\,\mathrm{mg\,l}^{(-1)}$ total dissolved solids and $9000\,\mathrm{mg\,l}^{(-1)}$ of chloroform-soluble material. Treatment of this effluent using *Yarrowia lipolytica* NCIM 3589, a marine hydrocarbon-degrading yeast isolated from Mumbai, India, gave a COD reduction of about 95% with a retention time of two days. Treatment with a chemical coagulant further reduced the COD and a consortium developed from garden soil clarified the effluent and adjusted the pH to between 6 and 7. The complete treatment reduced the COD content to $1500\,\mathrm{mg\,l}^{(-1)}$ which is a 99 per cent reduction from the original.

(D) **Anaerobic digestion:** In Lepar Hilier Palm Oil mill, Malaysia, at present there are two systems used in treating POME, firstly lagoon system second is the combination of lagoon and open digesting tanks. The Mills have 24,600 Ha of oil palm plantation which produces 54 t/Ha FFB. Annually, 259,890 tonnes of FFB is processed with the production of 300 to 4000 tonnes/month of CPO. The effluent produced has COD of 40,000 to 60,000 mg/l which has to be treated to the National standards of < 100 BOD before discharging into river. Anaerobic lagoon method ($30,000\,\mathrm{m}\,3 \times 4$ ponds) was adopted.

(E) **Activated sludge process:** The efficiency of the activated sludge process was evaluated by treating anaerobically digested and diluted raw POME obtained from Golden Hope Plantations, Malaysia (Vijayaraghavan et al. 2006). The treatment of POME was carried out at a fixed biomass concentration of

3900 ± 200 mg/L, whereas the corresponding sludge volume index was found to be around 105 ± 5 mL/g. The initial studies on the efficiency of the activated sludge reactor were carried out using diluted raw POME for varying the hydraulic retention time, viz: 18, 24, 30 and 36 h and influent COD concentration, viz: 1000, 2000, 3000, 4000 and 5000 mg/L, respectively. The results showed that at the end of 36 h of hydraulic retention time for the above influent COD, the COD removal efficiencies were found to be 83%, 72%, 64%, 54% and 42% whereas at 24 h hydraulic retention time they were 57%, 45%, 38%, 30% and 27%, respectively. The effectiveness of aerobic oxidation was also compared between anaerobically digested and diluted raw POME having corresponding CODs of 3908 and 3925 mg/L, for varying hydraulic retention time, viz: 18, 24, 30, 36, 42, 48, 54 and 60 h. The dissolved oxygen concentration and pH in the activated sludge reactor were found to be 1.8–2.2 mg/L and 7–8.5, respectively. The scum index was found to rise from 0.5% to 1.9% during the acclimatizing phase and biomass build-up phase.

18.2.5 Wastes to Energy and the Carbon Credit Schemes

In recent years, a novel closed tank methane recovery system and conversion to electricity generation was being planned for Lepar Hilier Palm Oil mill (Kyushu Institute of Technology 2003). Scientifically, it has been proved that anaerobic digestion of the bio-waste POME will produce biogas which is a mixture of CH_4 (65%) and CO_2(35%), which are Green House Gases (GHG). It has been shown that CH_4 emitted from this process has a good potential in the power generation using a gas engine. Biogas yield from POME is approximately 20 to 28 ($m^3 - CH_4/m^3$-Biogas). 1 m^3 biogas has the potential to generate approximately 1.8 kWh, which is about 25% power generation efficiency of its heat value. The power generated then can be supplied to Power Company by grid connection or can be consumed locally by the mill, small/medium scale industries or settlers in the residential areas. This proposal has the environmental component and carbon credit investment through reducing the GHG from reducing the emissions as well as saving from conventional power generation.

The Asian Palm Oil Company in Thailand is a producer of crude palm oil, with a production capacity of 40,000 tonnes of fresh palm per hour (total capacity is 50,000 tonnes). The company has 62 employees. The company converted its boilers to be able to utilize palm shells and fibre, rather than fuel oil, while the palm kernels are sold to manufacturers of low-grade palm oil. Using a 'Completely Stirred Tank Reactor "(CSTR) for energy generation With support from the Energy for Environment Foundation, the company also installed an anaerobic wastewater treatment system. The major piece of equipment is CSTR that uses microorganisms to digest the organic substances in the wastewater under anaerobic conditions. This process reduces BOD in the wastewater at the same time as it produces biogas. The biogas is then used to generate electricity. It was concluded from the experience that waste-to-energy practices require intensive technology development and high investment costs, and operating personnel must be qualified to handle the system (Waranusantikule 2003).

From the studies reviewed, POME management is a difficult task. A combination of treatment techniques should be used and there is need for adequate land availability. Biological treatment is the solution and the partially treated effluent and the sludge obtained makes an excellent soil additive. There is evidence that nitrogen may be lost if POME is stored for long periods in effluent ponds.

References

AdeOluwa O O (2005) Potential of Oil Palm Empty Fruit Bunch (FEB) as Fertilizer in Oil Palm (*Elaeis guineensis* jacq) Nurseries. A PhD thesis (supervised by G. O. Adeoye and M. K. C. Sridhar) submitted to the University of Ibadan, Ibadan, Nigeria, pp. 1–144

Agharan R (1984) E-Journal, www.geocities.com. Downloaded on 25th April, 2004

Aisueni N O and Omoti U (1999, Nov) The making of compost from empty oil palm bunch refuse. Books of Abstracts. Soil Science Society of Nigeria Conference, Benin. 21–25:48–49

Chin K K, Lee S W, and Mohammad H H (1996) A study of palm oil effluent treatment using a pond system. Water Sci. Tech. 34(11): 119–123

Corley R H V (1983) Potential productivity of tropical perennial crops. Experimental Agriculture 19: 217

Danmanhuri M A (1998). Hands-on experience in the production of empty fruit bunches (EFB) compost. Paper presented at the CETDEM Malaysian Organic Farming Seminar. Petaling, Jaya, Selangor, Malaysia, pp. 50–61. Dateline: 25/10/2002 09:45:09

Deraman M (1983) Carbon pellets prepared from fibres of oil palm empty fruit bunches:1.A Quantitative X-ray Diffraction Analysis PORIM Bull. Palm Oil Res. Inst. Malaysia 26: 70. Downloaded on 24th of April, 2004

Hamdan A B, Tarmizi A M, and Tayeh M D (1998) Empty fruit bunch mulching and nitrogen fertilizer amendment: The resultant effect of oil palm performance and soil properties. *PORIM Bull.* Palm Oil Res. Inst. Malaysia 37

Hartley C W (1988) The Oil Palm. Third Edition, Longmans Scientific and Technical London, pp. 1–761

Igwe J C and Onyegbado C C (2007) A Review of Palm Oil Mill Effluent (POME) Water Treatment. Global Journal of Environmental Research, 1(2): 54–62

Kittikun A H, Prasertsan P, Srisuwan G and Krause A (2000) Environmental Management of Palm Oil Mill. Internet Conference on Material Flow Analysis of Integrated Bio-Systems. March – October 2000. http://www.ias.unu.edu/proceedings/icibs/ic-mfa/kittikun/paper.html. Downloaded on 9th of March 2004.

Kolade O O, Coker A O, Sridhar M K C, and Adeoye G O (2006). Palm kernel waste management through composting and crop production. The Journal of Environmental Health Research, UK, 5(02): 81–85

Kouichi Miura (2001) Efficient Use of Oil Palm Wastes as Renewable Resource for Energy and Chemicals. Kyoto University, Japan, March

Kyushu Institute of Technology (2003) FELDA Lepar Hilir Palm Oil Mill, Biogas Project in Malaysia, Project Design Document. Matsushita Electric Industrial Co., Ltd.(National/Panasonic), EX CORPORATION, Ver. 2.1, April 2003, pp 1–43

Lim B (2000) The News Straits Times, Thursday. December 28. p. 6

Ma A N, Chean S A and Chow M C (1993) Current status of palm oil processing waste management. In: Waste Management in Malaysia: Current Status and Prospects for Bio Degradation, B.G. Yeoh et al., (Eds), pp. 111–136

Mutert E W and Fairhurst T H (1999) Oil palm – the great crop of South -East Asia: Potential, nutrition and management. Potash and Phosphate Institute http://www.fertilizer.org/ifa/ publicat/PDF/1999_biblio_45.pdf Downloaded on 24th of April, 2004

Onwueme I C and Sinha T D (1991) Field crop production in tropical Africa. CTA (The Technical Centre for Agricultural and Rural Co-operation), Ede, The Netherlands, pp. 1–480

Oswal N, Sarma P M, Zinjarde S S, Pant A (2002) Palm oil mill effluent treatment by a tropical marine yeast. Bioresource Technology, 85(1) 35–7

Poku K (2002) Small-scale palm oil processing in Africa. FAO Agric. Ser. Bull. 148

Sinnappa S (1978) Treatment Studies of Palm Oil Mill Wastes Water Pollution. Bangkok University Press Bangkok, Thailand, pp. 21–25.

Songeha S U (1974) National workshop on biogas technology. Proceedings in Kuala Lumpur. organized by the Malaysian working Group on Food Waste Materials, SIRM University of Malaysia, pp. 60–81

Sridhar M K C, Adeoye G O, and AdeOluwa O O (2001). Alternate nitrogen amendments for organic fertilizers, in optimizing nitrogen management in food and energy production and environmental protection: Proceedings of the 2nd International Nitrogen Conference on Science and Policy, The Scientific World, 1: 6

Suhaimi M and Ong H K (2001) Composting empty fruit bunches of oil palm. Malaysian Agricultural Research and Development Institute (MARDI), P.O. Box 12301, 50774 Kuala Lumpur, Malaysia. http://www.agnet.org/library/article/eb505a.html#eb505at4 2001-11-01

Thambirajah J J, Zulkifli M D, and Hashim M A (1995) Microbiological and biochemical changes during the composting of oil palm empty bunches; effect of nitrogen supplementation on the substrate. Bioresource Technology 52: 133–144

Vijayaraghavan K, Ahmad D, Mohd Aziz E B A (2006) Aerobic Treatment of Palm Oil Mill Effluent. Website accessed June 2008

Waranusantikule T (2003) Asia-Pacific Environmental Innovation Strategies (APEIS), Research on Innovative and Strategic Policy Options (RISPO) Good Practices Inventory: Waste-to-Energy Project by a Palm Oil Manufacturer. Thailand Environment Institute (TEI), tittaya@tei.or.th, 20 December 2003

Wood B J (1984) A Review on Current Methods of Dealing with Palm Oil Mill Effluent. Malaysian University Press, Kuala Lumpur, pp 19–32

WWF (2005) Agriculture and Environment: Palm Oil Better Management Practices: Reduce Fertiliser Use, Website updated November 13

Yusri A A, Rasol M, Mohammed O, Aziza Kume H T, and Hashimoto (1995) Biodegradation of oil palm empty bunch into compost by composite micro-organisms. Paper presented at the EU-ASEAN Conf. Combustion of solid and treated product. pp 8–9

Zaharah A R and Lim K C (2002) Malaysian Jour. of Soil Sci. (E-Journal). http://www.e-msss.com/mjss/abs04.htm. Down loaded on 2nd of May 2004

Zinatizadeh A A L, Mohamed A R, Abdullah A Z, Mashitah M D, Hasnain Isa M, and Najafpour G D (2006, October) Process modeling and analysis of palm oil mill effluent treatment in an up-flow anaerobic sludge fixed film bioreactor using response surface methodology (RSM). Water Research, 40(17): 3193–3208

Part IV
Enzymes Degrading Agro-Industrial Residues and Their Production

Chapter 19
Amylolytic Enzymes

Dhanya Gangadharan and Swetha Sivaramakrishnan

Contents

Abstract Amylolytic enzymes act on starch and related oligo- and polysaccharides. The recent wealth of information on the DNA sequence, structural analysis and catalytic mechanism led to the extensive research on starch hydrolyzing enzymes which led the concept of the alpha amylase family. Amylolytic enzymes are extensively used in starch liquefaction, paper industries, food, pharmaceutical and sugar industries which demands a specific hydrolysis profile. To fulfill the industrial requirements, the primary concern is the formulation of a simple indigenous and cost effective system for producing high titers of amylases. One alternative low cost and feasible production method is the use of agro-industrial residues as fermentation substrates. These residues represent one of the best reservoirs of fixed carbon in nature. Considerable research has been carried out in the effective utilization of these residues in large scale production of enzymes. This chapter gives a brief overview on the wide range of naturally occurring agricultural by products explored so far for the production of amylolytic enzymes.

D. Gangadharan (✉)
Biotechnology Division, National Institute for Interdisciplinary Science and Technology (NIIST), CSIR, Trivandrum-695 019, Kerala, India
e-mail: dhanya_28@yahoo.com

P. Singh nee' Nigam, A. Pandey (eds.), *Biotechnology for Agro-Industrial Residues Utilisation*, DOI 10.1007/978-1-4020-9942-7_19,

Keywords Amylases · Agro-industrial residues · Starch hydrolyzing enzymes · Oil cakes · Bran · Solid-state fermentation

19.1 Introduction

Starch represents one of the most abundant storage polysaccharides in nature and the most popular ingredient in food. It is composed exclusively of α-glucose units that are linked by α-1,4- or α-1,6-glycosidic bonds. The two high-molecular-weight components of starch are amylose (15–25%), a linear polymer consisting of α-1,4-linked glucopyranose residues with their molecular weights varying from hundreds to thousands, and amylopectin (75–85%), a branched polymer containing, in addition to α-1,4 glycosidic linkages, α-1,6-linked branch points occurring every 17–26 glucose units with molecular weight as high as 100 million (Bertoldo and Antranikian 2002). Because of the complex structure of starch, cells require an appropriate combination of enzymes for its depolymerization to oligosaccharides and smaller sugars, such as glucose and maltose. Amylolytic enzymes play an important role in the degradation of starch and are produced in bulk from microorganisms representing about 25–33% of the world enzyme market. Microbial enzymes are preferred for their stability over plant and animal enzymes which increases their spectrum of industrial applications. They also have the advantages of cost effectiveness, consistency, less time and space required for production and ease of process modification and optimization. The amylolytic enzymes find a wide spectrum of applications in food industry for production of glucose syrups, crystalline glucose, high fructose corn syrups, maltose syrups, reduction of viscosity of sugar syrups, reduction of haze formation in juices, solubilization and saccharification of starch for alcohol fermentation in brewing industries, retardation of staling in baking industry, in detergent industry used as an additive to remove starch based dirts, in paper industry for the reduction of viscosity of starch for appropriate coating of paper, in textile industry for warp sizing of textile fibers and in pharmaceutical industry they are used as a digestive aid (Sivaramakrishnan et al. 2006).

The vast research on whole genome sequencing and the accumulated protein sequence databases since the last two decades led to the study of a full range of starch hydrolyzing enzymes. The homology among alpha amylases from different origins was first studied by (Friedberg 1983). The detailed study of the amylolytic enzymes proved the existence of four highly conserved regions in eleven different α-amylases which is related to the catalytic and substrate binding sites (Nakajima et al. 1986). Thus the structural similarity and common catalytic mechanism among most of the amylases led to the concept for one enzyme family, 'the alpha amylase family'. The family included enzymes acting on α-glucosidase linkage to produce α-anomeric mono and oligosaccharides or form α-glucosidic linkages by transglycosylations, they possess four conserved regions containing the catalytic (Asp-206, Glu-230 and Asp-297) and substrate binding sites and a $(\beta/\alpha)_8$ or TIM barrel catalytic domain (Kuriki and Imanaka 1999). Thus the amylolytic and related enzymes have been classified into the families of glycoside hydrolases (GHs) and almost one hundred GH families have been reported. Amylolytic enzymes of microbial origin are divided into exo-acting, endo-acting, debranching and cyclodextrin producing enzymes.

19.2 Amylolytic Enzymes

19.2.1 Exo Acting Amylases – Glucoamylases and β-Amylases

Glucoamylases (1,4-α-D-glucan glucohydrolase, EC 3.2.1.3) catalyse hydrolysis of α-1,4 and α-1,6 glucosidic linkages to release β-D-glucose from the non-reducing ends of starch and related poly- and oligosaccharides. They have widely been reported to occur in a large number of microbes, including bacteria, yeast and fungi. Filamentous fungi, however, constitute the major source among all microorganisms and strains of genera *Aspergillus* and *Rhizopus* are mainly used for commercial production (Pandey 1995). β-amylases are known to be produced only by plants and certain bacteria mostly by several species of the genus *Bacillus*, including *B. polymyxa*, *B. cereus*, *B. megaterium*, and also by *Clostridium thermosulfurogenes* (Selvakumar et al. 1998). They hydrolyze α-1,4 bonds but cannot bypass α-1,6 linkages in amylopectin and glycogen and they produce maltose from amylose and maltose and a β-limit dextrin from amylopectin and glycogen.

19.2.2 Endo Acting Amylases

α-amylases (E.C. 3.2.1.1.) hydrolyse α-1,4 bonds and bypass α-1,6 linkages in amylopectin and glycogen. In spite of the wide distribution of amylases in microbes, animals and plants, microbial sources, namely fungal and bacterial amylases are preferred in industries. Among bacteria, *Bacillus* sp. is widely used for thermostable α-amylase production while fungi belonging to the genus *Aspergillus* are most common (Sivaramakrishnan et al. 2006). They are classified in two categories depending on the extent to which they hydrolyze starch. Liquefying α-amylases hydrolyze 30 to 40% of starch and saccharifying α-amylases hydrolyze 50 to 60%.

19.2.3 Debranching Amylases

Isoamylases and pullulanases are debranching enzymes that hydrolyze only α-1,6 linkages. On the basis of substrate specificity and product pattern, pullulanase (pullulan α-glucano-hydrolase; EC 3.2.1.41) have been classified into two groups: type I and type II. As they hydrolyze the α-glucosidase-resistant α-1,6 linkages in dextrins, they improve the starch saccharification rate and yield when used in combination with α-glucosidases. Many mesophilic (*Aerobacter aerogenes*, *B. macerans*, *B. acidopullulyticus* and *Bacillus* sp), thermophilic and hyperthermophilic bacteria and archae (*B.stearothermophilus*, *Clostridium thermosulfurogenes, Pyrococcus* and *Thermococcus* genus) have been reported to produce pullulanase (Gomes et al. 2003, Kunamneni and Singh 2006).

19.2.4 Cyclodextrinases

Cyclodextrin glycosyltransferase (α-1,4-D-glucan,α-4-D-(α − 1,4-D-glucano)-transferase, EC 2.4.1.19) produces a series of non-reducing cyclic dextrins (α-, β- and γ-cyclodextrins) from starch, amylose, and other polysaccharides. α-, β- and γ-cyclodextrins contain six, seven and eight glucose units, respectively, that are linked by α-1,4-bonds. *Thermococcus* sp., *B. coagulans, C. thermohydrosulfuricum* 39E, *B. sphaericus and* alkalophilic *Bacillus* sp are the most reported microorganisms producing these enzymes.

19.3 Production of Amylolytic Enzymes – Effective Utilisation of Agro Residues

Commercial production of enzymes is generally carried out by submerged (SmF) and solid state fermentation (SSF). The physico-chemical and nutritional requirements are unique for a particular microorganism. The composition and concentration of media and fermentation conditions greatly affect the growth and production of extracellular enzymes from microorganisms. The production of amylolytic enzymes in submerged fermentation employing synthetic media have been largely exploited (Tigue et al. 1995 and Hamilton et al. 1999) but is limited by the high cost of production. In the case of Smf the abundance of water gives more control of environmental factors such as temperature, oxygen concentration and pH and also provides ease in handling. However few reports have suggested agro residues as an alternative for synthetic basal media for the production amylase (Crueger and Crueger 2000, Hernandez et al. 2006, Gangadharan et al. 2008). Large scale production of enzymes would require formulation of a cost effective media and the commercial success of amylases is linked to the utilization of starchy biomass as an industrial raw material. The most inexpensive and highly energy rich substrates for fermentation is represented by agro-industrial residues.

The utilization of agro residues for the production of enzymes has gained renewed interest from researchers for the use of SSF as it solves solid waste disposal problem and also produce lesser waste water (Pandey 2003). Initially SSF was considered to be suitable for fungi and yeast considering the low water activity but there has been continous exploitation of bacterial cultures (Nampoothiri and Pandey 1996, Gangadharan et al. 2006 and Selvakumar 1999). Several naturally occurring agricultural by products such as wheat bran, coconut oil cake, groundnut oil cake, rice bran, wheat and paddy straw, sugar beet pulp, fruit pulps and peels, corn cobs, saw dust, maize bran, rice husk, soy hull, sago hampas, grape marc, coconut coir pith, banana waste, tea waste, cassava waste, aspen pulp, sweet sorghum pulp, apple pomace, peanut meal, cassava flour, wheat flour, corn flour, steamed rice, steam pre-treated willow, starch etc. could be used in one or the other industrial bioprocess for the production of value added products through SSF (Pandey et al. 2001).

Among the above substrates amylolytic enzyme production have been carried out mainly with wheat bran, rice bran, rice husk, oil cakes, tea waste, cassava, cassava bagasse, sugarcane bagasse (Mulimani et al. 2000). Banana waste, corn flour, saw-dust, soybean meal, sweet potato, potato, rice hull, sugar beet pulp have also been tried by some of the researchers. The utilization of agro residues such as wheat bran, molasses bran, maize meal, millet cereal, wheat flakes, barley bran, crushed maize, corncobs and crushed wheat have been exploited for the production alpha amylase by thermophilic fungus *Thermomyces lanuginosus* under solid state fermentation. Among the amylolytic enzymes commercial production of alpha amylase and glucoamylase utilizing agro residues are well studied and production of β amylases and pullulanase have been studied to a some extent. There are limited reports on the production of cyclodextrinases and research on the utilization of agro residues are yet to be explored in detail Thus the agro residues popularly employed for amylolytic enzyme production can be broadly classified as cereal brans, oil cakes and other starchy and non starchy substrates.

19.3.1 Cereals and Cereal Bran

Cereals are the fruits of cultivated grasses belonging to the genus *Gramineae.* The main cereals grown are wheat, rice, corn, barley, oats, sorghum and millet. Cereal grains contain 60–70% starch (Dendy and Dobraszczyk 2001). They have similar structure which include a hull (husks) and a kernel (caryopsis) and the kernel contains three components – bran, germ and endosperm. The bran is separated from cleaned and scoured cereals during milling. Among different agricultural by-products evaluated, wheat bran was found to be the best basal and standardized medium for optimal production of alpha amylase (Haq et al. 2003, Baysal et al. 2003, Balkan and Ertan 2007, Sivaramakrishnan et al. 2007). The strains of *Bacillus sp.* AS-1 and *Aspergillus sp.* AS-2 colonized well on the wheat bran based solid media and exhibited high production of α amylase and glucoamylase (Soni et al. 2003).

Table 19.1 Chemical composition of wheat bran and rice bran

Content	Wheat-bran (%)	Rice bran (%)
Moisture	6.4	12
Protein	16.4	16
Fat	6.8	22
Ash	6.5	10
Total dietary fibre	44.5	25
Starch	11.1	10–20

The physico- chemical properties of wheat bran (soft white) and rice bran are tabulated (Table 19.1). Corn gluten meal (CGM), a by-product of corn wet milling which are traditionally used for animal feed was found to be a promising substrate for the production of α amylase by *B. amyloliquefaciens* due to its high content of proteins ($\geq$ 60%), vitamin and other minerals (Tanyildizi et al. 2007). Agricultural raw starches such as pearl millet, rice, gram, hordium, corn and wheat starches at 1% levels were tested for the production of alpha amylase by *B. licheniformis* (Haq et al. 2005). Barley and oat brans are chiefly composed of beta glucans.Comparatively higher production was found in the case of pearl millet, which is represented by 56–65% starch, 20–22% amylose, free sugars ranging from 2.6–2.7% and a total protein content of 8–19% (Dendy and Dobraszczyk 2001). Spent brewing grain was found to be a good substrate for the production of α-amylase by *A.oryzae* under solid-state fermentation (Francis et al. 2003). Spent grain is the by product of breweries left after the grain (barley, corn, wheat, rice, and other grains) is fermented and the alcoholic solution drawn off. It is normally wet, with 80 to 85% moisture content and relatively high protein content (27–30%).

19.3.2 Oil Cakes

It is the solid residue that are usually extracted from various types of oily seeds like soya bean, pea nuts, linseed, cotton seed, cotton seed and sunflower by being pressed and removing the oil. They are valued for being rich in minerals and protein. They are rich in fibre and have high concentration of non-starch polysaccharides (NSP). Their chemical composition varies due to the differences in the extraction methods of oil. They are obtained by extraction of oil by means of a solvent from the expeller pressed oil cake. The meal may also be obtained directly from seeds after a preliminary treatment. The expeller pressed oil cake used for extraction are obtained by pressing clean and sound seeds. The meal are subjected to heat and steam treatment under controlled and regulated conditions so as to prevent denaturation of the protein and removal of traces of solvent. The material are in the form of either flakes or powder and are free from harmful constituents and castor cake or husk, rancidity, adulterants, insects or fungus infestation and from musty odour. The moisture content, crude protein and crude fibre (weight percent) of ground nut, cotton seed, linseed, mustard, sesame, coconut and safflower oil cakes have been tabulated (Table 19.2). Soya bean are also included under oil seeds and they are composed of oil and protein accounting to 60%, 35% carbohydrate and 5% ash.

The production of a thermostable pullulanase of *Thermoactinomyces thalpophilus* was studied in shake-flask cultures. Maximum production of pullulanase was obtained with 5% (w/v) soybean meal, 2% (w/v) yam starch (Odibo and Ob 1990). Oil cakes such as coconut oil cake (COC), sesame oil cake (SOC), groundnut oil cake (GOC), palm kernel cake (PKC) and olive oil cake (OOC) were screened to be used as substrate for the alpha amylase production by *A. oryzae* and they were also compared with wheat bran (WB). It was found that GOC and its combination with WB (1:1) resulted in higher enzyme titres (Ramachandran et al. 2004). Arasaratnam et al. have reported glucoamylase production by *A. niger* using rice bran and paddy husk as alternative substrates against wheat bran. Paddy husk was reported to enhance the nutrient utilization when mixed with the substrates like rice bran, corn flakes, soya flour and soy meal powder by *A. niger* CFTRI 1105 during SSF thereby increasing glucoamylase production (Arasaratnam et al. 2001).

19.3.3 Other Starchy and Non Starchy Substrates

Cassava (*Manihot esculenta* Crantz) is a root crop of tropical American origin, and the fourth most important staple crop in the tropics. Its starchy roots produce more raw starch per unit of land than any other staple crop. It is grown almost exclusively in arid and semiarid tropics, where it accounts for approximately 10 percent of the total caloric value of staple crops. Cassava starch is composed of unbranched amylose (20±5%) and branched amylopectin (20±5%. Cassava fibrous residue (CFR) contains about 10–15% crude fibre, 55–65% starch and very low ash content (1–1.2%) (dry weight basis) (Jyothi et al. 2005). Because of its low ash content, CFR could offer numerous advantages in comparison to other crop residues such as rice straw and wheat straw, which have 17.5% and 11.0% ash contents, respectively, for uses in bioconversion processes using microbial cultures (Pandey et al. 2000a, Pandey et al. 2000b). A study was carried out to investigate the α-amylase production by *B. subtilis* strain CM3 in SSF using CFR as the substrate (Swain and Ray 2007). Gonzalez et al. determined the optimal nutritional and operative conditions for amylolytic enzymes (α amylase and glucoamylase) production by *S. fibuligera* DSM-70554, cultured with cassava starch as the sole carbon source under different fermentation strategies, in order to improve cassava starch utilization. They described 97% degradation of cassava starch with a remaining 3% likely related to limit dextrins when grown under batch culture mode (Gonzalez et al. 2008).

Molasses, a by-product of sugar industry, is one of the cheapest sources of carbohydrates. Besides a large amount of sugar – 50% (sucrose 33.5%, invert sugar 21.2%), molasses contain nitrogenous substances (0.4–1.5%), vitamins such as thiamine (830μg per 100 g dry weight), pyridoxine (650μg per 100 g), folic acid (3.8μg per 100 g), biotin (120μg per 100 g), pantothenic acid (2140μg per 100 g), and trace elements (CaO 0.1–1.1%; MgO 0.03–0.1%; K_2O 2.6–5.0%) (Pandey 2003).

Table 19.2 Chemical composition of oil cakes

Oil cake	Moisture (%)	Crude protein (%)	Crude fat(%)	Crude fibre (%)
Groundnut	10	51	1.0	10
Cotton seed	8.0	40	8.0	10
Linseed	10	29	8.0	10
Mustard	10	35	8.0	9.0
Sesame	10	37	8.0	7.0
Coconut	10	21	8.0	12
Safflower	8.0	41	8.0	13

The enzyme titre and cost of α-amylase production by *Geobacillus thermoleovorans* using cane molasses and synthetic media has been compared. The enzyme titres were found to increase by 2.5 fold and cost reduced by nearly 22 fold when molasses was employed (Uma Maheswar Rao and Satyanarayana 2007). Cane molasses served as an excellent carbon and energy source for the economical production of glucoamylase by alginate-immobilized *Thermomucor indicae-seudaticae*, which was almost comparable with that in sucrose yeast-extract broth (Kumar and Satyanarayana 2007).

Potato is grown and consumed all over the world, and a large number of processed food industries market potato-based products. Although potato peel does not pose serious disposal and environmental problems, meaningful utilization of this nutrient-rich waste has not drawn much attention. Interestingly potato peel was found to be a superior substrate for solid state fermentation, compared to wheat bran, for the production of α-amylase by two thermophilic isolates of *B. licheniformis* and *B. subtilis* (Shukla and Kar 2006). Potato starch was found to be superior to other starch grains and tubers (amaranthus, wheat, sago, cassava, rice, maize etc) for the production of alpha amylase by *B. licheniformis* SPT 27 (DharaniAiyer 2004). High titres of β amylase production with 16.5% potato starch was reported in the case of *C. thermosulfurogenes* (Reddy et al. 2003). The composition of starch from *Amaranthus paniculatas* was reported to be 66.4%, which was utilized in the production of alpha amylase by *A. flavus* under SSF (Viswanathan and Surlikar 2001).

Brewery (BW) and meat processing (MPW) wastewaters, were used as a base of the culture media in the production of amylase by *A.niger* UO – 1 under submerged fermentation. BW contained (g/L): total sugars- 1.98, reducing sugars – 1.46, total nitrogen – 0.095, total phosphorous – 0.034 and MPW (g/L) was- total sugars – 1.82, reducing sugars – 0.99, total nitrogen – 0.172, total phosphorous – 0.028 (Hernandez et al. 2006). Tea waste is composed of approximately 19% crude protein, 5.4% calcium and 0.84% of phosphorous. They have been popularly used as cattle feed. SSF experiments with *A. niger* for the synthesis of glucoamylase production concluded tea waste, enriched with minerals, as a potential solid substrate (Selvakumar et al. 1998).

Banana is one the most consumed fruits in the world and India is one of the largest producing countries of this fruit. Each hectare of banana crop generates nearly 220 ton of plant residual waste that consists mainly of lignocellulose material and the waste disposal often causes serious environmental problem. The main residual wastes of the banana crop are leaves and pseudostem, both containing high levels of lignocelluloses (Shah et al. 2005). *A. oryzae*, produced amylase when banana fruit stalk was used as substrate in a solid state fermentation system. Banana waste has been exploited as an SSF substrate for α amylase production by *B. subtilis* (Krishna and Chandrasekaran 1996). An attempt was also made to utilize the food waste, kind of organic waste discharged from households, cafeterias and restaurants, which accounts for a considerable proportion of municipal solid garbage in China for the production of glucoamylase by *A. niger*. Wang et al. characterized the food waste and carbon content 53.68%, nitrogen 2.54%, reducing sugar 13.65%, total sugar 50.23%, starch 46.12%, crude protein – 15.56%, crude lipid – 18.06%, crude fiber – 2.26% (Wang et al. 2008).

19.4 Conclusion

The cost and availability of the substrates play an important role in the development of efficient processes. The feasibility of agricultural by products for the commercial production of amylolytic enzymes has been well explored. The effective use of these residues has served dual purpose of value addition and waste management. Eventhough wheat bran was given the prime position among agro residues as substrate, extensive research on the compositional analysis of the individual substrates has proved their high nutritive and productive value.

Abbreviations

SSF:	Solid state fermentation
SmF:	Submerged fermentation
CGM:	Corn gluten meal
NSP:	Non-starch polysaccharides (NSP)
COC:	Coconut oil cake
SOC:	Sesame oil cake
GOC:	Groundnut oil cake
PKC:	Palm kernel cake
OOC:	Olive oil cake
WB:	Wheat bran
CFR:	Cassava fibrous residue
BW:	Brewery wastewater
MPW:	Meat processing wastewater

References

Arasaratnam V, Mylvaganam K, Balasubramaniam K (2001) Glucoamylase production by *Aspergillus niger* in solid state fermentation with paddy husk as support. J Food Sci Technol 38:334–338

Balkan B, Ertan F (2007) Production of α-Amylase from *Penicillium chrysogenum* under solid-state fermentation by using some agricultural by-products. Food Technol Biotechnol 45(4):439–442

Baysal Z, Uyar F, Aytekin C (2003) Solid state fermentation for production of α-amylase by a thermotolerant Bacillus subtilis from hot-spring water. Process Biochem 38:1665–1668

Bertoldo C, Antranikian G (2002) Starch-hydrolyzing enzymes from thermophilic archaea and bacteria. Curr Opin Chem Biol 6:151–160

Crueger W, Crueger A, (2000). Substrates for industrial fermentation. In: Crueger W, Crueger A.(Eds.) Biotechnology, A textbook of Industrial Microbiology, Panima Publisher Corporation, New Delhi

Dendy AV, Dobraszczyk BJ (2001) Cereals and cereal products – Chemistry and technology. Aspen publication, Czech republic

Dharani Aiyer PV (2004) Effect of C:N ratio on alpha amylase production by *Bacillus licheniformis* SPT 27. Afr J Biotechnol 3(10) 519–522

Francis F, Sabu A, Nampoothiri KM et al. (2003) Use of response surface methodology for optimizing process parameters for the production of α-amylase by *Aspergillus oryzae.* Biochem Eng J 15:107–115

Friedberg F (1983) On the primary structure of amylases. FEBS Lett 152:139–140

Gangadharan D, Sivaramakrishnan S, Nampoothiri KM et al. (2008) Response surface methodology for the optimization of alpha amylase production by *Bacillus amyloliquefaciens* Bioresour Technol 99:4597–4602

Gangadharan D, Sivaramakrishnan S, Nampoothiri KM et al. (2006) Solid culturing of *bacillus amyloliquefaciens* for alpha amylase production. Food Technol Biotechnol 44(2): 269–274

Gomes I, Gomes J, Stenier W (2003) Highly thermostable amylase and pullulanase of the extreme thermophilic eubacterium *Rhodothermus marinus*: production and partial characterization. Bioresour Technol 90:207–214

Gonzalez CF, Farina JI, de Figueroa LIC (2008) Optimized amylolytic enzymes production in *Saccharomycopsis fibuligera* DSM-70554 An approach to efficient cassava starch utilization. Enzyme Microb Technol 42:272–277

Hamilton LM, Fogarty WM, Kelly CT (1999) Purification and properties of the raw starch degrading alpha amylase of *Bacillus sp.* IMD-434. Biotechnol Lett 21:111–115

Haq I, Ashraf H, Iqbal J et al. (2003) Production of alpha amylase by *Bacillus licheniformis* using an economical medium. Bioresour Technol 87:57–61

Haq I, Ashraf H, Qadeer MA et al. (2005) Pearl millet, a source of alpha amylase production by *Bacillus licheniformis.* Bioresour Technol 96:1201–1204

Hernandez MS, Rodrıguez MR, Guerra NP (2006) Amylase production by *Aspergillus niger* in submerged cultivation on two wastes from food industries. J Food Eng 7:93–100

Jyothi AN, Sasikiran K, Nambisan B et al. (2005) Optimization of glutamic acid production from cassava starch factory residues using *Brevibacterium divaricatum.* Process Biochem 40: 3576–3579

Krishna C, Chandrasekaran M (1996) Banana waste as substrate for α amylase production by *Bacillus sublitis* (CBTK 106) under solid state fermentation. Appl Microbiol Biotechnol 46:106–111

Kumar P, Satyanarayana T (2007) Economical glucoamylase production by alginate-immobilized *Thermomucor indicae-seudaticae* in cane molasses medium. Lett Appl Microbiol 45:392–397

Kunamneni A, Singh S (2006) Improved high thermal stability of pullulanase from a newly isolated thermophilic *Bacillus* sp. AN-7. Enzyme Microb Technol 39:1399–1404

Kuriki T,Imanaka T (1999) The concept of the α-amylase family: structural similarity and common catalytic mechanism. J Biosci Bioeng 87(5):557–565

Mulimani VH, Patil GN, Ramalingam (2000) α-Amylase production by solid state fermentation: a new practical approach to biotechnology courses. Biochem Edu 28:161–163

Nakajima R, Imanaka T, Aiba S (1986) Comparison of amino acid sequences of eleven different α-amylases. Appl Microbiol Biotechnol 23:355–360

Nampoothiri KM, Pandey A (1996) Solid state fermentation for l-glutamic acid production using *Brevibacterium* sp. Biotechnol Lett 16 (2):199–204

Odibo FJC, Ob SKC (1990) Optimum culture conditions for the production of the extracellular pullulanse of *Thermoactinomyces thalophilus.* Biol Wastes 32:1 9–15

Pandey A (1995) Glucoamylase research: An overview. Starch 47:439–445

Pandey A, (2003) Solid-state fermentation. Biochem Eng J 13:81–84

Pandey A, Soccol CR, Mitchell W (2000a) New development in solid state fermentation: I. Bioprocess and products. Process Biochem 35:1153–1169

Pandey A, Soccol CR, Nigam P (2000b) Biotechnological potential of agro-individual residues. II. Cassava bagasse. Bioresour Technol 74:81–87

Pandey A, Soccol CR, Rodriguez LJ et al. (2001) Solid-State Fermentation in Biotechnology. Asiatech, New Delhi

Ramachandran S, Patel AK, Nampoothiri KM et al. (2004) Alpha amylase from a fungal culture grown on oil cakes and its properties. Braz Arch Biol Technol 47(2): 309–317

Reddy RMP, Ramesh B, Mrudula S et al. (2003) Production of thermostable β-amylase by *Clostridium thermosulfurogenes* SV2 in solid-state fermentation: Optimization of nutrient levels using response surface methodology. Process Biochem 39:267–277

Selvakumar P, Ashakumary L, Pandey A (1998) Biosynthesis of glucoamylase from *Aspergillus niger* by solid state fermentation using tea waste as the basis of a solid substrate. Bioresour Technol 65:83–85

Selvakumar P, Pandey A (1999) Solid-state fermentation for the synthesis of inulinase from the strains of *Staphylococcus* sp. and *Kluyveromycesmarxianus*. Process Biochem 34(8):851–855

Shah MP, Reddy GV, Banerjee R, Babu PR et al. (2005) Microbial degradation of banana waste under solid state bioprocessing using two lignocellulolytic fungi (*Phylosticta* spp. MPS-001 and *Aspergillus* spp. MPS-002). Process Biochem 40:445–451

Shukla J, Kar R (2006) Potato peel as a solid state substrate for thermostable *a*-amylase production by thermophilic Bacillus isolates. World J Microb Biotechnol 22:417–422

Sivaramakrishnan S, Gangadharan D, Nampoothiri KM et al. (2006) α-amylases from microbial sources – An overview on recent developments. Food Technol Biotechnol 44 (2):173–184

Sivaramakrishnan S, Gangadharan D, Nampoothiri KM et al. (2007) Alpha amylase production by *Aspergillus oryzae* employing solid state fermentation. J Sci Ind Res India 66:621–626

Soni SK, Kaur A, J Kishore et al. (2003) A solid state fermentation based bacterial α-amylase and fungal glucoamylase system and its suitability for the hydrolysis of wheat starch. Process Biochem 39:185–192

Swain MR, Ray RC (2007) Alpha-amylase production by *Bacillus subtilis* CM3 in solid state fermentation using cassava fibrous residue. J Basic Microbiol 47:417–425

Tanyildizi MS, Ozer D, Elibol M (2007) Production of bacterial α-amylase by *B. amyloliquefaciens* under solid substrate fermentation. Biochem Eng J 37:294–297

Tigue MA, Kelly CT, Doyle EM, et al. (1995) The alkaline amylase of the alkalophilic *Bacillus* sp.IMD 370. Enzyme Microb Technol 17:570–573

Uma Maheswar Rao JL, Satyanarayana T (2007) Improving production of hyperthermostable and high maltose-forming α-amylase by an extreme thermophile *Geobacillus thermoleovorans* using response surface methodology and its application. Bioresour Technol 98:345–352

Viswanathan P, Surlikar NR (2001) Production of α amylase with *A. flavus* on Amaranthus grains by solid state fermentation. J Basic Microbiol 1:57–64

Wang Q, Wang X, Wang X (2008) Glucoamylase production from food waste by *Aspergillus niger* under submerged fermentation. Process Biochem 43:280–286

Chapter 20
Cellulolytic Enzymes

Reeta Rani Singhania

Contents

Abstract A world-wide interest has been emerged in the commercial potential of using cellulolytic enzymes to generate glucose feedstock using lignocellulosic biomass which could further be converted into value added products. Employing cellulolytic enzymes, for bio-ethanol production from biomass has led to the development of an environmentally safe and sustainable technology. Concept of bio-refinery is emerging to replace already existing petro-refinery as the later is supposed to exhaust in near future. In this chapter, importance of cellulolytic enzymes for biomass conversion and its production aspects have been discussed.

Keywords Cellulase · β-glucosidase · Bio-ethanol · Agro-industrial residues · Pre-treatment · *Tricoderma reesei*

R.R. Singhania (✉)
Biotechnology division, National Institute for Interdisciplinary Science and Technology, Trivandrum 695 019, Kerala, India
e-mail: reetasinghania_123@rediffmail.com

P. Singh nee' Nigam, A. Pandey (eds.), *Biotechnology for Agro-Industrial Residues Utilisation*, DOI 10.1007/978-1-4020-9942-7_20,

20.1 Introduction

Cellulolytic enzymes are the third most important industrial enzyme due to its versatile applications in various industries such as paper and pulp, textile and detergent industry. The resurgence in utilization of biomass for bio-ethanol and other value added organic compounds production has attracted major attention of researchers globally towards cellulases. Renewable plant resources could be used as a way to supplement hydrocarbon resources and meet increasing worldwide needs for consumer goods, which emphasizes on need of an environmentally safe and sound bio-refinery. Lignocellulosic biomass is more attractive for the purpose as it does not compete with food availability unlike starchy biomass. Cellulose is the most abundant and ubiquitous biopolymer on earth, considered to be almost inexhaustible raw material. At molecular level, it is a linear polymer of glucose composed of anhydroglucose units coupled to each other by β-1-4 glycosidic bonds. The number of glucose units in cellulose molecule varies from 250 to 10000 depending on the source and pre-treatment. Cellulose and hemicelluloses are the principle sources of fermentable sugars in lignocellulosic feedstock; however, nature has designed woody tissue for effective resistance to microbial attack. This is the reason that cellulose is relatively inaccessible not only to bigger molecules like protein but also to small molecules as water in some cases. There are crystalline and amorphous regions, in the polymer and also several types of surface irregularities exist. Due to the compact and stringent structure as well as its complex association with other component, very few reactive sites are available for enzyme attachment, which necessitates an appropriate pre-treatment method (Pandey and Soccol 2000). Suitable pre-treatment methods disrupt lignin coating and make the fibers accessible to enzyme action.

Cellulase is not a single enzyme but is a complex of three major types of cellulase: Cellobiohydrolase (CBH or 1,4-β-D-glucan cellobiohydrolase, EC 3.2.1.91), Endo β-glucanase (EG or endo-1,4 β-D glucan glucanohydrolase, EC 3.2.14) and β-glucosidase (BG, EC 3.2.1.21), which acts synergistically to produce oligosaccha-

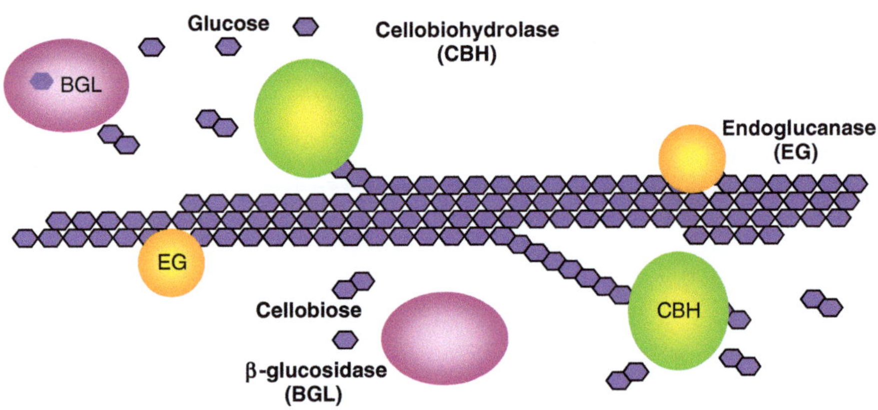

Fig. 20.1 Action of different component of cellulase

rides and glucose. Cellulase hydrolyzes cellulose and produce glucose, cellobiose and other oligo-saccharides as primary products. Bioconversion of cellulose can yield several useful bio-based products.

Figure 20.1 shows action of different component of cellulase, acting synergistically to hydrolyse cellulose. Endoglucanase acts first on amorphous cellulose fibers and attacks in between the chain randomly, to release small fibers with free reducing and non reducing ends. Then exoglucanase acts on free ends to release cellobiose, which is finally hydrolyzed by β-glucosidase to get the final end product as glucose.

20.2 Cellulase Production Employing Agro-Industrial Residues

Agro-industrial residues have been widely used for cellulase production employing variety of microorganisms, as are the rich source of cellulose. Majority of the reports on commercial production of cellulases available utilizes submerged fermentation because of ease of controlling the conditions. However, in nature, the growth and cellulose utilization of aerobic microorganisms elaborating cellulases probably resembles solid-state fermentation than a liquid culture. While the production cost in the crude fermentation by SmF was about $ 20/kg, by SSF it was only $ 0.2/kg if in situ fermentation was used. Ten fold reduction in production cost has been indicated by using SSF than SmF (Tengerdy 1996). Processing of agricultural wastes in SSF systems for cellulolytic enzyme production has been reviewed (Nigam and Singh 1996). They have emphasized with the appropriate technology, improved bioreactor design, and operation controls; SSF may become a competitive method for the production of cellulases. Detailed review of the application of SSF technology for cellulase production are available (Pandey et al. 1999, Xia and Cen 1999). Solid substrate fermentation can be proposed as a better technology for commercial production of cellulases considering the low cost input and ability to utilize naturally available sources of cellulose as substrate. *T reesei* has been exclusively studied as a microbial source of extracellular cellulase capable of hydrolyzing native cellulose. Solid-state fermentation is rapidly gaining interest as a cost effective technology for production of enzyme. The major technical limitation in fermentative production of cellulases remains the longer fermentation time and low productivity. Carbon sources in majority of commercial cellulase fermentations are cellulosic biomass including straw, spent hulls of cereals and pulses, rice or wheat bran, rice or wheat straw, sugarcane bagasse, water hyacinth, paper industry waste and other cellulosic biomass (Belghith et al. 2001 and Tengerdy 1996). Table 20.1 shows the various agro-industrial residues utilized by various microorganisms and the mode of cellulase production. Lactose, cellobiose, sophorose, gentibiose, are the known inducer of cellulase production but only lactose can be used for economically feasible industrial production.

SSF has been proved to be an effective technology for cellulase production for bio-ethanol applications. Though there is a limit to purify the cellulase, mainly because almost all the membranes are cellulosic in nature which serves as natural substrate for enzyme due to which the enzyme molecule get adhere onto the surface.

Table 20.1 Cellulase production-substrate, microorganisms and bioprocess employed

Substrate	Microorganisms	Method
Wheat bran/corn cob	*Aspergillus niger* NRRL3	SSF
Soybean industry residue	*Bacillus subtilis*	SSF
Banana waste	*B. subtilis*	SSF
Rice chaff/wheat bran(9:1)	Mixed culture *(T. reesei* and *A. niger)*	SSF
	A. niger	
Rice straw	*Neurospora crassa*	Smf-fed batch
Wheat straw	*Thermoascus auranticus*	SmF
Wheat straw/bran (8:2)	*Penicillium decumbans*	SSF
Sugar cane bagasse	*P. janthinellum*	SSF
	Trichoderma viride	SmF
	T. reesei	SmF
	H. grisea	SmF
Wheat bran	*Strptomyces drodowiczii*	SmF
Stream treated willow	*T. reesei*	SmF
Corn cob residue	*T. reesei* ZU 02	SmF
Corn stover residue	*T. reesei* ZU 02	SSF
	A. flavus	SSF
Saw dust/bagasse and corn cob	*P. chrysosporium*	SmF
Soy hull		SSF

Though, for hydrolyzing biomass, cellulase does not need high degree of purity. Concentrated crude extract can be used directly.

Even though *T. reesei, Penicillium, Aspergillus* and *Humicola* can hydrolyze native cellulose, the reaction may be sometime very slow due to recalcitrance of biomass. Very rarely cellulose can be found in pure state in nature, usually is embedded in matrix of lignin and is bound with hemicelluloses. It is necessary to remove lignin from cellulose with proper pre-treatment method to make cellulose accessible for the microorganisms. It is an important and a necessary step for commercial hydrolysis of lignocellulosic biomass.

20.3 Pre-Treatment

Cellulose and hemicelluloses are the principle source of C6 and C5 fermentable sugars in lignocellulosic feedstock; nature has designed woody tissue for effective resistance to microbial attack. It emphasizes on use of proper pre-treatment method for making these accessible for microorganisms. Pre-treatment of lignocellulosic biomass has been an active field for of research for several decades, and a wide variety of thermal, mechanical, chemical and biological pre-treatment approaches (and combinations thereof) have been investigated and reported in the scientific literature (McMillan 1994). It is important that the selected pre-treatment technology includes number of requirements;

1. Should improve the enzymatic accessibility of the lignocellulosic compound
2. Result in the minimum loss of the potential sugars

3. Prevent the formation of molecules which are inhibitory to microbial degradation or enzymatic action
4. Pre-treatment technology should be economically sound in order to make the overall process i.e. conversion of biomass to simple sugar a feasible technology

Pre-treatment involves delignification of the feedstock in order to make cellulose more accessible during hydrolysis. It results in separation of lignin and hemicellulose components from cellulose, as well as enlarges the inner surface area of fibers thus paving a way for enhanced enzymatic hydrolysis. Steam explosion, alkali and acid-pre-treatment are some of the common methods of pre-treatment. Steam explosion is most commonly used and alkali pre-treatment has been found to be better in lignin removal (Carrillo et al. 2005). Solid concentration is the key factor significantly affecting the process economics for a dilute acid pre-treatment/enzymatic hydrolysis based process. Solid loading of 30% have been also investigated for dilute acid pre-treatment. Still the relationship between enzymatic digestion and structural properties of pretreated material has to be explored for better understanding of the factors affecting cellulose hydrolysis.

20.4 Microbial Source of Cellulolytic Enzymes

Cellulolytic microbes are primarily cellulose degraders but generally do not utilize lipids or protein as energy source (Lynd et al. 2002). Many of them can utilize other carbohydrates in addition to cellulose but few anaerobic cellulolytic species have restricted carbohydrate range, limited to cellulose and their hydrolyzed product. Several fungi, bacteria and even actinomycetes have been involved in cellulase production. Certain fungi having characteristic ability to produce extracellular proteins in large amount have been studied extensively. Such fungal strains are most suited for the extracellular cellulase production. These organisms produce cellulases when grown on cellulose containing medium but in presence of easily utilizable sugars its production is inhibited. Major cellulase producers have been enlisted (Table 20.2). Fungi can grow and utilize agro-industrial residues better than other microbes as it closely resemble to their natural habitat. Several fungi have been extensively employed for commercial production of cellulases depending upon their ultimate application.

Filamentous fungi are well known as a cost effective resource for industrial cellulases. One of the most extensively studied fungi is *T. reesei* which is capable of hydrolyzing native cellulose (Reczey et al. 1996, Singhania et al. 2006, Singhania et al. 2007). Inspite of being a prolific natural producer of extracellular cellulases, it may not be the most effective cellulase system for use in biomass conversion processes that essentially require complete hydrolysis of the feedstock for economic viability. As a result of extensive studies of 40 years, more than 60 cellulolytic fungi have been reported, representing the soft-rot, brown-rot and white-rot fungi. Fungi such as *Humicola, Aspergillus* and *Penicillium*, bacteria such as *Cellulomonas, Pseudomonads* and actinomycetes such as, *streptomyces* are actively

Table 20.2 Major microorganisms employed in cellulase production

Major groups	Genus	Species
Fungi	Aspergillus	*A. niger*
		A. nidulans
		A. oryzae
		A. aculeatus
	Fusarium	*F. solani*
		F. fusosporium
	Humicola	*H. insolens*
		H. griesa
	Melanocarpus	*M. albomyces*
	Neorospora	*N. crassa*
	Phanerochaete	*P. chrysosporium*
	Penicillium	*P. brasilianum*
		P. occitanis
		P. decumbans
		P. purpurogenum
		P. janthinellum
	Talaromyces	*T. emersonii*
	Trichoderma	*T. reesei*
		T. harzianum
		T. longibrachiatum
Bacteria	Acidothermus	*A. cellulolyticus*
	Bacillus	*Bacillus* sp.
		B. subtilis
	Clostridium	*C. thermocellum*
		C. acitobutylicum
		C. cellulovorans
	Pseudomonas	*P. cellulose*
	Rhodothermus	*R. marinus*
	Cellulomonas	*C. fimi*
		C. uda
Actinomycetes	Streptomyces	*S. drodzowiczii*
		S. lividans
	Thermomonospora	*T. fusca*
		T. curvata

involved in cellulase production. Several other fungi are capable of utilizing cellulose but only few of them are capable of secreting extracellular cellulase complex, which could have practical applications in biomass hydrolysis. Besides *T. reesei; Penicillium, Aspergillus* and *Humicola* have the ability to hydrolyze native cellulose.

20.5 Regulation of Cellulases in Cellulolytic Microorganisms

Cellulase system of microbes can be generally regarded as complexed or noncomplexed. Non complexed cellulase systems from aerobic fungi and bacteria have components of cellulase system free and almost secretable. Cellulase production in cellulolytic microbes is tightly regulated by catabolite repression. The whole mechanism of induction and repression thus helps the microbes to save energy which is not desired when it has to be used as an industrial source of cellulases. The cellulolytic system of *Trichoderma* has been extensively studied. It is inducible and the repression is very strong when cellobiose accumulates in the medium beyond a threshold, though at basal levels it acts as an inducer. Since the CBH enzymes from *T. reesei* generate cellobiose, it tends to accumulate in the medium unless BGL (β-glucosidase) converts it into glucose immediately. In *T. reesei*, the BGL is produced in lowest quantities compared to the other two classes of enzymes, and it is slow acting, making BGL the rate limiting component in the hydrolysis of cellulose. The BGL from *T. reesei* is also subject to feed back inhibition by glucose and cellobiose. The commercial strains of fungi currently used for cellulase production, specifically *T. reesei* produces a cellulase mixture containing very little of BGL enzyme which is inhibited by excess glucose in the medium. Increasing the β-glucosidase production as well as its glucose tolerance is therefore highly desired for obtaining an enzyme preparation, which is efficient in biomass hydrolysis. There can be several approaches for this issue. This can be achieved by transforming the *T. reesei* with heterologus BGL genes that can produce efficient glucose tolerant enzyme. If the BGL gene expressed from a stronger promoter the yield can be increased several fold. Several foreign proteins have been successfully expressed in *Trichoderma* using the *cbh1* promoter to control their expression.

20.5.1 Beta-Glucosidase: Bottleneck in Cellulase Complex

Most of the cellulase complex is deficient in β-glucosidase or it contains small amount thereby making hydrolysis inefficient. Though *Aspergillus* and few *Penicillium* strains have been reported to produce high amount of β-glucosidase, isolation of potent β-glucosidase having glucose tolerance property producing fungal strain, is still the major bottleneck. Efficiency of cellulose complex can be increased by preparing a cocktail of enzyme supplementing β-glucosidase from heterogamous source (Sukumaran et al. 2009). Another strategy could be the over expression of BGL gene, which can be effected by use of the CBH1 promoter to drive its expression. The gene for glucose tolerant BGL can then be isolated and cloned into suitable vector and engineered further to express it from the CBH1 promoter. There have been limited reports of isolation of glucose tolerant BGL from filamentous fungi but there is lot of potential for isolating glucose tolerant BGL from unexplored filamentous fungi.

20.6 Application of Cellulase

Cellulases have occupied a major portion of enzyme market after amylase and protease. It has wide applications in various industries including, detergent industry, textile industry, paper and pulp industry and biofuel. The lignocellulosic plant biomass is renewable and can be used for producing several compounds which are currently being sourced from petroleum. This potential has led to the development of a "biorefinery" concept where plant biomass is the raw material for generating fuel and chemicals. With the imminent depletion of petroleum, renewable lignocellulosic feedstock could be one of the sources to supplement hydrocarbon resources and to meet the increasing worldwide needs for fuel, chemicals and materials. There is a compelling need for commercializing economically viable and environmentally safe biorefineries capable of making a whole range of bio-based products, especially ethanol for fuel applications, using carbohydrates and lignin present in the biomass.

Fuel ethanol production from lignocellulosic biomass is emerging as one of the most important technologies for sustainable production of renewable transportation fuels. In this chapter, bioethanol production has only been discussed as application of cellulolytic enzymes.

20.6.1 Bioethanol

The idea of generation of ethanol from lignocellulosic residues has been conceived by NREL (Northern renewable energy laboratory) in USA. In order to make it competitive with gasoline by the turn of the century, an extensive programme is going on with a strategy that will reduce the cost of bioconversion of biomass to ethanol, in countries like Canada, Denmark and Brazil. It was proposed to be done in two steps i.e. hydrolysis of lignocellulosic material into their monomers and thereby its further conversion into ethanol by fermentation. The crux of the technology lies in the cellulose hydrolysis and the latter one is well stabilized. The major requisite for exploitation is the efficient hydrolysis of the relatively recalcitrant cellulose to produce glucose which is the building block for all other chemicals and metabolites. While chemical methods do exist for biomass hydrolysis, they are inefficient and generate toxic byproducts and effluents that can create pollution. The most efficient method of biomass hydrolysis is through enzymatic saccharification where cellulases and hemicellulases are utilized. Significant research work has been done on cellulases and is still being carried out, with a major thrust on microbial cellulases.

Due to the apparent advantages of ethanol having high octane rating and also being a renewable alternative to existing transport fuels, there is now an increased interest in commercializing technologies for its production from inexpensive biomass (Schell et al. 2004). Most of the fuel ethanol produced in the world is currently sourced from starchy biomass or sucrose (molasses or cane juice), but the technology for ethanol production from non food plant sources is being developed rapidly so that large scale production will be a reality in the coming years. The process of converting low value biomass to ethanol via fermentation depends on the development of economically viable cellulolytic enzyme to achieve effective depolymerisation of the cellulosic content of the biomass. Reduction in cost of "biomass-ethanol" may also be achieved by efficient technologies for saccharification which includes the use of better "enzyme cocktails" and conditions for hydrolysis (Sukumaran et al. 2009). Cellulase preparation used in this process must hydrolyze crystalline cellulose completely, operate effectively at mild pH, withstand process stress and they need not be derived from microbes that are generally regarded as safe (GRAS). The ability to engineer cellulase systems in anticipation of each application is key to successful optimization and commercialization.

Agro-residues could be used as raw material for bioethanol production. Advances in industrial biotechnology, offers ample opportunities on economic utilization of agro-industrial residues. According to Indian scenario rice straw and sugarcane bagasse can be the probable feedstock for long term motive. A part of it can be used in the site itself for generating energy, but still a major part of it constitutes waste. Disposal of these residues itself is a major problem, causing pollution. Bioethanol production involves several steps starting from selection of proper feedstock, its pre-treatment, cellulase production, hydrolysis of feedstock using cellulases and finally fermentation of hydrolysate to obtain ethanol. This bioconversion of cellulose (enzymatic hydrolysis) is the costliest step in overall process which could be brought down by employing multifaceted approach as cheaper raw material for enzyme production, cheaper technology as solid-state fermentation, appropriate feedstock for bioconversion as well as appropriate pre-treatment method. Artificial cellulase preparation and engineering cellulases can help to modify cellulase to suit for the particular application. Expression cassettes, site directed mutagenesis and antisense technology have been successfully employed in designing cellulase. Potent cellulase gene from different filamentous fungi can be isolated, cloned and expressed in the host organism to get better combination or synergism. Enzyme cocktail can be prepared using cellulases from different sources to achieve maximum efficiency which otherwise is not possible due to lack of one or the other component of native cellulase. Cellulase from *T. reesei* can be supplemented by β-glucosidase from *A. niger* to overcome repression and feed back inhibition of β-glucosidase in *T. reesei* (Sukumaran et al. 2006).

20.7 Future Perspectives

Research has shed light into the mechanisms of microbial cellulase production and has led to the development of technologies for production and application of cellulose degrading enzymes. Lignocellulosic biomass is the potential source of biofuels besides biofertilizers, animal feed, chemicals and the raw material for paper industry. Exploitation of this renewable resource needs either chemical or biological treatment, and in the latter context cellulases have gained wide popularity over the past several decades. However use of the current commercial preparations of cellulase for bioconversion of lignocellulosic waste is economically not feasible.

Reduction in the cost of cellulase production and improvement in the efficiency of cellulase are major goals for future cellulase research. The former task may include optimizing growth conditions or processes, whereas, the latter required direct efforts in protein engineering and microbial genetics to improve the properties of the enzymes. Optimization of growth conditions and processes has been attempted to a large extent in improving cellulase production. Many of the commercial production of cellulases involve submerged fermentation technology and employ hyper producing mutants. In spite of several efforts directed at generating hyper producer by directed evolution, the cost of enzyme has remained high. Imparting desired features to enzyme by protein engineering are the area where cellulase research has to advance. Solid-state fermentation on lignocellulosic biomass particularly by using host/substrate specific microorganisms could be an alternative strategy. Filamentous fungi are capable of producing optimal cellulase complex for degradation of host lignocellulose (Tengerdy and Szakacs 2003). Enzyme complex prepared from same lignocellulosic material have been proved to perform better (Chahal et al. 1996). Mixed culture gives improved production and enzyme complexes with high efficiency. Thus, SSF may be considered as a cost effective means for large scale production of cellulases which probably would be several fold cheaper, compared to the commercial cellulase preparations.

Active research in the field of cellulase regulation has led to genetic improvement of cellulase production by various methods including over expressing cellulases from the *cbh1* promoter of *T. reesei* and generation of desired variation in the cellulase production profile of organism. Feedback inhibition of cellulase biosynthesis is another major problem to be encountered. *Trichoderma* and other cellulase producing microbes produces very little of β-glucosidase compared to other fractions of cellulolytic enzymes. Cellobiose is the potent inhibitor of CBH and EG and it gets accumulated due to low β-glucosidase production and thereby inhibits the hydrolysis of cellobiose into glucose. This issue has been addressed by various means like addition of exogenous β-glucosidases to remove cellobiose and engineering β-glucosidase genes into the organism so that it is overproduced. More and more research has been oriented towards genetic manipulation as process design and medium formulation has come to an age and future definitely requires controlled genetic intervention into the physiology of cellulase producers to improve production and thereby make cellulase production more cost effective. The major tasks ahead include overriding the feedback control by glucose and development of integrated bioprocesses for the production of cellulases.

20.8 Conclusion

Cellulase production with improved profile and efficiency is the crux in bioconversion of lignocellulosic biomass. Though, large number of research papers has been published and still publishing on all aspects of cellulase, its application for bioethanol production still has a long way to go to resume as economically feasible technology in near future. Improving its cellulose hydrolyzing efficiency and getting higher sugar concentration in the hydrolysate without any energy input, are still an area which has to be seriously addressed and solved. Protein engineering could serve to increase specific activities of enzyme as well as to increase process tolerance and stability. It needs a concerted effort in understanding the basic physiology of cellulolytic microbes and the utilization of this knowledge coupled with engineering principles to achieve a better utilization of lignocellulose, the most abundant natural resource.

References

Belghith H, Ellouz-Chaabouni S, Gargouri A (2001) Biostoning of denims by *Penicillium occitanis* (Pol6) cellulases. J Biotechnol 89:257–262

Carrillo F, Lis MJ, Colom X et al. (2005) Effect of alkali pre-treatment on cellulase hydrolysis of wheat straw: Kinetic study, Process Biochem 40:3360–3364

Chahal PS, Chahal DS, Simpson BK, Le GBB et al. (1996) Production of cellulose in solid-state fermentation with *Trichoderma reesei* MCG 80 on wheat straw. Appl Biochem Biotechnol 57/58 433–442

Lynd LR, Weimer PJ, van Zyl WH et al. (2002) Microbial Cellulase utilization. Fundam Biotechnol 66:506–577

McMillan JD (1994) Pre-treatment of lignocellulosic biomass. In: Himmel ME, Baker JO, Overend RP (ed) Enzymatic Conversion of Biomass for Fuels Production, ACS Symposium Series 566, Washington DC, American Chemical Society, pp. 292–324

Nigam P, Singh D (1996) Processing of agricultural wastes in solid state fermentation for cellulolytic enzyme production. J Sci Ind Res 55:457–467

Pandey A, Selvakumar P, Soccol CR et al. (1999) Solid state fermentation for the production of industrial enzymes. Curr Sci 77:149–162

Pandey A, Soccol CR (2000) Economic utilization of crop residues for value addition: A futuristic approach. J Sci Ind Res 59:12–22

Reczey K, Szengyel ZS, Eklund R et al. (1996) Cellulase production by *T. reesei*. Bioresour Technol 57:25–30

Schell DJ, Riley CJ, Dowe N et al. (2004) A bioethanol process development unit: initial operating experiences and results with a corn fiber feedstock. Bioresour Technol 91:179–188

Singhania RR, Sukumaran RK, Pillai A et al. (2006) Solid-state fermentation of lignocellulosic substrates for cellulase production by *Trichoderma reesei* NRRL 11460. Ind J Biotechnol 5:332–336

Singhania RR, Sukumaran RK, Pandey A (2007) Improved cellulase production by *Trichoderma reesei* RUT C30 under SSF through process optimization. Appl Biochem Biotechnol 142:60–70

Sukumaran RK, Singhania RR, Pandey A (2006) Microbial cellulases-Production, applications and challenges. J Sci Ind Res 64:832–844

Sukumaran RK, Singhania RR, Mathew G M, Pandey A et al. (2009) Cellulase production using biomass feed stock and its application in lignocellulose saccharification for bio-ethanol production. Renew Energ 34:421–424

Tengerdy RP (1996) Cellulase production by solid substrate fermentation. J Sci Ind Res 55:313–316

Tengerdy RP, Szakacs G (2003) Bioconversion of lignocelluloses in solid-state fermentation. Biochem Eng J 13:169–179

Xia L, Cen P (1999) Cellulase production by solid state fermentation on lignocellulosic waste from the xylose industry. Process Biochem 34:909–912

Chapter 21
Pectinolytic Enzymes

Nicemol Jacob

Contents

Abstract Pectic substances are prominent structural constituents of primary cell walls and middle lamella in non-woody plant tissues. Pectinases are a group of enzymes that contribute to the degradation of pectin by various mechanisms. In nature, pectinases are important for plants as they help in cell wall extension and fruit ripening. They have a significant role in maintaining ecological balance by causing decomposition and recycling of plant materials. The industrial applications of pectinolytic enzymes include fruit juice clarification, tissue maceration, wine clarification, plant fiber processing, oil extraction, coffee and tea fermentation etc. Microbial production of pectinolytic enzymes is mainly from filamentous fungi, yeasts and filamentous and non-filamentous bacteria and is produced in two different techniques viz; submerged fermentation (SmF) and solid-state fermentation (SSF). SSF permits the use of agricultural and agro-industrial residues as substrates for enzyme production. As these residues are renewable and in an abundant supply, they represent a potential low cost raw material for microbial enzyme production.

N. Jacob (✉)
Biotechnology Division, National Institute for Interdisciplinary Science and Technology (CSIR), Trivandrum 695019, Kerala, India
e-mail: nicemariacyril@yahoo.co.in

P. Singh nee' Nigam, A. Pandey (eds.), *Biotechnology for Agro-Industrial Residues Utilisation*, DOI 10.1007/978-1-4020-9942-7_21,

The significance of various agro-industrial residues as raw materials for pectinolytic enzyme production is highlighted in this article.

Keywords Pectin · D-Galacturonic acid · Pectinases · Fruit juice clarification · Wine clarification · Degumming of plant fibers

21.1 Introduction

Ecofriendly biotechnological processes seem to be very important as far as the modern society is concerned for which microbial enzymes are recognized as efficient tools. Pectinases are a group of enzymes that contribute to the degradation of pectin, which is a complex acidic polysaccharide present in the primary cell wall and middle lamella of higher plant tissues. The significance of these enzymes for the development of environment friendly industrial processes has already been established.

Pectinolytic enzymes can be applied in various industrial sectors wherever the degradation of pectin is favourable for a particular process. Pectinases can be divided into two groups based on the optimum activity pH of these enzymes: acidic and alkaline. Fruit juice clarification/extraction is one among the important applications of acidic pectinases (Rai et al. 2004; Sun et al. 2006; Lee et al. 2006; Sin et al. 2006). Fruit juices contain colloids that may lead to fouling problem during filtration process and these colloids are basically polysaccharides such as pectin and starch (Rai et al. 2004). Pre-treatment of juices with pectinases is performed to lower the amount of pectin present and to decrease the viscosity of the juice, which in turn accelerates the subsequent filtration process. Also, it helps to increase the clarity of the juice. Besides fruit juice clarification, pectinases can be used for fruit juice extraction to increase the juice yield. Wine processing industry also recognizes the importance of acidic pectinases (Roldan et al. 2006), where the enzyme can be applied at different stages. The addition of pectinases during crushing of the fruits increases the juice yield and also accelerates the release of anthocyanins into the juice. Pectinase treatment at the pre-fermentation or fermentation stage, settles out suspended particles. After fermentation, enzyme is added to the wine to increase its clarity and filtration rate (Kashyap et al. 2001). Tissue maceration is another important application of acidic pectinases in which organized tissue is transformed into a suspension of intact cells and it is significant in the food industry as well as in the field of biotechnology. The process can be applied for the liquefaction and saccharification of biomass, isolation of protoplasts (Balestri and Cinelli 2001) etc.

Alkaline pectinases are important for retting and degumming of plant fibers (Ossola and Galante 2004; Sharma et al. 2005) as an ecofriendly alternative to the traditional chemical processes. Plant fibers contain gum after decortication, which necessitates a degumming process, for textile purpose, in which the non-cellulosic gummy material is removed from the surface of the fibers. Pectinases have a leading role in the degumming of natural fibers by removing interlamellar pectin which acts as a cementing substance between the fibers. Alkaline pectinases are also applied in paper and pulp (Viikari et al. 2001; Ricard and Reid 2004) industry. Pre-treatment of pulp with pectinases is recommended to lower the cationic demand (Reid and Ricard 2000) and to decrease the cost of the process. Vegetable oil extraction can be augmented by applying enzymes which liquefy the structural cell wall components of oil containing crop which eliminates the use of carcinogenic organic solvents. Coffee and tea fermentation (Angayarkanni et al. 2002; Jayani et al. 2005) and treatment of pectic waste water are some other fields which utilize pectinolytic enzymes.

21.2 Pectin

The plant cell wall is a very dynamic structure as new material is constantly being laid down and old material degraded and removed. The cell wall lends strength and support to plants and the modern view of the plant cell wall is that it is a cellular compartment rather than a rigid and inert network involved in protection and structural support. Throughout plant development, the cell wall is subjected to many chemical and physical changes such as loosening during cell expansion and enzymatic degradation during fruit ripening (Mollet et al. 2003). The structural constituents of a young plant cell wall are cellulose, hemicellulose and pectic substances. The cellulose microfibrils provide strength to the cell wall, while hemicelluloses and pectic substances act as the cementing substance for the cellulose network.

Pectins or pectic substances contribute to complex physiological processes like cell growth and cell differentiation and so determine the integrity and rigidity of plant tissue. It is one of the most complex biomacromolecules in nature and it can be composed of as many as 17 different monosaccharides, with at least seven different polysaccharides. Pectic substances are the sole polysaccharides in the middle lamella responsible for cell cohesion. The texture of vegetables and fruits during growth, ripening and storage is strongly influenced by the type of pectin present. The structure of pectin present in fruits and vegetables depends on enzymatic and chemical modifications occurring during these processes. One of the most characteristic changes during the ripening of fleshy fruits is softening. The change is attributed to enzymatic degradation and solubilization of pectic substances.

21.2.1 Primary Structure of Pectin

The predominant structure of pectin consists of homogalacturonan (HG) which is an essentially unbranched molecule composed of poly α-1,4 D-galacturonic acid (PGA). The basic unit of PGA is shown in Fig. 21.1. The galacturonic acid (GA) residues can be methyl esterified at C-6 and some of the hydroxyl groups on C_2 or C_3 can be acetylated. Blocks of more than 10 unesterified GA residues generally yield pectin molecules, which are sensitive to calcium cross linking (Daas et al. 2001). Recent reports points to the fact that pectin is not an extended back bone consisting of homogalacturonan and rhamnogalacturonan regions, but rather a rhamnogalacturonan with neutral sugar and homogalacturonan side chains (Vincken et al. 2003). HG with β-D-xylose side chains is referred to as xylogalacturonan (XGA). Its pres-

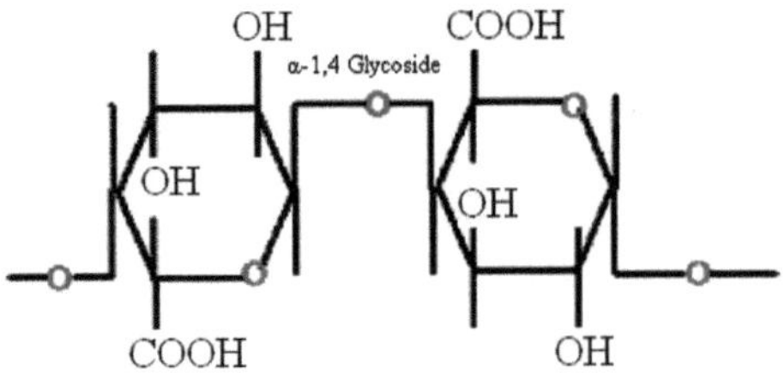

Fig. 21.1 The basic unit of pectin. Poly α-1,4 D-galacturonic acid is the basic unit of pectin. Blocks of this simple polymer alternate with hairy, non-gelling regions containing side chains with other unusual sugars

ence in plants seems to be confined to reproductive organs (fruits and seeds). HGs can contain clusters of four different side chains with very peculiar sugar residues such as apiose, aceric acid, Dha (3-deoxy-D-lyxo-2-heptulosaric acid) and Kdo (3-deoxy-D-manno-2-octulosonic acid). These side chains, together with the HG fragment of ~9 galacturonosyl residues to which they are attached, are referred to as rhamnogalacturonan II (RG-II) (Ridley et al. 2001). Two molecules of RG-II can complex with boron, forming a borate-diol ester, which can crosslink two HG molecules. Only the apifuranosyl residues of the 2-*O*-methyl-D-xylose containing side chains in each of the subunits of the dimer participate in the cross-linking (Ishii et al. 1999). The backbone of rhamnogalacturonan I (RG-I) is composed of as many as 100 repeats of the disaccharide, [→2)-α-L-Rhamnose-(1→4)-α-D-Galacturonic acid-(1→] (Albersheim et al. 1996). The rhamnosyl residues can be substituted at *O*-4 with neutral sugars. The side chains are mainly composed of galactosyl and/or arabinosyl residues. They can be single unit, but also polymeric such as arabinogalactan I (AG-I) and arabinan (50 glycosyl residues or more). AG-I is composed of a 1,4-linked β-D-galactose back bone. Arabinose residues can be attached to the *O*-3 of the galactosyl residues. The arabinans consist of a 1,5-linked α-L-arabinose back bone. Complexes of RG-I, AG-I and arabinan are often referred to as pectic hairy regions (HR), in which AG-I and arabinan are the hairs. Another type of arabinogalactan, arabinogalactan-II (AG-II), is mainly associated with proteins (arabinogalactan proteins or AGPs) and so it is unclear whether AG-II is part of the pectin complex. The major part of AGPs (>90 %) consists of polysaccharides. Even though the fine structure of the constituent polysaccharides of pectin is known in much detail, new structural details are added to the collection of elements already known, with the analysis of pectic polysaccharides from different plant sources (Vincken et al. 2003).

21.2.2 Secondary Structure of Pectin

Although there is a rather detailed picture on the primary structure of pectic constituent polysaccharides, surprisingly little is known on how they assemble into a macromolecular network (Vincken et al. 2003). Selective cleavage of the galacturonosyl residues of the RG-I back bone by treatment with lithium in ethylene diamine has been resulted in a collection of fragments in which neutral side chains are attached to rhamnitol at the reducing end (Lau et al. 1987). Also, oligosaccharides with small neutral sugar side chains attached to rhamnosyl residues have been found in the digests of pectic material obtained with either dilute acid or particular pectic enzymes. These observations suggest that the neutral sugar containing hairs are covalently linked to RG-I. There is also evidence that HG and RG-I are covalently attached to each other (Vincken et al. 2003). When plant tissues are treated with a rather crude mixture of enzymes including pectinases, cellulases and hemicellulases, the native pectic polysaccharides can be partially degraded and therefore the forthcoming material is referred to as modified hairy regions (MHR). A detailed structural characterization of the so called MHR has been extremely valuable in defining the various pectic subunits, and in providing clues on how the various subunits are connected to each other. The MHR has been subsequently fragmented at specific sites with novel enzymes. Based on these studies a tentative structure is put forward. Pectin is represented as one molecule containing smooth and hairy regions. These regions are assembled from three subunits: stretches of RG-I, containing single-unit galactosyl side chains, XGA and RG, containing long arabinan and AG-I side chains. The smooth regions are mainly composed of HG. The HR consist of RG-I, XGA, arabinan, AG-I and AG-II. Also, it is believed that XGA can be a continuation of HG and that the pectic back bone consists of regions with varying GA:Rhamnose ratios (Vincken et al. 2003). The schematic structure of pectin is given in Fig. 21.2.

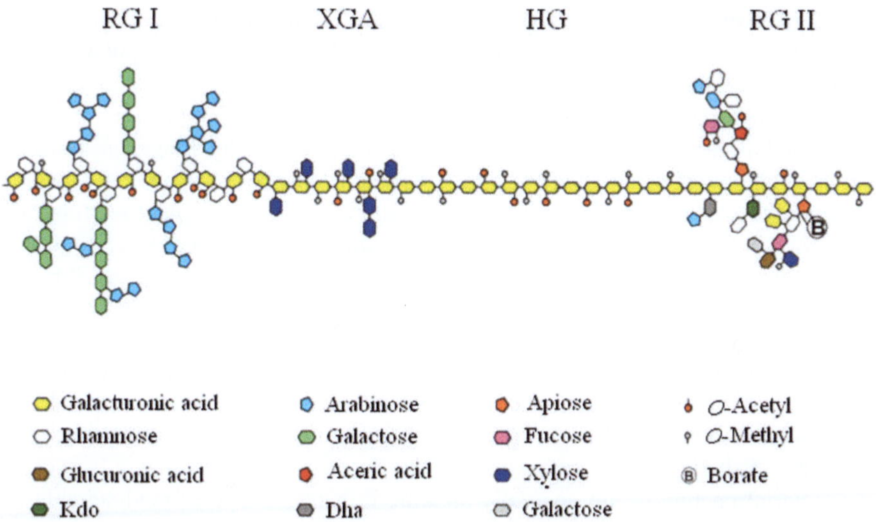

Fig. 21.2 Schematic structure of pectin. It is one of the most complex biomacromolecules in nature and it can be composed at least seven different polysaccharides. The major polysaccharides are HG, XGA, RG II and RG I and AG I, arabinan and AG II are side chains of RG I. The smooth regions are mainly composed of HG. The HR consist of RG-I, arabinan, AG-I, AG-II and XGA

21.3 Pectinases

Pectinases can be classified as esterases, eliminative depolymerases and hydrolytic depolymerases with respect to their role in the degradation of pectin. Pectinesterases or pectin methyl hydrolases catalyze hydrolytic removal of the methyl ester group of pectin, forming pectic acid. The target of the enzyme is a methyl ester group next to a non-esterified galacturonate unit. Depolymerases degrade the α-1,4 glycosidic linkages uniting adjacent GA residues of pectin. Among the depolymerases, hydrolytic depolymerases act by hydrolysis, whereas eliminative depolymerases effect depolymerization of pectin by transelimination, which results in galacturonide, with an unsaturated bond between C4 and C5. Hydrolytic depolymerases are classified into two, based on the substrate of preference. Polymethylgalacturonases act upon pectin, while polygalacturonases degrade PGA. Similarly, eliminative depolymerases or lyases are either polymethylgalacturonate lyases/pectin lyases or polygalacturonate lyases depending upon the substrate preferred. All the depolymerizing enzymes occur in two forms according to their site of action. Endoform of an enzyme catalyzes random cleavage of the substrate, whereas exoform of an enzyme catalyzes sequential cleavage of the substrate from the non-reducing end.

Pectinases are produced by plants, insects and saprophytic microorganisms such as bacteria and fungi. The major sources of plant pectinases are tomatoes and oranges (Torres et al. 2005). Among the animal sources, polygalacturonase obtained from whole-body extracts of the rice weevil, *Sitophilus oryzae* has been purified to apparent homogeneity (Shen et al. 1996). In contrast to plant and animal sources, microorganisms are attractive sources of enzymes because of their diversity and the possibility of yield enhancement by environmental and genetic manipulations and also due to the short life span of microorganisms.

Microbial production of pectinolytic enzymes is mainly from filamentous fungi, yeasts and filamentous and non-filamentous bacteria. Generally, fungal enzymes are acidic in nature, while alkaline enzymes are produced by bacterial strains. The most common source of commercial pectinolytic enzymes is the filamentous fungi especially *Aspergillus* sp. and is produced in two different techniques viz; submerged fermentation (SmF) and solid-state fermentation (SSF). The advantages of SSF over SmF include high volumetric productivity, relatively higher concentration of the products, less effluent generation, requirement for simple fermentation equipments etc. (Pandey et al. 1999). The growing condition in SSF approximates the natural habitat of filamentous fungi more closely than in liquid culture, so these microorganisms are able to grow well on solid substrate and excrete large quantities of enzyme (Castilho et al. 2000).

Bacterial pectinases are generally alkaline in nature and are found suitable for applications like plant fiber processing, treatment of pectic waste water, paper pulping etc. Bacterial pectinase production is usually carried out by SmF, since SSF is generally believed to be suitable for fungi which require a low water activity (0.6) compared to bacteria (0.95). High moisture content in the fermentation medium increases the possibility of contamination by other microbes. However, several researchers have reported enhanced enzyme production by bacterial strains under SSF, proving the fact that SSF process is a better option than SmF for bacterial strains also. Improved production of alkaline and thermotolerant pectinase has been reported by *Bacillus* sp. DT7 under SSF using wheat bran (Kashyap et al. 2003). An alkalophilic *Streptomyces* sp. RCK-SC, which is able to produce a thermostable alkaline pectinase, has been isolated from soil samples and in an immobilized cell system containing polyurethane foam, enzyme production has been enhanced by 32 % as compared to shake flask cultures, using wheat bran as solid substrate, at substrate-to-moisture ratio of 1:5 after 72 h of incubation (Kuhad et al. 2004).

21.3.1 Production of Pectinases Using Agro-Industrial Residues

SSF permits the use of agricultural and agro-industrial residues as substrates which are converted into bulk chemicals and fine products with high commercial value. Agro-industrial waste materials can be used both as source of energy for growth and as carbon for synthesis of cell biomass and other products (Mahmood et al. 1998). The selection of a substrate for enzyme production in an SSF process depends on several factors, mainly related with cost and availability of the substrate (Pandey et al. 1999). As agro-industrial residues are renewable and in an abundant supply (~3.5 billion tonnes/year), they represent a potential low cost raw material for microbial enzyme production (Robinson and Nigam 2003). Solid substrate not only supplies the nutrients to the microbial cultures growing in it but also serves as an anchorage for the cells. The ideal solid substrate is one that provides all the necessary nutrients for the microorganism. However, some of the nutrients in the solid substrate may be available in sub optimal concentrations, or even not present in the substrates. In such cases, it would be necessary to supplement them externally (Pandey et al. 1999) to enhance growth and subsequently enzyme production.

As there is an increasing demand for pectinases for various applications, its cost effective production using low value substrates is becoming important for industries. Agro-industrial residues like sugar beet pulp, citrus pulp, apple pomace, sugar cane bagasse, wheat bran, orange bagasse, grape pomace etc. are used for the production of pectinolytic enzymes by different fungi and bacteria. Many of the substrates used for SSF are unrefined and is of agricultural origin making complete characterization and exact reproducibility difficult (Mitchell and Lonsane 1992). However, a general picture of the composition of various agro industrial residues and their significance in pectinolytic enzyme production is presented below.

21.3.1.1 Sugarcane Bagasse

Sugarcane bagasse is the fibrous residue of cane stalks left over after the crushing and extraction of the juice from the sugar cane. It is a ligno-cellulosic residue (by-product) of the sugar industry and is almost completely used by the sugar factories themselves as fuel for the boilers. A sugar factory produces nearly 30 % of bagasse out of its total crushing. Several processes and products have been reported that utilize sugarcane bagasse as raw material. These include electricity generation, pulp and paper production, and products based on fermentation like animal feed, bioethanol, enzymes etc. (Pandey et al. 2000a). One of the successful attempts is the utilization of bagasse as raw material for the fermentative production of microbial enzymes employing potent microbial cultures.

Bagasse consists of approximately 50 % cellulose and 25 % each of hemicellulose and lignin. Chemically, bagasse contains about 50 % α-cellulose, 30 % pentosans and 2.4 % ash (Pandey et al. 2000a). Since the pectin content of sugar cane bagasse is low, it acts as only a solid inert support during SSF processes for pectinolytic enzyme production. The high fibre content is an advantage to be used as substrate for SSF, when it is used alone or with other low fiber substrates such as wheat bran, as it can increase the interparticle spacing with a possible increase in the aeration and nutrient diffusion, supporting enzyme production (Martins et al. 2002). Pectinases are inducible enzymes, and the inducers are GA, its polymer (pectin and polypectate) and structural relatives (mucic acid, tartonic acid and dulcitol) (Maldonado and Saad 1998). It is suggested to use a pectin rich inducer when sugar cane bagasse is used as the sole substrate. Sugar cane bagasse has been successfully used for the production of pectinases by *Aspergillus niger, Thermoascus aurantiacus,Moniliella* sp.,*Penicillium* sp. etc (Table 21.1).

Table 21.1 Utilization of agro-industrial residues for pectinolytic enzyme production

Microorganism	Substrate	Reference
Aspergillus niger	Sugar cane bagasse	Maldonado and Saad 1998
Aspergillus carbonarius	Wheat bran	Singh et al. 1999
Thermoascus aurantiacus	Wheat bran or orange bagasse or sugar cane bagasse	Martins et al. 2002
Bacillus sp. DT7	Wheat bran	Kashyap et al. 2003
Fusarium moniliforme	Wheat bran and orange pulp	Niture and Plant 2004
Moniliella sp.	Sugar cane bagasse, orange bagasse and wheat bran	Martin et al. 2004
Penicillium sp.	Sugar cane bagasse, orange bagasse and wheat bran	Martin et al. 2004
Aspergillus niger	Monosodium glutamate waste water and sugar beet pulp	Bai et al. 2004
Streptomyces sp. RCK-SC	Wheat bran	Kuhad et al. 2004
Penicillium viridicatum	Orange bagasse and wheat bran	Silva et al. 2005
Bacillus gibsoni	Sugar beet pulp	Li et al. 2005
Aspergillus niger	Apple pomace	Joshi et al. 2006
Aspergillus awamori	Grape pomace	Botella et al. 2007
Aspergillus niger	Wheat bran	Dinu et al. 2007
Streptomyces lydicus	Wheat bran	Jacob and Prema 2008

21.3.1.2 Sugar Beet Pulp

Sugar beet, alongside sugar cane, is the main source of sugar across the world. It is grown widely in Europe, North and South America, Asia and parts of North Africa and the crop is at the core of a multi-billion dollar global industry. After the sugar has been extracted, the remaining pulp contains very little sugar, but is valuable as a fiber and energy source. The pulp is composed of (% on dry basis) pectin, 28.7; cellulose, 20; hemicellulose, 17.5; protein, 9.0; lignin, 4.4; fat, 1.2; ash, 5.1 (Xue et al. 1992). Since the pectin content of beet pulp is high it can be used for the microbial production of pectinolytic enzymes without adding any pectinaceous materials as enzyme inducer. It has been used as raw material for pectinase production by *Aspergillus niger* (Bai et al. 2004). Sugar beet pulp has also been used as the carbon source as well as the pectinase inducer to produce extracellular alkaline pectinase, by *Bacillus gibsoni*, under SSF (Li et al. 2005).

21.3.1.3 Citrus Bagasse

Citrus bagasse is the biomass remaining after juice extraction from citrus fruits on an industrial scale. Citrus bagasse consists of peel, pulp and seeds, which correspond to half of the fresh fruit weight (Garzon and Hours 1992). Lemon and orange pulps are the major waste products released by citrus juice industry. Lemon pulps, essentially constituted of pectins, are similar to a homogenous cream from which no further juice can be mechanically recovered. The pulps can be disposed only after drying because pulps with high moisture content may undergo spontaneous fermentation causing environmental problems. The disposal cost of lemon pulps is high since it cannot be easily and cheaply dried to a stable product with moisture content lower than 10 %. In this respect, an alternative practical use of pulps is slurry state fermentation of the material for the production of pectinases (De Gregorio et al. 2002), as a value addition of the material and as an economical way of enzyme production.

Orange bagasse, composed of peel, seed and pulp, is the waste material released after juice extraction and is usually dried and marketed as a component of animal feed. Since the production cost of dried bagasse is high and its selling price is low, the process can be considered only as a waste disposal method (Mahmood et al. 1998). The pellet of orange bagasse contains 11.8 % fibre, 6.4 % protein, 63 % nitrogen, 6.7 % ash, 19 % total sugar (9 % reducing sugar) and 0.1 % pectin (Martins et al. 2002). The microbial cultures which are proved to be suitable for the production of pectinases utilizing orange bagasse as raw material are *Thermoascus aurantiacus*, *Penicillium* sp., *Moniliella* sp., *Penicillium viridicatum* etc. (Table 21.1).

21.3.1.4 Apple and Grape Pomaces

Pomace (ultimately from Latin *pomum*) or marc is the solid remains of grapes or other fruits after pressing for juice or oil. It is essentially the pulp, peel, seeds and stalks of the fruit after the oil, water or other liquid has been pressed out. One of the main agro-industrial by products, abundantly produced in Europe, is apple pomace, from the apple juice and cider industries. Apple pomace is often used to produce pectin. It has high moisture content ($\sim$80 %) which poses disposal problems for the pulping industry (Robinson and Nigam 2003). Apple pomace, composed of peel, seed and pulp, contains high levels of pectic substances and it could be used as a substrate for the microbial biosynthesis of pectinases. Apple pomace has been effectively used for the production of pectinases by *Aspergillus niger* (Joshi et al. 2006).

Grape pomace is produced in large qualities in wine production with the issue of disposal being an important environmental consideration. Grape pomace has traditionally been used to produce grape seed oil, a practice that continues to this day in small amounts, and Pomace brandy, such as grappa. Today, pomace is most commonly used as fodder or fertilizer. Some wineries will reuse the material as fertilizer while others are exploring options of selling the used pomace to biogas companies to be used in the creation of renewable energy. Grape pomace has been found to be suitable for pectinase production by *Aspergillus awamori* (Botella et al. 2007).

21.3.1.5 Wheat Bran

Wheat is an ancient grain and thought to have originated in southwestern Asia. It has been consumed as a food for more than 12,000 years. As it was looked upon as the Staff of Life, it played an important role of religious significance and was part of the sacred rituals of many cultures. As rice has been the dietary staple of Asia, wheat has served this role for many of other regions of the world. Wheat is a cereal grass of the Graminaceae (Poaceae) family and of the genus *Triticum* and it is estimated that approximately one-third of the world's people depend upon wheat for their nourishment. Today, the largest commercial producers of wheat include the Russian Federation, the United States, China, India, France and Canada.

Wheat bran is the hard outer layer of wheat grains, and consists of combined aleurone and pericarp. Along with germ, it is an integral part of whole grains, and is often produced as a by-product of milling in the production of refined grains. Wheat bran is composed of 8.12 % fibre, 15.7 % protein, 65 % nitrogen, 4.57 % ash and 16.7 % total sugar (5.22 % reducing sugar) (Martins et al. 2002).

Among the various agro-industrial residues, wheat bran however holds the key, as it is produced worldwide in enormous quantities as an important by-product of the cereal industry, and has most commonly been used as substrate for various biotechnological processes (Pandey et al. 1999). Moreover, as wheat bran is a cheap and readily available byproduct, the production of pectinase using wheat bran may be a cost-effective affair (Kashyap et al. 2003). Pectinolytic enzyme production using wheat bran is achieved using various microbial cultures such as *Aspergillus niger*, *Aspergillus carbonarius*, *Streptomyces* sp., *Streptomyces lydicus*, *Thermoascus auriantacus*, *Penicillium viridicatum*, *Fusarium moniliforme*, *Bacillus* sp., etc. as it is evident from Table 21.1. According to recent reports, the leading fungal pectinase producer is *Aspergillus carbonarius* with an enzyme yield of 480 U/g (Singh et al. 1999) under SSF using wheat bran as the substrate. The highest reported polygalacturonase production in SSF is from *Bacillus* sp. (23076 U/g) followed by *Streptomyces* sp. (4857 U/g) (Kapoor and Kuhad 2002; Kuhad et al. 2004) with wheat bran as the solid substrate. As it is evident from recent reports, wheat bran is the most potent solid substrate for pectinolytic enzyme production by both fungi and bacteria, with or without the presence of enzyme inducers.

21.3.1.6 Coffee Pulp and Husk

Coffee pulp and husk had been used previously for the production of pectinases since it contains considerable proportion of pectin in it. Coffee pulp or husk is a fibrous mucilagenous material acquired during the processing of coffee cherries by wet or dry process, respectively. The organic nature of the material makes it an ideal substrate for microbial processes for the production of value-added products (Pandey et al. 2000b).

21.3.2 Purification of Pectinases

An improved awareness of the properties of pectinases is important in commercialization of these enzymes in various fields. Enzyme inactivation and stability are considered to be the major constraints in the rapid development of biotechnological processes and the stability of pectinases is affected by both physical parameters (pH and temperature) and chemical parameters (inhibitors or activators). Stability studies provide valuable information regarding the structure and function of enzymes (Gummadi and Panda 2003). Purification of the enzyme is necessary before characterization because crude enzyme may contain different stabilizing components and so the properties may vary greatly. Purification of pectinases has been effected by combinations of different chromatography procedures (Celestino et al. 2006) or single step procedures like aqueous two phase systems (Lima et al. 2002) or affinity precipitation (Mondal et al. 2004).

When the raw material for enzyme production is an inert support, like sugar cane bagasse with inducers, the enzyme obtained will be less contaminated than in the case protein rich substrates like wheat bran. Purification requires more steps, when the enzyme is more contaminated, increasing the cost of the process. So the fermentation substrate should be selected carefully depending upon the end use of the enzyme and the growth requirements of a particular organism. If only crude enzyme is required for an application (fiber processing), purification is not necessary, but when it is applied for the quality improvement of a food material (fruit juice clarification), purification is compulsory.

21.4 Conclusions and Perspectives

Agro-industrial by-products can be successfully utilized for microbial pectinolytic enzyme production and as these residues are locally abundant low cost raw materials, enzyme production will be a cost effective affair. Also, the approach prevents the accumulation of these by-products as an environmental threat. Identification and bioconversion of new locally available agro-wastes is advantageous as it not only leads to the value addition of these residues, but also helps to keep the environment clean.

Abbreviations

AG I	–	Arabinogalactan-I
AG II	–	Arabinogalactan-II
AGPs	–	Arabinogalactan proteins
Dha	–	(3-deoxy-D-lyxo-2-heptulosaric acid)
GA	–	Galacturonic acid
HG	–	Homogalacturonan
HR	–	Hairy regions
Kdo	–	(3-deoxy-D-manno-2-octulosonic acid).
MHR	–	Modified hairy regions
PGA	–	Polygalacturonic acid
RG I	–	Rhamnogalacturonan-I
RG II	–	Rhamnogalacturonan-II
SmF	–	Submerged fermentation
SSF	–	Solid-state fermentation
XGA	–	Xylogalacturonan

References

Albersheim P, Darvill AG, O'Neill MA, Schols HA Voragen AGJ et al. (1996) An hypothesis: the same six polysaccharides are components of the primary cell walls of all higher plants. In: Visser J, Voragen AGJ (eds) Progress in Biotechnology 14: Pectins and Pectinases. Elsevier

Angayarkanni J, Palaniswamy M, Murugesan S, Swaminathan K et al. (2002) Improvement of tea leaves fermentation with *Aspergillus* spp pectinase. J Biosci Bioeng 94:299–303

Bai ZH, Zhang HX, Qi HY, Peng XW, Li BJ et al. (2004) Pectinase production by *Aspergillus niger* using waste water in solid state fermentation for eliciting plant disease resistance. Bioresour Technol 95:49–52

Balestri E, Cinelli F (2001) Isolation and cell wall regeneration of protoplasts from *Posidonia oceanica* and *Cymodocea nodosa*. Aquat Botany 70:237–242

Botella C, Diaz A, Ory I, Webb C, Blandino A et al. (2007) Xylanase and pectinase production by *Aspergillus awamori* on grape pomace in solid state fermentation. Process Biochem 42:98–101

Castilho LR, Medronho RA, Alves TLM et al. (2000) Production and extraction of pectinases obtained by solid-state fermentation of agro-industrial residues with *Aspergillus niger*. Bioresour Technol 71:45–50

Celestino SMC, Maria de Freitas S, Medrano FJ, Valle de Sousa M, Filho EXF et al. (2006) Purification and characterization of a novel pectinase from *Acrophialophora nainiana* with emphasis on its physicochemical properties. J Biotechnol 123:33–42

Daas PJH, Boxma B, Hopman AMCP, Voragen AGJ, Schols HA et al. (2001) Nonesterified galacturonic acid sequence homology of pectins. Biopolymers 58:1–8

De Gregorio A, Mandalari G, Arena N, Nucita F, Tripodo MM, Lo Curto RB et al. (2002) SCP and crude pectinase production by slurry-state fermentation of lemon pulps. Bioresour Technol 83:89–94

Dinu D, Nechifor MT, Stoian G, Costache M, Dinischiotu A et al. (2007) Enzymes with new biochemical properties in the pectinolytic complex produced by *Aspergillus niger* MIUG 16. J Biotechnol 131:128–137

Garzon CG, Hours RA (1992) Citrus waste: An alternative substrate for pectinase production in solid-state culture. Bioresour Technol 39:93–95

Gummadi SN, Panda T (2003) Purification and biochemical properties of microbial pectinases-a review. Process Biochem 38:987–996

Ishii T, Matsunaga T, Pellerin P, O'Neill MA, Darvill A, Albersheim P et al. (1999) The plant cell wall polysaccharide rhamnogalacturonan II self-assembles into a covalently cross-linked dimer. J Biol Chem 274:13098–13104

Jacob N, Prema P (2008) Novel process for the simultaneous extraction and degumming of banana fibers under solid-state cultivation. Braz J Microbiol 39:115–121

Jayani RS, Saxena S, Gupta R et al. (2005) Microbial pectinolytic enzymes: A review. Process Biochem 40:2931–2944

Joshi VK, Parmar M, Rana NS et al. (2006) Pectin esterase production from apple pomace in solid-state and submerged fermentations. Food Technol Biotechnol 44:253–256

Kapoor M, Kuhad RC (2002) Improved polygalacturonase production from *Bacillus* sp. MG-cp-2 under submerged (SmF) and solid-state (SSF) fermentation. Lett Appl Microbiol 34:317–322

Kashyap DR, Soni SK, Tewari R et al. (2003) Enhanced production of pectinase by *Bacillus* sp. DT7 using solid-state fermentation. Bioresour Technol 88:251–254

Kashyap DR, Vohra PK, Chopra S, Tewari R et al. (2001) Applications of pectinases in the commercial sector: A review. Bioresour Technol 77:215–227

Kuhad RC, Kapoor M, Rustagi R et al. (2004) Enhanced production of an alkaline pectinase from *Streptomyces* sp. RCK-SC by whole-cell immobilization and solid-state cultivation. World J Microbiol Biotechnol 20:257–263

Lau JM, McNeil M, Darvill AG Albersheim P et al. (1987) Treatment of rhamnogalacturonan I with lithium in ethylene diamine. Carbohydr Res 168:245–274

Lee WC, Yusof S, Hamid NSA, Baharin BS et al. (2006) Optimizing conditions for enzymatic clarification of banana juice using response surface methodology (RSM). J Food Eng 73: 55–63

Li Z, Bai Z, Zhang B, Xie H, Hu Q, Hao C, Xue W, Zhang H et al. (2005) Newly isolated *Bacillus gibsonii* S-2 capable of using sugar beet pulp for alkaline pectinase production. World J Microbiol Biotechnol 21:1483–1486

Lima AS, Alegre RM, Meirelles AJA et al. (2002) Partitioning of pectinolytic enzymes in polyethylene glycol/potassium phosphate aqueous two-phase systems. Carbohydr Polym 50:63–68

Mahmood AU, Greenman J, Scragg AH et al. (1998) Orange and potato peel extracts: Analysis and use as *Bacillus* substrates for the production of extracellular enzymes in continuous culture. Enzyme Microb Technol 22:30–137

Maldonado MC, Saad AMS (1998) Production of pectin esterase and polygalacturonase by *Aspergillus niger* in submerged and solid state systems. J Ind Microbiol Biotechnol 20:34–38

Martin N, De Souza SR, Da Silva R, Gomes E et al. (2004) Pectinase production by fungal strains in solid-state fermentation using agro-industrial bioproduct. Braz Arch Biol Technol 47:813–819

Martins ES, Silva D, Da Silva R, Gomes E et al. (2002) Solid-state production of thermostable pectinases from thermophilic *Thermoascus aurantiacus*. Process Biochem 37:949–954

Mitchell DA, Lonsane BK (1992) Definition, characteristics and potential. In: Doelle HW, Mitchell DA, Rolz CE et al. (eds) Solid Substrate Cultivation. Elsevier, Essex, UK

Mollet JC, Park SY Lord EM et al. (2003) Interaction of a stylar pectic polysaccharide and a basic protein (sca) mediates lily pollen tube adhesion. In: Voragen F, Schols H, Visser R et al. (eds) Advances in Pectin and Pectinase Research. Kluwer Academic Publishers, The Netherlands

Mondal K, Mehta P, Gupta MN et al. (2004) Affinity precipitation of *Aspergillus niger* pectinase by microwave-treated alginate. Protein Expr Purif 33:104–109

Ossola M, Galante YM (2004) Scouring of flax rove with the aid of enzymes. Enzyme Microb Technol 34:177–186

Pandey A, Selvakumar P, Soccol CR, Nigam P et al. (1999) Solid-state fermentation for the production of industrial enzymes. Curr Sci 77:146–162

Pandey A, Soccol CR, Nigam P, Brand D, Mohan R, Roussos S (2000b) Biotechnological potential of coffee pulp and coffee husk for bioprocesses. Biochem Eng J 6:153–162

Pandey A, Soccol CR, Nigam P, Soccol VT et al. (2000a) Biotechnological potential of agro-industrial residues. I. Sugarcane bagasse. Bioresour Technol 74:69–80

Rai P, Majumdar GC, DasGupta S, De S et al. (2004) Optimizing pectinase usage in pre-treatment of mosambi juice for clarification by response surface methodology. J Food Eng 64:397–403
Reid I, Ricard M (2000) Pectinase in paper making: Solving retention problems in mechanical pulp, bleached with hydrogen peroxide. Enzyme Microb Technol 26:115–123
Ricard M, Reid ID (2004) Purified pectinase lowers cationic demand in peroxide-bleached mechanical pulp. Enzyme Microb Technol 34:499–504
Ridley BL, O'Neill MA Mohnen D et al. (2001) Pectins: Structure, biosynthesis, and oligogalacturonide-related signaling. Phytochem 57:929–967
Robinson T, Nigam P (2003) Bioreactor design for protein enrichment of agricultural residues by solid state fermentation. Biochem Eng J 13:197–203
Roldan A, Palacios V, Penate X, Benitez T, Perez L et al. (2006) Use of *Trichoderma* enzymatic extracts on vinification of *Palomino fino* grapes in the sherry region. J Food Eng 75:375–382
Sharma HSS, Whiteside L, Kernaghan K et al. (2005) Enzymatic treatment of flax fibre at the roving stage for production of wet-spun yarn. Enzyme Microb Technol 37:386–394
Shen Z, Reese JC, Reeck GR et al. (1996) Purification and characterization of polygalacturonase from the rice weevil, *Sitophilus oryzae* (Coleoptera: Curculionidae). Insect Biochem Mol Biol 26:427–433
Silva D, Tokuioshi K, Martins ED, Da Silva R, Gomes E et al. (2005) Production of pectinase by solid-state fermentation with *Penicillium viridicatum* RFC3. Process Biochem 40:2885–2889
Sin HN, Yusof S, Hamid NSA, Abd Rahman R et al. (2006) Optimization of enzymatic clarification of sapodilla juice using response surface methodology. J Food Eng 73:313–319
Singh SA, Ramakrishna M, Rao AGA et al. (1999) Optimisation of downstream processing parameters for the recovery of pectinase from the fermented bran of *Aspergillus carbonarius*. Process Biochem 35:411–417
Sun Y, Wang Z, Wu J, Chen F, Liao X, Hu X et al. (2006) Optimising enzymatic maceration in pre-treatment of carrot juice concentrate by response surface methodology. Int J Food Sci Technol 41:1082–1089
Torres EF, Aguilar C, Esquivel JCC, Viniegra-Gonzalez G et al. (2005) Pectinases. In: Pandey A, Webb C, Soccol CR, Larroche C et al. (eds) Enzyme Technology. Asiatech Publishers, Inc., New Delhi
Viikari L, Tenkanen M, Suurnakki A et al. (2001) Biotechnology in the pulp and paper industry. In: Rehm HJ (ed) Biotechnology, vol. 10. Wiley, VCH, Singapore
Vincken J, Schols HA, Oomen RJFJ, Beldman G, Visser RGF, Voragen AGJ et al. (2003) Pectin-the hairy thing. In: Voragen F, Schols H, Visser R et al. (eds) Advances in Pectin and Pectinase Research. Kluwer Academic Publishers, The Netherlands
Xue M, Liu D, Zhang H, Qi H, Lei Z et al. (1992) A pilot process of solid state fermentation from sugar beet pulp for the production of microbial protein. J Ferment Bioeng 73:203–205

Chapter 22
Ligninolytic Enzymes

K.N. Niladevi

Contents

Abstract Ligninolytic enzymes are involved in the degradation of the complex and recalcitrant polymer lignin. This group of enzymes is highly versatile in nature and they find application in a wide variety of industries. The biotechnological significance of these enzymes has led to a drastic increase in the demand for these enzymes in the recent time. Production of enzymes/metabolites from microbial sources is a costly affair and the only alternate to minimize the production cost is the use of inexpensive raw materials. The utilization of agro-industrial residues in this as-

K.N. Niladevi (✉)
Biotechnology Division, National Institute for Interdisciplinary Science and Technology (CSIR), Trivandrum 695019, Kerala, India
e-mail: nilanandini@yahoo.co.in

P. Singh nee' Nigam, A. Pandey (eds.), *Biotechnology for Agro-Industrial Residues Utilisation*, DOI 10.1007/978-1-4020-9942-7_22,

pect is much appreciated due to their low cost and ease in availability. Adopting solid-state fermentation for enzyme production may add to the benefit of reducing the production costs. The studies have proved that huge quantities of lignocellulosic residues are available world wide for the production of ligninolytic enzymes. The current trend is to make use of every such locally available agro-industrial residue for enzymes production to meet the demand for the same from the industrial sectors.

Keywords Ligninolytic enzymes · Lignin peroxidase · Manganese peroxidase · Laccase · Agro-industrial residues · Solid-state fermentation

22.1 Introduction

The term lignin degrading enzymes encompasses mainly three oxidative enzymes; lignin peroxidase (LiP), manganese peroxidase (MnP) and laccase. The demand for these enzymes has increased in the recent years due to their potential applications in diverse biotechnological areas. Lignin degrading enzymes are widely used in the pollution abatement, especially in the treatment of industrial effluents that contains hazardous compounds like dyes, phenols and other xenobiotics. The role of these enzymes in textile dye decolourization and degradation of phenolic and non-phenolic aromatic compounds has been extensively studied (Wesenberg et al. 2003; Crecchio et al. 1995). The industrial preparation of paper requires separation and degradation of lignin in wood pulp. Pre-treatments of wood pulp with ligninolytic enzymes might provide milder and cleaner strategies of delignification that are also respectful of the integrity of cellulose (Kuhad et al. 1997). Ligninolytic enzymes are commonly used for biobleaching of kraft pulp and a laccase mediator system with the trade name Lignozym® is already in commercial application (Call and Mücke 1997). The ligninolytic enzymes also find application in the stabilization of wine and fruit juices (Minussi et al. 2002), denim washing (Pazarlioglu et al. 2005), cosmetic industry (Aaslyng et al. 1996) and Biosensors (Ferry and Leech 2005).

The ever increasing demand for these enzymes in the industrial sectors necessitates the production of enzymes from inexpensive raw materials. The ligninolytic enzymes can be produced from a wide variety of agro-industrial residues that are lignocellulosic in nature. It is a well known fact that majority of the agro-industrial residues are lignocellulosic in nature. The production of enzymes from agro-industrial residues is of great significance owing to two reasons; (i) the process helps in the reuse of industrial wastes and thereby reduces the problems arising out of their disposal and (ii) the enzymes produced from the residues can be utilized for various industrial applications.

22.2 Structure of Lignin

The significance of ligninolytic enzymes lies in the fact that they are capable of degrading the highly complex recalcitrant polymer, lignin. Lignin serves as the second major reservoir of fixed carbon sources in nature, next to cellulose, comprising 15% of the earth's biomass (Hammel 1992). True lignin is distributed widely but not universally throughout the plant kingdom. It is found in all vascular plants, where it is deposited in cell walls along with cellulose and hemicellulose. Lignin fills the spaces in the cell wall between different plant polysaccharides by covalent linking (Fig. 22.1) and thereby conferring mechanical strength to the cell wall and by extension the plant as a whole (Chabannes 2001). Lignin is synthesized in plants from the phenyl propanoid precursors such as coniferyl, synapyl, and *p*-coumaryl alcohols. These precursors (Fig. 22.2) are consisted of an aromatic ring and a 3-carbon side chain. Free radical copolymerization of these alcohols produces the heterogeneous, optically inactive, cross-linked, and highly polydisperse polymer (Kirk and Farrell 1987). In the lignin molecule the precursors form three types of subunits: hydroxyphenol- (H-type), guaiacyl- (G-type) and syringyl subunits (S-type). Unlike other biopolymers, lignin contains no readily hydrolysable bond recurring at periodic intervals along a linear backbone. Instead, lignin is a three

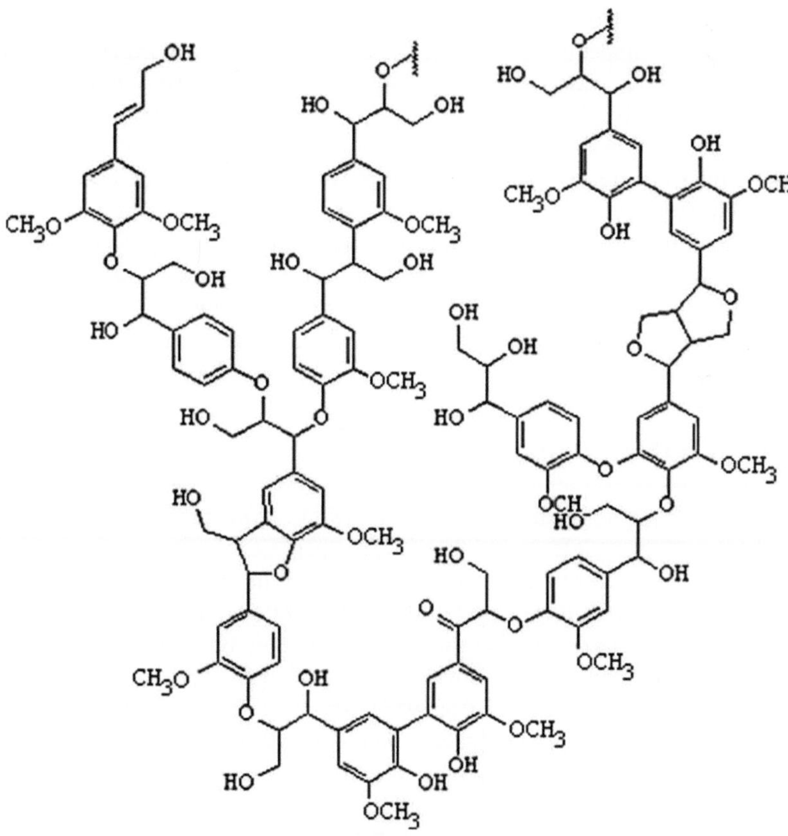

Fig. 22.1 Schematic structure of a lignin molecule
Source: www.research.uky.edu/.../green_energy.html

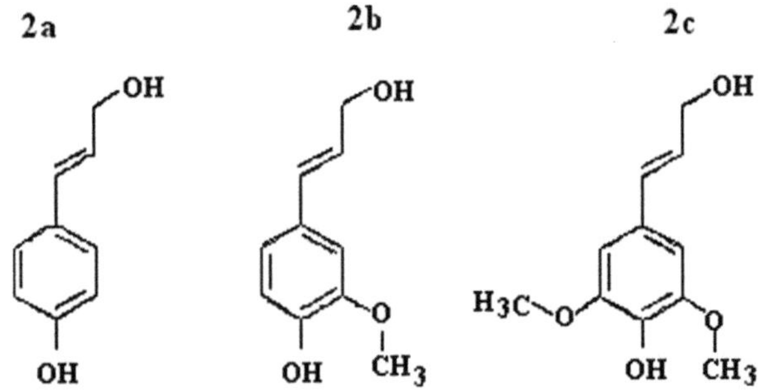

Fig. 22.2 Structure of the lignin precursors 2a. *p*-Coumaryl alcohol; 2b. Coniferyl alcohol; 2c. Sinapyl alcohol
Source: www.steve.gb.com/ science/ molecules.html

dimensional amorphous polymer containing many different stable C–C, C–O–C, β-O-4 linkages etc; the most common being the β-aryl ether (*β*-O-4) bond (Argyropoulos and Menachem 1997). It consists of an apparently random complex of phenolic and non- phenolic compounds.

The structural complexity of lignin makes it one of the most recalcitrant molecule and its breakdown involves multiple biochemical reactions, that has to take place more or less simultaneously; cleavage of inter monomeric linkages, demethylations, hydroxylations, side chain modifications and aromatic ring fission followed by dissimilation of the aliphatic metabolites produced (Vicuna 1988). Under natural conditions this task is achieved by the ligninolytic enzymes produced by different groups of microorganisms. The molecular mass of lignin is high, about 100 kDa or more, which prevents its uptake inside the microbial cell (Eriksson et al. 1990). Thus, the biological degradation of macromolecular lignin must occur through the activity of extracellular enzymes. The ability of these enzymes to act upon different phenolic and non-phenolic compounds in lignin is exploited in the commercial sector for different applications.

22.3 Lignin Degrading Enzymes

The extremely complex nature of lignin requires an array of oxidative enzymes to be involved in its complete degradation. Lignin peroxidase, manganese peroxidase and laccase are the major lignin degrading enzymes. The characteristics of these enzymes differ widely with the microbial source. The ability of an organism to produce one or more of these enzymes also varies greatly among different microbial groups. Apart from LiP, MnP and laccase, a wide range of other enzymes such as veratryl alcohol oxidase (Bourbonnais and Paice 1988), Aryl alcohol dehydrogenase, Quinone oxidoreductase, aromatic acid reductase, vanillate hydroxylase, dioxygenase, catalase (Leisola and Fietcher 1985; Buswell and Eriksson 1988) aromatic aldehyde oxidase (Deobald and Crawford 1989) and Glyoxal oxidase (Kersten 1990) are also believed to be involved in the tedious process of lignin degradation. However, these enzymes are of less importance as they could either act as mediators of lignin degradation by producing H_2O_2 required for the activity of peroxidases or catalyze the breakdown products of lignin degradation that is effected by the activities of enzymes like LiP, MnP and laccase.

Much attention has been paid in the recent time to commercialize these enzymes, especially laccases for different industrial purposes. Novozyme has commercialized laccases from *Trametes villosa* and *Trametes pubescens* (Novozym 51003), which is very efficient for lignin modification processes. The commercial laccase preparation Zylite (Zytex Pvt Ltd, Mumbai) is used for denim washing while the laccase mediator system Lignozym® finds application in bleaching processes. Laccase from *Aspergillus niger* has also been commercialized by Novo Nordisk under the trade name Denilite™, which is specifically meant for textile industry.

22.3.1 Lignin Peroxidase

Lignin peroxidase, commonly known as ligninase is one of the most important enzyme involved in the degradation of lignin. It was discovered in 1983 from the WRF, *P. chrysosporium*. Since then, this enzyme has been demonstrated in wide variety of organisms including brown rot fungi, soft rot fungi and filamentous bacteria. LiPs are oligomannose type glycoprotein with a molecular weight range of 38 KDa to 43 KDa (Schmidt et al. 1990). LiP is having relatively high redox potential, so the compounds with high redox potentials that are not oxidized by other enzymes are also oxidized by LiP. It is this particular character of LiP that makes it an important part of ligninolytic system. LiP can oxidize both phenolic and non-phenolic compounds. This enzyme employs free radical chemistry to cleave the propyl side chain of lignin substructures (Schoemaker et al. 1985) and have been shown to depolymerize lignin invivo (Hammel et al. 1993). LiPs have the unusual ability to cleave the recalcitrant nonphenolic units that comprise approximately 90% of lignin (Glenn et al. 1983; Tien and Kirk 1983). The reactions catalysed by LiP include Cα-Cβ cleavage of the propyl side chains of lignin and lignin models, hydroxylation of benzylic methylene groups, oxidation of benzyl alcohols to the corresponding aldehydes or ketones, phenol oxidation and even aromatic ring cleavage of non-phenolic lignin model compounds (Renganathan et al. 1985; Umezawa and Higuchi 1987; Chung and Aust 1995).

This heme peroxidase has a classical peroxidase catalytic mechanism for which H_2O_2 is required (Fig. 22.3). The native enzyme is oxidized by H_2O_2 and generates two-electron deficient compound I. Compound I can oxidize a compound and can be reduced to compound II, which is one electron deficient. A subsequent oxidation of another molecule by compound II returns the peroxidase to its native resting stage. When there is excess H_2O_2, it will combine with compound II of LiP, generating compound III, which is an inactive form of the enzyme. In many cases, the substrates are not directly accessible to heme of LiP and thus direct oxidation of substrate does not occur. In such cases involvement of redox mediator plays an important role. Veratryl alcohol is an excellent substrate for LiP and it acts as the redox mediator for indirect oxidation of other substrates. Veratryl alcohol stimulates oxidation by preventing enzyme inactivation (Valli et al. 1990) and it is oxidized by LiP to VA

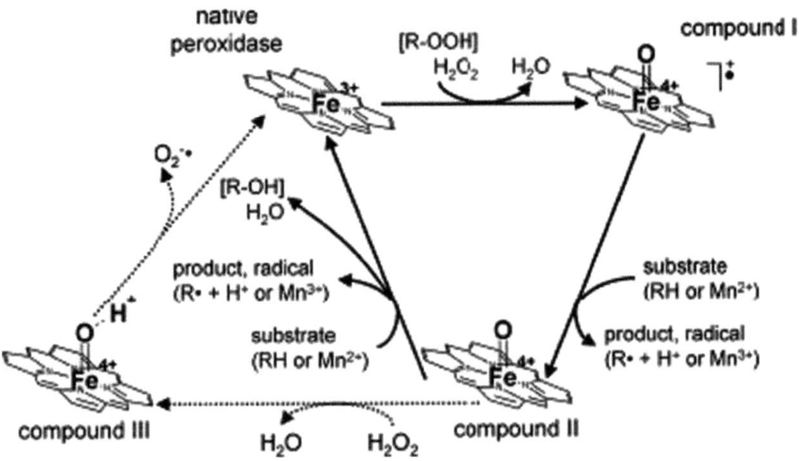

Fig. 22.3 Catalytic cycle of peroxidases

cation radical, which is a strong oxidant, and it acts as an electron transfer mediator in the catalytic reaction of LiP.

22.3.2 Manganese Peroxidase

Manganese peroxidase is another important enzyme produced by the lignin degraders. It is also a heme peroxidase and requires H_2O_2 for its activity. The redox potential of the MnP – Mn system is lower than that of LiP and normally it does not oxidize non-phenolic lignin models. However, it has been reported that the MnP from *Panus tigrinus* is able to degrade nonphenolic lignin model compounds (Maltseva et al. 1991). MnP shows a strong preference for Mn (II) as its reducing substrate (Glenn and Gold 1985). MnP oxidizes Mn^{2+} to Mn^{3+}, which is stabilized by organic acid chelators viz; oxalate, malonate, glyoxylate etc and acts in turn as a low molecular mass, diffusible, redox mediator that attacks organic molecule and oxidizes various compounds nonspecifically via hydrogen and one electron abstraction. The organic acids also facilitate the release of Mn (III) from the active site of the enzyme. The one electron oxidation of Mn (II) to Mn (III) in a multi step reaction cycle is as follows:

$$\text{MnP} + H_2O_2 \rightarrow \text{MnP compound I} + H_2O \text{ (Reaction 1)}$$
$$\text{MnP compound I} + \text{Mn (II)} \rightarrow \text{MnP compound II} + \text{Mn (III) (Reaction 2)}$$
$$\text{MnP compound II} + \text{Mn (II)} \rightarrow \text{MnP} + \text{Mn (III)} + H_2O\text{(Reaction3).}$$

This enzyme is mostly reported in WRF, where it is produced in combination with LiP or laccase.

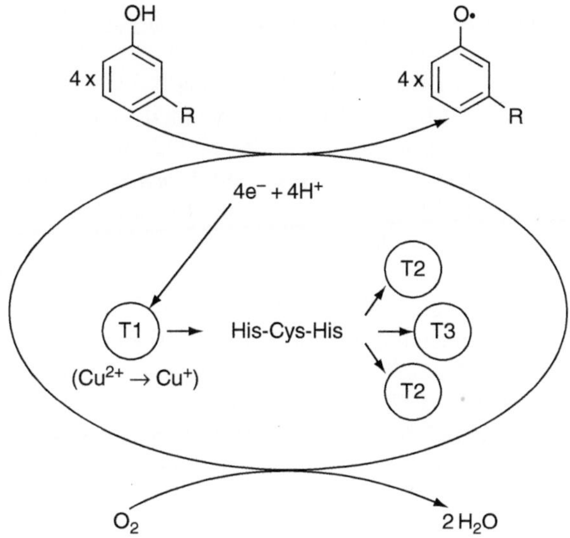

Fig. 22.4 Catalytic cycle of laccase

22.3.3 Laccase

Laccase is a polyphenol oxidase, which belongs to the family of blue multicopper oxidases. These enzymes catalyze the one-electron oxidation of four reducing-substrate molecules concomitant with the four-electron reduction of molecular oxygen to water (Piontek et al. 2002). Laccases oxidize a broad range of substrates, preferably phenolic compounds. In the presence of mediators, laccases exhibit an enlarged substrate range and are then able to oxidize compounds with a redox potential exceeding their own. Laccases differ from LiP and MnP in that it does not require H_2O_2 to oxidize its substrates. The active site of each laccase molecule has four copper ions: one type-1 (T1), one type-2 (T2) and two type-3 (T3) coppers (Fig. 22.4). The two copper ions at T3 site are antiferromagnetically coupled. The T_2 and T3 sites are arranged in a unique trinuclear cluster that is capable of binding oxygen, which is the final electron acceptor. Laccase activity is widely distributed among different groups of fungi, bacteria and also in few plants and insects.

22.4 Sources of Lignin Degrading Enzymes

Lignin degrading enzymes have been demonstrated in both plants and animals. However, the microorganisms remain the important source for lignin degrading enzymes. The production of ligninolytic enzymes from different microbial sources has been well documented.

Fungi are the most potent source of lignin degrading enzymes. The saprophytic fungi secrete these enzymes to degrade the lignin polymer. Among fungi, the white rot fungi are the best known producer of these enzymes, followed by the brown rots and the soft rots. The most studied lignin degrading system is that of *P. chrysosporium*. Lignin degradation by *P. chrysosporium is* a classical secondary metabolic activity induced particularly by nitrogen starvation. This organism secretes LiP, MnP and laccase for lignin degradation (Tien and Kirk 1983; Glenn and Gold 1985; Srinivasan et al. 1995). Different species of white rot fungus *Trametes* viz; *T. Versicolor, T. hirsuta,* and *T. ochracea* have been reported to be producing LiP and MnP along with laccase (Tomsovsky and Homolka 2003). There are also other fungal species such as *P. ostreatus* and *P. radiata* (Vares et al. 1995; Pradeep and Datta 2002) that secrete all the three major ligninolytic enzymes. There are many efficient delignifying fungi that secrete MnP as the sole extracellular peroxidase, including *Lentinula edodes* (Leatham 1986), *Bjerkandera adusta* (Wang et al. 2002), *Ceriporiopsis subvermispora* (Lobos et al. 1994), *Dichomitus squalens* (Perie et al. 1996) and *Rigidoporus lignosus* (Galliano et al. 1991). The commercial preparations of ligninolytic enzymes have been carried out mainly from fungal sources. The ligninolytic enzyme titer of fungi is much higher than that from the bacterial counterparts and this could be the main reason for the dependence on fungi for commercial enzyme production.

Among bacteria, the actinomycetes are the potent producers of ligninolytic enzymes. Extracellular lignin peroxidase has been demonstrated in different *Streptomyces* strains such as *S. viridosporus, S. chromofuscus* and*S. psammoticus* (Ramachandra et al. 1987; Pasti et al. 1990; Niladevi and Prema 2005). Apart from LiP, the *Streptomyces* strains remain as a good source for laccase too (Arias et al. 2003; Suzuki et al. 2003). Even though a large number of bacterial strains have been suggested in lignin degradation, the production of lignin degrading enzymes, especially lignin peroxidase is restricted to few strains of *Pseudomonas* (Yang et al. 2006) while laccase activity is established in bacteria like *Azospirillum lipoferum* and *Bacillus subtilis* (Givaudan et al. 1993; Martins et al. 2002). The details on bacterial production of MnP are scanty with only a few reports. Esposito et al. has reported the production of MnP by a *Streptomyces* strain CCT 4916 that was able to degrade the herbicide diuron by oxidative reactions (Esposito et al. 1998). MnP activity was also detected in the actinobacterium *S. psammoticus* (Niladevi and Prema, 2005).

22.5 Production of Ligninolytic Enzymes

Microorganisms are the best choice for the production of enzymes and other commercially important metabolites. This is because the micro and macro environment of the microorganisms can be controlled easily for enhancing the enzyme production as per the needs. Moreover, genetic manipulation is also easy with the microbial system. Production of ligninolytic enzymes from different fungal and bacterial sources has been carried out in both submerged fermentation (SmF) and solid-state fermentation (SSF) methods. However, the enzyme production by SSF is more prevalent, probably because of the advantages of SSF over SmF, which include higher product titers, lower wastewater output, reduced energy requirements, simpler fermentation media, etc (Pandey et al. 2001). Moreover, this technique offers the possibility of using by-products and wastes from food and agricultural industries as the raw material for enzyme production, making the process much more efficient from both economical and environmental standpoints. Solid-state fermentation is generally defined as the growth of microorganisms on solid materials in the absence or near absence of free water (Pandey et al. 2000a). In SSF, apart from supplying nutrients, the solid substrate also serves as an anchorage for the microorganisms, facilitating their growth and enzyme production.

22.6 Ligninolytic Enzyme Production from Agro-Industrial Residues

Selection of suitable substrate is one of the key factors determining the success of any fermentation process. The substrate must be easily available in the local area in surplus amount and of low cost to make the entire process cost effective. It is at this particular scenario the significance of agro-industrial residues blooms. The annual production of agro-industrial residues has been estimated to about 3.5 billion tonnes and they represent a potential low cost raw material for microbial enzyme production (Robinson and Nigam 2003). Lignocelluloses are the most abundant renewable organic matter on earth and they contribute to the majority of the agro-industrial residues available all over the world. The utilization of lignocelluloses for ligninolytic enzyme production has been studied extensively. Production of LiP, MnP and laccase from a wide variety of agro-industrial residues has been reported. Most of the works on ligninolytic enzyme production from agro-industrial residues deal with fungi (Rodriguez Couto and Sanroman 2005). The list of agro-industrial residues used for the production of ligninolytic enzymes by different microbial strains is given in Table 22.1.

22.6.1 Sugarcane Bagasse

The bagasse (or the crushed cane fibres), which results from the milling of sugar cane, is among the world's most widely used and available non wood fibres. It is used in the boilers for steam production which is used to power the process. The surplus bagasse is used in industry, to produce power, make paper, building materials, as a fuel and even as stock feeds. It contains about 50% cellulose, 25% hemicellulose and 25% lignin (Pandey et al. 2000b). The utilization of this lignocellulosic residue for the production of various industrially important enzymes has been a field of interest to the researchers. The fibrous nature of bagasse makes it more suitable for solid- state fermentation technique.

Sugar cane bagasse has been widely used for the production of ligninolytic enzymes in SSF. The production of laccase and manganese peroxidase in SSF by *Trametes versicolor* and laccase alone by *Flammulina velutipes* using bagasse has

Table 22.1 Microbial production of ligninolytic enzymes from agro-industrial residues

Support	Microbial strain	Ligniolytic enzymes	Reference
Bagasse	*T. versicolor, Flammulina velutipes Pleurotus ostreatus, Phanerochaete chrysosporium*	MnP, laccase laccase MnP, LiP, laccase	Pal et al. 1995; Pal et al. 1995 Pradeep and Datta 2002
Coffee pulp	*Streptomyces psammoticus Pleurotus ostreatus, P. pulmonarius*	Laccase MnP, laccase	Niladevi et al., 2007 Marnyye et al. 2002
Wheat bran	*Ganoderma sp. Fomes sclerodermeus*	Laccase MnP, laccase	Revankar et al. 2007 Papinutti et al. 2003
Wheat straw	*Phlebia radiata, Trametes versicolor Streptomyces cyaneus*	Lip, MnP, laccase MnP,laccase Laccase	Vares et al. 1995; Schlosser et al. 1997 Berrocal et al. 1997
Rice straw	*Streptomyces psammoticus*	Laccase	Niladevi et al. 2007
Orange peelings	*Trametes hirsuta*	Laccase	Rosales et al.,2007
Banana Skin	*Trametes pubescens*	Laccase	Osma et al., 2007
Rice bran	*Coriolus versicolor*	Laccase	Chawachart et al. 2004
Grape seeds	*T. hirsuta*	Laccase	Rodriguez-Couto et al., 2006
Kiwi fruit waste	*Trametes hirsuta*	Laccase	Rosales et al., 2005

been reported by Pal et al. (Pal et al. 1995). The authors have reported that a preferential degradation of non-condensed (syringyl-type) lignin units was observed during the fermentation of bagasse by these organisms. Bagasse has been successfully used for producing LiP, MnP and laccase in semi-solid-state fermentation also (Gonçalves et al. 1998). Bagasse powder has been found to have enhancing effect on ligninolytic enzyme production by *T. versicolor* in shake liquid culture (Masud Hossain and Anantharaman 2006).

22.6.2 Wheat Straw

Wheat is an annual agricultural crop grown for the grain that is a valuable food product. It is the staple food for majority of the human population and hence cultivated world wide. By-products from growing wheat have been used for many years for a variety of applications. The use, for example, of wheat straw as structural filler for mud bricks dates back several hundred years, for pulp back to 1827, and for building panels back into the early 1900s. Wheat straw is the stem and leaf of the wheat plant that is left after the harvest of grains. It consists of about 36% cellulose, 31% hemicellulose and 7% lignin. The current researchers have new approaches in utilizing the bulk quantity of wheat straw produced annually. This includes the production of ethanol and various enzymes from this agro-industrial residue.

Wheat straw is one of the best substrates for the production of ligninolytic enzymes. Laccase and MnP activity has been reported during the growth of *Trametes versicolor* on wheat straw (Schlosser et al. 1997). Wheat straw has served as a best substrate for the production of LiP, MnP and laccase from several other fungi too (Vares et al. 1995; Arora et al. 2002). Wheat straw has also been used for the production of ligninolytic enzymes under SSF by *Streptomyces* strains (Berrocal et al. 1997).

22.6.3 Rice Straw

Rice is the world's second largest cereal crop after wheat and produces the largest amount of crop residues, about 330 million metric tonnes. Ninety percent of the world's production is in developing countries of East and Southeast Asia (Van Soest 2006). A major portion of the rice straw is used as cattle feed and packaging material and not many other uses have been assigned to this agro-industrial waste. The production of ligninolytic enzymes from rice straw is of much relevance considering the higher lignin content of rice straw as compared to wheat straw (Rodriguez Couto and Sanroman 2005). However, the information on utilization of rice straw for this purpose is scanty. Production of all the three ligninolytic enzymes by *Streptomyces psammoticus* using rice straw in SmF has been reported (Niladevi and Prema, 2005) and it was the best substrate for laccase production in SSF as compared to other agro-industrial residues (Niladevi et al. 2007).

22.6.4 Wheat Bran

Wheat Bran is the portion of the grain immediately under the husk of the wheat kernel. During the milling process of the wheat grains for the production of wheat flour, large quantities of wheat bran is produced as by-product. Wheat bran is extensively used as feed for farm animals. Among the various agro-industrial residues, wheat bran however holds the key, as it is produced worldwide in enormous quantities as an important by-product of the cereal industry, and has most commonly been used as substrate for various biotechnological processes (Pandey et al. 1999). Similar to the production of other industrially important enzymes, wheat bran serves as one of the most suitable substrates for ligninolytic enzyme production. Wheat bran has been reported to be the best substrate for laccase production in SmF by *Ganoderma lucidum* (Songulashvili et al. 2007). Wheat bran has been used for the production of laccase by *Ganoderma* strain under SSF and very high laccase activity of 10,050 U/gds has been achieved (Revankar et al. 2007). Production of MnP and laccase from wheat bran has also been reported (Papinutti et al. 2003).

22.6.5 Rice Bran

Rice bran is a by-product of the rice milling process, and it contains various antioxidants that impart beneficial effects on human health. The major rice bran fraction contains 12%–13% oil and highly unsaponifiable components (4.3%). it also contains 4-hydroxy-3-methoxycinnamic acid (ferulic acid), which is also a component of the structure of non-lignified cell walls. The presence of lignin related compounds in rice bran is a factor that favours ligninolytic enzyme production. Laccase production has been carried out from rice bran in SmF and SSF using the basidiomycete fungus *Coriolus versicolor* and it was observed that rice bran was a better substrate for laccase production in both SmF and SSF as compared to other substrates like wheat bran and rice straw meal (Chawachart et al. 2004). However, it remains a fact that the nutritional requirement of the organism is the main factor that determines the suitability of a particular substrate.

22.6.6 Coffee Pulp

The processing of coffee cherries involves the removal of shell and mucilaginous part from the cherries. The processing is done by two methods: dry and wet processing. Depending upon the method of coffee cherries processing, i.e. wet or dry process, the solid residues (sub-products) obtained are termed as pulp or husk, respectively (Pandey et al. 2000c). Coffee pulp contains 50% carbohydrates, 10% proteins, 18% fibers, fat 2.5%, caffeine 1.3% and tannins 1.8–8.56% on dry weight basis (Elias 1979). Traditionally, coffee pulp had found only a limited application as fertilizers, livestock feed, compost, etc (Pandey et al. 2000c). In many of the coffee producing countries the disposal of coffee pulp causes significant environment problems and assigning new uses for this agro-industrial waste is gaining more attention.

The production of ligninolytic enzymes from coffee pulp could be an efficient method for the successful utilization of coffee pulp since, the presence of tannins and other phenolic compounds might facilitate lininolytic enzyme production. One of the popular uses of coffee pulp is for mushroom cultivation (Salmones et al. 2005). It has been established that the growth of white rot fungi on coffee pulp results in the production of various extracellular enzymes, including the ligninolytic enzymes such as laccase and mangaese peroxidase (Marnyye et al. 2002). Apart from white rot fungi, the production of laccase from coffee pulp (both SmF and SSF) by filamentous bacteria (*Streptomyces*) has also been reported (Niladevi et al. 2007; Niladevi and Prema 2008).

22.6.7 Fruit Wastes

Fruit wastes are another category of agro-industrial wastes that could be utilized successfully for the production of ligninolytic enzymes. It needs emphasis that the wastes generated from fruit industries are not widely used or accepted substrates for ligninolytic enzyme production. However, the available literature indicates that liginolytic enzymes could very well be produced from the residues obtained during the processing of different fruits and it is a step ahead of the conventional substrates those have been used commonly for the purpose.

Banana waste is one of the important fruit wastes available world wide as it is used as the common table fruit. The term banana waste includes mainly the leaf and pseudo stem that are commonly used for paper making. The banana peel, which is the waste from banana processing, is also termed as banana waste. Both these wastes can be utilized for the production of ligninolytic enzymes. Banana skin has been used as a support-substrate for the production of extracellular laccase by the white-rot fungus *Trametes pubescens* (Osma et al. 2007). Production of cellulolytic and ligninolytic enzymes from banana waste under SSF conditions has been investigated by Shah et al. and they have reported that the level of ligninolytic enzymes produced from banana waste was higher than the cellulolytic enzymes (Shah et al. 2005). The reports suggest the suitability of banana waste for ligninolytic enzymes production under SSF.

Grape seeds are the major waste from the grape processing units. The uitilization of grape seeds as support substrate for laccase production in laboratory-scale SSF bioreactors has been suggested (Rodriruez Couto et al. 2006). The lignin content of grape seeds is much higher (around 44%) than many of the other substrates and hence it can be utilized for the production of other ligninolytic enzymes like LiP and MnP.

The production of kiwi fruit in the world is around 926,008,000 kg per year. Part of this production is rejected because the kiwi fruit does not have the right shape. The possibility of utilizing rejected kiwi fruits and peelings for laccase production has been investigated (Rosales et al. 2005). The same group has also studied the utility of another fruit waste; the orange peelings, for laccase production under SSF (Rosales et al. 2007). The studies indicate that the same approach can be extended to other fruit wastes for ligninolytic enzymes production.

22.7 Purification of Ligninolytic Enzymes

The end use of the enzyme is the key factor that determines the extent of purity required. Many of the industrial applications of the ligninolytic enzymes (viz; dye decolourization, phenol degradation, biobleaching etc) require only crude preparations of enzyme and in such cases the enzyme purification is not obligatory. However in certain cases, purified enzymes are also used for application purposes (Kokol et al. 2007). Ion exchange and gel permeation chromatographic techniques are the commonly adopted methods for the purification of ligninolytic enzymes. The number of purification steps required depends largely upon the nature of the substrate from which the enzyme is produced and the mode of fermentation. In general, the enzymes produced by SSF are more contaminated than those produced by SmF and hence the enzyme purification becomes more tedious in SSF, partly depending on the nature of the substrate. Substrates like wheat and rice straw are deficient in protein that makes enzyme purification easier whereas, enzymes produced from cereal brans need more steps for purification. Purification and characterization of the ligninolytic enzymes (LiP, MnP and laccase) have been carried out from large numbers of microbial strains including the basidiomycetes and ascomycetes fungi, bacteria and actinobacteria. Purified ligninolytic enzymes with unusual properties like thermo stability, alkaline stability, salt tolerance etc are of much importance to the industries. Many reports are available on ligninolytic enzymes with novel properties (Nozomi et al. 2002; Hoshino et al. 2002; Suzuki et al. 2003; Niladevi et al. 2008) and these enzymes can be considered as an asset, owing to their significance in application sectors.

22.8 Conclusions

The concern towards minimizing the environmental problems is increasing all around and it necessitates the need for a green technology in every industrial sector. A large number of industries use agro-based natural materials as raw materials and generate considerable quantity of residues, which usually accumulates in the environment. The current trend in biotechnology research is oriented towards the utilization of these residues for the production of enzymes and other metabolites and great success has been attained by many researchers in this aspect, mostly with the aid of solid-state fermentation technology. In this scenario, bioconversion of lignocellulosic agro-industrial residues to ligninolytic enzymes is of immense importance while considering the demand for these enzymes in the global market.

References

Aaslyng D, Rorbaek K, Sorensen NH (1996) An enzyme for dying keratinous fibres. Int Pat Appl WO9719998

Argyropoulos DS, Menachem SB (1997) Lignin. In: Eriksson K (Ed.), Biotechnology in the Pulp and Paper Industry, Springer, Berlin, pp. 127–158

Arias ME, Arenas M, Rodriguez J, Soliveri J, Ball AS, Hernandez M (2003) Kraft pulp biobleaching and mediated oxidation of a non-phenolic substrate by laccase from *streptomyces cyaneus* CECT 3335. Appl. Environ. Microbiol. 69:1953–1958

Arora DS, Chander M, Gill PK (2002) Involvement of lignin peroxidase, manganese peroxidase and laccase in degradation and selective ligninolysis of wheat straw. Int. Biodeterior. Biodegrad. 50: 115–120

Berrocal MM, Rodriguez J, Ball AS, Perez–Leblic MI, Arias ME (1997) Solubilization and mineralization of [^{14}C] lignocellulose from wheat straw by *Streptomyces cyaneus* CECT 3335 during growth in solid state fermentation. Appl. Microbiol. Biotechnol. 48:379–384

Bourbonnais R, Paice MG (1988) Veratryl alcohol oxidases from the lignin-degrading basidiomycete *Pleurotus sajor-caju*. Biochem. J. 255:445–450

Buswell JA, Eriksson KE (1988) Methods in Enzymology 161, Wood WA, Kellogg SC (Eds), Academic press, p. 271

Call HP, Mücke I (1997) History, overview and applications of mediated lignolytic systems, especially laccase-mediator-systems (Lignozym-process). J. Biotechnol. 53:163–202

Chabannes M (2001) In situ analysis of lignins in transgenic tobacco reveals a differential impact of individual transformations on the spatial patterns of lignin deposition at the cellular and subcellular levels. Plant J. 28:271–282

Chawachart N, Khanongnuch C, Watanabe T, Lumyong S (2004) Rice bran as an efficient substrate for laccase production from thermotolerant basidiomycete *Coriolus versicolor* strain RC3. Fungal Divers. 15:23–32

Chung N, Aust SD (1995) Veratryl alcohol-mediated indirect oxidation of phenol by lignin peroxidase. Arch. Biochem. Biophys. 316:733–737

Claus H (2004) Laccases: Structure, reactions, distribution. Micron 35:93–96

Crecchio C, Ruggiero P, Pizzigallo MDR (1995) Polyphenol oxidases immobilized in organic gels: Properties and applications in the detoxification of aromatic compounds. Biotechnol. Bioeng. 48:585–591

Deobald LA, Crawford DL (1989) Lignin biotransformation by an aromatic aldehyde oxidase produced by *Streptomyces viridosporus* T7A. Appl. Biochem. Biotechnol 20/21:153–163

Elias LG (1979) In: J.E. Braham, R. Bressani (Eds.), Coffee Pulp: Composition, Technology and Utilization, Publication 108e, International Development Research Centre, Ottawa, Ont, pp. 17–24

Eriksson KE, Blanchette RA, Ander P (1990) Microbial and Enzymatic Degradation of Wood and Wood Components. Springer, Berlin Heidelberg, New York

Esposito E, Paulillo SM, Manfio GP (1998) Biodegradation of the herbicide diuron in soil by indigenous actinomycetes. Chemospere. 37:541–548

Ferry Y, Leech D (2005) Amperometric detection of catecholamine neurotransmitters using electrocatalytic substrate recycling at a laccase electrode. Electroanal. 17:2113–2119

Galliano H, Gas G, Seris JC, Boudet AU (1991) Lignin degradation by *Ridigoporus lignosus* involves synergistic action of two oxidizing enzymes: Mn peroxidase and laccase. Enzyme Microb. Technol. 13:478–482

Givaudan A, Effosse A, Faure D, Potier P, Bouillant ML, Bally R (1993) Polyphenol oxidase in *Azospirillum lipoferum* isolated from rice rhizosphere: evidence for laccase activity in non-motile strains of *Azospirillum lipoferum*. FEMS Microbiol. Lett. 108:205–210

Glenn JK, Gold MH (1985) Purification and characterization of an extracellular Mn (II)-dependent peroxidase from the lignin-degrading basidiomycete, *Phanerochaete chrysosporium*. Arch. Biochem. Biophys. 242:329–341

Glenn JK, Morgan MA, Mayfield MB, Kuwahara M, Gold MH (1983) An extracellular H_2O_2-requiring enzyme preparation involved in lignin biodegradation by the white rot basidiomycete *Phanerochaete chrysosporium*. Biochem. Biophysn. Res. Comm. 114: 1077–1083

Gonçalves AR, Esposito E, Benar P (1998) Evaluation of *Panus tigrinus* in the delignification of sugarcane bagasse by FTIR-PCA and pulp properties. J. Biotechnol. 66:177–185

Hammel KE (1992) Metal Ions in Biological Systems. Sigel H, Sigel A. Eds; Marcel Dekker, Inc, New York. 28:41

Hammel KE, Jensen KA, Mozuch MD, Landucci LL, Tien M, Pease EA (1993) Ligninolysis by a purified lignin peroxidase. J. Biol. Chem. 268:12274–12281

Hoshino F, Kajino T, Sugiyama H, Asami O, Takahashi H (2002) Thermally stable and hydrogen peroxide tolerant manganese peroxidase (MnP) from *Lenzites betulinus*. FEBS Lett. 530: 249–252

Kersten PJ (1990) Glyoxal oxidase of *Phanerochaete chrysosporium*: Its characterization and activation by lignin peroxidase. Proc. Natl. Acad. Sci. USA Biochem. 87:2936–2940

Kirk TK, Farrell RL (1987) Enzymatic "combustion": the microbial degradation of lignin. Ann. Rev. Microbiol. 41:465–505

Kokol V, Doliška A, Eichlerova I, Baldrian P, Nerud F (2007) Decolorization of textile dyes by whole cultures of *Ischnoderma resinosum* and by purified laccase and Mn-peroxidase. Enzyme Microb. Technol. 40:1673–1677

Kuhad RC, Singh A, Eriksson KEL. Microorganisms and enzymes involved in the degradation of plant fiber cell wall. In: Eriksson KEL, editor. Biotechnology in the Pulp and Paper Industry. Advances in Biochemical Engineering Biotechnology. Berlin: Springer Verlag; 1997. Chapter 2

Leatham GF (1986) Ligninolytic activities of *Lentinus edodes* and *Phanerochaete chrysosporium*. Appl. Microbiol. Biotechnol. 24:51–58

Leisola MSA, Fietcher A (1985) Advances in Biotechnological Processes, Mizrahi A, Van Wezel AL (Eds.), Alan R. Liss, New York, p. 59

Lobos S, Larraín J, Cullen D, Vicuna R (1994) Isoenzymes of manganese-dependent peroxidase and laccase produced by the lignin-degrading basidiomycete *Ceriporiopsis subvermispora*. Microbiol. 140:2691–2698

Maltseva OV, Niku-Paavola ML, Leontievsky AA, Myasoedova NM, Golovleva LA (1991) Ligninolytic enzymes of the white rot fungus *Panus tigrinus*. Biotechnol. Appl. Biochem. 13:291–302

Marnyye AV, Gerardo M, Jean-Michel S (2002) Waste-reducing cultivation of *Pleurotus ostreatus* and*Pleurotus pulmonarius* on coffee pulp: changes in the production of some lignocellulolytic enzymes. World J. Microbiol. Biotechnol. 18:201–207

Martins LO, Soares CM, Pereira MM, Teixeira M, Costa T, Jones GH, Henriques AO (2002) Molecular and biochemical characterization of a highly stable bacterial laccase that occurs as a structural component of the *Bacillus subtilis* endospore coat. J. Bio. Chem. 277: 18849–18859

Masud Hossain SK, Anantharaman N (2006) Activity enhancement of ligninolytic enzymes of *Trametes versicolor* with bagasse powder. Afr. J. Biotechnol. 5:189–194

Minussi RC, Pastore GM, Duran N (2002) Potential applications of laccase in the food industry. Trends Food Sci. Technol. 13:205–216

Niladevi KN, Jacob N, Prema P (2008) Evidence for a halotolerant-alkaline laccase in *Streptomyces psammoticus*: purification and characterization. Process Biochem. 43:654–660

Niladevi KN, Prema P (2005) Mangrove actinomycetes as the source of ligninolytic enzymes. Actinomycetol. 19:40–47

Niladevi KN, Prema P (2008) Effect of inducers and process parameters on laccase production by *Streptomyces psammoticus* and its application in dye decolourization. Bioresour. Technol. 99:4583–4589

Niladevi KN, Sukumaran RK, Prema P (2007) Utilization of rice straw for laccase production by *Streptomyces psammoticus* in solid-state fermentation. J. Ind. Microbiol. Biotechnol. 34: 665–674

Nozomi K, Tohru S, Keiichi K (2002) Purification and characterization of an alkaline manganese peroxidase from *Aspergillus terreus* LD-1. J. Biosci. Bioeng. 93:405–410

Osma JF, Toca Herrera JL, Rodriguez Couto S (2007) Banana skin: A novel waste for laccase production by *Trametes pubescens* under solid-state conditions. Application to synthetic dye decolouration. Dyes Pigments. 75:32–37

Pal M, Calvo AM, Terron MC, Gonzalez AE (1995) Solid-state fermentation of sugarcane bagasse with *Flammulina velutipes* and *Trametes versicolor* World J. Microbiol. Biotechnol 11: 541–545

Pandey A, Selvakumar P, Soccol CR, Nigam P (1999) Solid-state fermentation for the production of industrial enzymes. Curr Sci 77:146–162

Pandey A, Soccol CR, Nigam P, Brand D, Mohan R, Roussos S (2000c) Biotechnological potential of coffee pulp and coffee husk for bioprocesses. Biochem. Eng. J. 6:153–162

Pandey A, Soccol CR, Nigam P, Soccol VT (2000b) Biotechnological potential of agro-industrial residues. I. Sugarcane bagasse. Bioresour. Technol. 74:69–80

Pandey A, Soccol CR, Rodriguez-Leon JA, Nigam P (2001) Solid-State Fermentation in Biotechnology. Asiatech Publishers Inc., New Delhi, pp 221

Pandey A, Socool CR, Mitichell D (2000a) New developments in solid-state fermentation. I. Bioprocess and products. Process Biochem. 35:1153–1169

Papinutti VL, Diorio LA, Forchiassin F (2003) Production of laccase and manganese peroxidase by *Fomes sclerodermeus* grown on wheat bran. J. Ind. Microb. Biotechnol. 30:157–160

Pasti MB, Pometto III AL, Nuti MP, Crawford DL (1990) Lignin solubilizing ability of actinomycetes isolated from Termite (Termitidae) gut. Appl. Environ. Microbiol. 56: 2213–2318

Pazarlioglu NK, Sariijik M, Telefoncu A (2005) Laccase: Production by *Trametes versicolor* and application to denim washing. Process Biochem. 40:1673–1678

Perie FH, Sheng D, Gold MH (1996) Purification and characterization of two manganese peroxidase isozymes from the white-rot basidiomycete *Dichomitus squalens*. Biochem. Biophys. Acta. 1297:139–148

Piontek K, Antorini M, Choinowski T (2002) Crystal structure of a laccase from the fungus *Trametes versicolor* at 1.90 A° resolution containing a full complement of coppers. J. Biol. Chem. 277:37663–37669

Pradeep V, Datta M (2002) Production of ligninolytic enzymes for decolorization by cocultivation of white-rot fungi *Pleurotus ostreatus* and *Phanerochaete chrysosporium* under solid-state fermentation. Appl. Biochem. Biotechnol. 102:109–118

Ramachandra M, Pometto AL, Crawford DL (1987) Extracellular enzyme activities during lignocellulose degradation by *Streptomyces spp:* A comparative study of wild type and genetically manipulated strains. Appl. Environ. Microbiol. 53:2754–2760

Renganathan V, Miki K, Gold MH (1985) Multiple molecular forms of diarylpropane oxygenase, an H2O2-requiring, lignin-degrading enzyme from *Phanerochaete chrysosporium*. Arch. Biochem. Biophys. 241:304–314

Revankar MS, Desai KM, Lele SS (2007) Solid-state fermentation for enhanced production of laccase using indigenously isolated *Ganoderma* sp. Appl. Biochem. Biotechnol. 143:16–26

Robinson T, Nigam P (2003) Bioreactor design for protein enrichment of agricultural residues by solid state fermentation. Biochem. Eng. J. 13:197–203

Rodriguez Couto S, Lopez E, Sanroman MA (2006) Utilization of grape seeds for laccase production in solid-state fermentors. J. Food Eng. 74:263–267

Rodriguez Couto S, Sanroman MA (2005) Application of solid-state fermentation to ligninolytic enzyme production. Biochem. Eng. J. 22:211–219

Rosales E, Rodriguez Couto S, Sanroman MA (2005) Reutilisation of food processing wastes for production of relevant metabolites: Application to laccase production by *Trametes hirsuta.* J. Food Eng. 66:419–423

Rosales E, Rodriguez Couto S, Sanroman MA (2007) Increased laccase production by *Trametes hirsuta* grown on ground orange peelings. Enzyme Microb. Technol. 40:1286–1290

Salmones D, Mata G, Waliszewski KN (2005) Comparative culturing of *Pleurotus* spp. on coffee pulp and wheat straw: Biomass production and substrate biodegradation. Biores. Technol. 96:537–544

Schlosser D, Grey R, Fritsche W (1997) Patterns of ligninolytic enzymes in *Trametes versicolor*. Distribution of extra- and intracellular enzyme activities during cultivation on glucose, wheat straw and beech wood. Appl. Microbial. Biotechnol. 47:412–418

Schmidt B, Heimgartner U, Kozulic B, Leisola MSA (1990) Lignin peroxidases are oligomannose type glycoproteins. J. Biotechnol. 13:223–228

Schoemaker HE, Harvey PJ, Bowen RM, Palmer JM (1985) On the mechanism of enzymatic lignin breakdown. FEBS Lett. 183:2407–2412

Shah MP, Reddy GV, Banerjee R, Ravindra Babu P, Kothari IL (2005) Microbial degradation of banana waste under solid state bioprocessing using two lignocellulolytic fungi (*Phylosticta* spp. MPS-001 and *Aspergillus* spp. MPS-002). Process Biochem. 40:445–451

Songulashvili G, Elisashvili V, Wasser SP, Nevo E, Hadar Y (2007) Basidiomycetes laccase and manganese peroxidase activity in submerged fermentation of food industry wastes. Enzyme Microb. Technol. 41:57–61

Srinivasan C, D'Souza TM, Boominathan K, Reddy CA. (1995) Demonstration of laccase in the white rot basidiomycete *Phanerochaete chrysosporium* BKM-F1767. Appl. Environ. Microbiol. 61:4274–4277

Suzuki T, Endo K, Ito M, Tsujibo H, Miyamoto K, Inamori YA (2003) Thermostable laccase from *Streptomyces lavendulae* REN-7: Purification, Characterization, Nucleotide Sequence, And Expression. Biosci. Biotechnol. Biochem. 67:2167–2175

Tien M, Kirk TK (1983) Lignin-degrading enzyme from the hymenomycete *Phanerochaete chrysosporium* burds. Science (Washington) 221:661–662

Tomsovsky M, Homolka L (2003) Laccase and other ligninolytic enzyme activities of selected strains of *Trametes* spp. from different localities and substrates. Folia Microbiol. 48: 413–418

Umezawa T, Higuchi T (1987) Mechanism of aromatic ring cleavage of β-O-4 lignin substructure models by lignin peroxidase. FEBS Lett. 218:255–260

Valli K, Wariishi H, Gold MH (1990) Oxidation of monomethoxylated aromatic compounds by lignin peroxidase: role of veratryl alcohol in lignin biodegradation. Biochem. 29: 8535–8539

Van Soest PJ (2006) Rice straw, the role of silica and treatments to improve quality. Anim. Feed Sci. Technol. 130:137–171

Vares T, Kalsi M, Hatakka A (1995) Lignin peroxidases, manganese peroxidases and other ligninolytic enzymes produced by *Phlebia radiate* during solid state fermentation of wheat straw. Appl. Environ. Microbiol. 61:3515–3520

Vicuna R (1988) Bacterial degradation of lignin. Enzyme Microb. Technol. 10:646–655

Wang Y, Vazquez-Duhalt R, Pickard MA (2002) Purification, characterization, and chemical modification of manganese peroxidase from *Bjerkandera adusta* UAMH 8258. Curr. Microbiol. 45: 77–87

Wesenberg D, Kyriakides I, Agathos SN (2003) White-rot fungi and their enzymes for the treatment of industrial dye effluents. Biotechnol. Adv. 22: 161–187

Yang JS, Liu W, Ni JR (2006) Isolation, identification of lignin-degrading bacteria and purification of lignin peroxidase. Huan Jing Ke Xue. 27: 981–985

Part V
Bioconversion of Agro-Industrial Residues

Chapter 23
Anaerobic Treatment of Solid Agro-Industrial Residues

Michael Ward and Poonam Singh nee' Nigam

Contents

Abstract Solid agro-industrial wastes can be recycled in a cheaper eco-friendly bioconversion way using the anaerobic treatment method. With the depletion of fossil fuels the identification of bio-renewable fuel replacements is well underway. Biomethane derived from Biogas is one possible answer to this dilemma. Biogas production technology has been used for decades in developing countries for cheap

M. Ward (✉)
Centre for Vision Science, Queens University of Belfast, Royal Victoria Hospital, Belfast BT12 6BA, Northern Ireland, UK
e-mail: michaelward@hotmail.co.uk

P. Singh nee' Nigam, A. Pandey (eds.), *Biotechnology for Agro-Industrial Residues Utilisation*, DOI 10.1007/978-1-4020-9942-7_23,

production of fuel for heating and cooking from agro-residues, specifically animal manure. More developed countries have followed suit with production from a wide range of agro-wastes. With the convening of the EU, the 6th framework program and publication lead to the investigation of crops specifically for possible energy production. Varieties such as Maize, Sugar Beet and Wheat etc have been studied intensively. This route of investigation has not only applied to the EU but also across the world.

Keywords Agricultural residues · Anaerobic treatment · Biogas · Methanogenesis · Acetogenesis · Hydrolysis · Cattle manure · Biomethane

23.1 History of Anaerobic Digestion

The anaerobic digestion of agricultural and agro-industrial residues now has been studied in detail by researchers in various countries. This technology has been in practice on small rural scale or industrial scale utilizing locally produced waste industrial substrates and residual biomass. But the process of methanogenesis or methane production as a result of anaerobic digestion of organic matter was known much earlier than any publication on this subject came. Although Italian physicist Alessandro Volta (1745–1827) discovered methane as a type of combustible air, over 200 years ago. As we know it, Methanogenesis is the terminal step in the biodegradation of organic matter in many natural anaerobic (anoxic) habitats, such as swamps. Volta collected gas from marsh sediments and showed that it could burn into flames, still it was not until 1933 that the first publication highlighted an organism which produced methane as a catabolic end product (Thauer 1998). This publication by Stephenson and Stickland was pre-empted by the discovery of the enzyme "Hydrogenase" in a number of *Archaea* species two years previous. The enzyme allowed the micro-organism to thrive in an inorganic medium by use of formate as the sole carbon source. Due to the publication being the first to isolate, in pure culture, a methanogen and study the aforementioned hydrogenase, it has been marked as the beginning of the modern era for study of *Methanogenesis*.

Marjory Stephenson is heralded as the discoverer of hydrogenase and is recognized by her peers for her outstanding contribution to science as a whole but specifically Microbiology. Her lengthy contributions included cofounding of the Society for General Microbiology and her eventual Presidency (Postgate 1995).

23.2 Biogas Production from Agricultural Residues

Methanogenesis is methane production by methanogens via microbial decomposition of organic matter in anaerobic environments, such as river and lake beds. It is estimated that 1% of plant material formed per year by photosynthesis is remineralized via methane with an estimated 10^9 tons of combustible gas being produced (Thauer 1998). This is methane which is naturally produced and not harvested.

Instead it is oxidized, buried leading to methane deposits or more critically diffuses into the atmosphere. Atmospheric methane leads to an increase in the green house effect as it is a potent green house gas.

But it is important to note the process is not carried out solely by a single micro-organism but by syntrophic associations. There are a number of stages in the production of methane from agricultural residues. These stages are not always clearly defined. These stages include Hydrolysis, Acidogenesis, Acetogenesis and finally afore mentioned Methanogenesis. Typically, in a freshwater environment, plant material in the form of glucose from cellulose is completely decomposed to CO_2 and CH_4. This is carried out primarily by the fermentation of large complex polymers, such as carbohydrates or proteins, to Carbon Dioxide (CO_2), Hydrogen (H_2) acetate or formate, followed by the conversion of these substrates to methane. Methanogens have a wide range of optimum temperatures for gaseous production. But research has highlighted increased temperature not to be an advantage in production times and cost. The end product of these stages is termed *Biogas*.

The constituents of the Biogas are methane (CH_4), carbon dioxide (CO_2), making up approximately 90%. Other impurities such as hydrogen sulphide, nitrogen, hydrogen, methylmercaptans and oxygen complete the unrefined fuel source (Zinoviev et al. 2007). With increasing pressure to produce biofuels on a par with fossil based fuels, refinement and upgrading techniques allow the level of methane content within the gas to spike at around 97%. The refinement to almost total methane introduces the term Biomethane and increases the value in terms of energy of the end product. Processes involved can be Water Scrubbing, Pressure Swing Absorption Technologies, Chemical Absorption, Membrane Separation etc. These processes facilitate the introduction of Biomethane to the national gas grids for home use.

23.2.1 Hydrolysis of Agricultural Residues

The initial stage of depolymerisation of the residues is carried out by a large number of bacteria including obligate anaerobes such as *Clostridia* and facultative anaerobes such as *Streptococci* and *Enteric* bacteria (Nigam and Singh 2004). Complex polymers including carbohydrates, protein and lipids are broken down into monomers by enzymes produced by these micro-organisms. These enzymes include cellulase, amylase and protease. The hydrolases are either secreted from or anchored on the cell and the enzymes may be endohydrolases or exohydrolases (Parawira 2004). This extensive variety of hydrolyses allows for the degradation of organic particles to that of transportable molecules. Understanding of this stage is vital when dealing with solid residues where hydrolysis is usually the rate limiting step.

23.2.2 Acidogenesis of Hydrolysed Residues

The less complex end products of Hydrolysis are used as substrates by fermentive micro-organisms in this stage. These allow for the production of organic acids such as acetic, propionic, butyric and other short-chain fatty acids, alcohols, H_2, CO_2. Most of the products formed in the metabolism of glucose have, as an intermediate, pyruvic acid. Pyruvic acid fermentation can lead to the production of a number of C_1 and C_4 compounds such as Volatile Fatty Acids (VFAs) such as acetic, propionic and butyric acids. This is generally the fastest reaction in the anaerobic conversion from solid organic residues to liquid phase digestion. Due to the large numbers of species of bacteria involved in both stages, several organic acids and alcohols are produced. The concentration and proportion of individual VFAs produced in the acidogenic stage is important in the overall performance of the anaerobic digestion system since, acetic and butyric acids are the preferred precursors for methane formation (Hwang et al. 2001).

23.2.3 Acetogenesis

Obligate hydrogen-producing bacteria further degrade propionic, butyric and valric acids to acetate, formate, CO_2 and H_2. This intermediate depolymerisation is key, because prior to this degradation the acids are unable to be processed by methanogens. It is at this advanced point that definition between stages becomes blurred. With the breakdown to acetate, formate, CO_2 and H_2 occurring in both Acidogenic and Acetogenic stages. Acetogens are very slow growing and particularly sensitive to environmental change. Long lag periods have been of particular determent when it comes to the bacteria adjusting to new conditions (Xing et al. 1997). Acetogens need for low hydrogen partial pressure makes its syntrophic association with hydrogen-consuming methanogens ideal, especially when such reactions are thermodynamically unfavorable.

23.2.4 Methanogenesis

The final stage of this total process that of methane production in the form of Biogas. Utilization of mainly H_2/CO_2 and acetic acid form methane and CO_2. A limited number of other substrates can be used, such as methanol, methylamines, alcohols plus CO_2 and formate can also be used by methanogens.

Methanogens use a number of specialized enzymes in order to produce methane. Specifically reduction of CO_2 to methane utilizes Co-enzyme M (CoM) and Co-enzyme B (CoB). The reduction of CO_2 to the formyl level and consequent further reduction to methylene allows for the transfer of the methyl group to an enzyme containing CoM to create methyl-CoM. This methyl group consequently is removed by Co-enzyme F_{430} forming a Ni^{+2}–CH_3 complex. This complex finally is reduced by electrons from CoB and generates methane and a disulphide complex of CoM and CoB. The complex is reduced by hydrogen to generate CoM and CoB. This reaction allows for energy conservation in the process of Methanogenesis.

The microbiology and the biochemistry of methane synthesis have been described in detail by Nigam and Singh (2004).

23.3 Biogas Production Technology

Anaerobic installations for waste and residual biomass bioconversion are known as digesters. The primary process in biogas production is the digestion of organic material to soluble and gaseous products as a result of microbial metabolism. The advanced design for such installations presently in use are various types, such as two-phase, plug-flow, packed-bed and fluidized-bed digesters. The biomethanation plants and digesters types used for anaerobic digestion are significantly important for the overall economy of the process, these have been discussed in detail by Nigam and Singh (2004).

23.3.1 Small Scale Production – Rural

The technology for production of Biogas is not that of only recent development, quite the opposite. In developing countries such as India and rural China , other Asian and South American countries, the technology has been used for decades only on that of a small scale. In rural parts of India anaerobic digestion of manure in small digestion facilities produce what is known as "Gobar/Gober Gas". There have been approximations of over 2 million of these home facilities which provide energy for cooking or possible on-site electric generation. The importance of the facilities in countries which are otherwise totally dependant on outsourcing for energy need cannot be overstated. The design of the facility is that of an airtight circular pit made of concrete with a pipe connection. The manure is usually directed to the digester directly from the cattle shed. The cost efficiency and simplicity of process makes it one of the most environmentally sound energy sources for rural needs. Other facilities which can be operated on both small and large scale include plug-flow or covered lagoon digesters.

In Colombia experiments with diesel engines-generator sets partially fuelled by biogas demonstrated that biogas could be used for power generation, reducing electricity costs by 40% compared with purchase from the regional utility. These examples showcase potential on an obvious small scale, but it is that of large scale production that may redress the problems caused by fossil fuel dependence viewed not only in the obvious sky rocketing of cost, but also and more importantly environmental damage.

23.3.2 Large Scale Production – Industrial

In a number of European countries including Denmark and Sweden digester facilities are in place with the processing abilities of 25,000 to 100,000 gallons of manure per day. These facilities also handle other organic wastes collected from individual farms. Treatment of sewage stabilization and the removal of odour has in the past produced Biogas almost as a by-product with its use for reinvestment in treatment, but without optimization of the Methanogenesis (Nigam and Singh 2004).

Industrial sized digester facilities can have a number of different structures including One-Phase, Two-Phase, previously mentioned Plug-Flow, Batch Systems, Packed-Bed and Fluidized-Bed digesters (Yang et al. 2004). About 90% of the full-scale plants currently in use in Europe for anaerobic digestion of agro-residues rely on One-Phase systems (Vandevivere et al. 2002). This system is viewed by industrialists as more beneficial than Two-Phase or Batch-System due to reduce risk of malfunction and reduction of starting capital. But the trade off is that of a reduction in quality of methane percentage in the harvested Biogas. This is highlighted with the vast number of publications on the Multi-Phase systems which are viewed by researchers as the clear pathway for development. Research has provided data highlighting the limitations on the One-Phase system to be the delicate balancing acts between the stages leading to and including Methanogenesis. These include pH balance, substrate build up etc.

With Two-Phase systems a separation of the Hydrolysis and Acidogenesis from Methanogenesis stages allow for the elimination of the limitations of stability and control cause during One-Phase processing. But with this system comes the added costs of start up and maintenance. The advantages of a Two-Phase reactor over a One-Phase reaction are listed in Table 23.1. Batch Systems by design recirculate leachate from the feed stocks in order to continually disperse inoculants, nutrients and acids. Some Batch Systems operate the Methanogenesis stage at a higher temperature in order to decrease production times. But the failure for the system to gain recognition within the industry with little development of these facilities highlights its high use of end product in turn reducing its feasibility.

Table 23.1 Advantages of the two-stage system over the one-stage system when treating the same agro-industrial wastes. (Parawira 2004)

- Two-stage systems can treat three times the organic loading of a one-stage process, and therefore have shorter hydraulic retention time for rapidly degradable waste. The volumetric capacity of the two-stage system is theoretically higher than that of a single-stage system.
- Significantly higher biomass conversion efficiency and higher COD removal efficiency.
- Higher methane concentration (80–85%) in the biogas produced because the specific activity of methanogens is increased.
- Better process reliability, resilience and stability, especially with variable waste conditions and readily degradable waste, which causes unstable performance in one-stage systems.
- Physical separation of the acidogenic and methanogenic phases allows maintenance of appropriate densities of the acid- and methane-producing microbes enabling maximisation of their rates.
- The acid phase and methane phase can be started much more easily and quickly than in conventional, single-stage digesters.
- The acidification reactor can serve as a buffer system when the composition of the wastewater is variable and can help in the removal of compounds toxic to the methanogens.
- Based on information from full-scale operating systems, two-stage systems produce less and better quality Class A biosolids. This is the main reason for using the two-stage process.

23.4 Feed Stocks for Anaerobic Digestion

With vast numbers of grass species, crops, vegetable wastes and also live stock manure there are numerous feed stocks for production of Biogas. Primarily manure has been the main stay of these digesters in the developing world, but now also in industrialized countries (Tsai et al. 2003). With vast production of agro-wastes, such as corn husk, silage, plant stalks, fruit stones etc. there is another opportunity in which to not only provide waste disposal but environmentally beneficial energy production (Levin et al. 2006; Demirbas 2005). Research is also being directed toward specific growth of termed "Energy Crops" not for that of food consumption, but for that of specific use in digesters. Driven by the Biofuels Directive in Europe, countries such as Ireland have invested in feasibility studies into growth of these Energy Crops on arable lands once used in sugar production (Murphy and Power 2008), other studies include Reed Canary grass in Finland (Paavola et al. 2007).

23.4.1 Cattle Manure

Amon et al. carried out research highlighting levels of Biogas, methane yield and nutrient qualities obtained from Austrian Alpine dairy cows. Within this study the levels of milking were controlled along with the animal's dietary composition. The milk yield ranged from 11.2 to 29.2 liters (l) of milk per cow, with differing levels of hay, grass silage and maize silage comprising the feed viewed in Table 23.2. These studies were carried out at small laboratory scales using 1l eudiometer batch digesters.

The composition of the manure was determined in an effort to detect key constituents in the production of methane in the form of Biogas. The results highlighted manures with higher crude protein levels gave higher methane yield during digestion. Lignin, an integral part of the cell walls of plant cells, in the manure reduced the specific methane yield. This was highlighted in the higher feeding intensities and milk yield Table 23.3.

The treatment Dairy-3 yielded that of the highest methane content of 166.3 Nl $(\text{kg VS})^{-1}$. Previous research has highlighted more favorable biogas return derived from pig or poultry waste due to most of the biodegradable carbon in the cattle feed being digested in the animal's rumen and gut. With cattle feeds varying through out the year recorded levels of biogas production fluctuate. This highlights variation

Table 23.2 Use of agri-residues in Diet of Dairy cattle that delivered the manure for anaerobic digestion experiments. (Adapted from Amon et al. 2007)

Treatment	Milk yield [l/day]	Grass silage	Hay	Maize silage	Concentrate [kg Dry Mass]
Dairy 1	11.2	10.4	5.2	0	0
Dairy 2	11.2	6.4	5.4	5.8	0
Dairy 3	17.6	4.8	4.0	5.2	4.6
Dairy 4	16.0	10	5.0	0	5.8
Dairy 5	29.2	3.8	3.2	3.6	11.0
Dairy 6	29.2	6.2	3.0	0	10.0

Table 23.3 Effect of Composition of dairy cow manure on the yield of specific biogas and methane (Adapted from Amon et al. 2007)

Composition of manure [g (kg DM)$^{-1}$]					Gas yield[a] [Nl(kg/VS)]		
Treatment	DM[b]	XP	Cel	ADL	GE[MJ]	Biogas	Methane
Dairy 1	143.7	162.6	194.7	162.1	15.8	208.2	136.5
Dairy 2	128.8	154.3	227.3	128.2	17.3	213.1	131.8
Dairy 3	135.0	156.6	250.8	124.7	14.6	245.8	166.3
Dairy 4	159.6	150.6	164.1	183.3	19.3	222.5	143.1
Dairy 5	148.5	180.2	161.8	190.4	15.6	238.9	125.5
Dairy 6	157.3	296.5	210.1	121.7	16.8	267.7	159.2

DM = Dry Matter; XP = Crude Protein; Cel = Cellulose; ADL Lignin; GE = Gross Energy
[a] Nl = norm litre (273 K, 1.013 bar).
[b] [g (kg FM)$^{-1}$].

of crop feed sources have an effect on the production of the Biogas, with other research determining differing level of crop development adding to this problem (Amon et al. 2007).

23.4.2 Agricultural Residues (Case Study Turkey)

As with all but a select few countries, Turkey has a high dependence on importing to sustain its energy needs. Animal husbandry and agricultural development are both at high levels within the country, with both comes the obvious potential for production of renewable energy from agrowastes. This is highlighted with the extensive effort of honing biogas production projects since the 1960s (Demirbas 2005). Construction of a number of facilities has stemmed from this research. Tables 23.4 and 23.5 highlight the possible recoverable bio-energy from animals (1997) in Kilotonne of oil equivalent (Ktoe) and selected crops (2001) in Million tons of oil equivalent (Mtoe).

With the vast potential at its disposal Turkey is at the forefront among Organization for Economic Co-operation and Development (OECD) countries, with a total potential for 9.5 Mtoe from crop residues. With Biogas production being quotes

Table 23.4 Recoverable bioenergy potential of animal wastes (case study Turkey, 1997) (Adapted from Demirbas et al. 2006)

Animal	Number of animals[a]	Coefficient of conversion[b]	Energy potential (Kote)	Recoverable energy potential (Kote)
Sheep & Goats	75,095	0048	3604	1081
Donkey, horse, mule & camel	1370	0235	322	97
Poultry	311,500	0003	935	281
Cattle & Buffalo	12,121	0245	2970	891

[a] Thousand head.
[b] Kote per thousand head.

Table 23.5 Agricultural residue potential (case study Turkey, 2001) (Adapted from Demirbas et al. 2006)

Residue	Annual Production (million tons)	Energy Potential (Mtoe)
Wheat Straw	26.4	7.2
Barley Straw	13.5	3.9
Maize Stalk	4.2	1.2
Cotton Hull	2.9	0.9
Sunflower Head	2.7	0.8
Sugar Beet Waste	2.3	0.7
Hazelnut Shell	0.8	0.3
Oat Straw	0.5	0.2
Rye Straw	0.4	0.1
Rice Husk	0.3	0.1
Fruit Peel	0.3	0.1
Total	54.4	15.5

as 1.5–2 Mtoe of this total. The Turkish city of Izmit boasts the country's only waste-to-energy facility with a capacity of 5.4 MW. This highlights the shortfall in putting research into practice possibly due to further funding being necessary to facilitate the conclusive results being published.

23.4.3 Agro-Wastes: (Case Study Taiwan)

The 1970s energy crisis highlighted to the world that biomass utilization for energy production was a necessity. With a small island country such that of Taiwan importing more than 95% of its energy supplies, a collapse in distribution would be catastrophic. These possible dilemmas lead to the convening of the National Energy Conference (NEC). The aim of which was to boost the countries renewable energy supply to that of 3% of its total usage. Previously agro-residues had been used through processes as combustion for heat or electrical generation. But anaerobic digestion highlighted the possibility of gaseous product for a number of energy solutions including possible vehicle transport. Another aspect of Taiwanese industry is that of the every growing hog farms. With estimates of over 7 million animals these also highlighted a major possibility in feed stock for digesters.

23.4.3.1 Hog Manure

Hog manure and urine primarily posed the problem of environmental safety during disposal. By striving to identify a sustainable fuel lead to not only the production of Biogas, but also that of high grade fertilizers for rice crop growth. This industrial crop growth in turn produced agro-waste in the form of Rice Husk. Previous to Biofuel technology development, husk had been used industrially for combustion. But with this development production of hydrogen-rich synthesized gas has become possible. At the date of publication Biogas production based upon estimations of 7 million hogs could have reached the levels below:

Biogas production capacity: $6 : 4 \times 10^8 m^3$ = year (based on 0.25 m^3/day/head)
Electricity generation: $9 : 1 \times 10^8$ kW-h=year (based on 0.7 m^3 biogas/kW-h)
Equivalent electricity charge: US\$ $6 : 7 \times 10^7$ = year (based on US\$ 0.074/kW-h)
Equivalent methane mitigation: $2 : 5 \times 10^5$ metric ton=year (based on 60% methane in the biogas) (Tsai et al. 2004)

23.4.3.2 Rice Husk

A byproduct of the aforementioned Biogas production was that of a crop fertilizer. This was in turn used in rice crop productions which lead to the accumulation of rice husk. Again driven by the NEC Hydrogen gas production, in the form of Synthesis gas (Syngas), was investigated in the processing of these agro-wastes. This Syngas would in turn be converted to electrical power. Based upon preliminary evaluations by Tsai et al. the following figures for energy production and power saving were reached in favor of Syngas production:

Electricity generation: $4 : 5 \times 10^8$ kW-h=year (based on 3600 kcal/kg heating value, 30% energy efficiency)

Equivalent electricity charge saving: US\$ $3 : 3 \times 10^7$ = year (based on US\$ 0.074/kW-h)

23.4.4 Canadian Biomass Residues

As with most, if not all developed countries, Canada is almost totally dependant upon fossil fuel imports to sustain its energy needs. With not only an agricultural but forestry industry with the extent of the Canadian, a virtual untapped natural energy resource was being disposed of with no regard for is possibly enormous monetary and environmental benefit. This discarded waste biomass is also responsible of 10% of the countries green house gas emissions.

Throughout the country's provinces studies were carried out to discover the possible value in the production of methane and hydrogen as fuels, with a comparison of the two also undertaken. Taking into account residues such that of Agricultural residues, live stock manure agro-residues etc. the two gaseous fuels were compared. Potential methane yield from these feed stocks are summarised in Table 23.6.

Table 23.6 Potential methane production from a selection of residual biomass in Canada. (Adapted from Levin et al. 2007)

	Agricultural residues	Livestock manure	OMSW	Municipal biosolids
Total biomass (ODT/year)	1.29×10^7	1.61×10^7	1.50×10^7	3.87×10^5
Substrate used (t/year)	4.52×10^6	6.59×10^6	5.25×10^6	1.36×10^5
Volume methane (m^3/year)	2.10×10^9	2.31×10^9	2.06×10^9	6.45×10^7

Table 23.7 Comparison of total energy potentials from Methane and Hydrogen Biofuels (Levin et al. 2006)

Biomass	Energy recovery efficiency of methane[1]	Energy recovery efficiency of hydrogen[1]
Agricultural crop Residues	41.5	8.2
Livestock manure	36.4	4.9
OMSW	34.8	8.2
Municipal Biosolids	42.4	7.8

[1]Heating value of biofuels/theoretical energy value of biomass

In this study methane production was highlighted to be superior within the test study due to its higher energy yield than hydrogen (Table 23.7). But it is important to note these are preliminary studies and not on an industrial scale.

23.5 Cropgen

Funded by the European Union's (EU) 6th framework program involving 11 partners in 6 European countries, the research aimed to develop a biomass sustainable fuel source to be integrated into the existing energy infrastructure in the medium term. Long term aims are development of a renewable fuel economy via use of agro-residues and energy crops via anaerobic digestion.

From this body many different crops and plants have been studied as for their possible benefits over one another for biogas production. (Salter et al. 2007) highlights many values for methane potential from these feed stocks. It has been recorded that each crop or plant can be used at different stages of growth to yield greater amounts of methane. This has provided the opportunity of not only a mono-crop for production, but many crops selected to enhance a cropping system.

23.5.1 Biogas Production from Maize in Differing Vegetation Periods

As part of the Amon study Maize at three different developmental periods were harvested to identify the optimum for biogas production. These three stages were Milk, Wax and Full ripeness. Correlations between harvesting technology and biogas production yield were investigated. Analysis of the composition of the maize allowed estimates to be made as to the highest yield of biogas. Not only was the composition of the crop under scrutiny, but pre-storage and harvesting technique.

The study summated the crop to be stored as silage prior to digestion to increase yield, with late ripening varieties obtaining superior results. Optimum results were gained from whole maize crop again highlighting the possible production of energy crops solely for energy production.

23.5.2 Finnish Canary Grass

Another study stemming from the 6th framework program was that into Biogas production from canary grass. Investigation was undertaken to determine optimum harvesting periods to maximize gaseous production. The study used two plots of land from 2005–2006 encompassing three growth periods Generative, Flowering and Vegetative. The yield from each period was compared with the generative in both years reaping the greatest dividend 340–350 lCH_4/kg of volatile solid.

23.5.3 Biogas Production in Ireland

Decline of the sugar production industry in Ireland has lead to the availability of arable land previously used to grow sugar beet. Murphy and Power (2008) investigated the use of Wheat, Barley and Sugar Beet in crop rotation to produce methane. During three scenarios different crop variations were used as feed stocks for the digesters. These scenarios were:

Scenario 1: Wheat, Barley and Sugar Beet
Scenario 2: Wheat, Wheat and Sugar Beet
Scenario 3: Wheat alone

The afore mentioned crops have previously been shown to produce different amounts of Biogas with Wheat producing significantly more, but this is on a weight basis. On an area of crop production, i.e. larger area for crops to be produced, Sugar Beet is the superior crop. The study was limited to 48 Kilo Hectares (kha).

As a possible end product vehicle fuel was used to highlight the differences in each methane yield. Firstly the harvested Biogas would be cleansed of impurities to attain a 97% methane composition by removal of carbon dioxide to produce Biomethane. Finally by processes of compression, to 200 bar, the Biomethane could be used as transport fuel. Scenario 2 proved to be vastly superior with the production of Biomethane possibly allowing for the operating of 104, 591 cars or 3377 buses, converted to Biomethane as fuel, for an average year.

23.5.4 Biomethane Versus Bioethanol

The study also revealed a comparable view on the methane production against that of a paper previously written by the authors on Bioethanol production. In this comparison it was revealed gross energy output of the Biomethane system was 17% higher than that of the Bioethanol. Maybe of even greater significance was that of the cost of Biomethane production dipping to 76% of its counterpart. This suggested the production of Biomethane from such energy crops is superior to that of Bioethanol. Table 23.8 summates these findings.

Table 23.8 Comparison of Bioethanol and Biomethane from the crop rotation scenarios. (Murphy and Power 2008)

Scenario	$M_N^3/a \times 10^6$	PJ/a	€/l	€/MJ
	Biomethane			
1	106.49	3.9	0.93	0.025
2	118.19	4.33	0.90	0.027
3	90.48	3.31	0.83	0.023
Scenario	$l/a \times 10^6$	PJ/a	$€/M_N{}^3$	€/MJ
	Ethanol			
1	158.28	3.33	0.70	0.033
2	175.53	3.7	0.69	0.033
3	150.93	3.18	0.60	0.028

23.6 Concluding Remarks

With the obvious enormity in the gulf that shall be left by the loss of fossil fuels in 50, 100 or 150 years depending upon the literature, it is vital that a sustainable, reliable fuel source or sources be identified. A number of biofuels such as Biogas and Biomethane may well be the fuel that could fit bill. But as with all energy sources it is exactly that of fitting the bill that it shall be held accountable to. Even with the dwindling supplies of fossil fuels, it is still the cost of Biofuels compared with these which is the major point of contention.

Biogas has its uses in developing countries as a cheap energy sources especially on a small scale, but it is that of the larger scale production of possibly Biomethane which truly has the attention of organizations such as the EU (CROPGEN 2007). From such directives as the 6th Framework possible production of Biomethane could lead to the powering of cars, buses etc. It is of paramount importance to stress that this is a possibility that will only become a reality through hard work, stress, endeavor and funding which at times seems to have a limit. But at what point will that limit be surpassed pre or post fossil fuel?

References

Amon T, Amon B, Kryvoruchko V, Zollitsch W, Mayer K, Gruber L (2007) Biogas production from maize and airy cattle manure-Influence of biomass composition on the methane yield. Agriculture, Ecosystems and Environments 118:173–182

CROPGEN (2007). European Union 6th Framework Program

Demirbas A (2005) Potential applications of renewable energy sources, biomass combustion problems in boiler power systems and combustion related environmental issues. Prog. Eng. Comb. Sci. 31:171–192

Demirbas A, Pehlivan E, Altun T (2006) Potential evolution of Turkish agricultural residues as bio-gas, bio-char and bio-oil sources. International Journal of Hydrogen Energy. Elsevier 31: 613–620

Hwang S, Lee Y, Yang K (2001) Maximisation of acetic acid production in partial Acidogenesis of swine wastewater. Biotechnology and Bioengineering 75:521–529

Levin DB, Sparling R, Islam R, Cicek N (2006) Hydrogen production by *Clostridium thermocellum* 27405 from cellulosic biomass substrates. Int. J. Hydrogen Energ. 31:1496–1503

Levin D B, Zhu H, Beland M, Cicek N, Holbein B E (2007) Potential for hydrogen and ethane production from biomass residues in Canada. Bioresource Technology 98:654–660

Murphy J D, Power N (2008) Technical and economic analysis of biogas production in Ireland utilising three different crop rotations. Applied Energy 86:1–11

Nigam P, Singh D (2004) Biomethanogenesis In: Ashok Pandey (ed) Concise Encyclopedia of Bioresource Technology, The Haworth Press Inc., NY

Paavola T, Lehtomaki A, Seppala M, Rintala J (2007) Methane Production from Red Canary Grass. European Union 6th Framework Program

Parawira W (2004) Anaerobic Treatment of Agricultural Residues and Wastewater. University of Lund Department of Biotechnology

Postgate J (1995) Society for General Microbiology: Fifty Years On. Special publication of the Society for General Microbiology

Salter A, Delafield M, Heaven S, Gunton Z (2007) Anaerobic digestion of verge cuttings for transport fuel. Waste and Resource Management 160:105–112

Thauer R K (1998) Biochemistry of Methanogenesis: A Tribute to Marjory Stephenson. Microbiology 144:2377–2406

Tsai W T, Chou Y H, Chang Y M (2004) Progress in energy utilization from agrowastes in Taiwan. Renewable and Sustainable Energy Reviews 8:461–481

Vandevivere P, De Baere L, Verstraete W (2002) Types of anaerobic digester for solid wastes In: Mata-Alvarez J, editor. Biomethanization of the Organic Fraction of Municipal Solid Wastes. London: IWA Press; p. 112–140

Xing J, Criddle C, Hickey R (1997) Effects of a long-term periodic substrate perturbation on an anaerobic community. Water Research 31:2195–2204

Yang Y, Tsukahara K, Yagishita T, Sawayama S (2004) Performance of a fixed-bed reactor packed with carbon felt during anaerobic digestion of cellulose. Biosource Technology 94:197–201

Zinoviev S, Arumugam S, Miertus S (2007) Biofuel Production Technologies. Background paper, Dubrovik, Croatia

Chapter 24
Vermicomposting of Agro-Industrial Processing Waste

V.K. Garg and Renuka Gupta

Contents

Abstract Agro-industrial wastes- wastes from agriculture, food processing or any cellulose based industries- remain largely unutilized and often cause environmental

V.K. Garg (✉)
Department of Environmental Science and Engineering, Guru Jambheshwar University of Science and Technology, Hisar 125001, Haryana, India
e-mail: vinodkgarg@yahoo.com

P. Singh nee' Nigam, A. Pandey (eds.), *Biotechnology for Agro-Industrial Residues Utilisation*, DOI 10.1007/978-1-4020-9942-7_24,

problems like dispersing foul odors, occupying vast areas, ground and surface water pollution etc. These wastes could be converted into potential renewable source of energy, if managed sustainably and scientifically. In the last few decades, vermicomposting technology has been arising as a sustainable tool for the efficient utilization of the agro-industrial processing wastes and to convert them into value added products for land restoration practices. The product of the process, i.e., vermicompost is humus like, finely granulated and friable material which can be used as a fertilizer to reintegrate the organic matter to the agricultural soils. The usability of the process depends upon several factors like raw material, various process conditions- pH, temperature, moisture, aeration etc., type of vermicomposting system and earthworm species used. The chapter briefly discusses with the process technology of vermicomposting and the present state of research in the vermicomposting of agro-industrial processing wastes.

Keywords Vermicomposting · C:N ratio · Agricultural waste · *Eisenia fetida* · Vermiwash · Textile industry sludge

24.1 Introduction

In recent years, there is a marked trend towards the use of novel technologies, mainly based on biological processes, for recycling and efficient utilization of organic residues. With this, it is possible to conserve the available resources and to recover the natural products, and in some cases, to combat the disposal problems and minimize the pollution effects. Vermicomposting has been arising as an innovative and low cost biotechnology for the conversion of agro-industrial wastes into value added products, which can be utilized for improving the soil structure and fertility in organic farming. The product of the process, i.e., vermicompost is humus like, finely granulated and friable material which can be used as a fertilizer to reintegrate the organic matter to the agricultural soils. With the adverse impacts of agrochemicals on crop production, it is now well understood that optimum crop yield and maintenance of soil fertility can only be achieved by the balanced use of mineral fertilizers and organic manures. Vermicomposting of agro-industrial wastes can help mitigating these issues, as these residues, if utilized efficiently, represent a vast resource of plant nutrients. The process can be considered as a pre-treatment in order to obtain a stabilized material which may respond more efficiently and safely than the raw material to agricultural soils.

On the other hand, the traditional methods for disposal of these agro-industrial processing wastes such as burning in situ, open dumping, land filling are now-a-days experiencing unacceptability in many countries mainly due to scarcity of landfill sites and increasing public awareness about the impacts that land disposal and mass burning of the unsorted wastes can have on the environment. Vermicomposting technology can be a suitable tool to convert a significant proportion of these wastes into useful products, thereby reducing the associated environmental impacts.

This chapter describes in brief the process technology of vermicomposting as well as the present state of research in the vermicomposting of agro-industrial processing wastes.

24.2 Vermicomposting Technology

Vermicomposting is generally defined as the solid phase decomposition of organic residues in the aerobic environment by exploiting the optimum biological activity of earthworms and micro-organisms. The process depends upon the earthworms to fragment, mix and promote microbial activity in the organic waste material. The earthworms ingest organic solids and convert a portion of it into earthworm biomass and respiration products and egests peat like material termed as vermicompost (Loehr et al. 1985). As compared to the thermal composting, vermicomposting generates a product with lower mass, high humus content, processing time is lower, phytotoxicity is less likely, fertilizer value is usually greater, and an additional product (earthworms) which can have other uses is produced.

24.2.1 Basic Process

The vermicomposting process takes place in the mesophilic temperature range (35–40°C). The different phases during the process are as follows:

- Initial pre-composting phase: The organic waste is pre-composted for about 15 days before being fed to earthworms. During this phase, readily decomposable compounds are degraded and the potential volatile substances are eliminated which may be toxic to earthworms.
- Mesophilic phase: During this phase, earthworms, through their characteristic functions of breaking up organic matter, combine it with the soil particles and enhance microbial activities and condition organic waste materials for the formation of organic manures.
- Maturing and stabilization phase

In the vermicomposting process, the action of the earthworms is both physical/mechanical and biochemical. Physical participation in degrading the organic substances results in fragmentation, thereby increasing the surface area for further microbial colonization. Biochemical changes in organic matter decomposition are carried out through enzymatic digestion, enrichment by nitrogen excrement and transport of organic and inorganic material.

The passage of material through the earthworm intestine rapidly converts the locked up minerals of nitrogen, potassium, phosphorus, calcium etc. into the forms that are much more soluble and available to plants than the parent material. This is made possible by various enzymes present in their gut as well as enzymes of certain type of ingested microorganisms, viz., proteases, lipases, amylases, cellulases, chitinases etc which degrade the cellulosic and proteinaceous materials in organic waste (Hand et al. 1988). The earthworms seem to have developed mutualistic relationship with microorganisms ingested for decomposition of organic matter present in their food (Satchell 1983; Lattuad et al. 1999). Thus, the final quality of the vermicompost is the result of combined efforts taken by earthworms and the microorganisms. Also, it has been found that earthworms release coelomic fluids in which mucocytes, vacuolocytes, granulocytes and lymphocytes are present (Kale and Krishnamoorthy 1981) which kill the bacteria and the parasites present in the waste, thus, making the vermicompost odor and pathogen free. Significantly, the vermicompost is considered an excellent product of homogeneous and odor-less nature, has reduced levels of contaminants, rich in microbial population and tends to hold more nutrients over a longer period, without adversely impacting the environment.

24.2.2 Potential Agro-Industrial Residues

Virtually, any organic waste material of biological origin and biodegradable in nature may be used as a substrate material for the vermicomposting process, provided that it does not contain any material potentially toxic to earthworms. Since, the agro-industrial processing wastes are the byproduct or end product of the processing of agricultural materials; they offer potential opportunities to be used as a substrate for the earthworms and micro-organisms. The agro-industrial wastes are huge source of plant nutrients and their disposal means the ultimate loss of the resourceful material. At present, these wastes are either grossly underutilized or completely unutilized due to in situ burning in the fields or land disposal to the surrounding areas. These individually and cumulatively agro-industrial wastes could effectively be tapped for resource recovery through vermicomposting technology for use in sustainable land restoration practices. A wide variety of agro-industrial processing wastes explored for vermicomposting are encapsulated in Table 24.1.

Table 24.1 Potential agro-industrial processing wastes

Agricultural wastes
rice husk, cereal residues, wheat bran, millet straw etc.
Food processing waste
canning industry waste, breweries waste, dairy industry waste, sugar industry waste press mud and trash, wine industry waste, oil industry waste- non edible oil seed cake, coffee pulp, cotton waste etc.
Wood processing waste
Wood chips, wood shavings, saw dust
Other industrial wastes
fermentation waste, paper and cellulosic waste, vegetal tannery waste
Local organic products
Coco fiber dust, tea wastes, rice hulls etc.
Fruits and vegetable processing waste

24.2.3 Process Control

The optimization of vermicomposting process at a desired efficiency requires knowledge of control parameters that govern the continuity of the process. The following parameters are considered to be of great importance for an efficient process management as well as favorable for the effective growth and reproduction of the earthworms: (a) Temperature (b) Moisture (c) Aeration (d) pH (e) Carbon to nitrogen ratio (f) Food source

24.2.3.1 Temperature

Temperature is probably the most important factor affecting the metabolism, growth and reproduction of earthworms. It has been observed experimentally that, during vermicomposting process, most of earthworm species require moderate temperatures in the range of 10–35°C. At temperature in excess of 35°C, metabolic activity of earthworms begins to decline and sometimes, mortality also occurs. It is, therefore, desirable to maintain the temperature of the treatment process as constant as possible. Care should be taken especially when highly putrescible waste is processed as it generates lots of heat during initial stage of decomposition. In such conditions, to maintain the temperature within the optimal range, it is advisable to provide substrate biomass aeration for a few days before inoculation of earthworms. The key to successful vermicomposting lies in adding the materials to the surface of the piles or beds in thin, successive layers so that heating does not become excessive. Avoidance of such overheating, requires careful management. Earthworms are active and consume organic materials in a relatively narrow layer of 6–9 inches below the surface of compost or heap (Ismail 1997).

24.2.3.2 Moisture

Suitable moisture condition is an important requirement of earthworms. They require moisture in the range of 60–70% (Dominguez and Edwards 1997). The rmoisture content during vermicomposting process depends on many factors like physical status of the wastes, its porosity, type of vermicomposting system used etc. The excessive wet biomass may become anaerobic with consequent production of unpleasant odors, while lower moisture content in substrate biomass may dry up the earthworms. Optimum metabolic rates in the process can be achieved by using the suitable water content that does not restrict oxygen transfer and utilization in the feed mixture. The moisture content in the vermicomposting process is maintained through periodic sprinkling of water in the feed mixture.

24.2.3.3 Aeration

The adequate oxygen supply is the pre-requisite for a vermicomposting system. Factors such as high levels of fatty/oily substances in the feedstock or excessive moisture combined with poor aeration may render anaerobic conditions in vermicomposting system. Worms suffer severe mortality partly because they are deprived of oxygen and partly because of toxic substances (e.g. ammonia and other phytotoxic metabolites) produced under such conditions. Proper aeration could be achieved by periodic turning manually or by mechanical mixing of substrate biomass.

24.2.3.4 pH

The earthworms operates efficiently at pHs in much wide range of 5 to 9, however, a range of 7.5 to 8 is considered to be optimum. Although adjustment of pH in the starting biomass is rarely required, but, when the substrate biomass with high nitrogen is processed, pH should be conditioned appropriately. At high pH, release of ammonia may results in unpleasant odors from the initial feed mixtures. It is to be noted, that pH of the substrate undergoes considerable changes during vermicomposting process. In the initial stages, formation of CO_2 and organic acids lower the pH values, and as the process progresses, the pH value rises due to decomposition of proteins.

24.2.3.5 Carbon to Nitrogen Ratio

The carbon to nitrogen ratio of the organic waste affects substrate decomposition throughout vermicomposting process. Nevertheless, these two elements have to be not only available, but necessarily in a balanced ratio. The optimum C: N ratio for faster organic matter stabilization should be in the range of 25 to 30. Excessive nitrogen of a specific compostable substrate will allow rapid decomposition causing nitrogen loss through volatilization, while excessive carbon presence slows the biological activity. However, substrates with C: N ratios exceeding 40:1 decompose at a very slow rate; it is advisable to add some nitrogen supplement to ensure re-equilibrium of the C: N ratio and effective decomposition. For instance, the mixing of lignocellulosic residues (C: N ratio 100–200) with some bulking agent having low C: N ratio like sludge (10–15) allows the optimal condition for the biological transformation process. The C: N ratio keeps on changing during the vermicomposting process. The loss of carbon as CO_2 in the process of respiration and production of mucous and nitrogenous excrements by earthworms enhance the level of nitrogen, thus lowering the C: N ratio. It is to be noted that a decline in C: N ratio to < 20 indicate an advanced degree of organic matter stabilization and reflect a satisfactory degree of maturity of organic wastes (Senesi 1989).

24.2.3.6 Food Source

Concerning the biological aspects, it is necessary that initial feed material should be qualified physically, chemically and biologically. Following points should be taken care of before the onset of vermicomposting process:

(a) **Physical characteristics of Initial Feed Material:** Physical characteristics of the initial feed material has a marked influence on the stabilization process.

For example, the substrate biomass should always maintain adequate porosity to favor the movement of air, and, hence sufficient oxygen supply within the feed. Thus, prelim size reduction or shredding of the substrate material to such an extent should be done that represents a good compromise between the goal of increasing the surface area-to volume ratio of the substrate to enhance the microbial decomposition rates and the need of preventing exaggerated compaction because of excessive shredding of the initial substrate. Organic residues that are poorly structured and too wet (e.g. food residues, sludges) require mixing with proper bulking agents (e.g. saw dust, dried manure, wood chips etc.) to improve porosity and mitigate excessive moisture content.

(b) **Composition of Starting Material:** Composition of the starting material greatly influences the quality of the finished product. If the aim is the production of biofertilizers for the agricultural soils from the organic waste, it should not contain any non-biodegradable or toxic substance (e.g. inert materials, plastics, glass, metal objects, detergents, pharmaceuticals etc.), which pose a risk either directly to the earthworms or through their metabolic products. So, the source-separated organic fractions should, therefore, be selected for the vermicomposting process so that they will not contaminate the vermicompost produced and pollute the soil.

(c) **Salt content:** Worms are very sensitive to salts, preferring salt contents less than 0.5% in feed (Gunadi et al. 2002). If seaweed is used as a feed, then it should be rinsed so as to reduce salt content. If farmyard manures are to be used as feed material, they can be leached first to reduce the salt content. This is done by simply running water through the material for some time (Gaddie and Douglas 1975). If the manures are pre-composted outdoors, salts will not be a problem.

24.2.4 Major Steps in Vermicomposting Process

Vermicomposting process involves following steps:

- Collection of waste biomass.
- Segregation of biodegradable and non biodegradable waste biomass.
- Selection of biodegradable waste biomass for vermicomposting.
- Shredding of waste biomass to crush it into a homogeneous mixture and to increase surface area for biological action.
- Pre-composting to remove any volatile compounds potentially toxic to earthworms.
- Inoculation of earthworms and maintenance of proper conditions such as temperature, moisture, pH, aeration etc throughout the vermicomposting process.
- Screening and sorting of worms and cocoons after vermistabilization of waste.

During the decomposition, vermicomposting process can be marked complete by earthworms when the waste mixtures become fine, granulated and brown or dark brown in colour. Vermicompost can be collected in layer by layer so as to separate the intermingled earthworms. Then earthworms and cocoons can be separated from the remaining material by a dynamic separation method involving a sieve and photo/thermal stimulus.

24.3 Suitable Earthworm Species

The earthworm species should possess a few characteristics, in order to attain the objectives of verrmicomposting. These are as detailed below:

1. wide adaptability (tolerance) to environmental factors (capability to live in varying temperature and moisture conditions);
2. feeding preference and adaptability for wide range of organic material (high and rich organic matter);
3. high growth rate, low incubation period, high reproduction and cocoon production rate;
4. high consumption, digestion and assimilation rate for organic matter decomposition;
5. Easy to culture.

The vermicompost produced using different species of earthworms show variation in nutrient composition. So, the selection of the suitable species for particular vermicomposting application is important. It is well established that epigeic species of earthworms are used widely for the purpose of vermicomposting of different organic wastes (Ismail 2005). Among these, *Eudrilus eugeniae, Perionyx excavatus* and *Eisenia fetida* have a great potential as wastes decomposers (Hartenstein et al. 1979). These species are prolific breeders, maintaining a high reproduction rate under favorable, moisture and food availability. They show high metabolic activity and hence are particularly useful for vermicomposting. The vermicultural characteristics of these earthworm species have been given in Table 24.2.

Table 24.2 Vermicultural characteristics of some earthworms

Name of earth-worm	Optimum temp.(°C)	Age for cocoon production (weeks)	Upper limit of soil temp. tolerance (°C)	Vermi-stabilization time (weeks)	No. of young/ cocoons	Incubation period (weeks)	Av. size (g)
Eisenia fetida	18–25	5–9	25	6–8	2–4	3–4	0.5
Eudrilus eugeniae	20–25	7–10	30	3–4	2–3	4	1
Perionyx excavatus	25–30	15–18	30	4–5	1	4	1

Source: Dash and Senapati (1985).

24.3.1 *Eisenia fetida*

Eisenia fetida, popularly known as red wriggler, red worm, tiger worm etc is perhaps the most widely used earthworm for vermicomposting. The species has also in wide usages for various toxicological studies as test worm. Mature individuals can attain up to 1.5 g body weight. Each mature worm on average produces one cocoon every third day and from each cocoon emerge from 1 to 3 individuals on hatching within 23 days. Average life of a worm is 1–2 years.

24.3.2 *Eudrilus eugeniae*

Eudrilus eugeniae, a native of Equatorial West Africa, is commonly known as Night Crawler. It grows faster than other species accumulating mass at the rate of 12 mg day^{-1}. Mature individuals can attain body weight up to 4.3 g/individual. Maturity is attained over a period of 40 days, and, a week later, individuals commence cocoon production (on average one cocoon day^{-1}). Life span in laboratory has been estimated from 1 to 3 years. The temperature tolerance of *Eudrilus eugeniae* is lesser than that of *E. fetida.* This species is widely used as vermicomposting worm in tropical and sub-tropical regions.

24.3.3 *Perionyx excavatus*

This species is highly adaptable and can tolerate a wide range of moisture and quality of organic matter. Average growth rate of *Perionyx excavatus* is 3.5 mg day^{-1} and body weight (maximum) 600 mg. Maturity is attained within 21–22 days and reproduction commences by 24th day, with 1 to 3 hatchlings per cocoon. Scientists opine that species is amongst the best suited for vermicomposting in tropical climates.

24.4 Vermicompost Quality

Vermicompost is a peat like material containing most nutrients in plant available forms such as nitrates, phosphates, calcium, potassium, magnesium etc. It has high porosity, water holding capacity and high surface area that provides abundant sites for microbial activity and for the retention of nutrients. The plant growth regulators and other plant growth influencing materials i.e. auxins, cytokinins and humic substances etc. produced by the microbes have been found in vermicomposts (Atiyeh et al. 2002). The nutrients status of the vermicompost obtained from different organic materials is given in Table 24.3.

The vermicompost along with finely divided organic residues, partially digested by the earthworms, living microorganisms, nitrogenous and other excretory products of earthworm metabolism forms vermicast. The casts possess high moisture content and aerobic conditions and hence provide an extraordinary favorable microenvironment for wide range of decomposing microorganisms (Lee 1985). In fact, the most important role of earthworm on the soil may be the stimulation of the

Table 24.3 Chemical composition of vermicompost

Characteristics	Value
Organic carbon%	9.15 to 17.88
Total Nitrogen %	0.5 to 0.9
Phosphorus%	0.1 to 0.26
Potassium%	0.15 to 0.256
Sodium%	0.055 to 0.3
Calcium & magnesium (Meq/100 g)	22.67 to 47.6
Copper; mg kg^{-1}	2.0 to 9.5
Iron, mg kg^{-1}	2.0 to 9.3
Zinc, mg kg^{-1}	5.7 to 9.3
Sulphur, mg kg^{-1}	128.0 to 548.0

microbial activity in cast (Nowak 1975). Many investigators have shown that there is increased microbial population in earthworm casts than in the surrounding soil (Arthur 1965).

Various bacterial species of Pseudomonas, Micrococci, Acromobacter and fungal strains of *Aspergillus niger*, *Penicillium* sp were found in the vermicompost obtained from coir waste (Gobi et al. 2001). Rodriguez et al. (1996) investigated the bacterial pathogens of poultry litter in the vermicompost and their survival in the intestinal tract of earthworms cultivated. They found that the fecal coliforms like *Salmonella* and *Pseudomonas* species and other bacteria, which were present in poultry manure, were significantly reduced in the vermicompost prepared from it. They concluded that *E. fetida* presents an antibacterial response especially against gram negative bacteria starting from gizzard and continuing along intestine.

Vermicompost can not be described as being nutritionally superior to other organic manures but unique in the way in which it is produced, even right in the field and at low cost makes it very attractive for practical application. Various benefits of using vermicomposts are given below:

- Improves the physical structure and natural fertility of soil.
- Increases the water holding capacity of soil.
- Decreases the external inputs as chemical fertilizers reducing soil and water pollution.
- Helps in restoring the microbial population for nitrogen fixation and phosphate solubilization.
- Produces superior quality of food and yield is enhanced.

24.5 Vermiwash

In the vermicomposting process, the bed filled with organic wastes, bedding materials and earthworms is fitted with a drainage and collection system. Vermicomposting produces a leachate as a result of addition of moisture contents through the column of worm action. Draining of this water or leachate is important to prevent saturation of the vermicomposting unit and attraction of pests. The leachate so obtained is termed as vermiwash. It is beneficial in the sense that when collected it can be used as a liquid fertilizer as it contains large amounts of plant nutrients. It is a collection of excretory products and mucous secretion of the earthworms, along with the micronutrients from the organic molecules. Vermiwash, if collected properly, is clear and transparent, honey brown colored fluid. It should be noted; however, that plant bioassay test of vermiwash should be done prior to its use as foliar spray in order to explore the presence of pathogens and phytotoxic compounds. If used as fertilizer, the vermiwash is better diluted to avoid plant damage, but this automatically decreases its nutrient content so it has to be combined with other mineral fertilizers. Commercial formulations of liquid fertilizers are sometimes complemented with certain chemical compounds, such as polyoxyethylene tridecyl alcohol as dispersant and polyethylene nonylphenol as adherent, to increase nutrient availability for plants (Eibner et al. 1984).

24.6 Types of Vermicomposting Systems

Application of vermicomposting as an environmental biotechnology for the management of organic residues relies on the systems that lead to a satisfactory control of the process. These systems should be designed with the motto to achieve high decomposition rates within relatively short stabilization time. The type of the system to be adopted depends upon the land area available, the characteristics and amount of waste to be treated, estimated time required for the stabilization of substrate material and local climate. The common vermicomposting systems include **Windrow system, Beds and bins systems and Reactor system**.

Windrow systems are extensively used both in the open air and under cover, but require either a lot of land or large buildings. The two most common types of windrow include ***(a) Static pile windrows (batch)***: These windrows are simply piles of mixed bedding and feed that are inoculated with worms and allowed to stand until the processing is complete. These piles have a height of < 60 cm (at the time of settling).

(b) Top-fed windrows (continuous flow): These windrows are set up as a continuous-flow operation which means the bedding is placed first, then inoculated with worms, and then covered repeatedly with usually < 10 cm thick layers of food. Worms tend to consume food at the food/bedding interface, and then drop their castings near the bottom of the windrow. Unlike the batch windrows described above, these windrows require continuous feeding and are difficult to operate during winter due to excessive moisture content in the bed. In addition, if windrow covers are used, they must be removed and replaced every time the worms are fed, creating extra work for the operator. The advantages of top-feeding have mainly to do with the greater control the operator has over the worms' environment: since the food is added on a regular basis, the operator can easily assess conditions at the same time and modify such things as feeding rate, pH, moisture content, etc., as required.

This tends to result in a higher-efficiency system with greater worm production and reproduction.

In **Bed and bin system**, bins are used to breed and harvest the vermicompost and also in some cases beds are made on the ground for the same purpose. This method is labour intensive as vermicompost has to be separated manually.

The **Reactor systems** have raised beds with mesh bottom. Feed stocks are added daily in layers on top of the mesh. Finish vermicompost is collected by scrapping a thing layer just above the mesh, and then it falls into the chamber below. The reactor systems are rectangular boxes and not more than three meters in width. The worms are never disturbed in their beds – the material goes in the top, flows through the reactor (and the worms' guts), and comes out the bottom. For maximum efficiency these should be established under cover.

24.7 Vermicomposting of Different Agro-Industrial Wastes

Various agro-industrial wastes have been tested for their potential use as feed stock in vermicomposting process (Tables 24.4 and 24.5). A brief review of these studies is presented in this section

24.7.1 Sugar Industrial Waste

Sugar industry belongs to the most important agro-processing industries all over the world. During the production process, considerable amounts of byproducts such as (a) sugarcane bagasse (b) filter cake or press mud (c) sugarcane trash are produced. Sugarcane bagasse is the fibrous residue left over after the crushing and extraction of juice from the sugarcane stalk. Because of its high ligno-cellulose content and low ash content, sugarcane bagasse is efficiently utilized by the industry itself as a fuel to the boilers. In addition to this, bagasse has a significant market price to add the economical benefits to the industry.

The second byproduct, filter cake (or sometimes called as press mud) originates from the activated sludge process applied for the wastewater treatment in the sugar mills. For about 134 million tones of sugarcane crushed, 4.0 million tones of filter cake are generated (Yadav 1995). Filter cake has a great fertilizer value as it is a rich source of organic matter, organic carbon, sugar, cane wax, protein, enzymes, macronutrients (N, P and K), micronutrients (Zn, Fe, Cu, Mn etc) and microbes. Physically, filter cake is a soft and spongy, light weight, amorphous, dark brown to black material. It has been estimated that filter cake consists of about 1.0–3.1% N; 0.6–3.6%P and 0.3–1.8% K. Farmers are reluctant to apply it directly to the soils due to its mal-odor, immaturity and fear that its application may lead to crust formation, pH variation and pollution problem. Wax content of filter cake (8.15%) affects the soil properties by direct application and its high rate of application (up to 100 tones/acre) leads to soil sickness and water pollution. Filter cake generates

Table 24.4 Different Agro-industrial processing wastes which have been tested for vermicomposting in yesteryears

S. No.	Agro-industrial waste	Organic amendment	Earthworm species	Reference
1.	Filter cake, trash, bagasse	Cow dung	*Eudrilus eugeniae*	Sen and Chandra (2007)
2.	Distillery Sludge	Cow dung	*Perionyx excavatus*	Suthar and Singh (2008)
3.	Post harvest crop residues	Cattle shed manure	*Eudrilus eugeniae*	Suthar (2008)
4.	Solid textile mill sludge	(a) Cow dung, (b) Poultry droppings, (c) Biogas plant slurry, (d) Agricultural wastes	*Eisenia fetida*	Kaushik and Garg (2003) Garg et al. (2006) Kaushik and Garg (2004)
5.	Guar gum industry waste	–	*Perionyx excavatus*	Suthar (2006)
6.	Paper-pulp mill sludge	(a) *Mangifera indica*, (b) cow dung, (c) Saw dust	(a) *Eudrilus eugeniae,* (b) *Eisenia fetida,* (c) *Lampito mauritii*	Banu et al. (2001)
7.	Viticulture and winery waste	–	*Eisenia andrei*	Nogales et al. (2005)
8.	Wood chips (from platinum ore extraction process)	Sewage sludge	*Eisenia fetida*	Maboeta and Rensburg (2003)
9.	Olive pomace	Cattle manure	*Eisenia Andrei*	Plaza et al. (2008)
10.	Wheat straw	–	*Eisenia fetida*	Singh and Sharma (2002)
11.	Mustard residues and Sugarcane trash	Cow dung	*Eisenia fetida*	Bansal and Kapoor (2000)
12.	Filter cake	(a) Cow dung, (b) Horse dung	*Eisenia fetida*	Sangwan et al. (2007, 2008)
13.	Rubber leaf litter	–	(a) *Perionyx excavatus,* (b) *Eudrilus eugeniae,* (c) *Eisenia fetida*	Chaudhuri et al. (2003)
14.	Paper Waste	Cow dung	*Eisenia fetida*	Gupta and Garg (2009)

Table 24.5 Different Agro-industrial processing wastes which have been tested for vermicomposting in yesteryears

S. No.	Agro-industrial waste	Organic amendment	Earthworm species	Reference
1.	a) Paper-pulp mill sludge b) Dairy sludge	Cattle manure	*Eisenia andrei*	Elvira et al. (1998)
2.	Paper-pulp mill sludge	Brewery yeast	*Lumbricus terrestris*	Butt (1993)
3.	Paper-pulp mill sludge	Primary sewage sludge	*Eisenia andrei*	Elvira et al. (1996)
4.	Agricultural industrial waste	–	–	Viel et al. (1987)
5.	Coffee pulp	–	*Eisenia fetida*	Orozco et al. (1996)
6.	Lignocellulosic waste from	Municipal biosolids	*Eisenia andrei*	Benitez et al. (1999)
	Olive oil industry	–	–	–
7.	Paper-pulp mill sludge	(a) Pig slurry, (b) Poultry slurry, (c) Sewage sludge	*Eisenia andrei*	Elvira et al. (1997)
8.	Dairy sludge	(a) Cereal straw, (b) wood shavings	*Eisenia andrei*	Nogales et al. (1999)
9.	Cotton waste	Cattle dung	–	Zajonc and Sidor (1990)
10.	Lignocellulosic wastes	–	*Eisenia fetida*	Vinceslas-Akpa and Loquet (1996)

immense heat (65°C) and its natural decomposition takes long time and also does not remove the foul odor completely. The compost so obtained has lesser nutritive value and more compactness. However, the application of non-matured materials i.e. those with an incomplete stabilization of their organic fraction may lead to harmful effects to the soils.

Filter cake have been utilized significantly in the vermicomposting process for the production of stabilized and nutrient rich manure. Pre-treatment of filter cake is necessary prior to its utilization in vermiconversion processes to improve its palatability to the earthworms. It is kept in shade and turned for about 10–15 days till drying. This will reduce the foul odor as well as the volatile substances present in fresh filter cake.

The feasibility and adaptability of filter cake with various bulking agents like cow dung and horse waste have been tested earlier (Sangwan et al. 2008). It was hypothesized that different feed composition could affect the viability of process. The earthworms were fed to a given range of feed composition i.e. from 0% to 50% of filter cake amended with cow dung and horse waste agents in different set of experiments (composition given by vermireactor no. 1–6). The C: N ratio, which is the indicator of maturity of organic wastes, decreased with time in all the feed mixtures. A glimpse of Figs. 24.1 and 24.2 provides an insight of the degree of stabilization of organic matter in terms of reduction in C: N ratio. The C: N ratio is important because plants cannot assimilate mineral nitrogen unless this ratio is in the order of 20: 1 or less. Initial C: N ratios were in the range of 21.9–42.2 in filter cake amended with cow dung. It decreased significantly after 90 days of

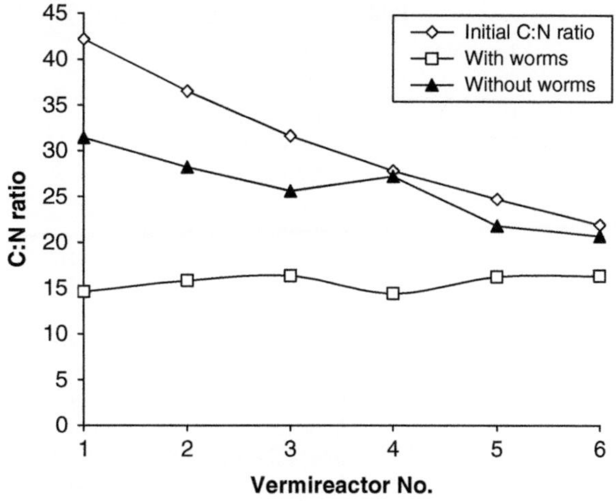

Fig. 24.1 Effect of Vermicomposting on C: N ratio in Filter cake amended with cow dung

worm activity and final C: N ratios were in the range of 14.4–16.3, depicting a high degree of stabilization (Fig. 24.1). In another set of experiment, initial C: N ratios for different feed mixtures in filter cake amended with horse dung were in the range of 22.5–34.9. While, the C: N ratios of vermicompost obtained were in the range of 15.8–18.1 (Fig. 24.2). Also, the C: N ratio in worm-inoculated units was lower than wormless units.

It was found that earthworms promote such microclimatic conditions in the vermireactors that increases the loss of organic carbon from substrates through microbial respiration. Addition of nitrogen in the form of mucus, nitrogenous excretory substances, growth stimulating hormones and enzymes from earthworms has also been reported (Tripathi and Bhardwaj 2004). These nitrogen rich substances were not originally present in feed and might have contributed additional nitrogen content.

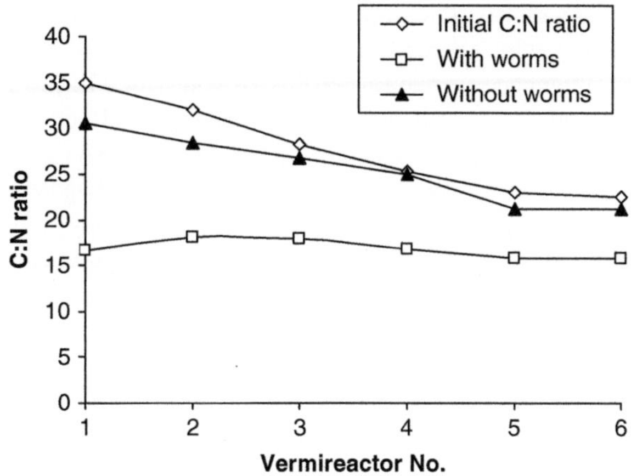

Fig. 24.2 Patterns of C: N ratio during vermicomposting of filter cake amended with horse dung

Table 24.6 The composition of Different potting media

Treatment no.	Composition
1.	Soil – control
2.	compost (10%) + soil
3.	compost (20%) + soil
4.	compost (30%) + soil
5.	compost (40%)+ soil
6.	commercial vermicompost (10%) + soil
7.	commercial vermicompost (20%) + soil
8.	commercial vermicompost (30%)+ soil
9.	commercial vermicompost (40%)+ soil
10.	filter cake (10%) & cow dung (90%) vermicompost + soil
11.	filter cake (20%) & cow dung (80%)vermicompost + soil
12.	filter cake(30%) & cow dung (70%)vermicompost + soil
13.	filter cake (40%) & cow dung (60%)vermicompost + soil
14.	filter cake (10%) & horse dung (90%)vermicompost + soil
15.	filter cake (20%) & horse dung (80%) vermicompost + soil
16.	filter cake (30%) & horse dung (70%)vermicompost + soil
17.	filter cake (40%) & horse dung (60%)vermicompost + soil

Further in the extension of these experiments, assessment of the quality of vermicompost produced from filter cake mixed with cow and horse waste on the growth and productivity of an ornamental plant, namely, marigold was done. A total of seventeen potting media were prepared by various combinations of vermicomposts, compost and soil. The composition of different potting media is presented in Table 24.6. The filter cake + cow dung and horse dung vermicomposts have higher manurial value and affects the growth and productivity of plants synergistically. Addition of vermicomposts in appropriate quantities had improved growth and flowering of plants including leaf areas, plant shoot biomass, root biomass, plant height (Fig. 24.3) and flower numbers (Fig. 24.4). The vermicompost addition also improved the physical, chemical and biological properties of the potting soil. The results also revealed that composts are lesser effective than vermicomposts. This shows that vermicomposts have a great potential in horticulture and agriculture in sustainable organic farming so that growth and productivity can be increased with maintenance of natural fertility of soil. But still prior to vermicompost application at large scale, further research involving its effects on other crops is desired.

The molecular structure of humic acids(HA) extracted from the vermicompost made from press mud, trash and sugarcane bagasse for 60 days was investigated by FT-IR and ^{13}C CP/MAS NMR spectroscopy (Sen and Chandra 2007). A rapid decrease in C: N ratio and lignocellulosic (lignin, cellulose and hemicellulose) content was observed during early phases of vermicomposting process. The solid state spectroscopy (FT-IR and ^{13}C NMR) is considered as the most powerful tool for examining the carbon composition of the organic material. The spectra of HA indicated a high rate of change in structure with increase in the alkyl C/O- alkyl C ratio during the process. Aromatic structures and carboxyl groups decreased after 40 days indicating extensive mineralization during final stages of vermicomposting.

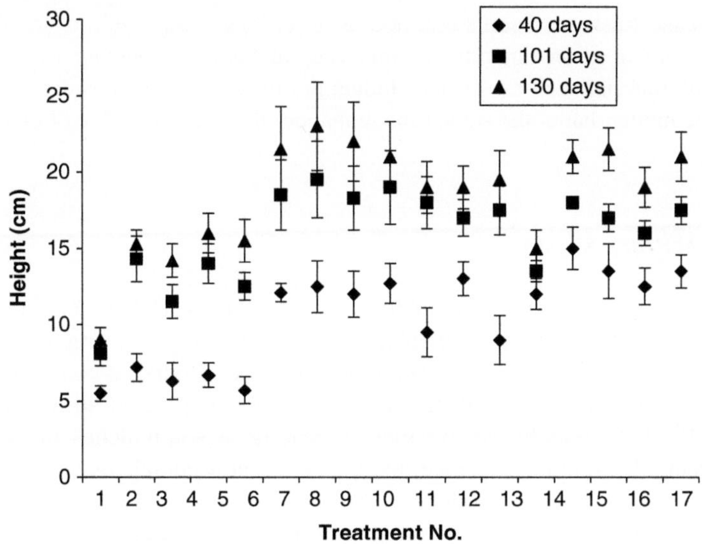

Fig. 24.3 Height of marigold plants grown in different treatments (treatment no. details in Table 24.6)

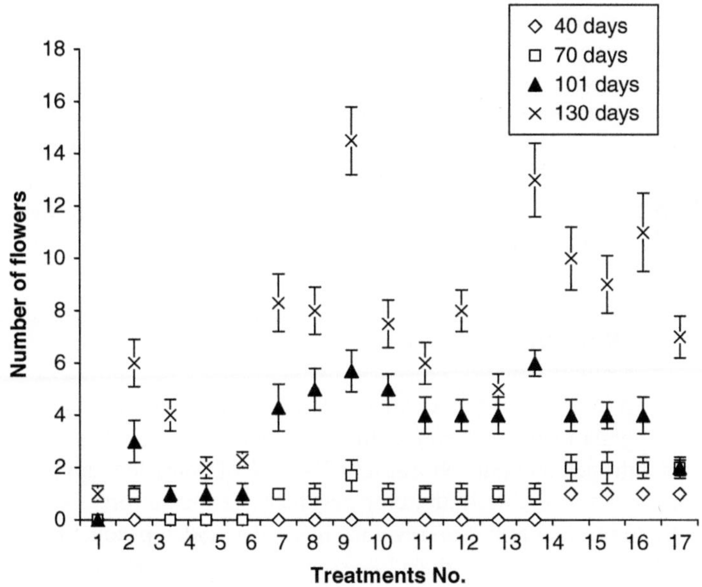

Fig. 24.4 Total number of flowers produced in different treatments (treatment no. details in Table 24.6)

Sugarcane trash has been suggested as an efficient soil conditioner. Moreover, sugarcane trash as an amendment material, in vermicomposting of some non-traditional material such as sewage sludge, not only supply a considerable amount of organic matter, but at the same time, enhances the nutritive value of end product.

24.7.2 Winery Waste

The major byproduct of winery industry is grape marc (GM), containing grape seeds, stalks and skins left over after crushing, draining and pressing stages of wine production. Grape marc is usually processed to produce alcohol and tartaric acid, resulting in formation of new lingo-cellulose by-product spent grape marc (SGM). The latter may be used as fuel for heating, as soil mulches and as organic amendments. In addition, wine lee, the material that accumulates in the bottom of wine fermentation tanks is produced. This by-product is used to make alcohol and tartarates, resulting in soild lees cakes (LC). Finally, the alcohol production from the above by-products leads to the generation of huge quantities of viscose, acidic wastewaters known as vinasse. Also, the main waste of viticulture activity is vine shoot (VS) produced during pruning of grapevines. The potential of these winery wastes (i.e. SGM, LC, VS, vinasse biosolids) in vermicomposting has been investigated using *Eisenia andrei* (Nogales et al. 2005). The evolution of the earthworm biomass and enzyme activity was tracked for 16 weeks of vermicomposting in a lab scale experiment. Changes in the hydrolytic enzymes and overall microbial activities during the vermicomposting process indicated the decomposition of winery wastes. The C: N ratio, conductivity, phytotoxicity was reduced, while humic materials, nutrient contents and pH were increased in all the cases, thereby, increasing the agronomic value of the winery waste. Thus, winery wastes show potential as raw substrate in vermicomposting technology and feasibility of such wastes in large scale studies can be explored.

24.7.3 Crop Residues

It is to be noted that a great percentage of the crop nutrient input is returned back in the form of plant residues during cultivation. For instance, it has been estimated that 30–35% nitrogen and phosphorus and 70–80% of potassium applied as the nutrient input remain available in the crop residues. Various efforts have been made to convert these nutrient rich crop residues into value added product i.e. vermicompost. The suitability of crop residues like soybean straw, wheat straw, maize stover, chickpea straw etc. for vermicomposting has been assessed. Vermicomposting of mustard residues and sugarcane trash mixed with cattle dung using the earthworm *Eisenia fetida* have been studied (Bansal and Kapoor 2000). There was an increase in mineral nitrogen after 90 days and microbial activity was also increased up to 60 days.

Preliminary studies were carried out on wheat straw to test the technical viability of an integrated system of composting, with bioinoculants and subsequent vermicomposting, to overcome the problem of lignocellulosic waste degradation, especially during the winter season (Singh and Sharma 2002). Wheat straw was pre-composted for 40 days by inoculating it with *Pleurotus sajor – caju, Trichoderma harzianun, Aspergillus niger* and *Azotobacter chroococcum* in different combinations. This was followed by vermicomposting for 30 days. The results indicated a significant decrease in cellulose, hemicellulose and lignin content during pre-decomposition and vermicomposting. The best quality compost was obtained when the substrate was treated with all the four bio-inoculums together followed by vermicomposting. Seasonal variations in vermicomposting of red gram pod husk, black gram pod husk, green gram pod husk, wheat straw, rice straw, sorghum straw, grass and parthenium waste and *Cicer arietinum* spiked with cattle dung were tested (Pulikeshi et al. 2003). The results revealed that seasonal changes directly affected the activities of *E. fetida* and indirectly the compostability of the waste.

The post-harvest residues of some crops, e.g. wheat (*Triticum aestivum*), millets (*Penniseum typhoides* and *Sorghum vulgare*), and a pulse (*Vigna radiata*) were utilized to recycle through vermicomposting by *Eudrilus eugeniae* (Suthar 2008). The crop residues were amended with animal dung; and four types of initial fed materials were prepared: (a) millet straw (*S. vulgare* + *Pennisenum typhoides* in equal quantity) + sheep manure (1:2 ratio) (MS), (b) pulse bran (*Vigna radiata*) + wheat straw (*Triticum aestivum*) + cow dung (1:1:2 ratio) (PWC), (c) mixed crop residues (mixing of all types crop residues, used in this study) + cow dung in 1:1 ratio (MCR + CD) and (d) cattle shed manure (CSM). The ready vermicompost obtained from MCR+CD vermibed showed the maximum increase in total N(143.4%), available P (111.1%) and exchangeable K (100.0%). The end product showed reduction in C: N ratio between the ranges of 60.7% (CSM) and 70.3% (MCR + CD). During experimentation, the maximum mortality for *E. eugeniae* was recorded in MS followed by CSM > PWC > MCR + CD. Results indicated that the C: N ratio of the substrate significantly influenced the growth parameters of *E. eugeniae*. This study clearly indicates that vermicomposting of crop residues and cattle dung can not only produce a value added product but also acts as good culture medium for large-scale production of earthworms.

24.7.4 Textile Industry Sludge

On an industrial scale the sludge resulting from the dyeing and printing operations of textile mills is managed through destructive methods: land filling practices and incineration. This sludge is characterized by high BOD, COD, sodium and other dissolved solids as well as micro-nutrients and heavy metals. Investigations have been made to transform textile mill sludge spiked with poultry droppings in to value added product, i.e., vermicompost by vermicomposting technology (Garg and Kaushik 2005). The growth and reproduction of *E. fetida* was monitored in a range of different feed mixtures for 77 days in the laboratory under controlled experimental conditions. The maximum growth was recorded in 100% cow dung (CD). Replacement of poultry droppings by cow dung in feed mixtures and vice versa had little or no effect on worm growth rate and reproduction potential. Worms grew and reproduced favourably in 70% poultry droppings (PD) + 30% solid textile mill sludge (STMS) and 60% PD + 40% STMS feed mixtures. Greater percentage of STMS in the feed mixture significantly affected the biomass gain and cocoon production. Net weight gain by earthworms in 100% CD was 2.9–18.2 fold higher than different STMS containing feed mixtures. The mean number of cocoon production was between 23.4 ± 4.65 (in 100% CD) and 3.6 ± 1.04 (in 50% PD + 50% STMS) cocoons earthworm^{-1} for different feed mixtures tested. Vermicomposting resulted in significant reduction in C: N ratio and increase in nitrogen and phosphorus contents. Total potassium, total calcium and heavy metals (Fe, Zn, Pb and Cd) contents were lower in the final product than initial feed mixtures. This demonstrated vermicomposting as an alternate technology for the recycling and environmentally safe disposal/management of textile mill sludge using an epigeic earthworm *E. fetida,* so avoiding its disposal in open dumps, agricultural fields etc.

24.7.5 Coir Pith

Coir pith is the byproduct of the coconut farm and the coir industry. A wide C: N ratio coupled with low N content, high lignin content and presence of soluble tannin related phenolic compounds (8–12%) cause slow decomposition of these agro-wastes. However, vermicomposting of coir pith has been successfully carried out by *Lumbricus rubellus* at pilot scale level (Kavian et al. 1998). Gobi et al. (2001) studied the vermicomposting of coir pith by *Eudrilus eugeniae* and found that the NPK values increased significantly from its original value after vermicomposting. The lignin and cellulose content were lesser than initial amount.

24.7.6 Cassava Roots

The peels of bitter cassava root, a major source of food carbohydrate in tropics, form toxic waste which is lethal to the soil invertebrates and inhibit the root growth of the plants. Investigations by Mba (1996) highlighted the ability of *Eudrilus eugeniae* to partially detoxify the wastes and convert the toxic cassava peels into valuable vermicompost. Further, in the field studies, it was found that cassava vermicompost enhanced cowpea aerial biomass production, but acidified the soil. Thus, the usefulness of the resources needs to be optimized in order to eliminate the toxin effects and increase the bio-fertilizing ability during vermicomposting. The optimization was done by adding three agricultural wastes, viz., poultry dropping, cow dung and guava leaves. Of the three, the guava leaves treatment increased the soil CEC, soil buffering capacity, eliminated the acidifying effect of cassava and promoted earthworms diversity and activities in cowpea plots.

24.7.7 Pulp and Paper Mill Sludge

The solid or semi-solid sludges obtained from the pulp and paper industry are great source of organic matter which can be effectively used in bioprocesses. However, two factors may limit the biooxidative processes: difficulty in degradation of structural polysaccharides and the low nitrogen content of the sludge. Both the problems could be solved by mixing these wastes with some nitrogen rich material acting as a natural inoculant of microbial populations.

The paper mill sludge was considered a suitable feed for *Lumbricus terrestris* (Butt 1993). This sludge had no deleterious effects on the earthworms, although worm growth rate was poor. The low level of nitrogen ($< 0.5\%$) was considered a limiting factor. By the addition of spent yeast from the brewing industry, the C:N ratio of this sludge could be adjusted according to the requirements. In one such feed mixture, containing paper mill waste and spent yeast from the brewing industry in 66:1 ratio, *Lumbricus terrestris* grew from the hatchling stage (50 mg biomass) to maturity (3–4 g biomass) with in 90 days with an acceptable low level of mortality. Kavian et al. (1998) studied bio-management of paper mill sludge using *Lumbricus rubellus* in order to convert solid effluents or semi-solid sludge from paper mill into a value added product, i. e., vermicompost using vermiculture biotechnology. The vermicomposting of paper mill sludge mixed with sewage sludge, pig slurry and poultry slurry in different ratios was studied by Elvira et al. (1997) by *E. andrei*. Solid paper-mill sludge mixed with sewage sludge in 3: 2 ratio resulted in the highest growth rate and the lowest mortality of *E. andrei*, whereas paper mill sludge mixed with pig slurry exhibited a high mortality. High mortality was not due to lack of food but the degradation process might have resulted in change of the environmental characteristics, the polysaccharides breakdown could modify the structure of the substrate so the water retention capacity decreases and this fact would increase worm mortality.

24.7.8 Coffee Pulp

The ability of *E. fetida* pre-adapted to coffee pulp was tested to transform coffee pulp into vermicompost under different experimental conditions in outdoor containers (Orozco et al. 1996). The results showed that C and N content were not affected by the depth of bed, whereas time affected both. After ingestion of pulp by the earthworms, an increase in available P, Ca, Mg but a decrease in K was detected.

24.7.9 Woodchips

Woodchips and sewage sludge that were produced as waste product by platinum mines were tested to vermicompost by *E. fetida* to examine the growth and reproductive success of the worms over 84 days for commercial use of vermitechnology (Maboeta and Rensburg 2003). The results revealed that there was no effect on worm growth but reproductive success decreased and aluminum, copper, nickel were bio-concentrated in the worms. Earthworms with an addition of microorganism inoculum (consisting of *Pseudomonas, Lactobacillus* and *Saccharomyces* spp) did not bio-concentrate any heavy metal in their body tissue and had a significantly higher reproductive success than their counterpart treatments without microbial inoculum. It was concluded that only economically feasible way to convert woodchip and sewage sludge from platinum mines would be with the addition of microorganism inoculate.

24.7.10 Oil Industry Waste

Olive oil mills use a two-phase centrifugation system for oil separation after pressing of olives. The process generates a semisolid waste, i.e., the two-phase olive pomace called "alperujo". Various chemical changes occurring in a mixture of two-phase olive pomace and cattle manure after vermicomposting with *E. andrei* for eight months were assessed (Plaza et al. 2008). Humic acid (HA)-like fractions were isolated from the substrate material before and after the vermicomposting process, and then analyzed for elemental and acidic functional group composition, by ultraviolet/visible, FT-IR and fluorescence spectroscopies. Prior to vermicomposting, the HA-like fractions contained a prevalent aliphatic character, large C contents, small O and acidic functional group contents. In addition to this, there was a marked presence of proteinaceous materials and polysaccharide-like structures, extended molecular heterogeneity and small degrees of aromatic ring polycondensation, polymerisation and humification. After vermicomposting, the total extractable C and HA–C contents in the bulk substrates increased, and the C and H contents, aliphatic structures, polypeptidic components and carbohydrates decreased in the HA-like fractions, whereas O and acidic functional group contents increased. Further, an adequate degree of maturity and stability was achieved after vermicomposting, and the HA-like fractions approached the characteristics typical of native soil HA. Vermicomposting was thus able to promote organic matter humification in the mixture olive pomace and cattle manure, thus enhancing the quality of these materials as soil organic amendments.

24.7.11 Food Industry Waste

Yadav and Garg (2008) investigated the effect of food industry sludge (FIS) amended with cow dung (CD) on the growth and fecundity of *E. fetida*. Nine waste mixtures of CD and increasing contents of FIS over a total amount of 150 g (100, 90, 80, 70, 60, 50, 40, 30 and 20% CD). It was inferred from the study that addition of 30% of FIS with CD had no adverse effect on the growth and fecundity of *E. fetida* (Fig. 24.5). There was a significant decrease in pH, total organic carbon and C: N ratio, but increase in nitrogen, potassium and phosphorus contents was recorded in the final vermicast than the initial feed mixture. After 84 day's of worm's activity, final C: N ratio was in the range of 9.58 to 23.33 in vermicompost and in the

Fig. 24.5 Growth rate per worm per day (mg) in different vermireactors containing Food Industry Sludge

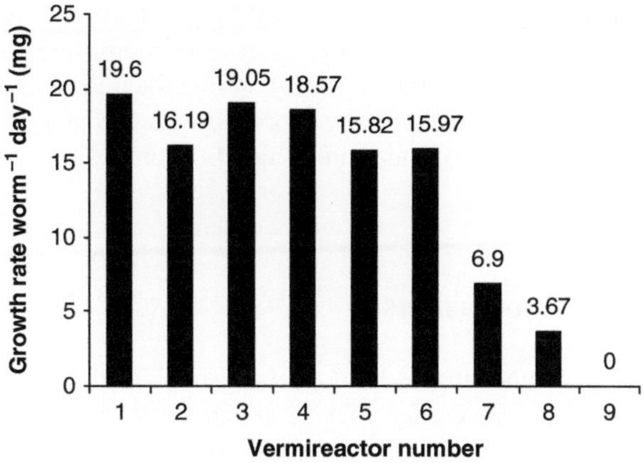

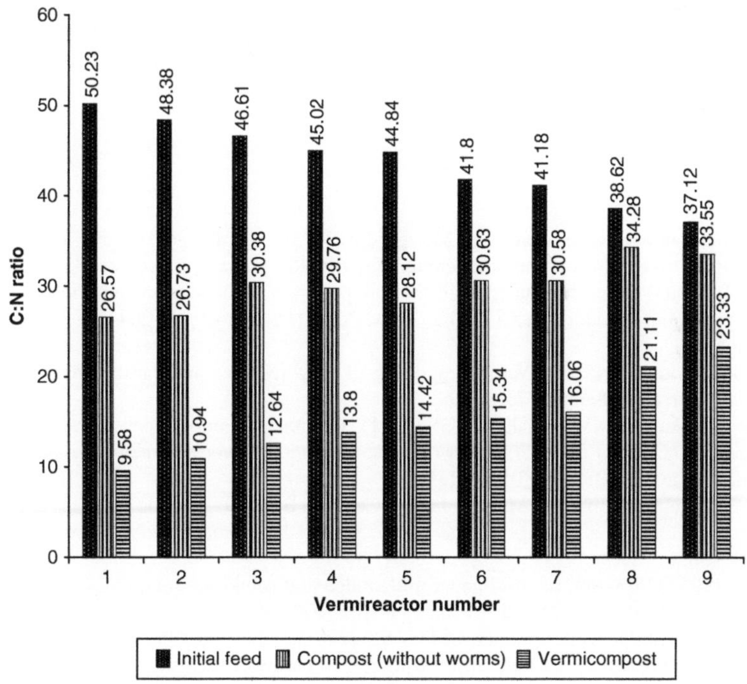

Fig. 24.6 C: N ratio of initial feed, Compost and Vermicompost in different vermireactors containing Food Industry sludge

range of 26.57–33.55 in compost (without worms). As evident from Fig. 24.6 that C: N ratio reduction was higher in vermicompost (37.14%–80.9%) than in compost (without worms) (9.61%–47%). A high degree of organic matter stabilization was achieved in all the vermireactors fed with FIS and CD. This demonstrates the role of earthworms in much more rapid decomposition and rates of mineralization of organic matter.

24.8 Conclusions

Vermicomposting technology is a suitable tool for efficient conversion of agro-industrial processing wastes, which serves as a rich source of plant nutrients. These waste materials are packed with a tremendous source of energy, protein and nutrients, which would otherwise be lost if they are disposed as such in the open dumps and landfills. Moreover, with the use of vermicompost as organic amendments in the agriculture, recycling of the nutrients back to the soil takes place, in turn, maintaining the sustainability of the ecosystem. Therefore, the vermicomposting technology has enormous potential in agro-industrial waste management in a sustainable and decentralized manner, as it yields rich organic fertilizer, safely disposes of organic waste and helps tackle environmental problems such as landfill and the expense of collecting and transporting this waste.

References

Atiyeh RM, Arancon NQ, Edwards CA, Metzger JD (2002) The influence of earthworm-processed pig manure on the growth and productivity of marigolds. Bioresour. Technol. 81:103–108.

Arthur DR (1965) Form and function in the interpretation of feeding in lumbricid worms, In: Carthy JD and Duddington CL (eds.) View Points in Biology, 4: 204–251, Butterworths, London.

Bansal S, Kapoor KK (2000) Vermicomposting of crop residues and cattle dung with *Eisenia foetida*. Bioresour. Technol. 73: 95–98.

Banu JR, Logakanthi S, Vijayalakshmi GS (2001) Biomanagement of paper mill sludge using an indigenous (*Lampito mauritii*) and two exotic (*Eudrilus eugineae* and *Eisenia foetida*) earthworms. J. Environ. Biol. 22(3):181–185.

Benitez E, Nogales R, Elvira C, Masciandaro G, Ceccanti B (1999) Enzyme activities as indicators of the stabilization of sewage sludge composting with *Eisenia foetida*. Bioresour. Technol. 67:297–303.

Butt KR (1993) Utilisation of solid paper-mill sludge and spend brewery yeast as feed for soil-dwelling earthworms. Bioresour. Technol. 44:105–107.

Chaudhuri PS, Pal TK, Bhattacharjee G, Dey SK (2003) Rubber leaf litters (*Hevea brasiliensis,* Var RRIM 600) as vermiculture substrate for epigeic earthworms, *Perionyx excavatus, Eudrilus eugeniae* and *Eisenia fetida*. Pedobiologia 47 (5 and 6):796–800.

Dominguez J, Edwards CA (1997) Effects of socking rate and moisture content on the growth and maturation of *Eisenia andrei* (Oligochaeta) in pig manure. Soil Biol Biochem 29:743–6.

Eibner R, Nolle HH, Shneider A (1984) *Procedimiento para preparer una dispersio'n fertilizante. Certificado de Invencio'n MX 6071.*

Elvira C, Goicoechea M, Sampdro L, Mato S, Nogales R (1996) Bioconversion of solid paper-pulp mill sludge by earthworms. Bioresour. Technol 75:173–177.

Elvira C, Sampedro L, Dominguez J, Mato, S (1997) Vermicomposting of wastewater sludge from paper-pulp industry with nitrogen rich materials. Soil Biol. Biochem. 29:759–762

Elvira C, Sampedro L, Benitez, E, Nogales R (1998) Vermicomposting of sludges from paper mill and dairy industries with *Eisenia andrei*: A pilot scale study. Bioresour. Technol. 63:205–211.

Gaddie RE (Sr.), Douglas DE (1975) Earthworms for Ecology and Profit. Volume 1: Scientific Earthworm Farming. Bookworm Publishing Company, Cal. 180pp.

Garg VK, Kaushik P, Dilbaghi N (2006) Vermiconversion of wastewater sludge from textile mill spiked with anaerobically digested biogas plant slurry employing *Eisenia foetida*. Ecotoxicol. Environ. Safety 65 (3): 412–419.

Garg VK, Kaushik, P (2005) Vermistablisation of solid textile mill sludge spiked with poultry droppings by an epigeic earthworm *Eisenia foetida*. Bioresour. Technol. 96: 1063–1071.

Gobi M, Balamurugan V, Vijayalakshmi GS (2001) Studies on the composting a fungi of coir waste by using *Phanerochaete chrysopsporium* and the earthworms, *Eudrilus eugineae*. J. Environ. Pollun 8(1): 45–48.

Gunadi B, Blount C, Edwards CA (2002) The growth and fecundity of *Eisenia foetida* (Savigny) in cattle solids pre-composted for different periods. Pedobiologia 46: 15–23.

Gupta R, Garg VK (2009) Vermiremediation and Nutrient recovery of non-recyclable paper waste employing *Eisenia fetida*. J of Hazard. Mat. 162:430–439.

Hand P, Hayes WA, Frankland JC, Satchell, JE (1988) The vermicomposting of cow slurry. Pedobiologia 31:199–209.

Hartenstein R, Neuhauser EF, Kaplan DL (1979) Reproductive potential of the earthworm *Eisenia foetida*. Oecologia 43:329–340.

Ismail SA (2005) The Earthworm Book. Other India Press, Goa, India.

Ismail SA (1997) Vermicology "Biology of Earthworms". Orient Longman Ltd. Chennai, India.

Kale RD, Krishnamoorthy RV (1981) What affects the abundance and diversity of earthworms in soils? Proc. Indian Acad. Sci. (Anim Sci.) 90:117–121.

Kaushik P, Garg VK (2003) Vermicomposting of mixed solid textile mill sludge and cow dung with epigeic earthworm *Eisenia foetida*. Bioresour. Technol. 90: 311–316.

Kaushik P, Garg VK (2004) Dynamics of biological and chemical parameters during vermicomposting of solid textile mill sludge mixed with cow dung and agricultural residues. Bioresour. Technol. 94:203–209.

Kavian MF, Ghatnekar SD, Kulkarni PR (1998) Conversion of coir pith into value added biofertiliser using *Lumbricus rubellus*. Indian J. Environ. Prot. 18 (5):354–358.

Lattuad C, Mora P, Garvin M, Locati S, Rouland C (1999) Enzymatic digestive capabilities in geophagus earthworms: Origin and activities of cellulytic enzymes. Pedobiologia 43(6): 842–850

Lee KE (1985) Earthworms: Their Ecology and Relationships With Soil and Land Use. Academic Press, Sydney.

Loehr RC, Neuhauser EF, Malecki MR (1985) Factors affecting the vermistabilization process. Water Res. 19:1311–1317.

Maboeta MS, Rensburg L (2003) Vermicomposting of industrially produced woodchips and sewage sludge utilizing *Eisenia fetida*. Ecotoxicol. Environ. Safety. 56: 265–270.

Mba CC (1996) Treated cassava peel vermicomposts enhanced earthworm activities and cowpea growth in field plots. Resour. Conserv. Recycl. 17:219–226.

Nogales R, Celia C, Benitez E (2005) Vermicomposting of winery wastes: A laboratory study.J. Environ. Sci. Health 40:59–673.

Nogales R, Elvira C, Benitez E, Thompson R, Gomez M (1999) Feasibility of vermicomposting dairy biosolids using a modified system to avoid earthworm mortality. J. Environ. Sci. Health B34:151–169.

Nowak E (1975) Population density of earthworms and some elements in their production in several grassland environments. Ekol. Pol. 23:459–491.

Orozco FH, Cegarra J, Trujillo LM, Roig A (1996) Vermicomposting of coffee pulp using the earthworm *Eisenia foetida*: effects on C and N contents and the availability of nutrients. Biol. Fertil. Soil 22:162–166.

Plaza C, Nogales R, Senesi N, Benitez E, Polo A. (2008) Organic matter humification by vermicomposting of cattle manure alone and mixed with two-phase olive pomace. Bioresour. Technol. 99:5085–5089.

Pulikeshi MB, Amoji SD, Shagoti UM, Biradar VA (2003) Seasonal variations in compostability and production of vermiprotein by *Eisenia fetida*. J. Environ. Biol. 24(2):165–171.

Rodriguez C, Viviana B, Finola M (1996) Bacteriology of poultry litter, compost and earthworm *Eisenia foetida*. Megadrilogica 6 (10):91–95.

Sangwan P, Kaushik CP, Garg VK (2008) Feasibility of utilization of horse dung spiked filter cake in vermicomposters using exotic earthworm *Eisenia foetida*. Bioresour. Technol. 99: 2442–2448

Satchell JE (1983) Earthworm Ecology from Darwin to Vermiculture. Chapman and Hall, London.

Sen B, Chandra TS (2007) Chemolytic and solid-state spectroscopic evaluation of organic matter transformation during vermicomposting of Sugar industry waste. Bioresour. Technol. 98: 1680–1683.

Senesi N (1989) Composted materials as organic fertilizers. Sci. Total Environ. 81/82: 521–524.

Singh A, Sharma S (2002) Composting of a crop residue through treatment with microorganisms and subsequent vermicomposting. Bioresour. Technol. 85:107–111.

Suthar S (2006) Potential utilization of guar gum industrial waste in vermicomposting production. Bioresour. Technol. 97:2474–2477.

Suthar S (2008) Bioconversion of post harvest crop residues and cattle shed manure into value added products using earthworm *Eudrilus eugeniae* Kinberg. Ecological Engineering 32: 206–214.

Suthar S, Singh S (2008) Feasibilty of vermicomposting in biostabilization of sludge from a distillery industry. Sci. Total Environ. 394:237–243.

Tripathi G, Bhardwaj P (2004) Comparative studies on biomass production, life cycles and composting efficiency of *Eisenia foetida* (Savigny) and *Lampito mauritii* (Kinberg). Bioresour. Technol. 92: 275–278.

Viel M, Sayag D, Andre L (1987) Optimization of agricultural, industrial waste management through in-vessel composting. In: de Bertoldi, M. (Ed.), Compost: Production, Quality and Use. Elseiver Appl. Sci. Essex, pp. 230–237.

Vinceslas-Akpa M, Loquet M (1997) Organic matter transformations in lignocellulosic waste products composted or vermicomposted (*Eisenia fetida andrei*): Chemical analysis and ^{13}C CPMAS NMR spectroscopy. Soil Biol. Biochem. 29: 751–758.

Yadav DV (1995) Recycling of sugar factory press mud in agriculture. In recycling of crop, animal, human and industrial wastes in agriculture. In: Tandon H L S. (ed.), Fertilizer development and consultation organization publication, New Delhi. 91–108.

Yadav A, Garg VK (2008). Effect of Feed composition on the growth and Fecundity of earthworms during vermicomposting. In Proceeding of Recent Advances in the Development of Sustainable Environment held at Haryana Engineering College, Jagadhari (Haryana) from May 3–4, 2008: 245–50.

Zajonc I, Sidor V (1990) Use of some wastes for vermicompost preparation and their influences on growth and reproduction of the earthworm *Eisenia foetida*. Pol'nohospodarstvo. 36:742–752.

Index

R

S

T